Transformer Engineering

POWER ENGINEERING

1. Power Distribution Planning Reference Book, *H. Lee Willis*
2. Transmission Network Protection: Theory and Practice, *Y. G. Paithankar*
3. Electrical Insulation in Power Systems, *N. H. Malik, A. A. Al-Arainy, and M. I. Qureshi*
4. Electrical Power Equipment Maintenance and Testing, *Paul Gill*
5. Protective Relaying: Principles and Applications, Second Edition, *J. Lewis Blackburn*
6. Understanding Electric Utilities and De-Regulation, *Lorrin Philipson and H. Lee Willis*
7. Electrical Power Cable Engineering, *William A. Thue*
8. Electric Systems, Dynamics, and Stability with Artificial Intelligence Applications, *James A. Momoh and Mohamed E. El-Hawary*
9. Insulation Coordination for Power Systems, *Andrew R. Hileman*
10. Distributed Power Generation: Planning and Evaluation, *H. Lee Willis and Walter G. Scott*
11. Electric Power System Applications of Optimization, James A. Momoh
12. Aging Power Delivery Infrastructures, *H. Lee Willis, Gregory V. Welch, and Randall R. Schrieber*
13. Restructured Electrical Power Systems: Operation, Trading, and Volatility, *Mohammad Shahidehpour and Muwaffaq Alomoush*
14. Electric Power Distribution Reliability, *Richard E. Brown*
15. Computer-Aided Power System Analysis, *Ramasamy Natarajan*
16. Power System Analysis: Short-Circuit Load Flow and Harmonics, *J. C. Das*
17. Power Transformers: Principles and Applications, *John J. Winders, Jr.*
18. Spatial Electric Load Forecasting: Second Edition, Revised and Expanded, *H. Lee Willis*
19. Dielectrics in Electric Fields, *Gorur G. Raju*
20. Protection Devices and Systems for High-Voltage Applications, *Vladimir Gurevich*
21. Electrical Power Cable Engineering: Second Edition, Revised and Expanded, *William A. Thue*
22. Vehicular Electric Power Systems: Land, Sea, Air, and Space Vehicles, *Ali Emadi, Mehrdad Ehsani, and John M. Miller*
23. Power Distribution Planning Reference Book: Second Edition, Revised and Expanded, *H. Lee Willis*
24. Power System State Estimation: Theory and Implementation, *Ali Abur and Antonio Gómez Expósito*
25. Transformer Engineering: Design and Practice, *S. V. Kulkarni and S. A. Khaparde*

ADDITIONAL VOLUMES IN PREPARATION

Transformer Engineering

Design and Practice

S. V. Kulkarni
S. A. Khaparde
Indian Institute of Technology, Bombay
Mumbai, India

CRC Press
Taylor & Francis Group
Boca Raton London New York

CRC Press is an imprint of the
Taylor & Francis Group, an informa business

Library of Congress Cataloging-in-Publication Data
A catalog record for this book is available from the Library of Congress.

ISBN: 0-8247-5653-3

This book is printed on acid-free paper.

Headquarters
Marcel Dekker 270 Madison Avenue, New York, NY 10016, U.S.A.
tel: 212-696-9000; fax: 212-685-4540

The publisher offers discounts on this book when ordered in bulk quantities. For more information, write to Special Sales/Professional Marketing at the headquarters address above.

Current printing (last digit):

10 9 8 7 6 5 4 3 2

PRINTED IN THE UNITED STATES OF AMERICA

Foreword

It is a great pleasure to welcome this new book from Prof. S. V. Kulkarni and Prof. S. A. Khaparde, and I congratulate them for the comprehensive treatment given in the book to nearly all aspects of transformer engineering.

Everyone involved in or with the subject area of this book, whether from academics or industry, knows that the last decade has been particularly dynamic and fast changing. Significant advances have been made in design, analysis and diagnostic techniques for transformers. The enabling factors for this technological leap are extremely competitive market conditions, tremendous improvements in computational facilities and rapid advances in instrumentation. The phenomenal growth and increasing complexity of power systems have put up tremendous responsibilities on the transformer industry to supply reliable transformers. The transformer as a system consists of several components and it is absolutely essential that the integrity of all these components individually and as a system is ensured. A transformer is a complex three-dimensional electromagnetic structure, and it is subjected to variety of stresses, viz. dielectric, thermal, electrodynamic, etc. In-depth understanding of various phenomena occurring inside the transformer is necessary. Most of these can now be simulated on computers so that suitable changes can be made at the design stage to eliminate potential problems.

I find that many of these challenges in the design and manufacture of transformers, to be met in fast changing market conditions and technological options, are elaborated in this book. There is a nice blend of theory and practice in almost every topic discussed in the text. The academic background of the authors has ensured that a thorough theoretical treatment is given to important topics. A number of landmark references are cited at appropriate places. The previous industry experience of S. V. Kulkarni is reflected in many discussions in the book. The various theories have been supported in the text by reference to actual practices.

For example, while deliberating on various issues of stray loss estimation and control, the relevant theory of eddy currents has been first explained. This theoretical basis is then used to explain various design and manufacturing practices established in the industry to analyze and minimize the stray losses in the windings and structural components. The design and manufacturing practices and processes have significant impact on the performance parameters of the transformers, and the same have been identified in the text while discussing various topics.

Wherever required, a number of examples and case studies are given which are of great practical value. The knowledge of zero-sequence characteristics of transformers is very important for utilities. It is essential to understand the difference between magnetizing and leakage zero-sequence reactances of the transformer. These two types of zero-sequence reactances are explained in the book for three-phase three-limb, three-phase five-limb and single-phase three-limb transformers with numerical examples. One may not find such a detailed treatment to zero-sequence reactances in the available literature. The effect of tank on the zero-sequence reactance characteristics is lucidly explained.

The discussions on the sympathetic inrush phenomenon, part-winding resonance, short-circuit withstand characteristics and noise reduction techniques should also be quite useful to the readers. With the increase in network complexity and severity of loads in some cases, the cooperation between the transformer manufacturers and users (utilities) is very critical. The design reviews with the involvement of users at various stages of contract should help in enhancing the reliability of transformers. I am happy to note that such areas of cooperation are identified at appropriate places in the text.

The book propagates the use of modern computational tools for optimization and quality enhancement of transformers. I know a number of previously published works of the authors in which Finite Element Method (FEM) has been applied for the stray loss control and insulation design of the transformers. The use of FEM has been aptly demonstrated in the book for various calculations along with some tips, which will be helpful to a novice in FEM.

The book is therefore a major contribution to the literature. The book will be extremely helpful and handy to the transformer industry and users. It will be also useful for teaching transformers to undergraduate and postgraduate students in universities. The thorough treatment of all-important aspects of transformer engineering given will provide the reader all the necessary background to pursue research and development activities in the domain of transformers.

It is anticipated that this book will become an essential reference for engineers concerned with design, application, planning, installation, and maintenance of power transformers.

H. Jin Sim, PE
VP, Waukesha Electric Systems
Past Chairman, IEEE Transformers Committee

Preface

In the last decade, rapid advancements and developments have taken place in the design, analysis, manufacturing and condition-monitoring technologies of transformers. The technological progress will continue in the forthcoming years. The phenomenal growth of power systems has put tremendous responsibilities on the transformer industry to supply reliable and cost-effective transformers.

There is a continuous increase in ratings of generator transformers and autotransformers. Further, the ongoing trend to go for higher system voltages for transmission increases the voltage rating of transformers. The increase in current and voltage ratings calls for special design and manufacturing considerations. Advanced computational techniques have to be used that should be backed up by experimental verification to ensure quality of design and manufacturing processes. Some of the vital design challenges are: stray loss control, accurate prediction of winding hot spots, short-circuit withstand capability and reliable insulation design. With the increase in MVA ratings, the weight and size of large transformers approach or exceed transport and manufacturing capability limits. Also, due to the ever-increasing competition in the global market, there are continual efforts to optimize the material content in transformers. Therefore, the difference between withstand levels (e.g., short circuit, dielectric) and corresponding operating stress levels is diminishing. Similarly, the guaranteed performance figures and actual test values are now very close. All these factors demand greater efforts from researchers and designers for accurate calculation of various stress levels and performance figures for the transformers. In addition, strict control of manufacturing processes is required. Manufacturing variations of components should be monitored and controlled.

Many of the standard books on transformers are now more than 10 years old. Some of these books are still relevant and widely referred for understanding the theory and operation of transformers. However, a comprehensive theoretical basis together with application of modern computational techniques is necessary to face the challenges of fast-changing and demanding conditions. This book is an effort in that direction. The principles of various physical phenomena occurring within a transformer are explained elaborately in the text, which could also be used in a course at the undergraduate or postgraduate level. Wherever required, adequate references have been cited so that readers can explore the phenomena in more depth. In fact, a large number of very useful references (more than 400) is one of the hallmarks of this book. Some of the references—classic sources that date back to the early part of the last century—explain many of the theories useful in transformer engineering. Some most recent works are also discussed to give readers a feel for the latest trends in transformer technology.

The first author worked in the transformer industry for 11 years before joining academia. He has vast experience in the design and development of transformers, from the small distribution range to 400 kV class 300 MVA ratings. He had ample opportunity to investigate problems in transformer operations and sites. A few case studies and site investigations in which he was actively involved have been incorporated at appropriate places in the text. Also, he found that some aspects of transformer engineering had not been given adequate treatment in the books available. Hence, the emphasis of this book is on these aspects: magnetizing asymmetry, zero-sequence reactance characteristics, stray losses and related theory of eddy currents, short-circuit forces and withstand, part winding resonance phenomena, insulation design, and design aspects of transformers for rectifier, furnace and HVDC applications. The book will be particularly useful to:

(1) Transformer designers and researchers engaged in optimization and quality-enhancement activities in today's competitive environment

(2) Utility engineers who would like to learn more about the system interaction aspects of transformers in an interconnected power system to improve on specifications and employ diagnostic tools for condition monitoring

(3) Undergraduate and postgraduate students who wish to integrate traditional transformer theory with modern computing practices

In Chapter 1, in addition to the transformer fundamentals, various types of transformers in a typical power system are explained along with their features.

There is a trend to use better materials to reduce core losses. Often the expected loss reduction is not obtained with these better grades. The design and manufacturing practices and processes that have significant impact on the core performance are highlighted in Chapter 2. The three-phase three-limb core has inherent magnetizing asymmetry that sometimes results in widely different no-load current and losses in three phases of the transformer during the no-load loss measurement by the three-wattmeter method. It is shown that one of the three wattmeters can have a negative reading depending on the magnitude of asymmetry between phases and the level of excitation. Although the inrush current phenomenon is well understood, the sympathetic inrush phenomenon—in which the magnetization level of a transformer is affected by energization of another interconnected transformer—is not well known. The factors influencing the phenomenon are elucidated in the chapter. The phenomenon was investigated by the first author in 1993 based on switching tests conducted at a site.

Chapter 3 is devoted to reactance of transformers, which can be calculated by either analytical or numerical methods. Procedures for the calculation of reactance of various types and configurations of windings, including zigzag and sandwich windings, are illustrated. The reactance for complex winding configurations can be easily calculated by the finite element method (FEM), which is the most widely used numerical method. The chapter gives exhaustive treatment of zero-sequence characteristics of the transformers. Procedures for calculation of the magnetizing zero-sequence and leakage zero-sequence reactances of the transformers are illustrated through examples (such a treatment is unusual in the published literature). The effect of the presence of delta winding on the zero-sequence reactance is also explained.

In order to accurately estimate and control the stray losses in windings and structural parts, an in-depth understanding of the fundamentals of eddy currents starting from the basics of electromagnetic fields is desirable. The treatment of eddy currents given in Chapter 4 is self-contained and useful for the conceptual understanding of the phenomena of stray losses in the windings and structural components of transformers described in Chapters 4 and 5, respectively. Stray losses in all the conducting components of the transformers have been given elaborate treatment. Different analytical and numerical approaches for their estimation are discussed and compared. A number of useful guidelines, graphs and equations are given that can be used by practicing engineers. A few interesting phenomena observed during the load-loss test of transformers are explained (e.g., the half turn effect). Various shielding arrangements for effective stray loss control are discussed and compared.

Failure of transformers due to short circuits is a major concern for transformer users. The success rate during actual short-circuit tests is far from satisfactory. The static force and withstand calculations are well established. Efforts are being made to standardize and improve the dynamic short-circuit calculations. A number of precautions (around 40) that can be taken at the specification, design and manufacturing stages of transformers for improvement in short-circuit withstand are elaborated in Chapter 6. The various failure mechanisms and factors that determine the withstand strength are explained.

Although methods for calculating impulse distribution are well established, failures of large transformers due to part-winding resonance and very fast transient overvoltages have attracted the attention of researchers. After an explanation of the methods for calculating series capacitances of commonly used windings, analytical and numerical methods for transient analysis are discussed in Chapter 7. The results of three different methods are presented for a typical winding. Methods for avoiding winding resonances are also explained.

Chapter 8 examines in detail the insulation design philosophy. Various factors that affect insulation strength are summarized. The formulae given for bulk oil and creepage withstand are very useful to designers. Procedures for the design of major and minor insulation systems are presented.

Chapter 9 deals with the thermal aspects of transformer design. After a description of the modes of heat transfer, various cooling systems are described. The insulation aging phenomenon and life expectancy are also discussed. A number of recent failures of large transformers have been attributed to the static electrification phenomenon, which is explained at the end of the chapter.

Various types of loads and tests that determine aspects of structural design are discussed in Chapter 10. Tank-stiffening arrangements are elaborated. This material has been scarce in the available literature. Because of increasing environmental concerns, many users are specifying transformers with lower noise levels. Different noise level reduction techniques are discussed and compared.

Chapter 11 is devoted to four special transformers: rectifier transformers, HVDC converter transformers, furnace transformers and phase-shifting transformers. Their design aspects and features, different from those of conventional distribution and power transformers, are enumerated.

The text concludes by identifying current research and development trends. The last chapter is intended to give pointers to readers desirous of pursuing research in transformers.

Even though the transformer is a mature product, there are still a number of design, manufacturing and power system interaction issues that continue to attract the attention of researchers. This book addresses many of these issues and provides leads to most of the remaining ones. It encompasses all the important aspects of transformer engineering including the recent advances in research and development activities. It also propagates the use of advanced computational tools such as FEM for optimization and quality enhancement of transformers.

S. V. Kulkarni
S. A. Khaparde

ACKNOWLEDGMENTS

We would like to thank our colleagues in the Electrical Engineering Department of the Indian Institute of Technology, Bombay, for their support and encouragement. In particular, we are grateful to Profs. R. K. Shevgaonkar, S. A. Soman, B. G. Fernandes, V. R. Sule, M. B. Patil, A. M. Kulkarni and Kishore Chatterjee, for their help in reviewing the book. Thanks are also due to Mr. V. K. Tandon, who suggested editorial corrections.

Research associates Mr. Sainath Bongane, Mr. Sachin Thakkar and Mr. G. D. Patil helped tremendously, and the excellent quality of the figures is due to their efforts. Ph.D. students Mr. G. B. Kumbhar, Mr. A. P. Agalgaonkar and Mr. M. U. Nabi also contributed in the refining of some topics in the book.

Previously, S. V. Kulkarni worked in Crompton Greaves Limited in the area of design and development of transformers. He sincerely acknowledges the rich and ample experience gained while working in the industry and is grateful to all his erstwhile senior colleagues. He would particularly like to express his sincere gratitude to Mr. C. R. Varier, Mr. T. K. Mukherjee, Mr. D. A. Koppikar, Mr. S. V. Manerikar, Mr. B. A. Subramanyam, Mr. G. S. Gulwadi, Mr. K. Vijayan, Mr. V. K. Lakhiani, Mr. P. V. Mathai, Mr. A. N. Kumthekar and Mr. K. V. Pawaskar for their support and guidance.

Many practical aspects of transformer technology are discussed in the book. Hence, it was essential to have those sections reviewed by practicing transformer experts. Mr. V. S. Joshi's valuable suggestions and comments on almost all the chapters resulted in refinement of the discussion on many practical points. Mr. K. Vijayan, with his expertise on insulation design, contributed significantly in refining Chapter 8. He also gave useful comments on Chapter 9. We thank Mr. G. S. Gulwadi for reviewing Chapters 1 and 8. Mr. V. D. Deodhar helped to improve Chapter 10.

We are also thankful to Dr. B. N. Jayaram for reviewing Chapter 7. The efforts of Dr. G. Bhat in improving Chapter 9 are greatly appreciated. Mr. V. K. Reddy contributed significantly to Chapter 10. Mr. M. W. Ranadive gave useful suggestions on some topics. We are grateful to Prof. J. Turowski, Mr. P. Ramachandran and Prof. L. Satish for constructive comments.

Ms. Rita Lazazzaro and Ms. Dana Bigelow of Marcel Dekker, Inc., constantly supported us and gave good editorial input.

Finally, the overwhelming support and encouragement of our family members is admirable. S. V. Kulkarni would like to particularly mention the sacrifice made and moral support given by his wife, Sushama.

Contents

Transformer Engineering

1

Transformer Fundamentals

1.1 Perspective

A transformer is a static device that transfers electrical energy from one circuit to another by electromagnetic induction without the change in frequency. The transformer, which can link circuits with different voltages, has been instrumental in enabling universal use of the alternating current system for transmission and distribution of electrical energy. Various components of power system, viz. generators, transmission lines, distribution networks and finally the loads, can be operated at their most suited voltage levels. As the transmission voltages are increased to higher levels in some part of the power system, transformers again play a key role in interconnection of systems at different voltage levels. Transformers occupy prominent positions in the power system, being the vital links between generating stations and points of utilization.

The transformer is an electromagnetic conversion device in which electrical energy received by primary winding is first converted into magnetic energy which is reconverted back into a useful electrical energy in other circuits (secondary winding, tertiary winding, etc.). Thus, the primary and secondary windings are not connected electrically, but coupled magnetically. A transformer is termed as either a step-up or step-down transformer depending upon whether the secondary voltage is higher or lower than the primary voltage, respectively. Transformers can be used to either step-up or step-down voltage depending upon the need and application; hence their windings are referred as high-voltage / low-voltage or high-tension / low-tension windings in place of primary / secondary windings.

***Magnetic circuit*:** Electrical energy transfer between two circuits takes place through a transformer without the use of moving parts; the transformer therefore has higher efficiency and low maintenance cost as compared to rotating electrical machines. There are continuous developments and introductions of better grades of core material. The important stages of core material development can be summarized as: non-oriented silicon steel, hot rolled grain oriented silicon steel, cold rolled grain oriented (CRGO) silicon steel, Hi-B, laser scribed and mechanically scribed. The last three materials are improved versions of CRGO. Saturation flux density has remained more or less constant around 2.0 Tesla for CRGO; but there is a continuous improvement in watts/kg and volt-amperes/kg characteristics in the rolling direction. The core material developments are spearheaded by big steel manufacturers, and the transformer designers can optimize the performance of core by using efficient design and manufacturing technologies. The core building technology has improved from the non-mitred to mitred and then to the step-lap construction. A trend of reduction of transformer core losses in the last few years is the result of a considerable increase in energy costs. The better grades of core steel not only reduce the core loss but they also help in reducing the noise level by few decibels. Use of amorphous steel for transformer cores results in substantial core loss reduction (loss is about one-third that of CRGO silicon steel). Since the manufacturing technology of handling this brittle material is difficult, its use in transformers is not widespread.

Windings: The rectangular paper-covered copper conductor is the most commonly used conductor for the windings of medium and large power transformers. These conductors can be individual strip conductors, bunched conductors or continuously transposed cable (CTC) conductors. In low voltage side of a distribution transformer, where much fewer turns are involved, the use of copper or aluminum foils may find preference. To enhance the short circuit withstand capability, the work hardened copper is commonly used instead of soft annealed copper, particularly for higher rating transformers. In the case of a generator transformer having high current rating, the CTC conductor is mostly used which gives better space factor and reduced eddy losses in windings. When the CTC conductor is used in transformers, it is usually of epoxy bonded type to enhance its short circuit strength. Another variety of copper conductor or aluminum conductor is with the thermally upgraded insulating paper, which is suitable for hot-spot temperature of about 110°C. It is possible to meet the special overloading conditions with the help of this insulating paper. Moreover, the aging of winding insulation material will be slowed down comparatively. For better mechanical properties, the epoxy diamond dot paper can be used as an interlayer insulation for a multi-layer winding. High temperature superconductors may find their application in power transformers which are expected to be available commercially within next few years. Their success shall depend on economic viability, ease of manufacture and reliability considerations.

***Insulation and cooling*:** Pre-compressed pressboard is used in windings as opposed to the softer materials used in earlier days. The major insulation (between windings, between winding and yoke, etc.) consists of a number of oil ducts formed by suitably spaced insulating cylinders / barriers. Well profiled angle rings, angle caps and other special insulation components are also used.

Mineral oil has traditionally been the most commonly used electrical insulating medium and coolant in transformers. Studies have proved that oil-barrier insulation system can be used at the rated voltages greater than 1000 kV. A high dielectric strength of oil-impregnated paper and pressboard is the main reason for using oil as the most important constituent of the transformer insulation system. Manufacturers have used silicon-based liquid for insulation and cooling. Due to non-toxic dielectric and self-extinguishing properties, it is selected as a replacement of Askarel. High cost of silicon is an inhibiting factor for its widespread use. Super-biodegradable vegetable seed based oils are also available for use in environmentally sensitive locations.

There is considerable advancement in the technology of gas immersed transformers in recent years. SF6 gas has excellent dielectric strength and is non-flammable. Hence, SF6 transformers find their application in the areas where fire-hazard prevention is of paramount importance. Due to lower specific gravity of SF6 gas, the gas insulated transformer is usually lighter than the oil insulated transformer. The dielectric strength of SF6 gas is a function of the operating pressure; the higher the pressure, the higher the dielectric strength. However, the heat capacity and thermal time constant of SF6 gas are smaller than that of oil, resulting in reduced overload capacity of SF6 transformers as compared to oil-immersed transformers. Environmental concerns, sealing problems, lower cooling capability and present high cost of manufacture are the challenges which have to be overcome for the widespread use of SF6 cooled transformers.

Dry-type resin cast and resin impregnated transformers use class *F* or *C* insulation. High cost of resins and lower heat dissipation capability limit the use of these transformers to small ratings. The dry-type transformers are primarily used for the indoor application in order to minimize fire hazards. Nomex paper insulation, which has temperature withstand capacity of 220°C, is widely used for dry-type transformers. The initial cost of a dry-type transformer may be 60 to 70% higher than that of an oil-cooled transformer at current prices, but its overall cost at the present level of energy rate can be very much comparable to that of the oil-cooled transformer.

***Design*:** With the rapid development of digital computers, the designers are freed from the drudgery of routine calculations. Computers are widely used for optimization of transformer design. Within a matter of a few minutes, today's computers can work out a number of designs (by varying flux density, core diameter, current density, etc.) and come up with an optimum design. The real benefit due to computers is in the area of analysis. Using commercial 2-D / 3-D field computation software, any kind of engineering analysis (electrostatic, electromagnetic, structural, thermal, etc.) can be performed for optimization and reliability enhancement of transformers.

***Manufacturing*:** In manufacturing technology, superior techniques listed below are used to reduce manufacturing time and at the same time to improve the product quality:

- High degree of automation for slitting / cutting operations to achieve better dimensional accuracy for the core laminations
- Step-lap joint for core construction to achieve a lower core loss and noise level; top yoke is assembled after lowering windings and insulation at the assembly stage
- Automated winding machines for standard distribution transformers
- Vapour phase drying for effective and fast drying (moisture removal) and cleaning
- Low frequency heating for the drying process of distribution transformers
- Pressurized chambers for windings and insulating parts to protect against pollution and dirt
- Vertical machines for winding large capacity transformer coils
- Isostatic clamping for accurate sizing of windings
- High frequency brazing for joints in the windings and connections

***Accessories*:** Bushings and tap changer (off-circuit and on-load) are the most important accessories of a transformer. The technology of bushing manufacture has advanced from the oil impregnated paper (OIP) type to resin impregnated paper (RIP) type, both of which use porcelain insulators. The silicon rubber bushings are also available for oil-to-air applications. Due to high elasticity and strength of the silicon rubber material, the strength of these bushings against mechanical stresses and shocks is higher. The oil-to-SF6 bushings are used in GIS (gas insulated substation) applications.

The service reliability of on load tap changers is of vital importance since the continuity of the transformer depends on the performance of tap changer for the entire (expected) life span of 30 to 40 years. It is well known that the tap changer failure is one of the principal causes of failure of transformers. Tap changers, particularly on-load tap changers (OLTC), must be inspected at regular intervals to maintain a high level of operating reliability. Particular attention must be given for inspecting the diverter switch unit, oil, shafts and motor drive unit. The majority of failures reported in service are due to mechanical problems related to the drive system, for which improvements in design may be necessary. For service reliability of OLTCs, several monitoring methods have been proposed, which include measurement of contact resistance, monitoring of drive motor torque / current, acoustic measurements, dissolved gas analysis and temperature rise measurements.

***Diagnostic techniques*:** Several on-line and off-line diagnostic tools are available for monitoring of transformers to provide information about their operating conditions. Cost of these tools should be lower and their performance reliability should be higher for their widespread use. The field experience in some of the monitoring techniques is very much limited. A close cooperation between manufacturers and utilities is necessary for developing good monitoring and diagnostic systems for transformers.

Transformer technology is developing at a tremendous rate. The computerized methods are replacing the manual working in the design. Continuous improvements in material and manufacturing technologies along with the use of advanced computational tools have contributed in making transformers more efficient, compact and reliable. The modern information technology, advanced diagnostic tools and several emerging trends in transformer applications are expected to fulfill a number of existing and future requirements of utilities and end-users of transformers.

1.2 Applications and Types of Transformers

Before invention of transformers, in initial days of electrical industry, power was distributed as direct current at low voltage. The voltage drop in lines limited the use of electricity to only urban areas where consumers were served with distribution circuits of small length. All the electrical equipment had to be designed for the same voltage. Development of the first transformer around 1885 dramatically changed transmission and distribution systems. The alternating current (AC) power generated at a low voltage could be stepped up for the transmission purpose to higher voltage and lower current, reducing voltage drops and transmission losses. Use of transformers made it possible to transmit the power economically hundreds of kilometers away from the generating station. Step-down transformers then reduced the voltage at the receiving stations for distribution of power at various standardized voltage levels for its use by the consumers. Transformers have made AC systems quite flexible because the various parts and equipment of the power system can be operated at economical voltage levels by use of transformers with suitable voltage ratio. A single-line diagram of a typical power system is shown in figure 1.1. The voltage levels mentioned in the figure are different in different countries depending upon their system design. Transformers can be broadly classified, depending upon their application as given below.

***a. Generator transformers*:** Power generated at a generating station (usually at a voltage in the range of 11 to 25 kV) is stepped up by a generator transformer to a higher voltage (220, 345, 400 or 765 kV) for transmission. The generator transformer is one of the most important and critical components of the power system. It usually has a fairly uniform load. Generator transformers are designed with higher losses since the cost of supplying losses is cheapest at the generating station. Lower noise level is usually not essential as other equipment in the generating station may be much noisier than the transformer.

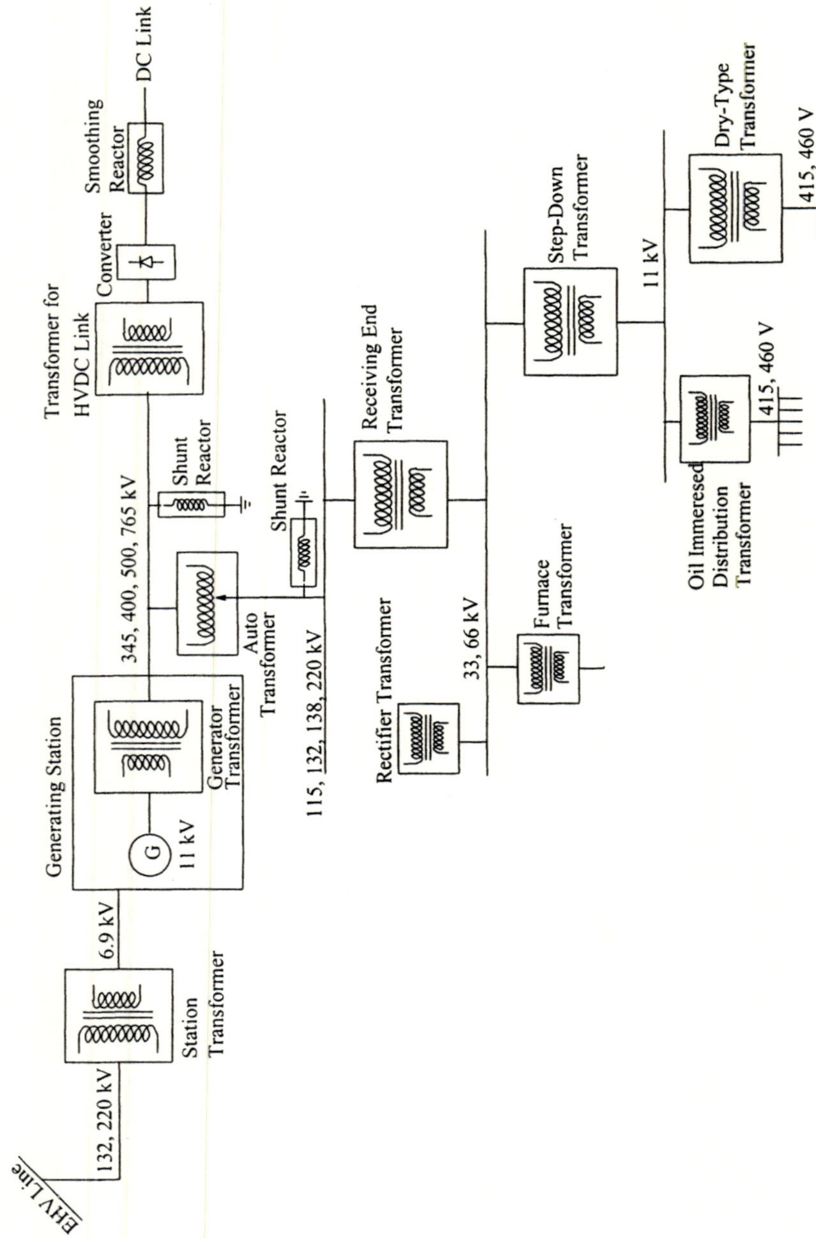

Figure 1.1 Different types of transformers in a typical power system

Generator transformers are usually provided with off-circuit tap changer with a small variation in voltage (e.g., ± 5%) because the voltage can always be controlled by field of the generator. Generator transformers with OLTC are also used for reactive power control of the system. They may be provided with a compact unit cooler arrangement for want of space in the generating stations (transformers with unit coolers have only one rating with oil forced and air forced cooling arrangement). Alternatively, they may also have oil to water heat exchangers for the same reason. It may be economical to design the tap winding as a part of main HV winding and not as a separate winding. This may be permissible since axial short circuit forces are lower due to a small tapping range. Special care has to be taken while designing high current LV lead termination to avoid any hot-spot in the conducting metallic structural parts in its vicinity. The epoxy bonded CTC conductor is commonly used for LV winding to minimize eddy losses and provide greater short circuit strength. Severe overexcitation conditions are taken into consideration while designing generator transformers.

***b. Unit auxiliary transformers*:** These are step-down transformers with primary connected to generator output directly. The secondary voltage is of the order of 6.9 kV for supplying to various auxiliary equipment in the generating station.

***c. Station transformers*:** These transformers are required to supply auxiliary equipment during setting up of the generating station and subsequently during each start-up operation. The rating of these transformers is small, and their primary is connected to a high voltage transmission line. This may result in a smaller conductor size for HV winding, necessitating special measures for increasing the short circuit strength. The split secondary winding arrangement is often employed to have economical circuit breaker ratings.

***d. Interconnecting transformers*:** These are normally autotransformers used to interconnect two grids / systems operating at two different system voltages (say, 400 and 220 kV or 345 and 138 kV). They are normally located in the transmission system between the generator transformer and receiving end transformer, and in this case they reduce the transmission voltage (400 or 345 kV) to the sub-transmission level (220 or 138 kV). In autotransformers, there is no electrical isolation between primary and secondary windings; some volt-amperes are conductively transformed and remaining are inductively transformed. Autotransformer design becomes more economical as the ratio of secondary voltage to primary voltage approaches towards unity. These are characterized by a wide tapping range and an additional tertiary winding which may be loaded or unloaded. Unloaded tertiary acts as a stabilizing winding by providing a path for the third harmonic currents. Synchronous condensers or shunt reactors are connected to the tertiary winding, if required, for reactive power compensation. In the case of an unloaded tertiary, adequate conductor area and proper supporting arrangement are provided for withstanding short circuit forces under asymmetrical fault conditions.

***e. Receiving station transformers*:** These are basically step-down transformers reducing transmission / sub-transmission voltage to primary feeder level (e.g., 33 kV). Some of these may be directly supplying an industrial plant. Loads on these transformers vary over wider limits, and their losses are expensive. The farther the location of transformers from the generating station, the higher the cost of supplying the losses. Automatic tap changing on load is usually necessary, and tapping range is higher to account for wide variation in the voltage. A lower noise level is desirable if they are close to residential areas.

***f. Distribution transformers*:** Using distribution transformers, the primary feeder voltage is reduced to actual utilization voltage (~ 415 or 460 V) for domestic / industrial use. A great variety of transformers fall into this category due to many different arrangements and connections. Load on these transformers varies widely, and they are often overloaded. A lower value of no-load loss is desirable to improve all-day efficiency. Hence, the no-load loss is usually capitalized with a high rate at the tendering stage. Since very little supervision is possible, users expect the least maintenance on these transformers. The cost of supplying losses and reactive power is highest for these transformers.

Classification of transformers as above is based on their location and broad function in the power system. Transformers can be further classified as per their specific application as given below. In this chapter, only main features are highlighted; details of some of them are discussed in the subsequent chapters.

***g. Phase shifting transformers*:** These are used to control power flow over transmission lines by varying the phase angle between input and output voltages of the transformer. Through a proper tap change, the output voltage can be made to either lead or lag the input voltage. The amount of phase shift required directly affects the rating and size of the transformer. Presently, there are two types of design: single-core and two-core design. Single-core design is used for small phase shifts and lower MVA / voltage ratings. Two-core design is normally used for bulk power transfer with large ratings of phase shifting transformers. It consists of two transformers, one associated with the line terminals and other with the tap changer.

***h. Earthing or grounding transformers*:** These are used to get a neutral point that facilitates grounding and detection of earth faults in an ungrounded part of a network (e.g., the delta connected systems). The windings are usually connected in the zigzag manner, which helps in eliminating third harmonic voltages in the lines. These transformers have the advantage that they are not affected by a DC magnetization.

i. Transformers for rectifier and inverter circuits: These are otherwise normal transformers except for the special design and manufacturing features to take into account the harmonic effects. Due to extra harmonic losses, operating flux density in core is kept lower (around 1.6 Tesla) and also conductor dimensions are smaller for these transformers. A proper de-rating factor is applied depending upon the magnitudes of various harmonic components. A designer has to adequately check the electromagnetic and thermal aspects of design. For transformers used with HVDC converters, insulation design is the most challenging design aspect. The insulation has to be designed for combined AC-DC voltage stresses.

j. Furnace duty transformers: These are used to feed the arc or induction furnaces. They are characterized by a low secondary voltage (80 to 1000 V) and high current (10 to 60 kA) depending upon the MVA rating. Non-magnetic steel is invariably used for the LV lead termination and tank in the vicinity of LV leads to eliminate hot spots and minimize stray losses. High current bus-bars are interleaved to reduce the leakage reactance. For very high current cases, the LV terminals are in the form of U-shaped copper tubes of certain inside and outside diameters so that they can be cooled by oil / water circulation from inside. In many cases, a booster transformer is used along with the main transformer to reduce the rating of tap-changers.

k. Freight loco transformers: These are mounted on the locomotives within the engine compartment itself. The primary voltage collected from an overhead line is stepped down to an appropriate level by these transformers for feeding to the rectifiers, whose output DC voltage drives the locomotives. The structural design should be such that it can withstand vibrations. Analysis of natural frequencies of vibration is done to eliminate possibility of resonance.

l. Hermetically sealed transformers: This construction does not permit any outside atmospheric air to get into the tank. It is completely sealed without any breathing arrangement, obviating need of periodic filtration and other normal maintenance. These transformers are filled with mineral oil or synthetic liquid as a cooling / dielectric medium and sealed completely by having an inert gas, like nitrogen, between the coolant and top tank plate. The tank is of welded cover construction, eliminating the joint and related leakage problems. Here, the expansion of oil is absorbed by the inert gas layer. The tank design should be suitable for pressure buildup at elevated temperatures. The cooling is not effective at the surface of oil, which is at the highest temperature. In another type of sealed construction, these disadvantages are overcome by deletion of the gas layer. The expansion of oil is absorbed by the deformation of the cooling system, which can be an integral part of the tank structure.

m. Outdoor and indoor transformers: Most of the transformers are of outdoor duty type, which have to be designed for withstanding atmospheric pollutants. The creepage distance of bushing insulator gets decided according to the pollution level. The higher the pollution level, the greater the creepage distance required from the live terminal to ground. Contrary to the outdoor transformers, an indoor transformer is designed for installation under a weatherproof roof and/or in a properly ventilated room. Standards define the minimum ventilation required for an effective cooling. Adequate clearances are kept between the walls and transformer to eliminate the possibility of higher noise level due to reverberations.

There are many more types of transformers having applications in electronics, electric heaters, traction, etc. Some applications have significant impact on the design of transformers. The duty (load) of transformers can be very onerous. For example, current density in transformers with frequent motor starting duty has to be lower to take care of high starting current of motors, which can be of the order of 6 to 8 times the full load current.

Shunt and series reactors are very important components of the power system. Design of reactors, which have only one winding, is similar to transformers in many aspects. Their special features are given below.

***n. Shunt Reactors*:** These are used to compensate the capacitive VARs generated during low loads and switching operations in extra high voltage transmission networks, thereby maintaining the voltage profile of a transmission line within desirable limits. These are installed at a number of places along the length of the line. They can be either permanently connected or switched type. Use of shunt reactors under normal operating conditions may result in poor voltage levels and increased losses. Hence, the switched-in types are better since they are connected only when the voltage levels are required to be controlled. When connected to the tertiary windings of a large transformer, they become cost-effective. Voltage drop in high series reactance between HV and tertiary windings must be accounted for when deciding the voltage rating of tertiary connected shunt reactors. Shunt reactors can be of core-less (air-core) or gapped-core (magnetic circuit with non-magnetic gaps) design. The flux density in the air-core reactor has to be smaller as the flux path is not well constrained. Eddy losses in the winding and stray losses in the structural conducting parts are higher in this type of reactor. In contrast, the gapped-core reactor is more compact due to higher permissible flux density. The gap length can be suitably designed to get a desired reactance value. Shunt reactors are usually designed to have a constant impedance characteristics up to 1.5 times the rated voltage to minimize the harmonic current generation under over-voltage conditions.

***o. Series Reactors*:** These reactors are connected in series with generators, feeders and transmission lines for limiting fault currents under short circuits. Series reactors should have linear magnetic characteristics under fault conditions. They are designed to withstand mechanical and thermal effects of short circuits. Series reactors used in transmission lines have a fully insulated winding since both its ends should be able to withstand the lightning impulse voltages. The value of series reactance has to be judiciously selected because a higher value reduces the power transfer capability of the line. The smoothing reactors used in HVDC transmission system, connected between the converter and DC line, smoothen the DC voltage ripple.

1.3 Principles and Equivalent Circuit of a Transformer

1.3.1 Ideal transformer

A transformer works on the principle of electromagnetic induction, according to which a voltage is induced in a coil if it links a changing flux. Figure 1.2 shows a single-phase transformer consisting of two windings, wound on a magnetic core and linked by a mutual flux ϕ_m. Transformer is in no-load condition with primary connected to a source of sinusoidal voltage of frequency f Hz. Primary winding draws a small excitation current, i_0 (instantaneous value), from the source to set up the mutual flux ϕ_m in the core. All the flux is assumed to be contained in the core (no leakage). The windings 1 and 2 have N_1 and N_2 turns respectively. The instantaneous value of induced electromotive force in the winding 1 due to the mutual flux is

$$e_1 = N_1 \frac{d\phi_m}{dt} \tag{1.1}$$

Equation 1.1 is as per the circuit viewpoint; there is flux viewpoint also [1], in which induced voltage (counter electromotive force) is represented as $e_1 = -N_1(d\phi_m / dt)$. The elaborate explanation for both the viewpoints is given in [2]. If the winding is assumed to have zero winding resistance,

$$v_1 = e_1 \tag{1.2}$$

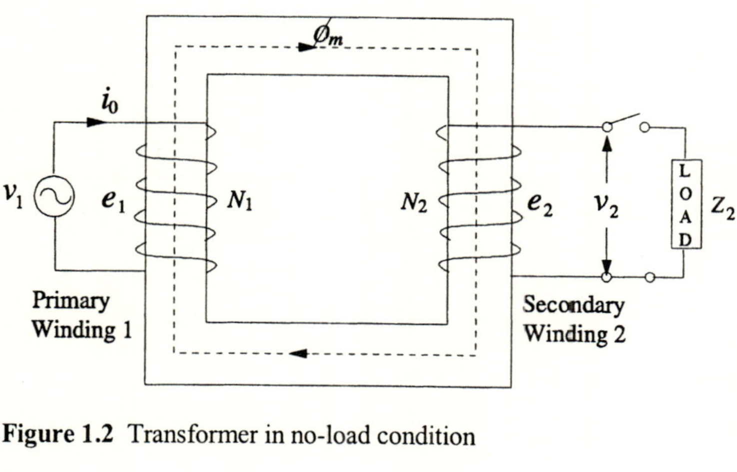

Figure 1.2 Transformer in no-load condition

Since v_1 (instantaneous value of the applied voltage) is sinusoidally varying, the flux ϕ_m must also be sinusoidal in nature varying with frequency *f*. Let

$$\phi_m = \phi_{mp} \sin \omega t \tag{1.3}$$

where ϕ_{mp} is the peak value of mutual flux ϕ_m and $\omega = 2\pi f$ rad/sec. After substituting the value of ϕ_m in equation 1.1, we get

$$e_1 = N_1\, \omega\, \phi_{mp} \cos \omega t \tag{1.4}$$

The r.m.s. value of the induced voltage, E_1, is obtained by dividing the peak value in equation 1.4 by $\sqrt{2}$,

$$E_1 = 4.44\, \phi_{mp}\, f\, N_1 \tag{1.5}$$

Equation 1.5 is known as *emf equation* of a transformer. For a given number of turns and frequency, the flux (and flux density) in a core is entirely determined by the applied voltage.

The voltage induced in winding 2 due to the mutual flux ϕ_m is given by

$$e_2 = N_2 \frac{d\phi_m}{dt} \tag{1.6}$$

The ratio of two induced voltages can be derived from equations 1.1 and 1.6 as

$$e_1/e_2 = N_1/N_2 = a \tag{1.7}$$

where a is known as ratio of transformation. Similarly, r.m.s. value of the induced voltage in winding 2 is

$$E_2 = 4.44\, \phi_{mp}\, f\, N_2 \tag{1.8}$$

The exciting current (i_0) is only of magnetizing nature (i_m) if B-H curve of core material is assumed without hysteresis and if eddy current losses are neglected. The magnetizing current (i_m) is in phase with the mutual flux in the absence of hysteresis. Also, linear magnetic (B-H) characteristics are assumed.

Now, if the secondary winding in figure 1.2 is loaded, secondary current is set up as per Lenz's law such that the secondary magnetomotive force (mmf), i_2N_2, opposes the mutual flux ϕ_m tending to reduce it. In an ideal transformer $e_1 = v_1$, because for a constant value of the applied voltage, induced voltage and corresponding mutual flux must remain constant. This can happen only if the primary draws more current (i_1') for neutralizing the demagnetizing effect of secondary ampere-turns. In r.m.s. notations,

$$I_1' N_1 = I_2 N_2 \tag{1.9}$$

Thus, the total primary current is a vector sum of the no-load current (i.e., magnetizing component, I_m, since core losses are neglected) and the load current (I_1'),

$$\bar{I}_1 = \bar{I}_1' + \bar{I}_m \tag{1.10}$$

For an infinite permeability magnetic material, magnetizing current is zero. Equation 1.9 then becomes

$$I_1 N_1 = I_2 N_2 \tag{1.11}$$

Thus, for an ideal transformer when its no-load current is neglected, primary ampere-turns are equal to secondary ampere-turns. The same result can also be arrived at by applying Ampere's law, which states that the magnetomotive force around a closed path is given by

$$\oint H \bullet dl = i \tag{1.12}$$

where i is the current enclosed by the line integral of the magnetic field intensity H around the closed path of flux ϕ_m,

$$\oint H \bullet dl = \oint (B/\mu) \bullet dl = i_1 N_1 - i_2 N_2 \tag{1.13}$$

If the relative permeability of the magnetic path is assumed as infinite, the integral of magnetic field intensity around the closed path is zero. Hence, in the r.m.s. notations,

$$I_1 N_1 - I_2 N_2 = 0 \tag{1.14}$$

which is the same result as in equation 1.11.

Thus, for an ideal transformer (zero winding resistance, no leakage flux, linear B-H curve with an infinite permeability, no core losses), it can be summarized as,

$$\frac{E_1}{E_2} = \frac{V_1}{V_2} = \frac{N_1}{N_2} = \frac{I_2}{I_1} \tag{1.15}$$

and

$$V_1 I_1 = V_2 I_2 \tag{1.16}$$

Schematic representation of the transformer in figure 1.2 is shown in figure 1.3. The polarities of voltages depend upon the directions in which the primary and secondary windings are wound. It is common practice to put a dot at the end of the windings such that the dotted ends of the windings are positive at the same time, meaning that the voltage drops from the dotted to unmarked terminals are in phase. Also, currents flowing from the dotted to unmarked terminals in the windings produce an mmf acting in the same direction in the magnetic circuit.

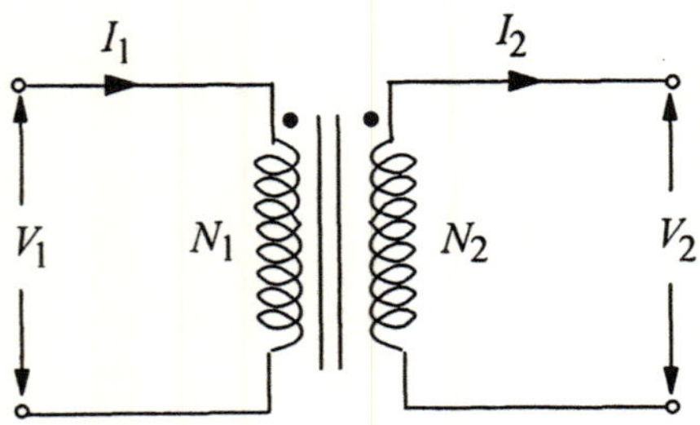

Figure 1.3 Schematic representation of transformer

If the secondary winding in figure 1.2 is loaded with an impedance Z_2,

$$Z_2 = V_2/I_2 \tag{1.17}$$

Substituting from equation 1.15 for V_2 and I_2,

$$Z_2 = \frac{(N_2/N_1)\,V_1}{(N_1/N_2)\,I_1} \tag{1.18}$$

Hence, the impedance as referred to the primary winding 1 is

$$Z_2' = \frac{V_1}{I_1} = (N_1 / N_2)^2 Z_2 \tag{1.19}$$

Similarly, any impedance Z_1 in the primary circuit can be referred to the secondary side 2 as

$$Z_1' = (N_2 / N_1)^2 Z_1 \tag{1.20}$$

It can be summarized from equations 1.15, 1.16, 1.19 and 1.20 that for an ideal transformer, voltages are transformed in ratio of turns, currents in inverse ratio of turns and impedances in square of ratio of turns, whereas the volt-amperes and power remain unchanged.

The ideal transformer transforms direct voltage, i.e., DC voltages on primary and secondary sides are related by turns ratio. This is not a surprising result because for the ideal transformer, we have assumed infinite core material permeability with linear (non-saturating) characteristics permitting core flux to rise without limit under a DC voltage application. When a DC voltage (V_{d1}) is applied to the primary winding with the secondary winding open-circuited,

$$V_{d1} = N_1 (d\phi_m / dt) \tag{1.21}$$

Thus, $(d\phi_m / dt)$ is constant (flux permitted to rise with time without any limit) and is equal to (V_{d1} / N_1). Voltage at the secondary of the ideal transformer is

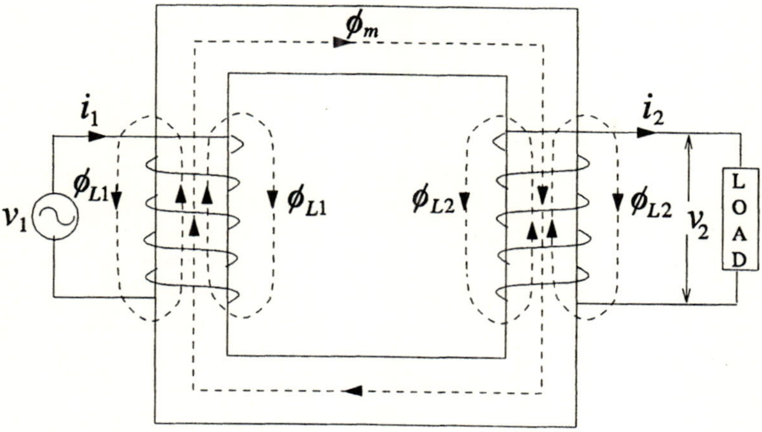

Figure 1.4 Practical transformer

$$V_{d2} = N_2(d\phi_m / dt) = (N_2 / N_1)\, V_{d1} \tag{1.22}$$

However, for a practical transformer, during the steady-state condition, current has a value of V_{d1}/R_1, and the magnetic circuit is driven into saturation reducing eventually the value of induced voltages E_1 and E_2 to zero (in saturation there is hardly any change in the flux even though the current may still be increasing till the steady state condition is reached). The current value, V_{d1}/R_1, is quite high, resulting in damage to the transformer.

1.3.2 Practical transformer

Analysis presented for the ideal transformer is merely to explain the fundamentals of transformer action; such a transformer never exists and the equivalent circuit of a real transformer shown in figure 1.4 is now developed.

Whenever a magnetic material undergoes a cyclic magnetization, two types of losses, eddy and hysteresis losses, occur in it. These losses are always present in transformers as the flux in their ferromagnetic core is of alternating nature. A detailed explanation of these losses is given in Chapter 2.

The hysteresis loss and eddy loss are minimized by use of a better grade of core material and thinner laminations, respectively. The total no-load current, I_0, consists of magnetizing component (I_m) responsible for producing the mutual flux ϕ_m and core loss component (I_c) accounting for active power drawn from the source to supply eddy and hysteresis losses. The core loss component is in phase with the induced voltage and leads the magnetizing component by 90°. With the secondary winding open-circuited, the transformer behaves as a highly inductive circuit due to magnetic core, and hence the no-load current lags the applied voltage by an angle slightly less than 90° (I_m is usually much greater than I_c). In the equivalent circuit shown in figure 1.5, the magnetizing component is represented by the inductive reactance X_m, whereas the loss component is accounted by the resistance R_c.

Let R_1 and R_2 be the resistances of windings 1 and 2, respectively. In a practical transformer, some part of the flux linking primary winding does not link the secondary. This flux component is proportional to the primary current and is responsible for a voltage drop which is accounted by an inductive reactance X_{L1} (leakage reactance) put in series with the primary winding of the ideal transformer. Similarly, the leakage reactance X_{L2} is added in series with the secondary winding to account for the voltage drop due to flux linking only the secondary winding. One can omit the ideal transformer from the equivalent circuit, if all the quantities are either referred to the primary or secondary side of the transformer. For example, in equivalent circuit of figure 1.5 (b), all quantities are referred to the primary side, where

$$X_{L2}^{'} = X_{L2}(N_1 / N_2)^2 \tag{1.23}$$

$$R_2^{'} = R_2(N_1 / N_2)^2 \tag{1.24}$$

This equivalent circuit is a passive lumped-T representation, valid generally for sinusoidal steady-state analysis at power frequencies.. For higher frequencies, capacitive effects must be considered, as discussed in Chapter 7. For any transient analysis, all the reactances in the equivalent circuit should be replaced by the corresponding equivalent inductances.

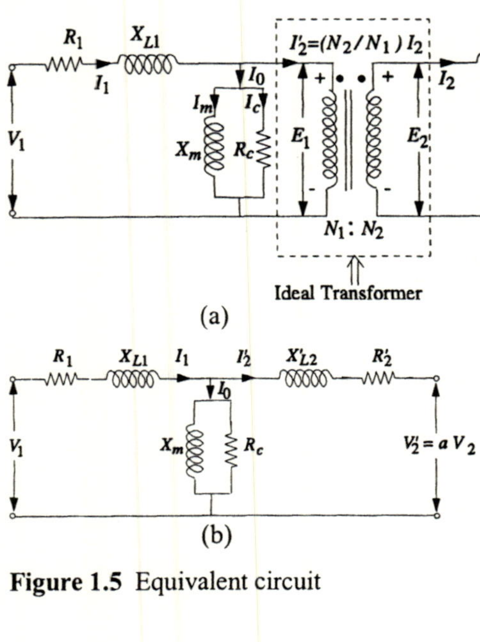

Figure 1.5 Equivalent circuit

While drawing a vector diagram, it must be remembered that all the quantities in it must be of same frequency. Actually, the magnetization (B-H) curve of core material is of non-linear nature, and it introduces higher order harmonics in the magnetizing current for a sinusoidal applied voltage of fundamental frequency. In the vector diagram, however, a linear B-H curve is assumed neglecting harmonics. The aspects related to the core magnetization and losses are dealt in Chapter 2. For figure 1.5 (a), the following equations can be written:

$$V_1 = E_1 + (R_1 + jX_{L1})\, I_1 \tag{1.25}$$

$$V_2 = E_2 - (R_2 + jX_{L2})\, I_2 \tag{1.26}$$

Vector diagrams for primary and secondary voltages / currents are shown in figure 1.6. The output terminal voltage V_2 is taken as a reference vector along x-axis. The load power factor angle is denoted by θ_2. The induced voltages are in phase and lead the mutual flux ϕ_{mr} (r.m.s. value of ϕ_m) by 90° in line with equations 1.1 and 1.6. The magnetizing component (I_m) of no-load current (I_0) is in phase with ϕ_{mr}, whereas the loss component I_c leads ϕ_{mr} by 90° and is in phase with the induced voltage E_1. The core loss is given as

$$P_c = I_c\, E_1 \tag{1.27}$$

or

$$P_c = I_c^2\, R_c \tag{1.28}$$

The mutual reactance X_m is

$$X_m = \frac{E_1}{I_m} \tag{1.29}$$

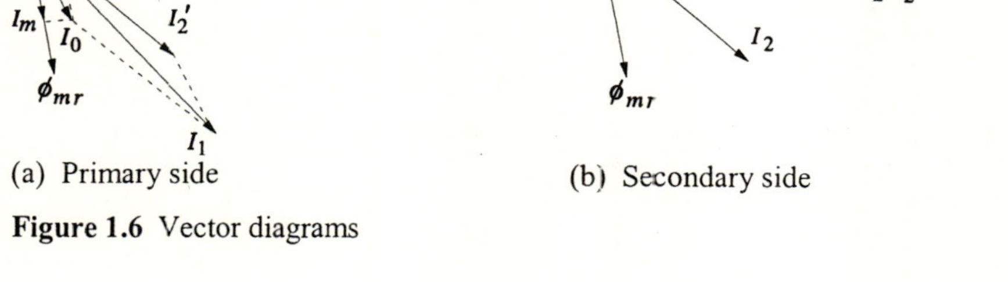

(a) Primary side (b) Secondary side

Figure 1.6 Vector diagrams

The magnitude of secondary current referred to the primary (I_2') is same as that of secondary current (I_2), since the turns of primary and secondary windings are assumed equal. There is some phase shift between the terminal voltages V_1 and V_2 due to the voltage drops in the leakage impedances. The voltage drops in the resistances and leakage reactances have been exaggerated in the vector diagram. The voltage across a winding resistance is usually around 0.5% of the terminal voltage for large power transformers, whereas the voltage drop in leakage impedance depends on the impedance value of the transformer. For small distribution transformers (e.g., 5 MVA), the value of impedance is around 4 to 7% and for power transformers it can be anywhere in the range of 8 to 20% depending upon the regulation and system protection requirements. The lower the percentage impedance, the lower the voltage drop. However, the required ratings of circuit breakers will be higher.

1.3.3 Mutual and leakage inductances

The leakage flux ϕ_{L1} shown in figure 1.4 is produced by the current i_1 in winding 1, which only links winding 1. Similarly, the leakage flux ϕ_{L2} is produced by the current i_2 in winding 2, which only links winding 2. The primary leakage inductance is

$$L_{L1} = N_1 \frac{d\phi_{L1}}{di_1} \tag{1.30}$$

Differential reluctance offered to the path of leakage flux is

$$\Re_{L1} = N_1 \frac{di_1}{d\phi_{L1}} \tag{1.31}$$

Equations 1.30 and 1.31 give

$$L_{L1} = \frac{N_1^2}{\Re_{L1}} \tag{1.32}$$

Similarly, the leakage inductance of secondary winding is

$$L_{L2} = \frac{N_2^2}{\Re_{L2}} \tag{1.33}$$

Let us derive the expression for mutual inductance M. Using equation 1.6,

$$e_2 = N_2 \frac{d\phi_m}{dt} = N_2 \frac{d\phi_m}{di_m} \frac{di_m}{dt} = M_{21} \frac{di_m}{dt} \tag{1.34}$$

where,

$$M_{21} = N_2 \frac{d\phi_m}{di_m} = N_2 \frac{N_1}{\Re_m} \tag{1.35}$$

M_{21} represents flux linkages in the secondary winding due to magnetizing current (i_m) in the primary winding divided by the current i_m. Reluctance offered to the path of mutual flux (ϕ_m) is denoted by $\Re_m$. Similarly,

$$M_{12} = N_1 \frac{N_2}{\Re_m} \tag{1.36}$$

Thus, the mutual inductance M is given by

$$M = M_{12} = M_{21} = \frac{N_1 N_2}{\Re_m} \tag{1.37}$$

Let $\Re_1$ represent the total reluctance of parallel paths of two fluxes, viz. leakage flux ϕ_{L1} of winding 1 and mutual flux ϕ_m. Also let

$$k_1 = \frac{\Re_1}{\Re_m} \tag{1.38}$$

The self inductance L_1 of winding 1, when $i_2 = 0$, is

$$L_1 = \frac{N_1^2}{\Re_1} = \frac{N_1^2}{\Re_1} \frac{\Re_m}{\Re_m} \frac{N_2}{N_2} = \frac{aM}{k_1} \tag{1.39}$$

Similarly let

$$k_2 = \frac{\Re_2}{\Re_m} \tag{1.40}$$

and we get

$$L_2 = \frac{M}{ak_2} \tag{1.41}$$

Hence,

$$M = \sqrt{k_1 k_2}\sqrt{L_1 L_2} = k\sqrt{L_1 L_2} \tag{1.42}$$

Coefficient $k = \sqrt{k_1 k_2}$ is a measure of coupling between the two windings. From definitions of k_1 and k_2 (e.g., $k_1 = \frac{\Re_1}{\Re_m} = \frac{1/\Re_m}{1/\Re_1} = \frac{1/\Re_m}{1/\Re_m + 1/\Re_{L1}}$), it is clear that $0 \le k_1 \le 1$ and $0 \le k_2 \le 1$, giving $0 \le k \le 1$. For $k = 1$, windings are said to have perfect coupling with no leakage flux, which is possible only for the ideal transformer.

1.3.4 Simplified equivalent circuit

Since the no-load current and voltage drop in the leakage impedance are usually small, it is often permissible to simplify the equivalent circuit of figure 1.5 (b) by doing some approximations. Terminal voltages (V_1, V_2) are not appreciably different than the corresponding induced voltages, and hence a little error is caused if the no-load current is made to correspond to the terminal voltage instead of the induced voltage. For example, if the excitation branch (consisting of X_m in parallel R_c) is shifted to the input terminals (excited by V_1), the approximate equivalent circuit will be as shown in figure 1.7 (a). If we totally neglect the no-load excitation current, since it is much less as compared to the full load current, the circuit can be further simplified as shown in figure 1.7 (b). This simplified circuit, in which a transformer is represented by the series impedance of Z_{eq1}, is considered to be sufficiently accurate for modeling purpose in power system studies. Since R_{eq1} is much smaller than X_{eq1}, a transformer can be represented just as a series reactance in most cases.

1.4 Representation of Transformer in Power System

As seen in the previous section, ohmic values of resistance and leakage reactance of a transformer depend upon whether they are referred on the LV side or HV side. A great advantage is realized by expressing voltage, current, impedance and volt-amperes in per-unit or percentage of base values of these quantities. The per-unit quantities, once expressed on a particular base, are same when referred to either side of the transformer. Thus, the value of per-unit impedance remains same on either side obviating the need for any calculations by using equations 1.19 and 1.20. This approach is very handy in power system calculations, where a large number of transformers, each having a number of windings, are present.

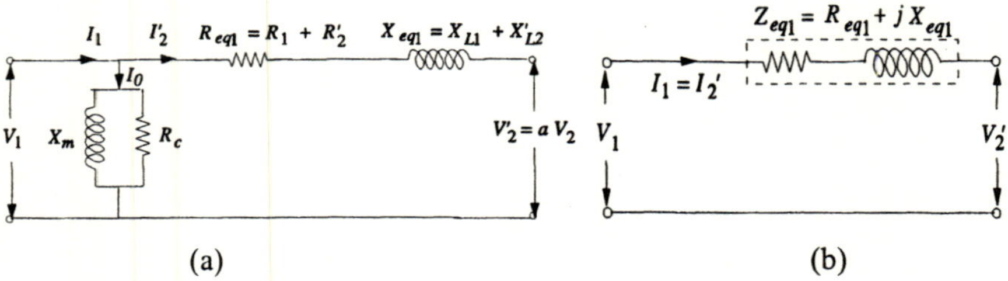

Figure 1.7 Simplified equivalent circuit

For a system, the per-unit values are derived by choosing a set of base values for various quantities. Although the base values can be chosen arbitrarily, it is preferable to use the rated quantities of a device as the base values. The per-unit quantity (p.u.) is related to the base quantity by the following relationship:

$$Per\text{-}unit\ quantity = \frac{Actual\ value\ of\ the\ quantity}{Base\ value\ of\ the\ quantity} \tag{1.43}$$

The actual and base values must be expressed in the same unit. Usually, base values of voltage and volt-amperes are chosen first, from which other base quantities are determined. The ba·- values of voltages on the LV side and HV side are denoted by V_{bL} and V_{bH} respectively. The corresponding values of base currents for the LV side and HV side are I_{bL} and I_{bH} respectively. If rated voltage of LV winding is taken as a base voltage (V_{bL}) for the LV side,

$$Per\text{-}unit\ rated\ voltage\ on\ LV\ side = \frac{Rated\ voltage\ of\ LV\ side}{V_{bL}} = 1.0 \tag{1.44}$$

Hence, the per-unit values of rated quantities are equal to unity when rated quantities are chosen as the base quantities. The per-unit quantities are ratios and dimensionless, which are to be multiplied by 100 to get the percentage (%) values.

The value of base impedance on the LV side is,

$$Z_{bL} = \frac{V_{bL}}{I_{bL}} = \frac{V_{bL}^2}{V_{bL}\ I_{bL}} = \frac{V_{bL}^2}{(VA)_b} \tag{1.45}$$

where $(VA)_b$ denotes base volt-amperes. Similarly for the HV side,

$$Z_{bH} = \frac{V_{bH}^2}{(VA)_b} \tag{1.46}$$

For the simplified equivalent circuit of figure 1.7, the equivalent total resistance referred to the primary (LV) side can be expressed in per-unit notation as,

$$(R_{eq1})_{pu} = \frac{R_{eq1}}{Z_{bL}} \tag{1.47}$$

If R_{eq2} is the total equivalent resistance of the windings referred to the secondary (HV) winding, it follows from the equations 1.24 and 1.47 that

$$(R_{eq1})_{pu} = \frac{R_{eq2}(N_1/N_2)^2}{Z_{bL}} = \frac{R_{eq2}}{Z_{bL}(N_2/N_1)^2} = \frac{R_{eq2}}{Z_{bH}} = (R_{eq2})_{pu} = (R_{eq})_{pu} \tag{1.48}$$

Similarly, it can easily be verified that the per-unit values of impedance calculated on the LV and HV sides are equal.

The per-unit impedance can be expressed as

$$(Z_{eq})_{pu} = (Z_{eq1})_{pu} = \frac{Z_{eq1}}{Z_{bL}} = \frac{Z_{eq1}\, I_{bL}}{Z_{bL}\, I_{bL}} = \frac{Z_{eq1}\, I_{bL}}{V_{bL}} \tag{1.49}$$

Thus, $(Z_{eq})_{pu}$ denotes the per-unit value of leakage impedance voltage drop on the LV (or HV) side. For example, if 1000/100 V transformer has $(Z_{eq})_{pu}$ of 0.1, the voltage drop across the equivalent leakage impedance referred to the LV side is 0.1 times 100 volts, i.e., 10 volts; the corresponding voltage drop on the HV side is 100 volts ($= 0.1 \times 1000$). Similarly,

$$(R_{eq})_{pu} = (R_{eq1})_{pu} = \frac{R_{eq1}}{Z_{bL}} = \frac{R_{eq1}\, I_{bL}\, I_{bL}}{Z_{bL}\, I_{bL}\, I_{bL}} = \frac{R_{eq1}\, I_{bL}^2}{V_{bL}\, I_{bL}} \tag{1.50}$$

Thus, the per-unit value of resistance $(R_{eq})_{pu}$ is a ratio of ohmic loss at the rated current to the rated volt-amperes. For example, $(R_{eq})_{pu}$ of 0.02 for 50 kVA, 1000/100 V transformer means that the total ohmic loss at the rated current is 0.02 times (2% of) 50 kVA, i.e., 1000 watts.

Another advantage of using per-unit system is that the impedances of transformers of the same type (irrespective of their ratings) lie usually within a small known range of per-unit values although the ohmic values may be widely different.

For large power transformers, base voltage is usually expressed in kV and base volt-amperes in MVA. Hence, the base impedance on either side can be calculated as

$$Z_{bL} = \frac{(\mathrm{kV})_{bL}^2}{(\mathrm{MVA})_b} \quad \text{and} \quad Z_{bH} = \frac{(\mathrm{kV})_{bH}^2}{(\mathrm{MVA})_b} \tag{1.51}$$

For a three-phase transformer, the total three-phase MVA and line-to-line kV are taken as the base values. It can be shown that when ohmic value of impedance is transferred from one side to other, the multiplying factor is the ratio of squares of line-to-line voltages of both sides irrespective of whether transformer connection is star-star or star-delta [3].

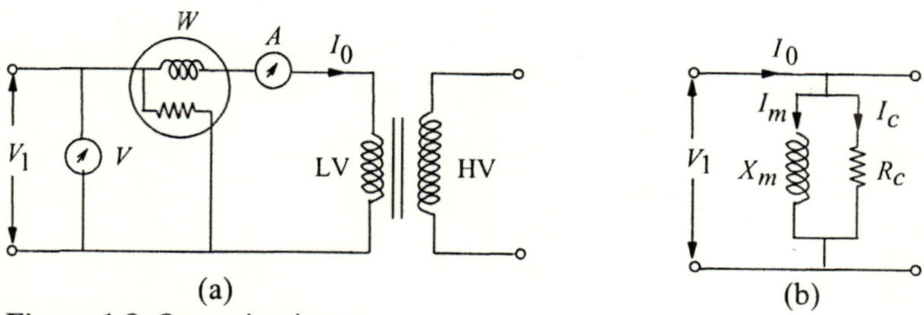

Figure 1.8 Open circuit test

1.5 Open Circuit and Short Circuit Tests

Parameters of the equivalent circuit can be determined by open circuit (no-load) and short circuit (load) tests. Open circuit test determines the parameters of shunt branch of the equivalent circuit of figure 1.5. The circuit diagram for conducting the test is shown in figure 1.8. The rated voltage is applied to one winding and other winding is kept open (usually LV winding is supplied, while HV is kept open for ease of testing and availability of supply). Since the no-load current is a very small percentage of the full load current, which can be in the range of 0.2 to 2% (for large power transformers, e.g., above 300 MVA, no-load current can be as small as about 0.2%), the voltage drop in LV resistance and leakage reactance is negligible as compared to the rated voltage ($\therefore V_1 \cong E_1$ in figure 1.5). The input power measured by a wattmeter consists of the core loss and primary winding ohmic loss. If the no-load current is 1% of full load current, ohmic loss in primary winding resistance is just 0.01% of the load loss at rated current; the value of winding loss is negligible as compared to the core losses. Hence, the entire wattmeter reading can be taken as the total core loss. The equivalent circuit of figure 1.5 (b) gets simplified to that shown in figure 1.8 (b).

The no-load (core loss) P_c measured by the wattmeter is expressed as

$$P_c = V_1 I_0 \cos\theta_0 \tag{1.52}$$

From the measured values of P_c, V_1 and I_0, the value of no-load power factor can be calculated from equation 1.52 as

$$\cos\theta_0 = \frac{P_c}{V_1 I_0} \tag{1.53}$$

With reference to the vector diagram of figure 1.6, the magnetizing component (I_m) and the core loss component (I_c) of the no-load current (I_0) are

$$I_c = I_0 \cos\theta_0 \tag{1.54}$$

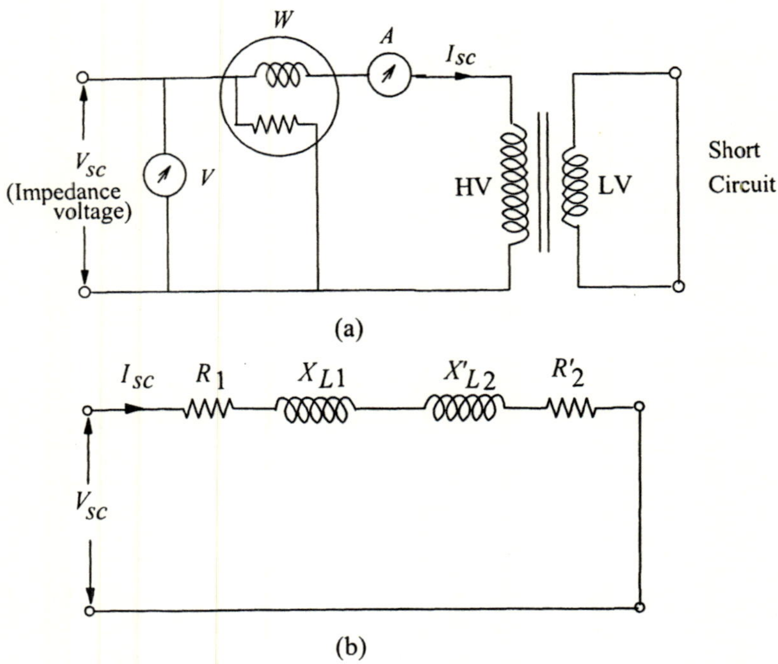

Figure 1.9 Short circuit test (with LV terminals short-circuited)

$$I_m = I_0 \sin\theta_0 \tag{1.55}$$

The shunt branch parameters of the equivalent circuit can be estimated as

$$R_c = \frac{P_c}{I_c^2} \quad \text{or} \quad R_c = \frac{V_1^2}{P_c} \tag{1.56}$$

$$X_m = \frac{V_1}{I_m} \tag{1.57}$$

These values are with reference to the LV side, since the measuring instruments are placed on the LV side. If required, they can be referred to the HV side by using the operator a^2. The value of magnetizing reactance is very high as compared to the leakage reactance. For a no-load current of 0.2% (and with the assumption that $I_0 \cong I_m$), the value of X_m is 500 per-unit.

A short circuit test is done to measure the load loss and leakage impedance of a transformer. In this test, usually the LV winding is short-circuited and voltage is applied to the HV winding in order to circulate the rated currents in both the windings; the voltage required to be applied is called as the impedance voltage of the transformer. For a transformer having 10% leakage impedance, voltage required to circulate the rated current is 10% of the rated voltage. The circuit diagram for short circuit test is shown in figure 1.9 (a), in which LV winding (secondary winding 2) is short-circuited. For an applied voltage of 10%, assuming for the equivalent circuit of figure 1.5 (b) that the primary and referred secondary leakage impedances are equal, 5% of voltage appears across the shunt excitation branch. With a no-load current of 2% at rated voltage, the current in the shunt branch for a 5% voltage is just 0.1% of rated current (assuming linear B-H curve). Hence, the shunt branch can be neglected giving the simplified circuit of figure 1.9 (b) for the short circuit test. Since the core loss varies approximately in the square proportion of the applied voltage, with 5% voltage across the shunt excitation branch, it is just 0.25% of the core loss at the rated voltage. Hence, almost the entire loss measured by the wattmeter is the load loss of the transformer.

Equivalent circuit parameters $R_{eq1}(=R_1+R_2^{'})$, $X_{eq1}(=X_{L1}+X_{L2}^{'})$ and $Z_{eq1}(=R_{eq1}+jX_{eq1})$ can now be determined from the measured quantities of power (P_L), voltage (V_{SC}) and current (I_{SC}) as

$$Z_{eq1}=\frac{V_{SC}}{I_{SC}} \tag{1.58}$$

$$R_{eq1}=\frac{P_L}{I_{SC}^2} \tag{1.59}$$

$$X_{eq1}=\sqrt{Z_{eq1}^2-R_{eq1}^2} \tag{1. 60}$$

R_{eq1} is the equivalent AC resistance referred to the primary (HV) winding and accounts for the losses in DC resistance of windings, eddy losses in windings and stray losses in structural parts. It is not practically possible to apportion parts of stray losses to the two windings. Hence, if the resistance parameter is required for each winding, it is usually assumed that $R_1=R_2^{'}=(1/2)\ R_{eq1}$. Similarly it is assumed that $X_{L1}=X_{L2}^{'}$, although it is strictly not true. Since the value of % *R* is much smaller than % *Z*, practically percentage reactance (% *X*) is taken to be the same as percentage impedance (% *Z*). This approximation may not be true for very small distribution transformers.

1.6 Voltage Regulation and Efficiency

Since many electrical equipments and appliances operate most effectively at their rated voltage, it is necessary that the output voltage of a transformer is within narrow limits when the magnitude and power factor of loads vary. Voltage regulation is an important performance parameter of a transformer that determines the quality of electricity supplied to consumers. The voltage regulation for a specific load is defined as a change in magnitude of secondary voltage after removal of the load (primary voltage being held constant) expressed as a fraction of the secondary voltage corresponding to the no-load condition.

$$\text{Regulation (p.u.)} = \frac{V_{2oc} - V_2}{V_{2oc}} \tag{1.61}$$

where V_2 is the secondary terminal voltage at a specific load and V_{2oc} is the secondary terminal voltage when the load is removed. For the approximate equivalent circuit of a transformer (figure 1.7 (b)), if all the quantities are referred to the secondary side, the voltage regulation for a lagging power factor load is given as [4]

$$\text{Regulation (p.u.)} = \frac{I_2 R_{eq2} \cos\theta_2 + I_2 X_{eq2} \sin\theta_2}{V_{2oc}} + \frac{1}{2}\left(\frac{I_2 X_{eq2} \cos\theta_2 - I_2 R_{eq2} \sin\theta_2}{V_{2oc}}\right)^2 \tag{1.62}$$

where R_{eq2} and X_{eq2} are the equivalent resistance and leakage reactance of the transformer referred to the secondary side respectively. The secondary load current (I_2) lags behind the secondary terminal voltage (V_2) by an angle θ_2. Under the rated load conditions, with the rated values taken as base quantities,

$$\text{Regulation (p.u.)} = \varepsilon_r \cos\theta_2 + \varepsilon_x \sin\theta_2 + \frac{1}{2}(\varepsilon_x \cos\theta_2 - \varepsilon_r \sin\theta_2)^2 \tag{1.63}$$

where $\varepsilon_r = \dfrac{I_2 R_{eq2}}{V_{2oc}}$ represents the per-unit resistance drop and $\varepsilon_x = \dfrac{I_2 X_{eq2}}{V_{2oc}}$ represents the per-unit leakage reactance drop. For a leading power factor load (I_2 leads V_2 by an angle θ_2),

$$\text{Regulation (p.u.)} = \varepsilon_r \cos\theta_2 - \varepsilon_x \sin\theta_2 + \frac{1}{2}(\varepsilon_x \cos\theta_2 + \varepsilon_r \sin\theta_2)^2 \tag{1.64}$$

The square term is usually small and may be neglected, simplifying equations 1.63 and 1.64 as

$$\text{Regulation(p.u.)} = \varepsilon_r \cos\theta_2 \pm \varepsilon_x \sin\theta_2 \tag{1.65}$$

The efficiency of a transformer, like any other device, is defined as the ratio of useful output power to input power.

$$\eta = \frac{output\ power}{input\ power} \tag{1.66}$$

The percentage efficiency of a transformer is in the range of 95 to 99%. For large power transformers with low loss designs, the efficiency can be as high as 99.7%. There is a possibility of error if the efficiency is determined from the measured values of output and input powers, as the wattmeter readings may have an error of about 1%. Hence, it is a more accurate approach if the efficiency is determined using the measured values of losses by the open circuit and short circuit tests. The efficiency is then given as

$$\eta = \frac{output\ power}{output\ power + losses} \tag{1.67}$$

$$\therefore \eta = \frac{V_2 I_2 \cos\theta_2}{V_2 I_2 \cos\theta_2 + P_c + P_L} = \frac{V_2 I_2 \cos\theta_2}{V_2 I_2 \cos\theta_2 + P_c + I_2^2 R_{eq2}} \tag{1.68}$$

Although the load power factor has some effect on the mutual flux ϕ_m and hence the core loss, the effect is insignificant, allowing us to assume that the core loss is constant at all the load conditions. Hence, for the assumed constant values of P_c and V_2 (secondary terminal load voltage also varies with load, but the variation is too small to be accounted in efficiency calculations), the condition for maximum efficiency, at a given load power factor, can be derived by differentiating the expression for η with respect to I_2 and equating it to zero:

$$\frac{d\eta}{dI_2} = \frac{d}{dI_2}\left[\frac{V_2 I_2 \cos\theta_2}{V_2 I_2 \cos\theta_2 + (P_c + I_2^2 R_{eq2})}\right] = 0 \tag{1.69}$$

Solving it further, we get

$$P_c = I_2^2 R_{eq2} \tag{1.70}$$

Thus, the maximum efficiency occurs at a load at which variable load loss equals the constant core (no-load) loss. Further,

$$\frac{I_2^2}{I_{2FL}^2} = \frac{P_c}{I_{2FL}^2\ R_{eq2}} \tag{1.71}$$

where I_{2FL} is the full-load (rated) current and $I_{2FL}^2 R_{eq2}$ is the load loss at the rated load conditions. Therefore, the per-unit load at which the maximum efficiency occurs is

$$(per\text{-}unit\ load)_{\eta_{max}} = \frac{I_2}{I_{2FL}} = \sqrt{\frac{P_c}{(P_L)_{FL}}} \tag{1.72}$$

The value of maximum efficiency can be found out by substituting the value of I_2 from equation 1.72 in equation 1.68. Similarly, it can easily be shown that the maximum efficiency, for a given load, occurs at unity power factor ($\cos\theta = 1$).

The rating of transformers is expressed in volt-amperes and not in watts because heating (temperature) determines the life of the transformers. Hence, the rated output is limited by the losses, which depend on the voltage (no-load loss) and the current (load loss), and are almost unaffected by the load power factor. The amount of heat depends on the r.m.s. values of current and voltage and not on the power factor. Hence, the power delivered through a transformer may not be a unique value. The rating of a transformer is therefore not expressed in power rating (watts) but by the one which indicates the apparent power (volt-amperes) that it can deliver.

Example 1.1

A single-phase transformer is designed to operate at 220/110 V, 60 Hz. What will be effect on the transformer performance if frequency reduces by 5% to 57 Hz and voltage increases by 5% to 231 volts?

Solution:

The emf equation of a transformer is given by

$$V_1 = 4.44\ \phi_{mp}\ f\ N_1$$

Now $\phi_{mp} = B_{mp}\ A_c$

where B_{mp} = peak value of flux density in core (wb/m^2)

A_c = core cross-sectional area in m^2

Hence, for a given number of turns (N_1) and core area (A_c),

$$\frac{(231)_{57\,Hz}}{(220)_{60\,Hz}} = \frac{4.44\ (B_{mp})_{57\,Hz} \times A_c \times 57 \times N_1}{4.44\ (B_{mp})_{60\,Hz} \times A_c \times 60 \times N_1}$$

$$\frac{(B_{mp})_{57\,Hz}}{(B_{mp})_{60\,Hz}} = \frac{231 \times 60}{220 \times 57} = 1.10$$

Thus, with the reduced frequency and higher applied voltage, the flux density in the core increases resulting in higher no-load current, core losses and noise. This example shows that a transformer designer has to select the operating peak flux density in the core depending on the overfluxing conditions (simultaneous over-voltage and under-frequency) specified by the user.

Example 1.2

Tests on 31.5 MVA, 132/33 kV star/delta 3-phase transformer gave following results (loss values given are for three phases):

- Open circuit test: 33 kV, 5.5 A, 21 kW
- Short circuit test: 13.2 kV, 137.8 A, 100 kW

Calculate: a) equivalent circuit parameters referred to LV side
b) efficiency at full load and half of full load with unity power factor
c) regulation at full load with 0.8 power factor lagging

Solution:

a) Unless otherwise stated, the specified values of voltages should be taken as line-to-line values. The equivalent circuit parameters are on per-phase basis, and hence all the quantities in 3-phase balanced system are converted into the per-phase values.

The open circuit test is performed on the LV side with the application of rated voltage of 33 kV. For a delta connected LV winding, line and phase voltages are the same; per-phase LV current is line current divided by $\sqrt{3}$. Hence we get,

$$\text{Per-phase no-load excitation current } = \frac{5.5}{\sqrt{3}} = 3.2 \text{ A}$$

$$\text{Per-phase core loss } = \frac{21000}{3} = 7000 \text{ W}$$

$$\text{Core-loss component } = I_c = I_0 \cos\theta_0 = \frac{P_c}{V_1} = \frac{7000}{33000} = 0.21 \text{ A}$$

$$\text{Magnetizing component } = I_m = \sqrt{I_0^2 - I_c^2} = \sqrt{3.2^2 - 0.21^2} = 3.19 \text{ A}$$

Values of core-loss resistance and magnetizing reactance referred to the LV side are

$$R_c = \frac{V_1}{I_c} = \frac{33000}{0.21} = 157 \text{ k}\Omega$$

$$X_m = \frac{V_1}{I_m} = \frac{33000}{3.19} = 10.34 \text{ k}\Omega$$

$$\text{Base impedance on the LV side } = \frac{\text{kV}^2}{\text{MVA}} = \frac{33^2}{(31.5/3)} = 103.7\ \Omega$$

Thus, the per-unit value of X_m is 100 $(=10.34\ \text{k}\Omega/103.7\ \Omega)$.

Now the short circuit test is performed on the HV side (applied voltage of 13.2 kV is a fraction of rated HV voltage of 132 kV).

Per-phase applied voltage to the star-connected HV winding $= \dfrac{13200}{\sqrt{3}} = 7621$ V

For star connection, phase current is equal to line current (=137.8 A).
Now, ohmic value of the leakage impedance and resistance referred to the HV side can be calculated as

$$Z_{eqH} = \frac{V_{sc}}{I_{sc}} = \frac{7621}{137.8} = 55.3\ \Omega$$

$$\text{Per phase value of load-loss} = \frac{100000}{3}\ \text{W}$$

$$R_{eqH} = \frac{P_L}{I_{sc}^2} = \frac{100000/3}{137.8^2} = 1.755\ \Omega$$

$$\therefore X_{eqH} = \sqrt{Z_{eqH}^2 - R_{eqH}^2} = \sqrt{55.3^2 - 1.755^2} = 55.27\ \Omega$$

Now, the equivalent circuit quantities calculated on the HV side can be referred to the LV side using the transformation ratio a,

$$a = \frac{N_1}{N_2} \cong \frac{V_{pL}}{V_{pH}}$$

where V_{pL} and V_{pH} are phase voltages of LV and HV windings respectively.

$$\therefore R_{eqL} = a^2 R_{eqH} = \left[\frac{33000}{132000/\sqrt{3}}\right]^2 \times 1.755 = 0.329\ \Omega$$

Similarly,

$$X_{eqL} = a^2 X_{eqH} = \left[\frac{33000}{132000/\sqrt{3}}\right]^2 \times 55.27 = 10.36\ \Omega$$

In per-unit quantities,

$$\varepsilon_r = (R_{eqL})_{pu} = \frac{R_{eqL}}{Z_{bL}} = \frac{0.329}{103.5} = 0.0032$$

$$\varepsilon_x = (X_{eqL})_{pu} = \frac{X_{eqL}}{Z_{bL}} = \frac{10.36}{103.5} = 0.1$$

We could have directly found the value of $(R_{eqL})_{pu}$ by using line-to-line and 3-phase quantities using equation 1.50 as

$$(R_{eqL})_{pu} = \frac{100 \times 10^3}{31.5 \times 10^6} = 0.0032$$

Similarly we can directly infer the value of $(X_{eqL})_{pu}$, which is almost equal to $(Z_{eqL})_{pu}$, from the short circuit test results because it equals the fraction of rated voltage applied to circulate the rated current,

$$(X_{eqL})_{pu} = (X_{eqH})_{pu} = (X_{eq})_{pu} = \frac{13.2 \times 10^3}{132 \times 10^3} = 0.1$$

b) Efficiency can be worked out either by using per-phase or 3-phase quantities. The percentage efficiency at full load and unity power factor $(\cos\theta = 1)$ can be calculated by using equation 1.68 as

$$\eta_{FL} = \frac{31.5 \times 10^6 \times 1}{31.5 \times 10^6 \times 1 + (21 \times 10^3 + 100 \times 10^3)} \times 100 = 99.62\ \%$$

and at half the full load,

$$\eta_{0.5FL} = \frac{0.5 \times 31.5 \times 10^6 \times 1}{0.5 \times 31.5 \times 10^6 \times 1 + (21 \times 10^3 + (0.5)^2 \times 100 \times 10^3)} \times 100 = 99.71\ \%$$

The maximum efficiency occurs at a load of 45.8% $(= \sqrt{21/100} \times 100)$ as per equation 1.72.

c) Regulation at full load and 0.8 power factor lagging is calculated from equation 1.63 as

$$\text{Regulation} = (\varepsilon_r \cos\theta_2 + \varepsilon_x \sin\theta_2) + \frac{1}{2}(\varepsilon_x \cos\theta_2 - \varepsilon_r \sin\theta_2)^2$$

$$= (0.0032 \times 0.8 + 0.1 \times 0.6) + \frac{1}{2}(0.1 \times 0.8 - 0.0032 \times 0.6)^2$$

$$= 0.066$$

Example 1.3

A 500 kVA single-phase transformer is designed to have a resistance of 1% with its maximum efficiency occurring at 250 kVA load. Find efficiency of the transformer when it is supplying full load at 0.8 power factor lagging.

Solution:

The percentage resistance is given as

$$\% R = \frac{Load\ loss\ at\ full\ load}{Base\ VA} \times 100$$

$$\therefore \text{Load loss at full load} = \frac{1}{100}\left[500 \times 10^3\right] = 5000\ \text{W}$$

Now from equation 1.72, for the same terminal voltage,

$$\frac{I_{\eta_{max}}}{I_{FL}} = \frac{250}{500} = \sqrt{\frac{P_c}{5000}}$$

$$\therefore P_c = \text{No-load loss} = 1250\ \text{W}$$

Total loss at full load = 1250 + 5000 = 6250 W

Efficiency at full load and 0.8 power factor =

$$\frac{output\ power}{output\ power + losses} \times 100 = \frac{500 \times 0.8}{500 \times 0.8 + 6.25} \times 100 = 98.5\%$$

Although the efficiency of a transformer is given by the ratio of output power to input power, there are some specific applications of transformer in which its performance cannot be judged only by this efficiency. Distribution transformers, for example, supply a load which varies over a wide range throughout the day. For such transformers, the parameter *all-day efficiency* is of more relevance and is defined as

$$\eta_{all-day} = \frac{(output)_{\text{kWh}}}{(input)_{\text{kWh}}} = \frac{(output)_{\text{kWh}}}{(output)_{\text{kWh}} + (losses)_{\text{kWh}}} \tag{1.73}$$

The output and losses are computed for a period of 24 hours using the load cycle. No-load losses are constant (independent of load); hence it is important to design distribution transformers with a lower value of no-load losses so that a higher value of all day energy efficiency is achieved.

1.7 Parallel Operation of Transformers

For supplying a load in excess of the rating of an existing transformer, two or more transformers may be connected in parallel with the existing transformer. It is usually economical to install another transformer in parallel instead of replacing the existing transformer by a single larger unit. The cost of a spare unit in the case of two parallel transformers (of equal rating) is also lower than that of a single large transformer. In addition, it is preferable to have a parallel transformer for the reason of reliability. With this, at least half the load can be supplied with one transformer out of service. For parallel connection of transformers, primary windings of the transformers are connected to source bus-bars and secondary windings are connected to the load bus-bars. There are various conditions that must be fulfilled for the successful parallel operation of transformers. These are as follows:

1. The line voltage ratios of the transformers must be equal (on each tap): If the transformers connected in parallel have slightly different voltage ratios, then due to the inequality of induced emfs in the secondary windings, a circulating current will flow in the loop formed by the secondary windings under the no-load condition, which may be much greater than the normal no-load current. The current will be quite high as the leakage impedance is low. When the secondary windings are loaded, this circulating current will tend to produce unequal loading on the two transformers, and it may not be possible to take the full load from this group of two parallel transformers (one of the transformers may get overloaded).

2. The transformers should have equal per-unit leakage impedances and the same ratio of equivalent leakage reactance to the equivalent resistance (X/R): If the ratings of both the transformers are equal, their per-unit leakage impedances should be equal in order to have equal loading of both the transformers. If the ratings are unequal, their per-unit leakage impedances based on their own ratings should be equal so that the currents carried by them will be proportional to their ratings. In other words, for unequal ratings, the numerical (ohmic) values of their impedances should be in inverse proportion to their ratings to have current in them in line with their ratings.

A difference in the ratio of the reactance value to resistance value of the per-unit impedance results in a different phase angle of the currents carried by the two paralleled transformers; one transformer will be working with a higher power factor and the other with a lower power factor than that of the combined output. Hence, the real power will not be proportionally shared by the transformers.

3. The transformers should have the same polarity: The transformers should be properly connected with regard to their polarity. If they are connected with incorrect polarities then the two emfs, induced in the secondary windings which are in parallel, will act together in the local secondary circuit and produce a short circuit.

The previous three conditions are applicable to both single-phase as well as three-phase transformers. In addition to these three conditions, two more conditions are essential for the parallel operation of three-phase transformers:

4. The transformers should have the same phase sequence: The phase sequence of line voltages of both the transformers must be identical for parallel operation of three-phase transformers. If the phase sequence is an incorrect, in every cycle each pair of phases will get short-circuited.

5. The transformers should have the zero relative phase displacement between the secondary line voltages: The transformer windings can be connected in a variety of ways which produce different magnitudes and phase displacements of the secondary voltage. All the transformer connections can be classified into distinct vector groups. Each vector group notation consists of first an uppercase letter denoting HV connection, a second lowercase letter denoting LV connection, followed by a *clock number* representing LV winding's phase displacement with respect to HV winding (at 12 o'clock). There are four groups into which all possible three-phase connections can be classified:

Group 1: Zero phase displacement (Yy0, Dd0, Dz0)
Group 2: 180° phase displacement (Yy6, Dd6, Dz6)
Group 3: -30° phase displacement (Yd1, Dy1, Yz1)
Group 4: +30° phase displacement (Yd11, Dy11, Yz11)

In above notations, letters y (or Y), d (or D) and z represent star, delta and zigzag connections respectively. In order to have zero relative phase displacement of secondary side line voltages, the transformers belonging to the same group can be paralleled. For example, two transformers with Yd1 and Dy1 connections can be paralleled. The transformers of groups 1 and 2 can only be paralleled with transformers of their own group. However, the transformers of groups 3 and 4 can be paralleled by reversing the phase sequence of one of them. For example, a transformer with Yd11 connection (group 4) can be paralleled with that having Dy1 connection (group 3) by reversing the phase sequence of both primary and secondary terminals of the Dy1 transformer.

References

1. Say, M. G. *The performance and design of alternating current machines*, 2nd edition, Sir Isaac Pitman and Sons, London, 1955.
2. Toro, V. D. *Principles of electrical engineering,* 2nd edition, Prentice Hall, New Delhi, 1977.
3. Stevenson, W. D. *Elements of power system analysis,* 4th edition, McGraw-Hill, Tokyo, 1982, pp. 138-162.
4. MIT Press, *Magnetic circuits and transformers,* 14th edition, John Wiley and Sons, New York, 1962, pp. 259-406.

2

Magnetic Characteristics

The magnetic circuit is one of the most important active parts of a transformer. It consists of laminated iron core and carries flux linked to windings. Energy is transferred from one electrical circuit to another through the magnetic field carried by the core. The iron core provides a low reluctance path to the magnetic flux thereby reducing magnetizing current. Most of the flux is contained in the core reducing stray losses in structural parts. Due to on-going research and development efforts [1] by steel and transformer manufacturers, core materials with improved characteristics are getting developed and applied with better core building technologies. In the early days of transformer manufacturing, inferior grades of laminated steel (as per today's standards) were used with inherent high losses and magnetizing volt-amperes. Later on it was found that the addition of silicon content of about 4 to 5% improves the performance characteristics significantly, due to a marked reduction in eddy losses (on account of the increase in material resistivity) and increase in permeability. Hysteresis loss is also lower due to a narrower hysteresis loop. The addition of silicon also helps to reduce the aging effects. Although silicon makes the material brittle, it is well within limits and does not pose problems during the process of core building. Subsequently, the cold rolled manufacturing technology in which the grains are oriented in the direction of rolling gave a new direction to material development for many decades, and even today newer materials are centered around the basic grain orientation process. Important stages of core material development are: non-oriented, hot rolled grain oriented (HRGO), cold rolled grain oriented (CRGO), high permeability cold rolled grain oriented (Hi-B), laser scribed and mechanically scribed. Laminations with lower thickness are manufactured and used to take advantage of lower eddy losses. Currently the lowest thickness available is 0.23 mm, and the popular thickness range is 0.23 mm to 0.35 mm for power transformers. Maximum thickness of lamination used in small transformers can be as high as 0.50 mm. The lower the thickness of laminations, the higher core building time is required since the number of laminations for a given core area increases. Inorganic coating (generally glass film and phosphate layer) having thickness of 0.002 to 0.003 mm is provided on both the surfaces of laminations, which is sufficient to withstand eddy voltages (of the order of a few volts).

Since the core is in the vicinity of high voltage windings, it is grounded to drain out the statically induced voltages. If the core is sectionalized by ducts (of about 5 mm) for the cooling purpose, individual sections have to be grounded. Some users prefer to ground the core outside tank through a separate bushing. All the internal structural parts of a transformer (e.g., frames) are grounded. While designing the grounding system, due care must be taken to avoid multiple grounding, which otherwise results into circulating currents and subsequent failure of transformers. The tank is grounded externally by a suitable arrangement. Frames, used for clamping yokes and supporting windings, are generally grounded by connecting them to the tank by means of a copper or aluminum strip. If the frame-to-tank connection is done at two places, a closed loop formed may link appreciable stray leakage flux. A large circulating current may get induced which can eventually burn the connecting strips.

2.1 Construction

2.1.1 Types of core

A part of a core, which is surrounded by windings, is called a limb or leg. Remaining part of the core, which is not surrounded by windings, but is essential for completing the path of flux, is called as yoke. This type of construction (termed as core type) is more common and has the following distinct advantages: viz. construction is simpler, cooling is better and repair is easy. Shell-type construction, in which a cross section of windings in the plane of core is surrounded by limbs and yokes, is also used. It has the advantage that one can use sandwich construction of LV and HV windings to get very low impedance, if desired, which is not easily possible in the core-type construction. In this book, most of the discussion is related to the core-type construction, and where required reference to shell-type construction has been made.

The core construction mainly depends on technical specifications, manufacturing limitations, and transport considerations. It is economical to have all the windings of three phases in one core frame. A three-phase transformer is cheaper (by about 20 to 25%) than three single-phase transformers connected in a bank. But from the spare unit consideration, users find it more economical to buy four single-phase transformers as compared to two three-phase transformers. Also, if the three-phase rating is too large to be manufactured in transformer works (weights and dimensions exceeding the manufacturing capability) and transported, there is no option but to manufacture and supply single-phase units. In figure 2.1, various types of core construction are shown.

In a single-phase three-limb core (figure 2.1 (a)), windings are placed around the central limb, called as main limb. Flux in the main limb gets equally divided between two yokes and it returns via end limbs. The yoke and end limb area should be only 50% of the main limb area for the same operating flux density. This type of construction can be alternately called as single-phase shell-type transformer. Zero-sequence impedance is equal to positive-sequence impedance for this construction (in a bank of single-phase transformers).

Sometimes in a single-phase transformer windings are split into two parts and placed around two limbs as shown in figure 2.1 (b). This construction is sometimes adopted for very large ratings. Magnitude of short-circuit forces are lower because of the fact that ampere-turns/height are reduced. The area of limbs and yokes is the same. Similar to the single-phase three-limb transformer, one can have additional two end limbs and two end yokes as shown in figure 2.1 (c) to get a single-phase four-limb transformer to reduce the height for the transport purpose.

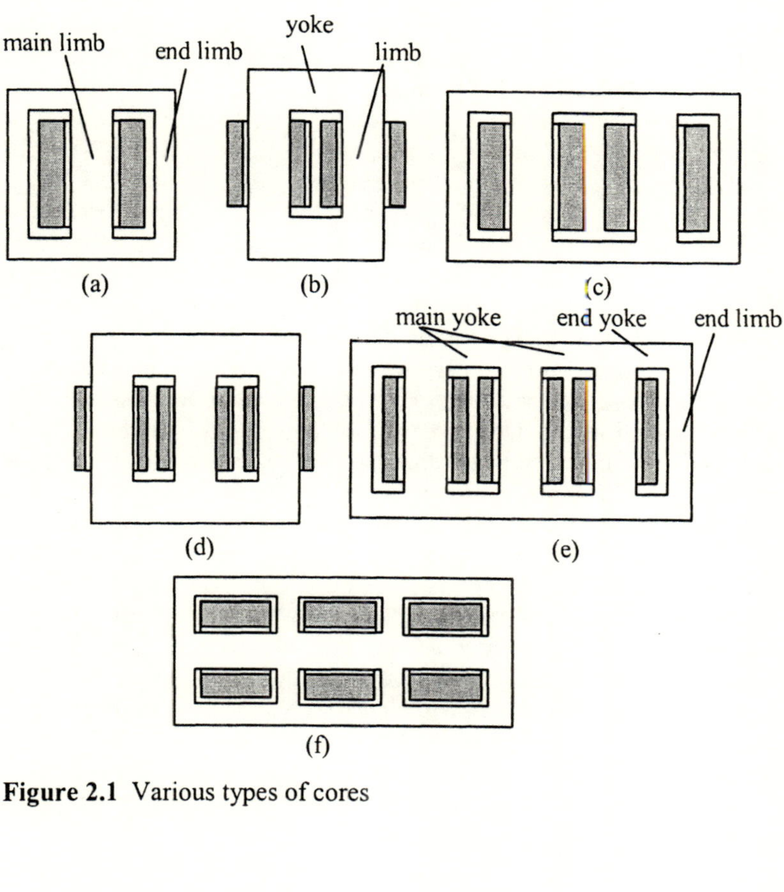

Figure 2.1 Various types of cores

The most commonly used construction, for small and medium rating transformers, is three-phase three-limb construction as shown in figure 2.1 (d). For each phase, the limb flux returns through yokes and other two limbs (the same amount of peak flux flows in limbs and yokes). In this construction, limbs and yokes usually have the same area. Sometimes the yokes are provided with a 5% additional area as compared to the limbs for reducing no-load losses. It is to be noted that the increase in yoke area of 5% reduces flux density in the yoke by 5%, reduces watts/kg by more than 5% (due to non-linear characteristics) but the yoke weight increases by 5%. Also, there may be additional loss due to cross-fluxing since there may not be perfect matching between lamination steps of limb and yoke at the joint. Hence, the reduction in losses may not be very significant. The provision of extra yoke area may improve the performance under over-excitation conditions. Eddy losses in structural parts, due to flux leaking out of core due to its saturation under over-excitation condition, are reduced to some extent [2, 3]. The three-phase three-limb construction has inherent three-phase asymmetry resulting in unequal no-load currents and losses in three phases; the phenomenon is discussed in section 2.5.1. One can get symmetrical core by connecting it in star or delta so that the windings of three phases are electrically as well as physically displaced by 120 degrees. This construction results into minimum core weight and tank size, but it is seldom used because of complexities in manufacturing.

In large power transformers, in order to reduce the height for transportability, three-phase five-limb construction depicted in figure 2.1 (e) is used. The magnetic length represented by the end yoke and end limb has a higher reluctance as compared to that represented by the main yoke. Hence, as the flux starts rising, it first takes the path of low reluctance of the main yoke. Since the main yoke is not large enough to carry all the flux from the limb, it saturates and forces the remaining flux into the end limb. Since the spilling over of flux to the end limb occurs near the flux peak and also due to the fact that the ratio of reluctances of these two paths varies due to non-linear properties of the core, fluxes in both main yoke and end yoke / end limb paths are non-sinusoidal even though the main limb flux is varying sinusoidally [2, 4]. Extra losses occur in the yokes and end limbs due to the flux harmonics. In order to compensate these extra losses, it is a normal practice to keep the main yoke area 60% and end yoke / end limb area 50% of the main limb area. The zero-sequence impedance is much higher for the three-phase five-limb core than the three-limb core due to low reluctance path (of yokes and end limbs) available to the in-phase zero-sequence fluxes, and its value is close to but less than the positive-sequence impedance value. This is true if the applied voltage during the zero-sequence test is small enough so that the yokes and end limbs are not saturated. The aspects related to zero-sequence impedances for various types of core construction are elaborated in Chapter 3. Figure 2.1 (f) shows a typical 3-phase shell-type construction.

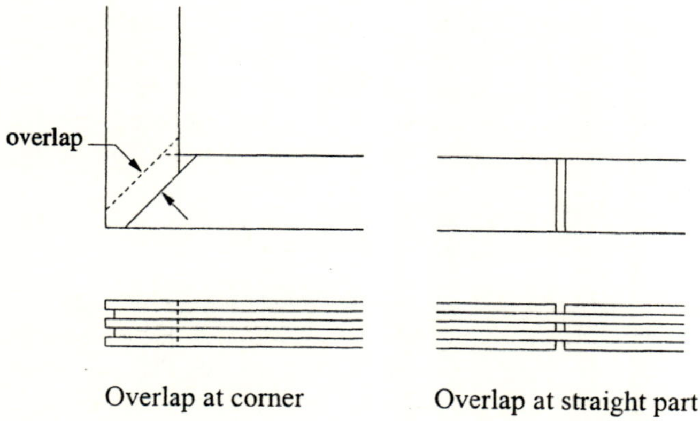

Figure 2.2 Overlapping at joints

2.1.2 Analysis of overlapping joints and building factor

While building a core, the laminations are placed in such a way that the gaps between the laminations at the joint of limb and yoke are overlapped by the laminations in the next layer. This is done so that there is no continuous gap at the joint when the laminations are stacked one above the other (figure 2.2). The overlap distance is kept around 15 to 20 mm. There are two types of joints most widely used in transformers: non-mitred and mitred joints (figure 2.3). Non-mitred joints, in which the overlap angle is 90°, are quite simple from the manufacturing point of view, but the loss in the corner joints is more since the flux in the joint region is not along the direction of grain orientation. Hence, the non-mitred joints are used for smaller rating transformers. These joints were commonly adopted in earlier days when non-oriented material was used.

In case of mitred joints the angle of overlap (α) is of the order of 30° to 60°, the most commonly used angle is 45°. The flux crosses from limb to yoke along the grain orientation in mitred joints minimizing losses in them. For air-gaps of equal length, the excitation requirement of cores with mitred joints is $\sin\alpha$ times that with non-mitred joints [5].

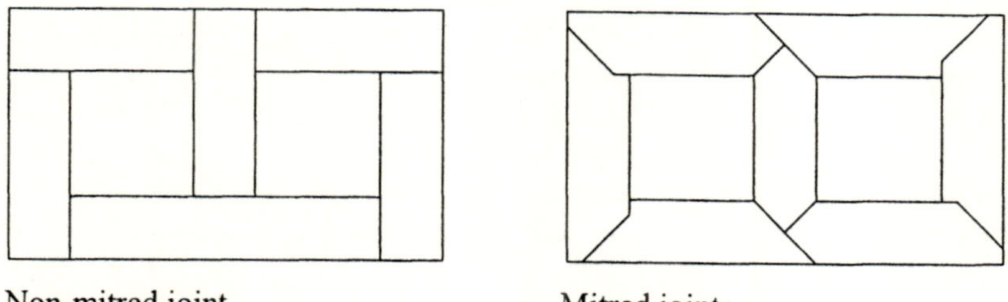

Figure 2.3 Commonly used joints

Better grades of core material (Hi-B, scribed, etc.) having specific loss (watts/kg) 15 to 20% lower than conventional CRGO material (termed hereafter as CGO grade, e.g., M4) are regularly used. However, it has been observed that the use of these better materials may not give the expected loss reduction if a proper value of *building factor* is not used in loss calculations. It is defined as

$$Building\ \ factor = \frac{Built\ \ transformer\ \ core\ \ loss\ \ (watts\,/\,kg)}{Material\ \ Epstein\ \ core\ \ loss\ \ (watts\,/\,kg)} \tag{2.1}$$

The building factor generally increases as grade of the material improves from CGO to Hi-B to scribed (domain refined). This is a logical fact because at the corner joints the flux is not along the grain orientation, and the increase in watts/kg due to deviation from direction of grain orientation is higher for a better grade material. The factor is also a function of operating flux density; it deteriorates more for better grade materials with the increase in operating flux density. Hence, cores built with better grade material may not give the expected benefit in line with Epstein measurements done on individual lamination. Therefore, appropriate building factors should be taken for better grade materials using experimental / test data.

Single-phase two-limb transformers give significantly better performances than three-phase cores. For a single-phase two-limb core, building factor is as low as 1.0 for the domain refined grade (laser or mechanically scribed material) and slightly lower than 1.0 for CGO grade [6]. The reason for such a lower value of losses is attributed to lightly loaded corners and spatial redistribution of flux in limbs and yokes across the width of laminations. Needless to say, the higher the proportion of corner weight in the total core weight, the higher are the losses. Also the loss contribution due to the corner weight is higher in case of 90° joints as compared to 45° joints since there is over-crowding of flux at the inner edge and flux is not along the grain orientation while passing from limb to yoke in the former case. Smaller the overlapping length better is the core performance; but the improvement may not be noticeable. It is also reported in [6, 7] that the gap at the core joint has significant impact on the no-load loss and current. As compared to 0 mm gap, the increase in loss is 1 to 2% for 1.5 mm gap, 3 to 4% for 2.0 mm gap and 8 to 12% for 3 mm gap. These figures highlight the need for maintaining minimum gap at the core joints.

Lesser the laminations per lay, lower is the core loss. The experience shows that from 4 laminations per lay to 2 laminations per lay, there is an advantage in loss of about 3 to 4%. There is further advantage of 2 to 3% in 1 lamination per lay. As the number of laminations per lay reduces, the manufacturing time for core building increases and hence most of the manufacturers have standardized the core building with 2 laminations per lay.

A number of works have been reported in the literature, which have analyzed various factors affecting core losses. A core model for three-phase three-limb transformer using a lumped circuit model is reported in [8]. The length of equivalent air gap is varied as a function of the instantaneous value of the flux in the laminations. The anisotropy is also taken into account in the model. An analytical solution using 2-D finite difference method is described in [9] to calculate spatial flux distribution and core losses. The method takes into account magnetic anisotropy and non-linearity. The effect of overlap length and number of laminations per lay on core losses has been analyzed in [10] for wound core distribution transformers.

Joints of limbs and yokes contribute significantly to the core loss due to cross-fluxing and crowding of flux lines in them. Hence, the higher the corner area and weight, the higher is the core loss. The corner area in single-phase three-limb cores, single-phase four-limb cores and three-phase five-limb cores is less due to smaller core diameter at the corners, reducing the loss contribution due to the corners. However, this reduction is more than compensated by increase in loss because of higher overall weight (due to additional end limbs and yokes).

Building factor is usually in the range of 1.1 to 1.25 for three-phase three-limb cores with mitred joints. Higher the ratio of window height to window width, lower is the contribution of corners to the loss and hence the building factor is lower.

Single-phase two-limb and single-phase three-limb cores have been shown [11] to have fairly uniform flux distribution and low level of total harmonic distortion as compared to single-phase four-limb and three-phase five-limb cores.

Step-lap joint is used by many manufacturers due to its excellent performance figures. It consists of a group of laminations (commonly 5 to 7) stacked with a staggered joint as shown in figure 2.4. Its superior performance as compared to the conventional mitred construction has been analyzed in [12, 13]. It is shown [13] that, for a operating flux density of 1.7 T, the flux density in the mitred joint in the core sheet area shunting the air gap rises to 2.7 T (heavy saturation), while in the gap the flux density is about 0.7 T. Contrary to this, in the step-lap joint of 6 steps, the flux totally avoids the gap with flux density of just 0.04 T, and gets redistributed almost equally in laminations of other five steps with a flux density close to 2.0 T. This explains why the no-load performance figures (current, loss and noise) show a marked improvement for the step-lap joints.

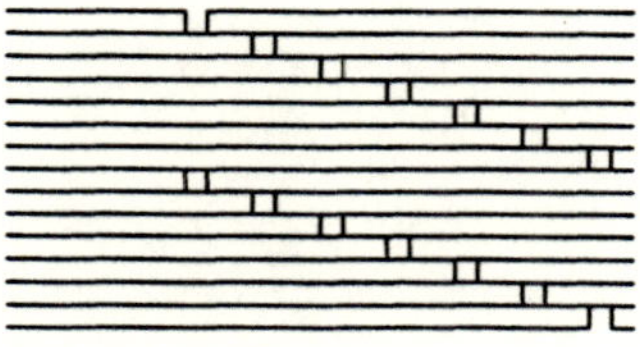

Step-lap joint

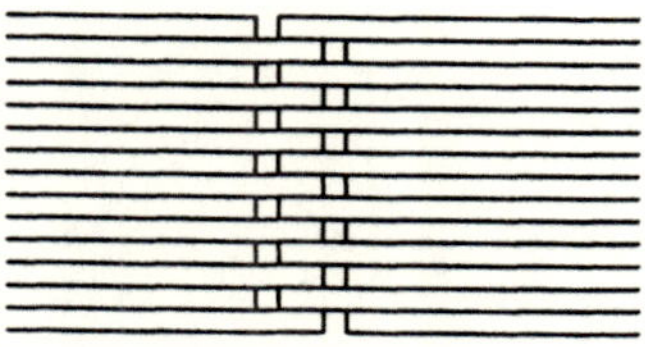

Conventional mitred joint

Figure 2.4 Step-lap and conventional joint

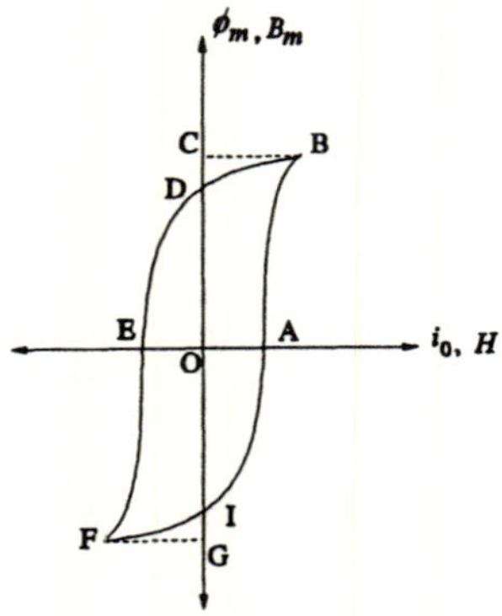

a. Hysteresis loop

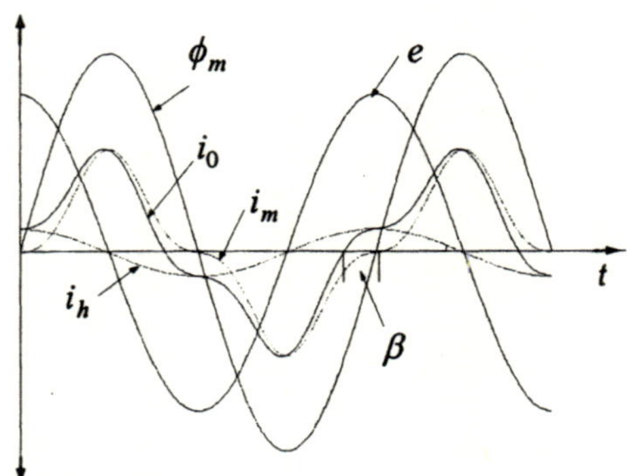

b. Waveforms

Figure 2.5 Hysteresis loss

2.2 Hysteresis and Eddy Losses

Hysteresis and eddy current losses together constitute the no-load loss. As discussed in Chapter 1, the loss due to no-load current flowing in the primary winding is negligible. Also, at the rated flux density condition on no-load, since most of the flux is confined to the core, negligible losses are produced in the structural parts due to near absence of the stray flux. The hysteresis and eddy losses arise due to successive reversal of magnetization in the iron core with sinusoidal application of voltage at a particular frequency f (cycles/second).

Eddy current loss, occurring on account of eddy currents produced due to induced voltages in laminations in response to an alternating flux, is proportional to the square of thickness of laminations, square of frequency and square of effective (r.m.s.) value of flux density.

Hysteresis loss is proportional to the area of hysteresis loop (figure 2.5(a)). Let e, i_0 and ϕ_m denote the induced voltage, no-load current and core flux respectively. As per equation 1.1, voltage e leads the flux ϕ_m by 90°. Due to hysteresis phenomenon, current i_0 leads ϕ_m by a hysteresis angle (β) as shown in figure 2.5 (b). Energy, either supplied to the magnetic circuit or returned back by the magnetic circuit is given by

$$Energy = \int e\, i_0\, dt = \int N \frac{d\phi_m}{dt}\, i_0\, dt = \int N\, i_0\, d\phi_m \tag{2.2}$$

If we consider quadrant I of the hysteresis loop, the area OABCDO represents the energy supplied. Both induced voltage and current are positive for path AB. For path BD, the energy represented by the area BCD is returned back to the source since the voltage and current are having opposite signs giving a negative value of energy. Thus, for the quadrant I the area OABDO represents the energy loss; the area under hysteresis loop ABDEFIA represents the total energy loss termed as the hysteresis loss. This loss has a constant value per cycle meaning thereby that it is directly proportional to frequency (the higher the frequency (cycles/second), the higher is the loss). The non-sinusoidal current i_0 can be resolved into two sinusoidal components: i_m in-phase with ϕ_m and i_h in phase with *e*. The component i_h represents the hysteresis loss.

The eddy loss (P_e) and hysteresis loss (P_h) are thus given by

$$P_e = k_1\ f^2\, t^2\, B_{rms}^2 \tag{2.3}$$

$$P_h = k_2\ f\ B_{mp}^n \tag{2.4}$$

where

t is thickness of individual lamination

k_1 and k_2 are constants which depend on material

B_{rms} is the rated effective flux density corresponding to the actual r.m.s. voltage on the sine wave basis

B_{mp} is the actual peak value of the flux density

n is the *Steinmetz* constant having a value of 1.6 to 2.0 for hot rolled laminations and a value of more than 2.0 for cold rolled laminations due to use of higher operating flux density in them.

In r.m.s. notations, when the hysteresis component (I_h) shown in figure 2.5 (b) is added to the eddy current loss component, we get the total core loss current (I_c). In practice, the equations 2.3 and 2.4 are not used by designers for calculation of no-load loss. There are at least two approaches generally used; in one approach the building factor for the entire core is derived based on the experimental / test data, whereas in the second approach the effect of corner weight is separately accounted by a factor based on the experimental / test data.

$$No\ load\ loss = W_t \times K_b \times w \tag{2.5}$$

or

$$No\ load\ loss = (W_t - W_c) \times w + W_c \times w \times K_c \tag{2.6}$$

where,

w is watts/kg for a particular operating peak flux density as given by lamination supplier (Epstein core loss),

K_b is the building factor,

W_c denotes corner weight out of total weight of W_t, and

K_c is factor representing extra loss occurring at the corner joints (whose value is higher for smaller core diameters).

2.3 Excitation Characteristics

Excitation current can be calculated by one of the following two methods. In the first method, magnetic circuit is divided into many sections, within each of which the flux density can be assumed to be of constant value. The corresponding value of magnetic field intensity (H) is obtained for the lamination material (from its magnetization curve) and for the air gap at joints. The excitation current can then be calculated as the total magnetomotive force required for all magnetic sections (n) divided by number of turns (N) of the excited winding,

$$I_0 = \frac{\sum H_n l_n}{N} \tag{2.7}$$

where l is length of each magnetic section.

It is not practically possible to calculate the no-load current by estimating ampere-turns required in different parts of the core to establish a given flux density. The calculation is mainly complicated by the corner joints. Hence, designers prefer the second method, which uses empirical factors derived from test results. Designers generally refer the VA/kg (volt-amperes required per kg of material) versus induction (flux density) curve of the lamination material. This VA/kg is multiplied by a factor (which is based on test results) representing additional excitation required at the joints to get VA/kg of the built core. In that case, the no-load line current for a three-phase transformer can be calculated as

$$No\ load\ current = \frac{VA/kg\ of\ built\ core \times core\ weight\ in\ kg}{\sqrt{3} \times rated\ voltage} \tag{2.8}$$

Generally, manufacturers test transformers of various ratings with different core materials at voltage levels below and above the rated voltage and derive their own VA/kg versus induction curves.

As seen from figure 2.5 (b), excitation current of a transformer is rich in harmonics due to non-linear magnetic characteristics. For CRGO material, the usually observed range of various harmonics is as follows. For the fundamental component of 1 per-unit, 3rd harmonic is 0.3 to 0.5 per-unit, 5th harmonic is 0.1 to 0.3 per-unit and 7th harmonic is about 0.04 to 0.1 per-unit. The harmonics higher than the 7th harmonic are of insignificant magnitude. The effective value of total no-load current is given as

$$I_0 = \sqrt{I_1^2 + I_3^2 + I_5^2 + I_7^2 + ----} \tag{2.9}$$

In above equation, I_1 is the effective (r.m.s.) value of the fundamental component (50 or 60 Hz) whereas I_3, I_5 and I_7 are the effective values of 3rd, 5th and 7th harmonics respectively. The effect of higher harmonics of diminishing magnitude have a small influence on the effective value of resultant no-load current (40% value of 3rd harmonic increases the resultant by only about 8%). Since the no-load current itself is in the range of 0.2 to 2% of the full load current, harmonics in no-load current do not appreciably increase the copper loss in windings except during extreme levels of core saturation. The harmonic components of current do not contribute to the core loss if the applied voltage is sinusoidal.

If the current harmonic components are modified or constrained, flux density in the core gets modified. For example, if the third harmonic current is suppressed by isolating the neutral, the flux density will be flat-topped for a sinusoidal current as shown in figure 2.6 (hysteresis is neglected for simplicity). For this case, the flux can be expressed as

$$\phi_m = \phi_{mp1}\sin\omega t + \phi_{mp3}\sin 3\omega t + \phi_{mp5}\sin 5\omega t + \phi_{mp7}\sin 7\omega t + -- \qquad (2.10)$$

where ϕ_{mp1}, ϕ_{mp3}, -- etc. represent the peak values of fundamental and harmonic components. The induced voltage per turn is

$$e = \frac{d\phi}{dt} = \omega\begin{bmatrix}\phi_{mp1}\cos\omega t + 3\phi_{mp3}\cos 3\omega t + 5\phi_{mp5}\cos 5\omega t \\ +7\phi_{mp7}\cos 7\omega t + ----\end{bmatrix} \qquad (2.11)$$

The induced voltage as seen in figure 2.6 is peaky in nature with pronounced third harmonic component (only the third harmonic component is shown for clarity). Thus, even a small deviation of flux from the sinusoidal nature introduces appreciable harmonic components in voltages (15% third harmonic component in flux results into 45% of third harmonic component in the voltage). This results in increase of eddy losses but hysteresis loss reduces as the maximum value of flux density is reduced. The net effect on the total core loss will depend on the relative changes in eddy and hysteresis losses.

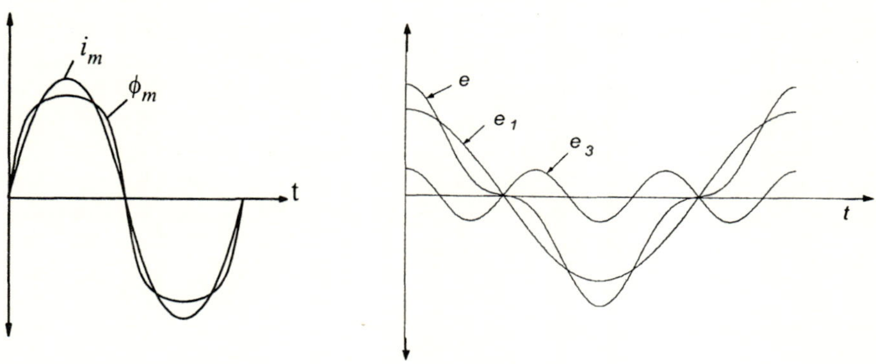

Figure 2.6 Waveforms of flux and voltage for sinusoidal magnetizing current

2.4 Over-Excitation Performance

The choice of operating flux density of a core has a very significant impact on the overall size, material cost and performance of a transformer. For the currently available various grades of CRGO material, although losses and magnetizing volt-amperes are lower for better grades, viz. Hi-B material (M0H, M1H, M2H), laser scribed, mechanical scribed, etc., as compared to CGO material (M2, M3, M4, M5, M6, etc.), the saturation flux density has remained same (about 2.0 T). The peak operating flux density (B_{mp}) gets limited by the over-excitation conditions specified by users. The slope of B-H curve of CRGO material significantly worsens after about 1.9 T (for a small increase in flux density, relatively much higher magnetizing current is drawn). Hence, the point corresponding to 1.9 T can be termed as knee-point of the B-H curve. It has been seen in example 1.1 that the simultaneous over-voltage and under-frequency conditions increase the flux density in the core. Hence, for an over-excitation condition (over-voltage and under-frequency) of α %, general guideline can be to use operating peak flux density of $[1.9/(1+\alpha/100)]$. For the 10% continuous over-excitation specification, B_{mp} of 1.73 T $[=1.9/(1+0.1)]$ can be the upper limit. For a power system, in which a voltage profile is well maintained, a continuous over-excitation condition of 5% is specified. In this case, B_{mp} of 1.8 T may be used as long as the core temperature and noise levels are within permissible limits; these limits are generally achievable with the step-lap core construction.

When a transformer is subjected to an over-excitation, core contains an amount of flux sufficient to saturate it. The remaining flux spills out of the core. The over-excitation must be extreme and of a long duration to produce damaging effect in the core laminations. The laminations can easily withstand temperatures in the region of 800°C (they are annealed at this temperature during their manufacture), but insulation in the vicinity of core laminations, viz. press-board insulation (class A: 105°C) and core bolt insulation (class B: 130°C) may get damaged. Since the flux flows in air (outside core) only during the part of a cycle when core gets saturated, the air flux and exciting current are in the form of pulses having high harmonic content which increases the eddy losses and temperature rise in windings and structural parts. Guidelines for permissible short-time over-excitation of transformers are given in [14, 15]. Generator transformers are more susceptible for overvoltages due load rejection conditions and therefore need special design considerations.

2.5 No-Load Loss Test

Hysteresis loss is a function of average voltage or maximum flux density, whereas eddy loss is a function of r.m.s. voltage or r.m.s. flux density. Hence, the total core loss is a function of voltage wave-shape. If the sine-wave excitation cannot be ensured during the test, the following correction procedure can be applied to derive the value of no-load loss on the sine wave basis [16, 17]. When a voltmeter corresponding to the mean value is used, reading is proportional to the maximum value of flux density in the core. Hence, if the applied non-sinusoidal voltage has the same maximum flux density as that of the desired sine-wave voltage, hysteresis loss will be measured corresponding to the sine wave. The r.m.s. value may not be equal to r.m.s. value of desired sine wave; hence eddy loss has to be corrected by using a factor K_e,

$$K_e = \frac{(actual\ r.m.s.\ voltage\ \ of\ applied\ voltage)^2}{(r.m.s.\ voltage\ \ on\ sine\ wave\ basis)^2}$$

True core loss of transformer (P_c) on the sine wave basis is then calculated from the measured loss (P_m) as

$$P_c = \frac{P_m}{P_h^{'} + K_e P_e^{'}} \tag{2.12}$$

where $P_h^{'}$ and $P_e^{'}$ are hysteresis and eddy loss fractions of the total core loss respectively. The following values are usually taken for these two fractions,

$P_h^{'} = 0.5$ and $P_e^{'} = 0.5$ for cold rolled steel

$P_h^{'} = 0.7$ and $P_e^{'} = 0.3$ for hot rolled steel

The calculation as per equation 2.12 is recommended in ANSI Standard C57.12.90-1999. For highly distorted waveforms (with multiple zero crossings per period), a correction which can be applied to this equation is given in [18].

As per IEC 60076-1 (Edition 2.1, 2000), the test voltage has to be adjusted according to a voltmeter responsive to the mean value of voltage but scaled to read the r.m.s. voltage of a sinusoidal wave having the same mean value (let the reading of this voltmeter be V_1). At the same time, a voltmeter responsive to the r.m.s. value of voltage is connected in parallel with the mean value voltmeter and let its reading be V. The test voltage wave shape is satisfactory if the readings V_1 and V are within 3% of each other. If the measured no-load loss is P_m then the corrected no-load loss (P_c) is given as

$$P_c = P_m(1 + d) \tag{2.13}$$

where $d = \dfrac{V_1 - V}{V_1}$ (usually negative)

The method given in [19] allows the determination of the core loss from the measured data under non-sinusoidal excitation without artificial separation of the hysteresis and eddy current losses. Harmonic components are taken into account. The computed results are compared with the IEC method.

A voltage regulator with a large capacitor bank is better than a conventional rotating machine source from the point of view of getting as sinusoidal voltage as possible for core loss measurements.

The no-load loss test and the calculation of parameters of shunt branch of the equivalent circuit of a transformer have been elaborated in Chapter 1. Now, special topics / case studies related to the no-load test are discussed.

2.5.1 Asymmetrical magnetizing phenomenon

Unlike in a bank of three single-phase transformers having independent magnetic circuits, a three-phase three-limb transformer has interlinked magnetic circuit. The excitation current and power drawn by each phase winding are not the actual current and power required by the corresponding magnetic sections of the core. The current drawn by each phase winding is determined by the combination of requirements of all the three core branches. Consider a three-phase three-limb core shown in figure 2.7. Let the magnetomotive force required to produce instantaneous values of fluxes (ϕ_r, ϕ_y and ϕ_b) in the path between points P_1 to P_2 for the phase windings (r, y and b) be $NI_r^{'}$, $NI_y^{'}$ and $NI_b^{'}$ respectively. There is an inherent asymmetry in the core as the length of magnetic path of winding y between the points P_1 and P_2 is less than that of windings r and b. Let the actual currents drawn be I_r, I_y and I_b.

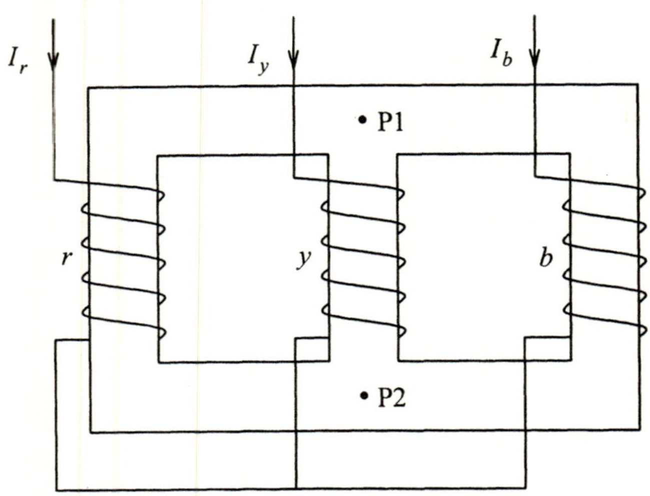

Figure 2.7 Three-phase three-limb core with Y connected primary

The following equations can be written:

$$NI_r - NI_y = NI_r' - NI_y' \tag{2.14}$$

$$NI_y - NI_b = NI_y' - NI_b' \tag{2.15}$$

$$NI_b - NI_r = NI_b' - NI_r' \tag{2.16}$$

For a Y-connected winding (star connected without grounded neutral),

$$I_r + I_y + I_b = 0 \tag{2.17}$$

It follows from equations 2.14 to 2.17 that

$$I_r = I_r' - (1/3)\,(I_r' + I_y' + I_b') = I_r' - I_z \tag{2.18}$$

$$I_y = I_y' - (1/3)\,(I_r' + I_y' + I_b') = I_y' - I_z \tag{2.19}$$

$$I_b = I_b' - (1/3)\,(I_r' + I_y' + I_b') = I_b' - I_z \tag{2.20}$$

where I_z is the zero-sequence component of the currents required to establish the required magnetomotive forces,

$$I_z = (1/3)\,(I_r' + I_y' + I_b') \tag{2.21}$$

Higher the magnetizing asymmetry, higher is the magnitude of I_z. The magnetomotive force, NI_z, is responsible for producing a zero-sequence leakage flux in the space outside core between points P_1 and P_2 [20]. The magnitude of this zero-sequence leakage flux is quite small as compared to the mutual flux in the core. For convenience, the reluctance of the magnetic path of winding y between points P_1 and P_2 is taken as half that of windings r and b. For sinusoidal applied voltages, fluxes are also sinusoidal, and the excitation current required then contains harmonics due to non-linear magnetic characteristics. Thus, the required excitation currents in three-phases can be expressed as (harmonics of order more than 3 are neglected)

$$I_r' = I_c \cos\omega t + I_1 \sin\omega t - I_3 \sin 3\omega t \tag{2.22}$$

$$I_y' = 0.5 I_c \cos(\omega t - 120^0) + 0.5 I_1 \sin(\omega t - 120^0) - 0.5 I_3 \sin 3(\omega t - 120^0) \tag{2.23}$$

$$I_b' = I_c \cos(\omega t - 240^0) + I_1 \sin(\omega t - 240^0) - I_3 \sin 3(\omega t - 240^0) \tag{2.24}$$

where I_c is the core loss component, and a negative sign is taken for third harmonic components [21] to get a peaky nature of the excitation current (for a sinusoidal flux, excitation current is peaky in nature due to non-linear magnetic characteristics). Substituting these expressions in equation 2.21,

$$I_z = 0.167 I_c \cos(\omega t + 60^0) + 0.167 I_1 \sin(\omega t + 60^0) - 0.833 I_3 \sin(3\omega t) \quad (2.25)$$

After substituting this expression for I_z and expressions for $I_r^{'}$, $I_y^{'}$ and $I_b^{'}$ from equations 2.22 to 2.24 in equations 2.18 to 2.20, the actual excitation currents drawn are

$$I_r = 0.928 I_c \cos(\omega t - 9^0) + 0.928 I_1 \sin(\omega t - 9^0) - 0.167 I_3 \sin(3\omega t) \quad (2.26)$$

$$I_y = 0.667 I_c \cos(\omega t - 120^0) + 0.667 I_1 \sin(\omega t - 120^0) + 0.333 I_3 \sin(3\omega t) \quad (2.27)$$

$$I_b = 0.928 I_c \cos(\omega t + 129^0) + 0.928 I_1 \sin(\omega t + 129^0) - 0.167 I_3 \sin(3\omega t) \quad (2.28)$$

The condition that the sum of 3rd harmonic currents in three phases has to be zero (since the neutral is isolated) is satisfied by above three equations. The essence of the mathematical treatment can be understood by the vector diagrams of fundamental and third harmonic components shown in figure 2.8. The magnitudes of I_r and I_b are almost equal and these are greater than the magnitude of I_y. The current I_y, though smallest of all the three currents, is higher than the current required to excite middle phase alone ($I_y > I_y^{'}$). The currents in the outer limbs are slightly less than that needed to excite outer limbs alone ($I_r < I_r^{'}$ and $I_b < I_b^{'}$). In actual practice, the currents I_r and I_b may differ slightly due to minor differences in the characteristics of their magnetic paths (e.g., unequal air gap lengths at corner joints). The third harmonic component drawn by phase *y* is greater than that of phases r and b.

Since the applied voltage is assumed to be sinusoidal, only the fundamental component contributes to the power. The power corresponding to phase r will be negative if I_z is large enough to cause the angle between V_r and I_r to exceed 90°. Negative power is read in one of the phases during the no-load loss test for transformers whose yoke lengths are quite appreciable as compared to limb heights increasing the asymmetry between the middle and outer phases.

It has been proved in [22] that for a length of central limb between points P_1 and P_2 equal to half that of outer limbs (reluctance of central limb is half that of outer limbs) in figure 2.7,

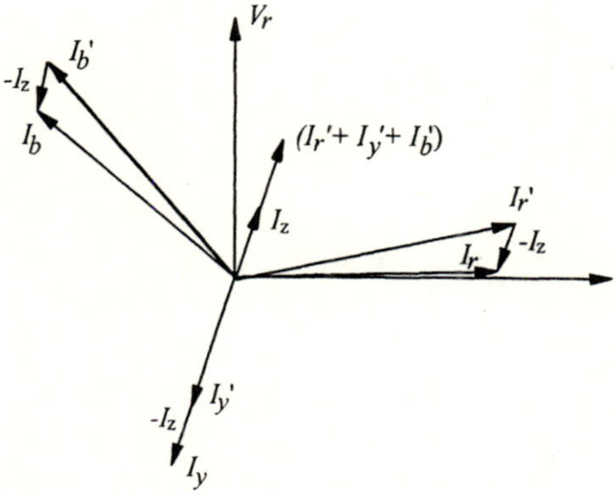

a. Fundamental components

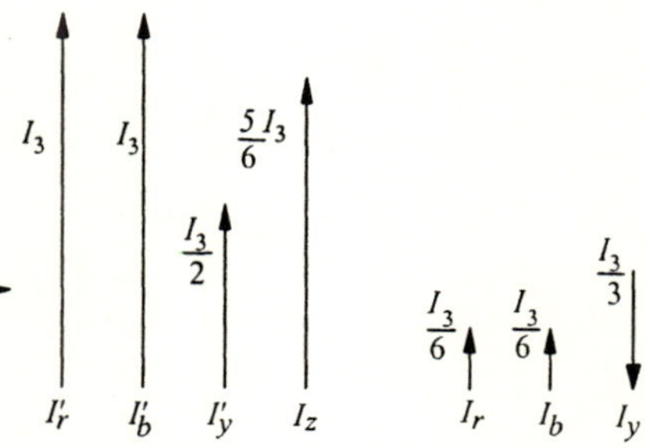

b. Third harmonic components

Figure 2.8 Magnetizing asymmetry

$$I_r : I_y : I_b = 1 : 0.718 : 1 \tag{2.29}$$

The effect of change in excitation is illustrated for r phase in figure 2.9. During no-load loss test, losses are generally measured at 90%, 100% and 110% of the rated voltage. The magnetizing component of excitation current is more sensitive to the increase in flux density as compared to the core loss component. Consequently as the voltage is increased, the no-load power factor decreases. The value of I_z also increases and hence the possibility of reading negative power increases with the increase in applied voltage. When the angle between V_r and I_r is 90°, the r phase wattmeter reads zero, and if it exceeds 90° the wattmeter reads negative.

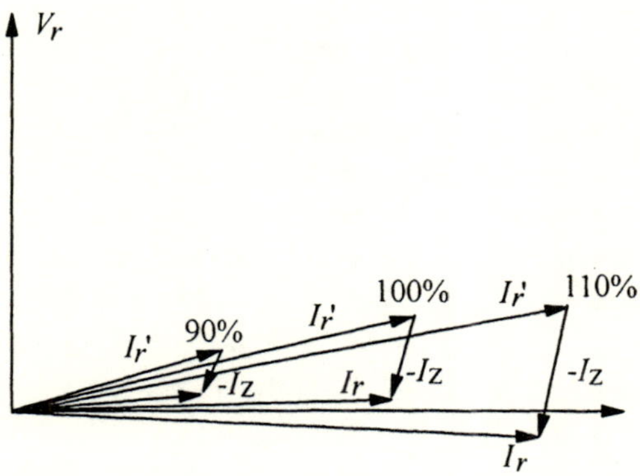

Figure 2.9 Effect of excitation level

The magnetizing asymmetry phenomenon described above has been analyzed by using mutual impedances between 3 windings in [23]. It is shown that phase currents and powers are balanced if mutual impedances Z_{ry}, Z_{yb} and Z_{br} are equal. These impedances are function of number of turns and disposition of windings, winding connections within a phase and more importantly on dimensions and layout of the core. These mutual impedances, which are unbalanced in three-phase three-limb core ($Z_{ry} = Z_{yb} \neq Z_{br}$), redistribute the power shared between the three phases. The form of asymmetry occurring in the phase currents and powers is different for three-limb and five-limb cores. It is reported that there is star point displacement in a five-limb transformer, which tends to reduce the unbalance caused by the inequality of mutual impedances.

Similar analysis can be done for a delta connected primary winding, for which the measured line current is the difference between currents of the corresponding two phases. It can be proved that [24] when the delta connected winding is energized, for Yd1 or Dy11 connection, line current drawn by r phase is higher than that drawn by y and b phases, which are equal ($I_{r-L} > I_{y-L} = I_{b-L}$). For Yd11 or Dy1 connection, the line current drawn by b phase is higher than that drawn by r and y phases, which are equal ($I_{b-L} > I_{y-L} = I_{r-L}$). It should be noted that, for the delta connected primary winding also, the magnetic section corresponding to y phase requires least magnetizing current, i.e., ($I_y^{'} < I_r^{'}, I_b^{'}$), but the phasor addition of two phase currents results into a condition that line current I_{y-L} equals the current of one of the outer phases.

2.5.2 Magnetic balance test

This test is performed at works or site as an investigative test to check the healthiness of windings and core. In this test, a low voltage (say, 230 V) is applied to a winding of one phase with all other windings kept open circuited. Voltages induced in the corresponding windings of other two phases are measured. When a middle phase (y) is excited, voltage induced in r and b phases should be in the range of 40 to 60% of the applied voltage. Ideally it should be 50% but due to difference in reluctance of the magnetic paths corresponding to r and b phases (on account of minor differences in air gaps at joints, etc.), some deviation from the expected values need not be considered as abnormal. When r (or b) phase is excited, one may get y-phase induced voltage as high as 90% and the voltage induced in b (or r) phase as low as 10% for a healthy core. The addition of r.m.s. voltages induced in unexcited phases need not necessarily be equal to the voltage applied to the excited phase due to non-linear characteristics of the magnetic circuit and the harmonics present in the fluxes of the unexcited limbs.

The results of the magnetic balance test should be taken as indicative ones and some other test (e.g., no-load loss test at rated voltage in manufacturer's works) should be performed to confirm the conclusions. The magnetic balance test can be done at various stages of manufacturing, viz. before and after connections, before final tests, before dispatch; these test results can be used for comparison with those done at any subsequent time to check whether any problem is developed in the core and windings. If the voltages measured do not fall in the expected range, a problem in core or windings can be suspected. Suppose there is a turn-to-turn fault in r phase. When a low voltage is applied to y phase winding, instead of getting almost equal induced voltages in r and b phase windings, a much higher voltage is obtained in b phase winding as compared to that of r phase, indicating a fault in winding of phase r. A high fault current circulating in the faulty section, opposes the magnetizing flux thereby reducing the induced voltage in the faulty phase. For the test, the core should be demagnetized because a slight magnetization (e.g., after resistance measurement) can give erratic results. The demagnetization can be achieved by a repeated application of variable AC voltage which is slowly reduced to zero.

2.5.3 Trouble-shooting by no-load loss test

Detection and location of turn-to-turn fault can be done by the results of no-load loss test. Suppose, it is suspected that during impulse testing a particular winding has failed. The turn-to-turn fault may not result in appreciable change in the transfer function (impedance) of the winding and hence there is no appreciable disturbance noticed in the recorded impulse waveforms. The fault in the suspected winding can be confirmed by doing a no-load loss test. Therefore, it is usually recommended to do the no-load loss test after all high voltage dielectric tests for detecting any developed fault in the windings. No-load loss value shoots up for a fault between turns. In order to locate the exact position of a fault, the parallel conductors are electrically separated at both ends, and then resistance is measured between all the points available $(1, 1', 2, 2', 3, 3')$ as shown in figure 2.10 for a winding with 3 parallel conductors.

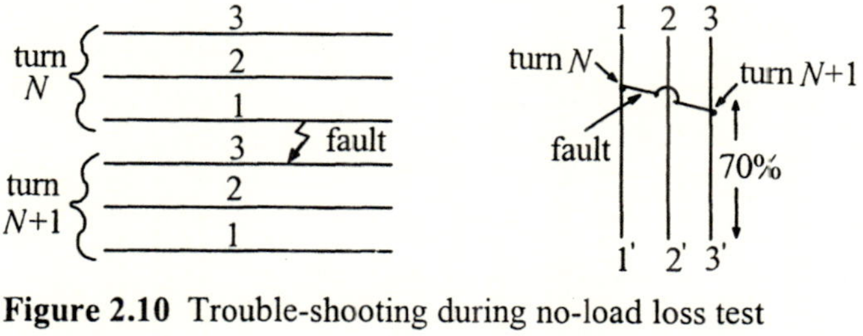

Figure 2.10 Trouble-shooting during no-load loss test

Let us assume that each of the parallel conductors is having a resistance of 0.6 ohms. If the fault is at a location 70% from the winding bottom between conductor 1 of one turn and conductor 3 of next turn, then the measured values of resistances between $1-3$ and $1'-3'$ will be 0.36 ohms ($2 \times 0.3 \times 0.6$) and 0.84 ohms ($2 \times 0.7 \times 0.6$) respectively. A voltage corresponding to one turn circulates very high currents since these are limited only by above resistances (reactance in path is negligible). The increase in no-load loss corresponds approximately to the loss in these two resistance paths due to circulating currents.

2.5.4 Effect of impulse test on no-load loss

A slight increase of about few % in the no-load loss is sometimes observed after impulse tests due to partial breakdown of interlaminar insulation (particularly at the edges) resulting into higher eddy loss. The phenomenon has been analyzed in [25], wherein it is reported that voltages are induced in core by electrostatic as well as electromagnetic inductions. The core loss increase of an average value of less than 2% has been reported. It is further commented that the phenomenon is harmful to the extent that it increases the loss and that the loss will not increase at site. Application of an adhesive at the edges can prevent this partial and localized damage to the core during the high voltage tests.

2.6 Impact of Manufacturing Processes on Core Performance

For building cores of various ratings of transformers, different lamination widths are required. Since the lamination rolls are available in some standard widths from material suppliers, slitting operation is required to get the required widths. It is obvious that most of the times a full width cannot be utilized and the scrap of leftover material has to be minimized by a meticulous planning exercise. A manufacturer having a wide product range, generally uses the leftover of large transformer cores for the cores of small distribution transformers.

The next operation is that of cutting the laminations in different shapes (e.g., mitred joint in figure 2.3). Finally, the corner protrusions of the built core are cut because they are not useful (do not carry the flux), and they may contribute to the noise level of the transformer due to their vibrations.

In a bolted yoke construction, which ensures rigidity of the core, holes are punched in the yoke laminations. There is distortion of flux at the position of holes as shown in figure 2.11.

Figure 2.11 Effect of yoke bolts

This leads to an increase in core loss. Hence, many manufactures use boltless yoke construction, in which special clamping arrangement consisting of fiber-glass / stainless steel bands is adopted. The boltless yoke construction results in better utilization of core material and reduction in core loss. Small guiding holes are needed to facilitate the placement of laminations and core-building. In order to strengthen the bonding of laminations, suitable epoxy resin is applied at the edges of yoke laminations, which also results in reduction of noise level.

The processes of slitting, cutting and punching result in edges having burrs, which not only worsen the stacking factor but also result into shorting of adjacent laminations (due to damage of insulation coating) increasing eddy losses. The upper limit of acceptable burr level is about 20 microns. A lower burr level improves the stacking factor of the core and reduces loss. Higher the stacking factor of laminations (which may be about 0.97 to 0.98), higher is the core area obtained leading to a more cost effective design. This staking factor, which is decided by lamination coating and burr, is different than the core space factor (which of the order of 0.88 to 0.90). The core space factor is defined as the actual core area obtained divided by the core circle area. The burrs can be removed by passing the laminations through a de-burring process. A thin coating of varnish may be applied at the edges to cover up the scratches formed during the de-burring process.

All the above processes and multiple handling of laminations result into development of mechanical strains inside the laminations, disturbing the original grain orientation and thereby causing increase in core loss. This effect can be mitigated by annealing the laminations at a temperature of about 800°C under inert gas atmosphere. If state-of-the-art lamination slitting and cutting machines having high degree of automation are used, handling of laminations is reduced substantially and hence annealing is not considered necessary.

Core limbs are generally made of a large number of steps in order to get a maximum core area for a given core diameter leading to an optimum design. Yokes on the other hand may have lesser number of steps to provide better axial support to the windings. The mismatch in number of limb and yoke steps may result into some extra loss at the corner joints.

One question which is many times asked by production and planning departments is: Can materials of different grades be mixed to overcome the problem of non-availability of a particular grade in sufficient quantity? Needless to say, one can mix only a better grade material, which can give marginal reduction in the no-load loss. Actually, the grade mixing should not be encouraged. In exceptional cases, however, mixing of a better grade may be allowed to minimize shop inventory problems.

2.7 Inrush Current

2.7.1 Theory

If it were possible to switch on a transformer exactly at an instant of a voltage wave which corresponds to the actual flux density in the core at that instant, it would not have resulted in any transient. In actual practice, a transient phenomenon in the form of inrush current is unavoidable since the instant of switching cannot be easily controlled and the instant of switching favourable to one phase is not favourable to other two phases.

When a transformer is switched off, the excitation current follows the hysteresis curve to zero, whereas the flux density value changes to a non-zero value B_r as evident from the hysteresis loop in figure 2.5 (point D). For a residual flux density of $+B_r$, a maximum inrush current is drawn when a transformer is switched on at the instant when the applied voltage is zero as shown in figure 2.12. If transformer was not switched off, excitation current (i) and flux density would have followed the dotted curves. As per the *constant flux linkage theorem*, magnetic flux in an inductive circuit cannot change suddenly; the flux just after closing the switch (at $t = 0^+$) must remain equal to the flux just before closing the switch (at $t = 0^-$). Hence, the flux density, instead of starting from the negative maximum value ($-B_{mp}$), starts from $+B_r$ and reaches the peak positive value of ($B_r + 2B_{mp}$) driving the core into saturation.

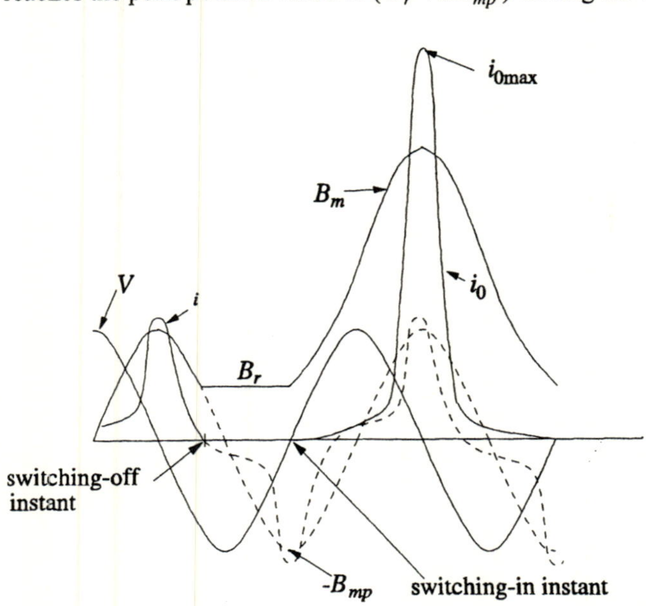

Figure 2.12 Case of maximum inrush current

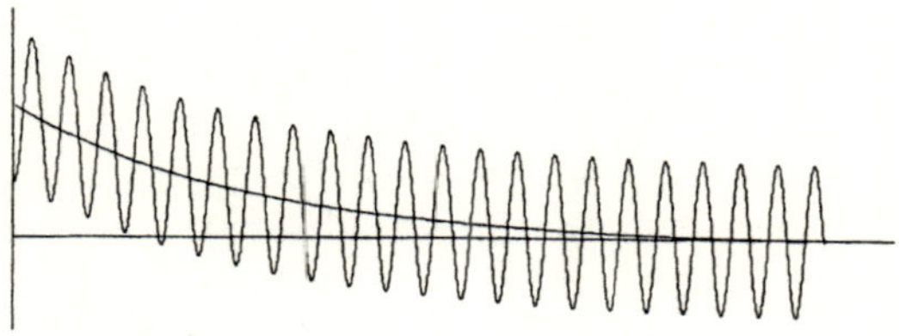

Figure 2.13 Flux wave

For an applied sinusoidal voltage, the flux density is sinusoidal and magnetizing current is peaky in nature. The same result can be obtained by solving the following governing equation:

$$V_p \sin(\omega t + \theta) = i_0 R_1 + N_1 \frac{d\phi_m}{dt} \tag{2.30}$$

where

V_p = peak value of the applied voltage

θ = angle at which voltage is switched on

i_0 = instantaneous value of magnetizing current

ϕ_m = instantaneous value of flux at any time t

R_1 = primary winding resistance

N_1 = primary winding turns

The solution of the equation is quite straightforward when linear magnetic characteristics are assumed. The solution is obtained by using the initial conditions that at $t = 0$, $\phi_m = \pm\phi_r$,

$$\phi_m = (\phi_{mp} \cos\theta \pm \phi_r)\, e^{\frac{-R_1}{L_1}t} - \phi_{mp} \cos(\omega t + \theta) \tag{2.31}$$

For $\theta = 0$ and residual flux of $+\phi_r$, the waveform of flux (flux density) is shown in figure 2.13. It can be observed from equation 2.31 and the flux waveform that the flux wave has a transient DC component, which decays at a rate determined by the ratio of resistance to inductance of primary winding (R_1 / L_1), and a steady-state AC component ($-\phi_{mp} \cos(\omega t + \theta)$). A typical waveform of an inrush current is shown in figure 2.14 for a phase switched on at the most unfavourable instant (i.e., at zero crossing of the applied voltage wave).

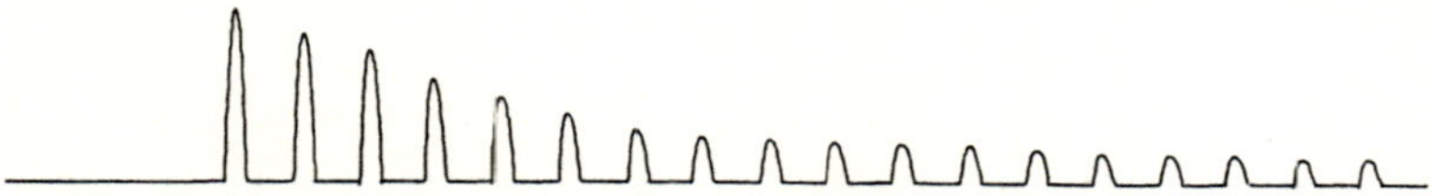

Figure 2.14 Typical inrush current waveform

It can be observed that the current waveform is completely offset in first few cycles with wiping out of alternate half cycles because the flux density is below saturation value for these half cycles (resulting in very small current value). Hence, the inrush current is highly asymmetrical and has a predominant second harmonic component which is used by differential protection schemes to restrain relays from operating.

Time constant (*L/R*) of the circuit is not constant; the value of *L* changes depending on the extent of core saturation. During the first few cycles, saturation is high and *L* is low. Hence, initial rate of decay of inrush current is quite high. As the losses damp the circuit and saturation drops, *L* increases slowing down the decay. Hence, the decay of inrush current starts with a high initial rate and progressively reduces; the total phenomenon lasts for few seconds. Smaller transformers have higher rates of decay. In general, transformers having higher losses (lower efficiency) have higher decay rates of inrush current [26].

While arriving at equation 2.31, linear magnetic characteristics are assumed, which is a major approximation. Accurate procedures for calculation of inrush currents for single-phase transformers is given in a number of references [27, 28] in which non-linear magnetic characteristics are elaborately represented. For estimation of inrush current in three-phase transformers, the analysis is more involved [29, 30, 31]. A method of calculation of inrush currents in harmonic domain for single-phase and three-phase transformers using operational matrices is given in [32]. Inrush currents of transformers and associated overvoltages in HVDC systems are dealt in [33], wherein AC system impedance, which is generally inductive (that of generators, transformers and transmission lines), is shown to resonate with filters on AC bus-bars (which act as lumped capacitance below fifth harmonic). If the resulting resonance frequency of the combined AC system and filters is equal to or close to a harmonic component of inrush current of the same frequency, overvoltages occur.

2.7.2 Estimation of magnitude of first peak

A transformer user is generally interested in knowing the maximum value of inrush current and the rate of decay of inrush current. If the saturation flux density of core material is 2.03 T, the flux of magnitude ($2.03 \times A_c$) is contained in the core, where A_c is the net core area. Rest of the flux spills out of the core, whose path is predominantly in air. The ampere-turns required to produce the air flux are so large that they can be also assumed to produce 2.03 T in the core. For the worst instant of switching it can be written that

$$\phi_{air} = \mu_0 \, H \, A_w = 2\phi_{mp} + \phi_r - 2.03\, A_c \tag{2.32}$$

where A_w is the mean area enclosed by a winding turn. Hence, the maximum inrush current (first peak), $i_{0\max}$, drawn by the energized winding with N_1 turns and h_w height can be calculated for a single-phase transformer from the following equation (H is magnetic field intensity) as

$$H = \frac{N_1\, i_{0\max}}{h_w} \tag{2.33}$$

$$\therefore i_{0\max} = \frac{h_w\, H}{N_1} = \frac{(2\phi_{mp} + \phi_r - 2.03\, A_c)\, h_w}{\mu_0\, A_w\, N_1} = \frac{(2B_{mp} + B_r - 2.03)\, A_c\, h_w}{\mu_0\, A_w\, N_1} \tag{2.34}$$

For three-phase transformers, calculation of inrush current needs more explanation. Let us consider the following three cases:

1) If it is a delta connected primary, each of the phases is independently connected to the network, and the inrush phenomenon corresponding to flux of each phase takes place as in the case of a single-phase transformer. This results in the same value of phase inrush current as that of the single-phase transformer. But in terms of line currents, inrush is less severe. Under normal operating conditions, line current is $\sqrt{3}$ times the phase current. During the inrush condition, only one phase is having large inrush current (the phase which gets switched at the worst or near worst instant of voltage switching); hence the line current is almost equal to the phase current. Hence, the per-unit line inrush current of three-phase transformer with delta connected primary is 0.577 times ($1/\sqrt{3}$ times) the corresponding inrush current of a single-phase transformer.

2) For a bank of three single-phase transformers having independent magnetic circuits, with star connected primary and delta connected secondary, the current distribution expressed in terms of the maximum inrush current ($i_{0\max}$) of single-phase transformer is shown in figure 2.15. It is assumed that phase a has the maximum transient inrush current.

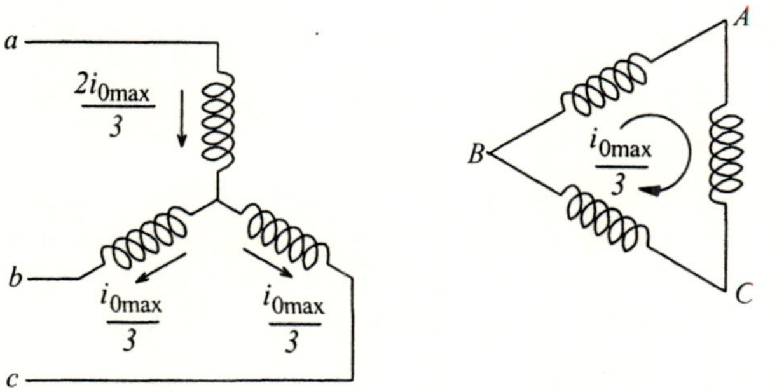

Figure 2.15 Inrush in Y-delta bank of transformers

Since the delta allows the flow of zero-sequence currents, it holds the neutral voltage at a stable value and maintains the normal line-to-neutral voltages across the phases. The presence of delta also ensures full single-phase transient in the phase that has maximum inrush transient (phase a in this case). Two-thirds of the required single-phase inrush current $(2\, i_{0\max}/3)$ flows in phase a on the star side and the remaining one-third flows on the delta side. Hence, the maximum inrush current in this case is two-thirds that of single-phase transformer. The phases b and c do not get magnetized since currents in them are equal and opposite on the star and delta sides.

3) A three-phase three-limb transformer, in which the phases are magnetically interlinked, can be treated as consisting of three independent single-phase transformers [17, 29] under inrush transients. Hence, for a star connected primary winding, the inrush phenomenon is similar to that of Case 2, irrespective of whether secondary is star or delta winding. Maximum inrush current is approximately equal to two-thirds of that corresponding to single-phase operation of one limb.

2.7.3 Estimation of decay pattern

The equation 2.34 is an approximate formula giving maximum possible inrush current. The operating engineers may be interested in knowing inrush current peak values for the first few cycles or the time after which inrush current reduces to a value equal to the rated current. The procedures for estimating inrush current peaks for first few cycles are given in [34, 35]. The procedures are generally applicable for some tens of initial cycles.

Example 2.1

Calculate inrush current peaks for first 5 cycles for a 31.5 MVA, 132/33 kV, 50 Hz, Yd1 transformer, when energized from 132 kV winding having 920 turns, mean diameter of 980 mm and height of 1250 mm. The peak operating flux density is 1.7 T for core area of 0.22 m^2. The sum of system and winding resistances is 0.9 ohms.

Solution:

The transformer is assumed to be energized at the instant when voltage is at zero value. It is also assumed that the residual flux is in the same direction as that of the initial flux change, thus giving a maximum possible value of inrush current. After the core saturation, the inrush current gets limited by air core reactance, X_s, which can be calculated by the fundamental formula.

Step 1:

$$X_s = \frac{\mu_0 N^2 A_w}{h_w} \times 2 \times \pi \times f \tag{2.35}$$

N = number of turns of excited winding = 920 turns

A_w = area inside the mean turn of excited winding = $\frac{\pi}{4}(mean\ diameter)^2$

$= \frac{\pi}{4}(0.980)^2 = 0.754\ \text{m}^2$

h_w = height of energized winding = 1.25 m

$$\therefore X_s = \frac{4\pi \times 10^{-7} \times 920^2 \times 0.754}{1.25} \times 2 \times \pi \times 50 = 202 \text{ ohms}$$

Step 2:

Now angle θ is calculated [34], which corresponds to the instant at which the core saturates,

$$\theta = K_1 \cos^{-1}\left\{\frac{B_s - B_{mp} - B_r}{B_{mp}}\right\} \tag{2.36}$$

where B_s = saturation flux density = 2.03 Tesla

B_{mp} = peak value of designed steady-state flux density in the core = 1.7 T

B_r = residual flux density = 0.8 × B_{mp} = 1.36 T

(For cold rolled material, maximum residual flux density is usually taken as 80% of the rated peak flux density, whereas for hot rolled material it can be taken as 60% of the rated peak flux density)

K_1 = correction factor for saturation angle = 0.9

$$\therefore \theta = 0.9 \times \cos^{-1}\left\{\frac{2.03 - 1.7 - 1.36}{1.7}\right\} = 2.0 \text{ radians}$$

Step 3:

The inrush current peak for the first cycle is calculated as [34],

$$i_{0\max} = \frac{K_2 V\sqrt{2}}{X_s}(1 - \cos\theta) \tag{2.37}$$

where V = r.m.s. value of applied alternating voltage

K_2 = correction factor for the peak value = 1.15

$$\therefore i_{0\max} = \frac{1.15 \times (132000/\sqrt{3})\sqrt{2}}{202}(1 - \cos 2.0) = 869 \text{ amperes}$$

The current calculated by equation 2.34, for a core area of 0.22 m^2 is

$$i_{0\,\max} = \frac{(2\times 1.7 + 1.36 - 2.03)\times 0.22\times 1.25}{4\pi\times 10^{-7}\times 0.754\times 920} = 861 \text{ amperes,}$$

which is very close to that calculated by the more accurate method.

Step 4:

After having calculated the value of inrush current peak for the first cycle, the residual flux density at the end of first cycle is calculated. The residual component of flux density reduces due to losses in the circuit and hence is a function of damping provided by the transformer losses. The new value of residual flux density is calculated as [34]

$$B_r(new) = B_r(old) - B_{mp}\times\frac{K_3 R}{X_s}\left[2(\sin\theta - \theta\cos\theta)\right] \tag{2.38}$$

where R = sum of transformer winding resistance and system resistance
= 0.9 ohms

K_3 = correction factor for the decay of inrush = 2.26

$$\therefore\ B_r(new) = 1.36 - 1.7\times\frac{2.26\times 0.9}{202}\left[2(\sin 2.0 - 2.0\cos 2.0\right] = 1.3\text{ T}$$

Now steps 2, 3 and 4 are repeated to calculate the peaks of subsequent cycles. The inrush current peaks for the first 5 cycles are: 869 A, 846 A, 825 A, 805 A and 786 A on single phase basis. Since it is a Y-delta connected three-phase three-limb transformer, actual line currents are approximately two-thirds of these values (579 A, 564 A, 550 A, 537 A and 524 A).

The inrush of magnetizing current may not be harmful to a transformer itself (although repeated switching on and off in short period of time is not advisable). Behavior of transformer under inrush condition continues to attract attention of researchers. The differences in forces acting on the windings during inrush and short circuit conditions are enumerated in [36]. Inrush may result in the inadvertent operation of the overload and differential relays, tripping the transformer out of the circuit as soon as it is switched on. Relays which discriminate between inrush and fault conditions are commonly used. Some of their features are:

1) Differential relay with second harmonic restraint, which makes use of the fact that the inrush current has a predominant second harmonic component which is used to prevent the relay from operating.

2) Differential relay with reduced sensitivity to the inrush current by virtue of higher pick-up for the offset wave plus a time delay to override high initial peaks of the inrush current.

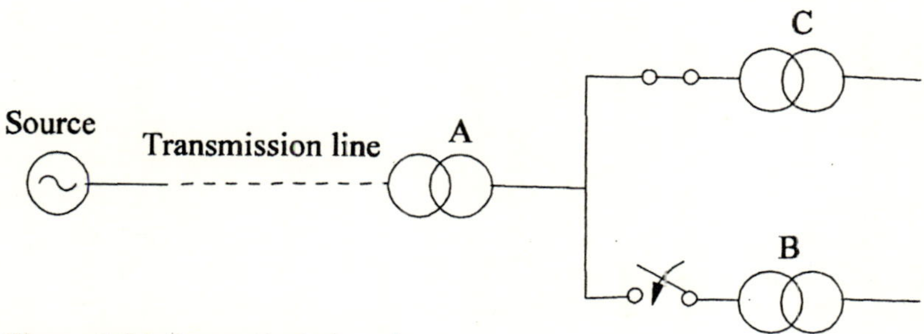

Figure 2.16 Sympathetic inrush

A technique of discriminating inrush current and internal fault is described in [37], which uses wavelet transforms and neural networks. The ability of wavelet transforms to extract information from the transient signals simultaneously in time and frequency domains is used for the discrimination. Large inrush may cause an excessive momentary dip in the supply voltage affecting operation of other interconnected electrical equipment. Switching on of a particular transformer in an interconnected network can affect already energized transformers as explained below.

2.7.4 Sympathetic inrush phenomenon

It has long been known that transient magnetizing inrush currents, sometimes reaching magnitudes as high as six to eight times the rated current, flow in a transformer winding when switched on to an electric power network. It has not been generally appreciated, however, that the other transformers, already connected to the network near the transformer being switched, may also have a transient magnetizing current of appreciable magnitude at the same time. In order to understand how energizing of a transformer in a network affects the operating conditions of other transformers connected to the same network, consider a network as shown in figure 2.16.

When transformer B is switched on to the network already feeding similar transformers (C) in the neighbourhood, the transient magnetizing inrush current of the switched-on transformer also flows into these other transformers and produces in them a DC flux which gets superimposed on their normal AC magnetizing flux. This gives rise to increased flux density and corresponding higher magnetizing currents in these other transformers in the neighborhood [17, 38, 39]. This sympathetic inrush current in these other transformers is less than their own inrush current when energized. Depending on the magnitude of decaying DC component, this sympathetic (indirect) inrush phenomenon leads to an increased noise level of these connected transformers due to higher core flux density for the transient period. It may also lead to mal-operation of protective equipment of these transformers. The phenomenon of increase in noise level of an upstream power transformer during the energization of a downstream distribution transformer (fed by the power transformer) has been analyzed in [40] supported by noise level measurements done during switching tests at site.

Let us now analyze the case of parallel transformers shown in figure 2.17 (a). The transformers may or may not be paralleled on the secondary side. The DC component of inrush current of the transformer being energized flows through the transmission line resistance (between source and transformer) producing a DC voltage drop across it. The DC voltage drop forces the already energized transformer towards / into saturation in opposite direction of the transformer which is being switched on, resulting in a buildup of magnetizing current in the already energized transformer; this rate of buildup is same as the rate at which DC component of magnetizing current is decreasing in the transformer being switched on. When the two parallel transformers are similar and magnitudes of DC components of currents in both the transformers become equal, there is no DC component in the line feeding both the transformers. However, there is a DC component circulating in the loop circuit between them, whose rate of decay is very slow due to high inductance and small resistance of windings of the two transformers. The waveforms of currents are shown in figure 2.17 (b).

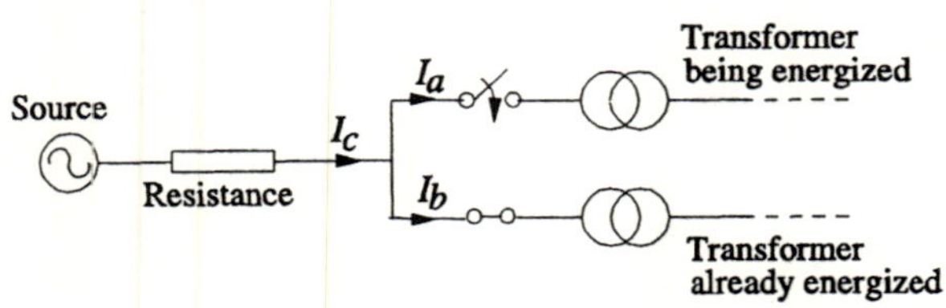

(a)

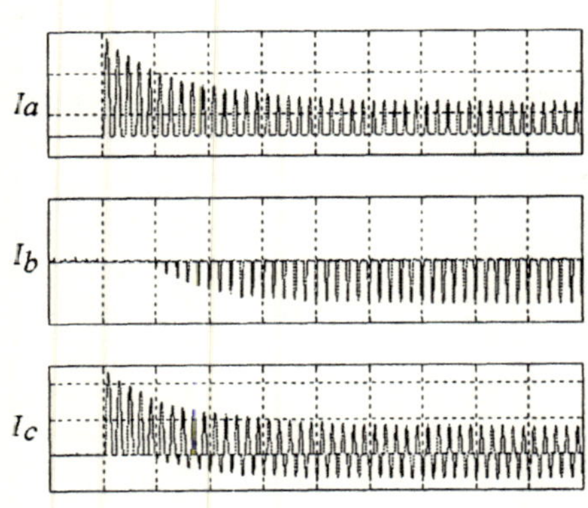

(b)

Figure 2.17 Inrush current in parallel transformers

Since the line current feeding the transformers becomes symmetrical (waveform I_c) devoid of the second harmonic component, differential relaying with second harmonic constraint is provided to each transformer separately instead of protecting them as a unit [41]. The phenomenon is more severe when transformers are fed from a weak system (transformers connected to a common feeder with a limited fault level and high internal resistance).

2.7.5 Factors affecting inrush phenomenon

Various factors affecting the inrush current phenomenon are now summarized:

A. Switching-on angle (α)

Inrush current decreases when switching-on angle (on the voltage wave) increases. It is maximum for $\alpha = 0°$ and minimum for $\alpha = 90°$.

B. Residual flux density

Inrush current is significantly aggravated by residual flux density, which depends upon core material characteristics and the power factor of the load at interruption when a transformer was switched off. The instant of switching-off has an effect on residual flux density depending upon the type of load [17]. The total current is made up of the magnetizing current component and load current component. The current interruption generally occurs at or near zero of the total current waveform. The magnetizing current passes through its maximum value before the instant at which total current is switched off for no load, lagging load and unity power factor load conditions, resulting in maximum value of residual flux density as per B-H curve of figure 2.5. For leading loads, if the leading component is less than the magnetizing component, at zero of the resultant current the magnetizing component will have reached the maximum value resulting in the maximum residual. On the contrary, if the leading current component is more than the magnetizing component, the angle between maximum of the magnetizing current and zero of the resultant current will be more than 90°. Hence, at the interruption of the resultant current, the magnetizing component will not have reached its maximum resulting in a lower value of residual flux density.

Residual flux density also depends on the core material. Its maximum value is usually taken as about 80% and 60% of the saturation value for cold rolled and hot rolled materials respectively. It is also a function of joint characteristics. Hence, its value for a core with the mitred joint is different than that with the step-lap joint.

C. Series resistance

The resistance of line between the source and transformer has a predominant effect on the inrush phenomenon. Due to the damping effect, series resistance between the transformer and source not only reduces the maximum initial inrush current but also hastens its decay rate. Transformers near a generator usually have a longer inrush because of low line resistance. Similarly, large power transformers tend to have a long inrush as they have a large inductance as compared to the system resistance.

Consider a series circuit of two transformers, T feeding T1 as shown in figure 2.18. When transformer T1 is energized, transformer T experiences sympathetic inrush. Resistance between T and T1 contributes mainly to the decay of inrush of T1 (and T) [42] and not the resistance on the primary side of T.

In case of parallel transformers (figure 2.17), the sympathetic inrush phenomenon experienced by the transformer already energized is due to the coupling between the transformers on account of DC voltage drop in the transmission line feeding them. Hence, the higher the transmission line resistance the higher is the sympathetic inrush [43].

D. *Inrush under load*

If a transformer is switched on with load, the inrush peaks are affected to some extent by the load power factor. When it is switched on under heavy load (large secondary current) with the power factor close to unity, the peak value of inrush current is smaller, and as the power factor reduces (to either lagging or leading), the inrush current peak is higher [27].

2.7.5 Mitigation of inrush current

During the inrush phenomenon, inrush current in the saturated core condition is limited by the air-core reactance of the windings and hence it is usually lower than the peak short-circuit current due to faults. Since transformers are designed to withstand mechanical effects of short circuit forces, inrush currents may not be considered to be dangerous, although they may unnecessarily cause operation of protective equipment like relays and fuses.

One of the natural ways of reducing inrush current is to switch-in transformers through a closing resistor. The rated voltage is applied through a large resistor so that the voltage at the transformer terminals is lower than the rated value (e.g., 50%) reducing the inrush current. The resistor is subsequently by-passed to apply full voltage to the transformer. Such a scheme with pre-insertion (closing) resistors is recommended in [33, 44] to suppress the inrush currents in transformers. The closing resistor should be small enough to allow passage of the normal magnetizing current.

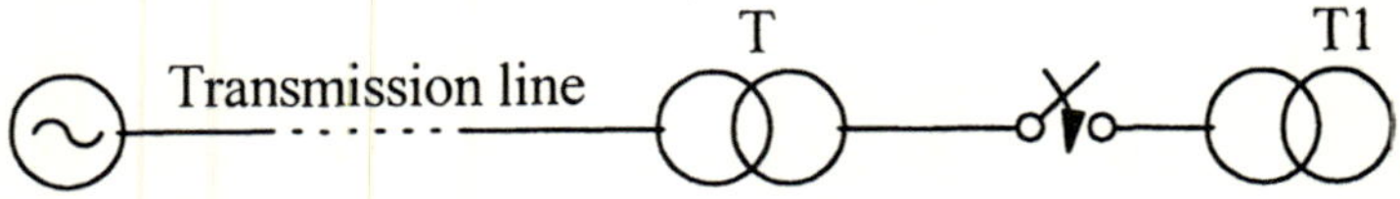

Figure 2.18 Sympathetic inrush in series connection

If possible, a transformer should be switched from its high voltage winding, which is usually the outer winding in core-type transformers and therefore has a higher air core reactance resulting in a lower value of inrush current.

Since residual flux is one of the main reasons for high inrush currents, any attempt to reduce it helps in mitigating the inrush phenomenon. When a transformer is being switched off, if a capacitor of suitable size is connected across it [26], a damped oscillation will result, causing an alternating current to flow in the transformer winding. The amplitude of current decreases with time, gradually reducing the area of the traversed hysteresis loop, eventually reducing both current and residual flux to zero. For small transformers, a variable AC source can be used to demagnetize the core. The applied voltage can be slowly reduced to zero for demagnetization.

Various schemes of controlled closing at favourable instants have been proposed in [45]. In these methods, each winding is closed when the prospective and dynamic (transient) core fluxes are equal resulting in an optimal energization without core saturation or inrush transients.

2.8 Influence of Core Construction and Winding Connections on No-Load Harmonic Phenomenon

The excitation current is a small percentage of the rated current in transformers. With the increase in rating of transformers, generally the percentage no-load current reduces. The harmonics in the excitation current may cause interference with communication systems and result into inadvertent tripping of protective equipment. Due to non-linear magnetic characteristics, it can be said that:

1) for a sinusoidal applied voltage, the flux is sinusoidal and magnetizing current is peaky in nature with a pronounced third harmonic component.

2) if the magnetizing current is constrained to have the sinusoidal nature, the flux wave will be flat-topped. The induced voltages in windings will be peaky in nature with a pronounced third harmonic component.

Harmonic phenomenon in a three-phase transformer depends on the type of magnetic circuit (separate or inter-linked) and the type of winding connections (star / delta / zigzag). Let us consider the following common cases:

1) ***Yy connection (isolated neutral)***: Since the neutral is isolated, third harmonic currents cannot flow in phases and lines making the magnetizing current almost sinusoidal (if higher harmonics are neglected). This results into a flat-topped flux wave. For transformers having independent magnetic circuits (bank of single-phase transformers), there is a low reluctance magnetic path available (in the form of end limbs) for the third harmonic flux. Due to the corresponding induced third harmonic voltages in three phases, the neutral gets shifted. Since the third harmonic voltages are at thrice the fundamental frequency, the neutral

voltage oscillates at thrice the fundamental frequency causing fluctuations in the line to neutral voltages. For the third harmonic flux of $x\,\%$ of the fundamental component, the third harmonic voltage is $3\,x\,\%$ of the fundamental frequency voltage as explained in Section 2.3. This increases voltage stresses due to higher resultant r.m.s. line to neutral voltages. Hence, Yy transformers with independent magnetic circuits are usually not preferred. The behavior of three-phase five-limb transformers is similar to that of bank of single-phase transformers since the path (provided by end yokes and end limbs) is available for the third harmonic flux (as long as third harmonic fluxes are not high enough to cause saturation of this magnetic path). These disadvantages of Yy connection are to a large extent overcome in three-phase three-limb transformers. The third harmonic flux, which flows in the same direction in all the three phases, has to return from one yoke to another through the surrounding non-magnetic paths having high reluctance. This reduces the third harmonic flux and the associated effects such as neutral instability.

For the same reason, a moderate single-phase load can be taken between line and neutral in three-phase three-limb transformers without undue unbalancing of phase voltages. The disadvantage, however, is that the flux returning through paths outside the core causes additional stray losses in the structural parts.

It should be noted that line-to-line voltage is free of third harmonic components in the case of Y connected winding because these components present in phase to neutral voltages get cancelled in the line to line voltage. Sometimes a tertiary delta winding is provided with a Yy transformer so that the third harmonic currents can flow in the closed delta making the flux and voltage almost sinusoidal.

2) *Yy connection with neutral*: If the system and transformer neutrals are grounded, the third harmonic voltages will practically disappear due to the fact that there is a path available for the third harmonic currents to flow. This connection is equivalent to application of an independent excitation to each phase. The main disadvantage of this connection is that the third harmonic currents cause interference in communication circuits running parallel to power lines. If a tertiary delta winding is provided, the third harmonic ground currents will be reduced but not completely eliminated; the current shared by the tertiary winding depends on the relative value of impedances offered by the two paths.

3) *Yd or Dy connection*: Due to the presence of a delta connected winding, these connections are free of third harmonic voltage problems associated with the Y connections. The neutral is also stabilized permitting a moderate single-phase load from line to neutral. For a delta connected primary winding under a no-load condition, although phases carry third harmonic currents, they get cancelled in lines (this is not strictly true because the magnitude of third harmonic current components in three phases are not necessarily equal due to the magnetizing asymmetry as described in Section 2.5.1).

2.9 Transformer Noise

Transformers located near a residential area should have sound level as low as possible. A low noise transformer is being increasingly specified by transformer users; noise levels specified are 10 to 15 dB lower than the prevailing levels mentioned in the international standards (e.g., NEMA-TR1: Sound levels in transformers and reactors, 1981). The design and manufacture of a transformer with a low sound level require in-depth understanding of sources of noise. Core, windings and cooling equipment are the three main sources of noise. The core is the most important and significant source of the transformer noise, which is elaborated in this chapter. The other two sources of noise are discussed in Chapter 10 along with the noise reduction techniques

The core vibrates due to magnetic and magnetostrictive forces. Magnetic forces appear due to non-magnetic gaps at the corner joints of limbs and yokes. The force per unit cross-sectional area can be given as

$$\sigma = \frac{B_m^2}{2\mu_0} = \frac{B_{mp}^2}{4\mu_0}\left(1+\cos 2\omega t\right) \tag{2.39}$$

where B_{mp} = peak value of flux density in the gap between corresponding laminations of yoke and limb

μ_0 = permeability of free space

ω = fundamental angular frequency

These magnetic forces depend upon the kind of interlacing between the limb and yoke; these are highest when there is no overlapping (continuous air gap). The magnetic forces are smaller for 90° overlapping, which further reduce for 45° overlapping. These are the least for the step-lap joint due to reduction in the value of flux density in the overlapping region at the joint.

The forces produced by the magnetostriction phenomenon are much higher than the magnetic forces in transformers. Magnetostriction is a change in configuration of magnetizable material in a magnetic field, which leads to periodic changes in the length of material. An alternating field sets the core in vibration. This vibration is transmitted, after some attenuation, through the oil and tank structure to the surrounding air. This finally results in a characteristic hum. The magnetostriction phenomenon is characterized by the coefficient of magnetostriction ε,

$$\varepsilon = \frac{\Delta l}{l} \tag{2.40}$$

where l and Δl are length of lamination sheet and its change respectively. The coefficient ε depends on the instantaneous value of flux density according to the expression [46, 47]

$$\varepsilon_{(t)} = \sum_{\upsilon=1}^{n} K_{\upsilon} B^{2\upsilon} \tag{2.41}$$

where B = instantaneous value of flux density

K_{υ} = coefficient which depends on level of magnetization, type of lamination material and its treatment

With the increasing exponent (order number υ), the coefficients K_{υ} usually are decreasing. The magnetostriction force is given by

$$F = \varepsilon_{(t)} E A \tag{2.42}$$

where E is the modulus of elasticity in the direction of force and A is the cross-sectional area of a lamination sheet. The previous two equations indicate that the magnetostriction force varies with time and contains even harmonics of the power frequency (120, 240, 360, --- Hz for 60 Hz power frequency). Therefore, the noise also contains all harmonics of 120 Hz. The amplitude of core vibration and noise increase manifold if the fundamental mechanical natural frequency of the core is close to 120 Hz. The natural frequencies of the core can be calculated approximately by analytical / empirical formulae or by the more accurate Finite Element Method.

A typical magnetostriction curve is shown in figure 2.19. The change in dimension is not linearly proportional to the flux density. The value of the magnetostriction can be positive or negative, depending on the type of the magnetic material, and the mechanical and thermal treatments. Magnetostriction is generally positive (increase in length by a few microns with increase in flux density) for CRGO material at annealing temperatures below 800°C, and as the annealing temperature is increased (≥ 800°C), it can be displaced to negative values [48]. The mechanical stressing may change it to positive values. Magnetostriction is minimum along the rolling direction and maximum along the 90° direction.

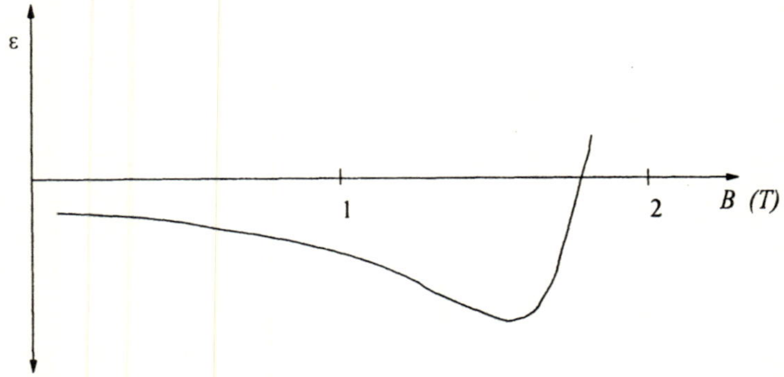

Figure 2.19 Magnetostriction curve

Most of the noise transmitted from a core comes principally from the yoke region because the noise from the limb is effectively damped by windings (copper and insulation material) around the limb. Hence, empirical / semi-empirical formulae used by manufacturers for the noise level calculations have the yoke flux density as a predominant factor. The quality of yoke clamping has a significant influence on the noise level. Apart from the yoke flux density, other factors which decide the noise level are: limb flux density, type of core material, leg center (distance between the centers of two adjacent phases), core weight, frequency, etc. The higher the flux density, leg centers, core weight and frequency of operation, the higher is the noise level.

The noise level is closely related to the operating peak flux density and core weight. The change in noise level as a function of these two factors can be expressed as [49]:

$$\Delta L = 10\ Log_{10}\left[\left(\frac{B_2}{B_1}\right)^8 \left(\frac{W_2}{W_1}\right)^{1.6}\right] \tag{2.43}$$

If core weight is assumed to change with flux density approximately in inverse proportion, for a flux density change from 1.6 T to 1.7 T, the increase in noise level is 1.7 dB [$\approx 64\ Log_{10}(1.7/1.6)$]. Hence, one of the ways of reducing noise is by designing transformer at lower operating flux density. For a flux density reduction of 0.1 T, the noise level reduction of about 2 dB is obtained. This method results into an increase of material content and it may be justified economically if the user has specified a lower no-load loss, in which case the natural choice is to use a lower flux density. The use of step-lap joint gives much better noise reduction (4 to 5 dB). The noise performance of step-lap joint is compared with that of mitred joint in [50]. Some manufacturers also use yoke reinforcement (leading to reduction in yoke flux density); the method has the advantage that copper content does not go up since the winding mean diameters do not increase. Bonding of laminations by adhesives and placing of anti-vibration / damping elements between the core and tank can give further reduction in the noise level. The use of Hi-B / scribed material can also give a reduction of 2 to 3 dB. When a noise level reduction of the order of 15 to 20 dB is required, some of these methods are necessary but not sufficient, and the methods involving changes in structural design are adopted (which are discussed in Chapter 10).

In a gapped core shunt reactor, the vibration is quite high as compared to a transformer due to forces between every two magnetic packets (sections) separated by a non-magnetic gap of few tens of millimeters. The magnetic field creates pulsating forces across these air gaps which can be calculated by equation 2.39. Hence, reactor cores are designed as very stiff structures to eliminate excessive vibrations. The non-magnetic gaps are created and supported by placing a non-magnetic material such as stone spacers or ceramic blocks having a high modulus of elasticity. The dimensional stability and core tightness can be further ensured by placing epoxy impregnated polyester material and fiberglass cloth of 2 to 3 mm between the last limb packet and top / bottom yoke. The material gets hardened after getting heated during the processing stage (heating and vacuum cycles) and bonds the yoke with limb packet. Care should also be taken that the fundamental mechanical natural frequency of vibration of reactor core structure is more and sufficiently away from twice the power frequency.

References

1. Baehr, R. Up-to-date possibilities for further improvement of power transformer magnetic circuit features, *International Summer School of Transformers, ISST'93*, Lodz, Poland, 1993.
2. Koppikar, D. A., Kulkarni, S. V., Khaparde, S. A., and Arora, B. A modified approach to overfluxing analysis of transformer, *International Journal of Electrical Power & Energy Systems*, Vol. 20, No. 4, 1998, pp. 235-239.
3. Koppikar, D. A., Kulkarni, S. V., and Khaparde, S. A. Overfluxing simulation of transformer by 3D FEM analysis, *Fourth Conference on EHV Technology*, IISc Bangalore, India, 1998, pp. 69-71.
4. Bean, R. L., Chackan, N., Moore, H. R., and Wentz, E. C. *Transformers for the electric power industry*, McGraw-Hill, New York, 1959, p. 103.
5. Karsai, K., Kerenyi, D., and Kiss, L. *Large power transformers*, Elsevier Publication, Amsterdam, 1987, p. 41.
6. Girgis, R. S., teNijenhuis, E. G., Gramm, K., and Wrethag, J. E. Experimental investigations on effect of core production attributes on transformer core loss performance, *IEEE Transactions on Power Delivery*, Vol. 13, No. 2, April 1998, pp. 526-531.
7. Lofler, F., Pfutzner, H., Booth, T., Bengtsson, C., and Gramm, K. Influence of air gaps in staked cores consisting of several packages, *IEEE Transactions on Magnetics*, Vol. 30, No. 3, March 1994, pp. 913-915.
8. Elleuch, M. and Poloujadoff, M. New transformer model including joint air gaps and lamination anisotropy, *IEEE Transactions on Magnetics*, Vol. 34, No. 5, September 1998, pp. 3701-3711.
9. Mechler, G. F. and Girgis, R. S. Calculation of spatial loss distribution in stacked power and distribution transformer cores, *IEEE Transactions on Power Delivery*, Vol. 13, No. 2, April 1998, pp. 532-537.
10. Olivares, J. C., Kulkarni, S. V., Canedo, J., Driesen, J., and Escarela, R. Impact of joint design parameters on transformer losses, *International Journal of Power and Energy Systems*, Vol. 23, No. 3, 2003, pp. 151-157.
11. teNyenhuis, E. G., Mechler, G. F., and Girgis, R. S. Flux distribution and core loss calculation for single phase and five limb three phase transformer core designs, *IEEE Transactions on Power Delivery*, Vol. 15, No. 1, January 2000, pp. 204-209.

12. Loffler, F., Booth, T., Pfutzner, H., Bengtsson, C., and Gramm, K. Relevance of step-lap joints for magnetic characteristics of transformer cores, *Proceedings IEE – Electrical Power Applications*, Vol. 142, No. 6, November 1995, pp. 371-378.
13. Mechler, G. F. and Girgis, R. S. Magnetic flux distributions in transformer core joints, *IEEE Transactions on Power Delivery*, Vol. 15, No. 1, January 2000, pp. 198-203.
14. Alexander, G. W., Corbin, S. L., and McNutt, W. J. Influence of design and operating practices on excitation of generator step-up transformers, *IEEE Transactions on Power Apparatus and Systems*, Vol. PAS-85, No. 8, August 1966, pp. 901-909.
15. Gantner, J. and Birch, F. H. Transformer overfluxing protection, *Electra*, No. 31, Report of Working Group 01 of CIGRE Study Committee No. 34 (Protection), pp. 65-74.
16. Camilli, G. A flux voltmeter for magnetic tests, *AIEE Transactions*, 1926, pp. 721-728.
17. Blume, L. F., Boyajian, A., Camilli, G., Lennox, T. C., Minneci, S., and Montsinger, V. M. *Transformer engineering*, John Wiley and Sons, New York, and Chapman and Hall, London, 1951.
18. Arseneau, R. and Moore, W. J. M. A method for estimating the sinusoidal iron losses of a transformer from measurements made with distorted voltage waveforms, *IEEE Transactions on Power Apparatus and Systems*, Vol. PAS-103, No. 10, October 1984, pp. 2912-2918.
19. Tran-Quoc, T. and Pierrat, L. Correction of the measured core losses under distorted flux, *IEEE Transactions on Magnetics*, Vol. 33, No. 2, March 1997, pp. 2045-2048.
20. Walker, C. S. The excitation requirements of 3-phase core-type 3-legged Y-connected transformers, *AIEE Transactions*, December 1957, pp. 1113-1119.
21. Kemp, P. *Alternating current waveforms*, Chapman and Hall Limited, Second Edition, London, 1952.
22. Brailsford, F. Current and power relationships in the measurement of iron losses in a three-limb transformer core, *Proceedings IEE*, Vol. 101, Pt. 2, 1954, pp. 409-416.
23. Parion J. I. Magnetizing asymmetry in three-phase core-type transformers, *Electrical Energy*, July 1957, pp. 330-336.
24. Nene, A. S., Kulkarni, S. V., and Vijayan, K. Asymmetrical magnetizing phenomenon of three phase core type transformers, *International Conference on Transformers*, *TRAFOTECH-1994*, Bangalore, India, January 1994, pp. IV 21-26.
25. McWhirter, J. H., Specht, T. R., and Albright, W. D. Effect of impulse testing on transformer iron loss, *AIEE Transactions*, February 1957, pp. 1275-1279.

26. Blume, L. F., Camilli, G., Farnham, S. B., and Peterson, H. A. Transformer magnetizing inrush currents and influence on system operation, *AIEE Transactions*, Vol. 63, 1944, pp. 366-375.
27. Lin, C. E., Cheng, C. L., Huang, C. L., and Yeh, J. C. Investigation of magnetizing inrush current in transformers- Part I: Numerical simulation, *IEEE Transactions on Power Delivery*, Vol. 8, No. 1, January 1993, pp. 246-254.
28. Finzi, L. A. and Mutschler, W. H. The inrush of magnetizing current in single-phase transformers, *AIEE Transactions*, Vol. 70, 1951, pp. 1436-1438.
29. Yacamini, R. and Abu-Nasser, A. The calculation of inrush currents in three-phase transformers, *Proceedings IEE*, Vol. 133, Pt. B, No. 1, January 1986, pp. 31-40.
30. Macfadyen, W. K., Simpson, R. R. S., Slater, R. D., and Wood, W. S. Method of predicting transient current patterns in transformers, *Proceedings IEE*, Vol. 120, No. 11, November 1973, pp. 1393-1396.
31. Nakra, H. L. and Barton, T. H. Three phase transformer transients, *IEEE PES Winter Meeting*, New York, 1974, Paper T-74-243-2.
32. Rico, J. J., Acha, E., and Madrigal, M. The study of inrush current phenomenon using operational matrices, *IEEE Transactions on Power Delivery*, Vol. 16, No. 2, April 2001, pp. 231-237.
33. Yacamini, R. and Abu-Nasser, A. Transformer inrush currents and their associated overvoltages in HVDC schemes, *Proceedings IEE*, Vol. 133, Pt. C, No. 6, September 1986, pp. 353-358.
34. Specht, T. R. Transformer magnetizing inrush current, *AIEE Transactions*, Vol. 70, 1951, pp. 323-328.
35. Holcomb, J. E. Distribution transformer magnetizing inrush current, *AIEE Transactions*, December 1961, pp. 697-702.
36. Adly, A. A. Computation of inrush current forces on transformer windings, *IEEE Transactions on Magnetics*, Vol. 37, No. 4, July 2001, pp. 2855-2857.
37. Mao, P. L. and Aggarwal, R. K. A novel approach to the classification of the transient phenomenon in power transformers using combined wavelet transform and neural network, *IEEE Transactions on Power Delivery*, Vol. 16, No. 4, October 2001, pp. 654-660.
38. Elmore, W. A. *Protective relaying: theory and applications*, Marcel Dekker, New York, 1994, p. 146.
39. Rudenburg, R. *Transient performance of electric power system,* First Edition, McGraw-Hill Book Company, Inc., 1950, pp. 640-641.
40. Kulkarni, S. V. Influence of system operating conditions on magnetizing inrush phenomenon of transformer, *International Conference on Transformers*, *TRAFOTECH-94,* Bangalore, India, January 1994, pp. VI 19-23.

41. Mason, C. R. *Art and science of protective relaying,* John Wiley & Sons, Inc., New York, 1956, pp. 259-261.
42. Bronzeado, H. and Yacamini, R. Phenomenon of sympathetic interaction between transformers caused by inrush transients, *Proceedings IEE - Science, Measurement and Technology*, Vol. 142, No. 4, July 1995, pp. 323-329.
43. Hayward, C. D. Prolonged inrush currents with parallel transformers affect differential relaying, *AIEE Transactions*, Vol. 60, 1941, pp. 1096-1101.
44. Holmgrem, B., Jenkins, R. S., and Riubrugent, J. Transformer inrush current, *CIGRE 1968*, Paper No. 12-03.
45. Brunke, J. H. and Frohlich, K. J. Elimination of transformer inrush currents by controlled switching- Part I: Theoretical considerations, *IEEE Transactions on Power Delivery*, Vol. 16, No. 2, April 2001, pp. 276-285.
46. Krondl, M. and Kronauer, E. Some contributions to the problem of transformer noise, *Bulletin Oerlikon*, No. 356, pp. 1-15.
47. Majer, K. Evaluation of transformer noise and vibrations, *International Summer School of Transformers, ISST'93*, Lodz, Poland, 1993.
48. Foster, S. L. and Reiplinger, E. Characteristics and control of transformer sound, *IEEE Transactions on Power Apparatus and Systems*, Vol. PAS-100, No. 3, March 1981, pp. 1072-1075.
49. Brandes, D. Noise levels of medium power transformers, *Elektrizitatswirtschaft*, No. 11, 1977.
50. Weiser, B., Pfutzner, B., and Anger, J. Relevance of magnetostriction and forces for the generation of audible noise of transformer cores, *IEEE Transactions on Magnetics*, Vol. 36, No. 5, September 2000, pp. 3759-3777.

3

Impedance Characteristics

The leakage impedance of a transformer is one of the most important specifications that has significant impact on its overall design. Leakage impedance, which consists of resistive and reactive components, has been introduced and explained in Chapter 1. This chapter focuses on the reactive component (leakage reactance), whereas Chapters 4 and 5 deal with the resistive component. The load loss (and hence the effective AC resistance) and leakage impedance are derived from the results of short circuit test. The leakage reactance is then calculated from the impedance and resistance (Section 1.5 of Chapter 1). Since the resistance of a transformer is generally quite less as compared to its reactance, the latter is almost equal to the leakage impedance. Material cost of the transformer varies with the change in specified impedance value. Generally, a particular value of impedance results into a minimum transformer cost. It will be expensive to design the transformer with impedance below or above this value. If the impedance is too low, short circuit currents and forces are quite high, which necessitate use of lower current density thereby increasing the material content. On the other hand, if the impedance required is too high, it increases the eddy loss in windings and stray loss in structural parts appreciably resulting into much higher load loss and winding / oil temperature rise; which again will force the designer to increase the copper content and / or use extra cooling arrangement. The percentage impedance, which is specified by transformer users, can be as low as 2% for small distribution transformers and as high as 20% for large power transformers. Impedance values outside this range are generally specified for special applications.

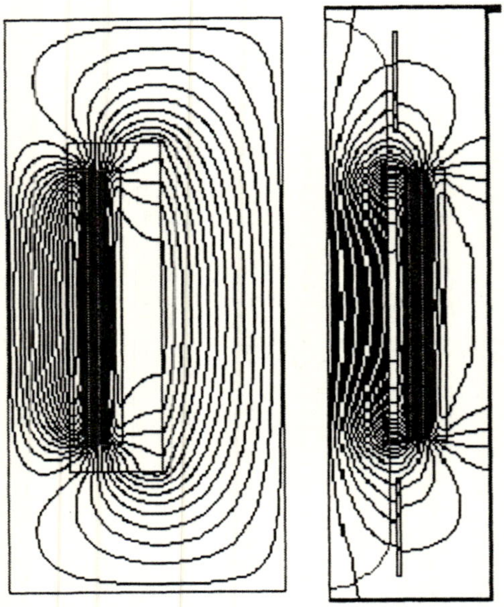

(a) (b)

Figure 3.1 Leakage field in a transformer

3.1 Reactance Calculation

3.1.1 Concentric primary and secondary windings

Transformer is a three-dimensional electromagnetic structure with the leakage field appreciably different in the core window cross section (figure 3.1 (a)) as compared to that in the cross section perpendicular to the window (figure 3.1 (b)). For reactance ($\cong$ impedance) calculations, however, values can be estimated reasonably close to test values by considering only the window cross section. A high level of accuracy of 3-D calculations may not be necessary since the tolerance on reactance values is generally in the range of $\pm 7.5\%$ or $\pm 10\%$.

For uniformly distributed ampere-turns along LV and HV windings (having equal heights), the leakage field is predominantly axial, except at the winding ends, where there is fringing (since the leakage flux finds a shorter path to return via yoke or limb). The typical leakage field pattern shown in figure 3.1 (a) can be replaced by parallel flux lines of equal length (height) as shown in figure 3.2 (a). The equivalent height (H_{eq}) is obtained by dividing winding height (H_w) by the Rogowski factor K_R (<1.0),

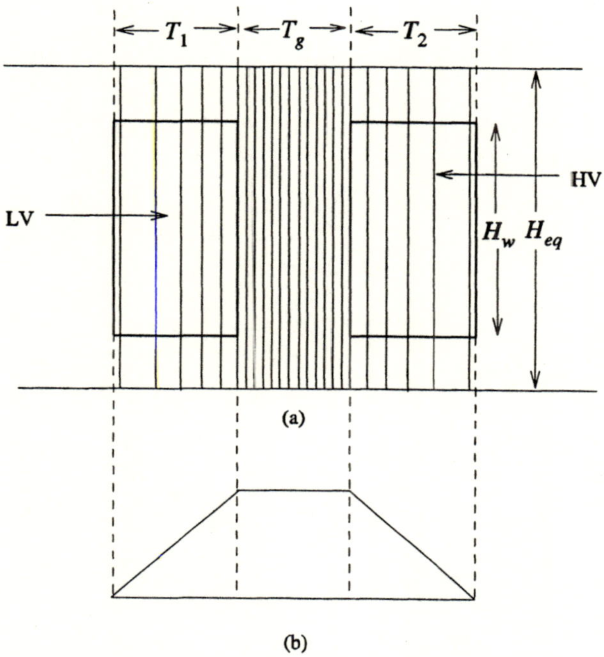

Figure 3.2 (a) Leakage field with equivalent height
(b) Magnetomotive force or flux density diagram

$$H_{eq} = \frac{H_w}{K_R} \tag{3.1}$$

The leakage magnetomotive (mmf) distribution across the cross section of windings is of trapezoidal form as shown in figure 3.2 (b). The mmf at any point depends on the ampere-turns enclosed by a flux contour at that point; it increases linearly with the ampere-turns from a value of zero at the inside diameter of LV winding to the maximum value of one per-unit (total ampere-turns of LV or HV winding) at the outside diameter. In the gap (T_g) between LV and HV windings, since flux contour at any point encloses full LV (or HV) ampere-turns, the mmf is of constant value. The mmf starts reducing linearly from the maximum value at the inside diameter of the HV winding and approaches zero at its outside diameter. The core is assumed to have infinite permeability requiring no magnetizing mmf, and hence the primary and secondary mmfs exactly balance each other. The flux density distribution is of the same form as that of the mmf distribution. Since the core is assumed to have zero reluctance, no mmf is expended in the return path through it for any contour of flux. Hence, for a closed contour of flux at a distance x from the inside diameter of LV winding, it can be written that

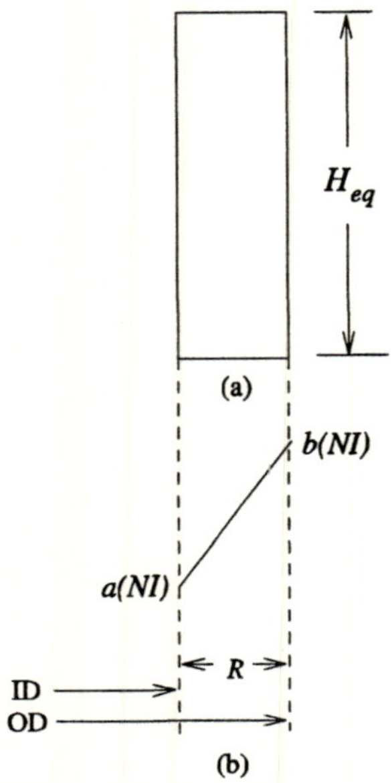

Figure 3.3 (a) Flux tube
(b) MMF diagram

$$\left(\frac{B_x}{\mu_0}\right)H_{eq} = (NI)_x \tag{3.2}$$

or

$$B_x = \frac{\mu_0 (NI)_x}{H_{eq}} \tag{3.3}$$

For deriving the formula for reactance, let us derive a general expression for the flux linkages of a flux tube having radial depth R and height H_{eq}. The ampere-turns enclosed by a flux contour at the inside diameter (ID) and outside diameter (OD) of this flux tube are $a(NI)$ and $b(NI)$ respectively as shown in figure 3.3, where NI are the rated ampere-turns. The general formulation is useful when a winding is split radially into a number of sections separated by gaps. The r.m.s. value of flux density at a distance x from the ID of this flux tube can now be inferred from equation 3.3 as

$$B_x = \frac{\mu_0}{H_{eq}}\left[\left(a + \frac{b-a}{R}x\right)NI\right] \tag{3.4}$$

The flux linkages of an incremental flux tube of width dx placed at x are

$$d\psi = N_x \phi_x = N_x B_x A \tag{3.5}$$

where A is the area of flux tube given by

$$A = \pi(ID + 2x)\,dx \tag{3.6}$$

Substituting equations 3.4 and 3.6 in equation 3.5,

$$d\psi = \left\{\left(a + \frac{b-a}{R}x\right)N\right\}\left\{\frac{\mu_0}{H_{eq}}\left[\left(a + \frac{b-a}{R}x\right)NI\right]\right\}\left[\pi(ID + 2x)dx\right] \tag{3.7}$$

Hence, the total flux linkages of the flux tube are given by

$$\psi = \int_0^R d\psi = \frac{\mu_0 \pi N^2 I}{H_{eq}} \int_0^R \left(a + \frac{b-a}{R}x\right)^2 (ID + 2x)dx \tag{3.8}$$

After integration and a few arithmetic operations, we get

$$\psi = \frac{\mu_0 \pi N^2 I}{H_{eq}} \frac{R}{3}\left[(a^2 + ab + b^2)ID + \frac{(a^2 + ab + b^2)3R}{2} - \frac{2a^2 + ab}{2}R\right] \tag{3.9}$$

The last term in square bracket can be neglected without introducing an appreciable error to arrive at a simple formula for the regular design use.

$$\therefore \psi = \frac{\mu_0 \pi N^2 I}{H_{eq}} \frac{R}{3}(a^2 + ab + b^2)\left[ID + \frac{3R}{2}\right] \tag{3.10}$$

The term $\left[ID + \frac{3R}{2}\right]$ can be taken to be approximately equal to the mean diameter (D_m) of the flux tube (for large diameters of windings / gaps with comparatively lower values of their radial depths).

$$\therefore \psi = \frac{\mu_0 \pi N^2 I}{H_{eq}} \frac{R}{3}(a^2 + ab + b^2)\,D_m \tag{3.11}$$

Now, let

$$ATD = \frac{R}{3}(a^2 + ab + b^2)\,D_m \tag{3.12}$$

which corresponds to the area of Ampere-Turn Diagram. The leakage inductance of a transformer with n flux tubes can now be given as

$$L = \frac{\sum_{k=1}^{n}\psi}{I} = \frac{\mu_0 \pi N^2}{H_{eq}} \sum_{k=1}^{n} ATD \tag{3.13}$$

and the corresponding expression for the leakage reactance X is

$$X = 2\pi f \frac{\mu_0 \pi N^2}{H_{eq}} \sum_{k=1}^{n} ATD \tag{3.14}$$

For the base impedance of Z_b, the formula for percentage leakage reactance is

$$\begin{aligned} \%X &= \frac{X}{Z_b} = \frac{IX}{V} \times 100 = 2\pi f \frac{\mu_0 \pi IN^2}{H_{eq} V} \sum_{k=1}^{n} ATD \times 100 \\ &= 2\pi f \frac{\mu_0 \pi (NI)}{H_{eq}(V/N)} \sum_{k=1}^{n} ATD \times 100 \end{aligned} \tag{3.15}$$

where V is rated voltage and term (V/N) is volts/turn of the transformer. Substituting $\mu_0 = 4\pi \times 10^{-7}$ and adjusting constants so that the dimensions used in the formula are in units of centimeters (H_{eq} in cm and $\sum ATD$ in cm^2),

$$\%X = 2.48 \times 10^{-5} f \frac{(Ampere\ Turns)}{H_{eq}(Volts/Turn)} \sum_{k=1}^{n} ATD \tag{3.16}$$

After having derived the general formula, we will now apply it for a simple case of a two winding transformer shown in figure 3.2. The constants a and b have the values of 0 and 1 for LV, 1 and 1 for gap, and 1 and 0 for HV respectively. If D_1, D_g and D_2 are the mean diameters and T_1, T_g and T_2 are the radial depths of LV, gap and HV respectively, using equation 3.12 we get

$$\sum ATD = \frac{1}{3}(T_1 \times D_1) + (T_g \times D_g) + \frac{1}{3}(T_2 \times D_2) \tag{3.17}$$

The value of H_{eq} is calculated by equation 3.1, for which the Rogowski factor K_R is given by

$$K_R = 1 - \frac{1 - e^{-\pi H_w/(T_1 + T_g + T_2)}}{\pi H_w/(T_1 + T_g + T_2)} \tag{3.18}$$

For taking into account the effect of core, a more accurate but complex expression for K_R can be used as given in [1]. For most of the cases, equation 3.18 gives sufficiently accurate results.

For an autotransformer, transformed ampere-turns should be used in equation 3.16 (difference between turns corresponding to HV and LV phase voltages multiplied by HV current) and the calculated impedance is multiplied by the auto-factor,

$$K_A = 1 - \frac{V_{LV}}{V_{HV}} \tag{3.19}$$

where V_{LV} and V_{HV} are the rated line voltages of LV and HV windings respectively.

3.1.2 Sandwich windings

The reactance formula derived in the previous section can also be used for sandwich windings in core-type or shell-type transformers with slight modifications. Figure 3.4 shows a configuration of such windings with two sections. The mean diameter of windings is denoted by D_m. If there are total N turns and S sections in windings, then remembering the fact that reactance is proportional to the square of turns, the reactance between LV and HV windings corresponding to any one section (having N/S turns) is given by

$$\%X_S = 2.48 \times 10^{-5} f \frac{NI}{H_{eq}(V/N)} \frac{1}{S^2} \sum_{k=1}^{n} ATD \tag{3.20}$$

where

$$\sum_{k=1}^{n} ATD = \frac{1}{3}(T_1 \times D_m) + (T_g \times D_m) + \frac{1}{3}(T_2 \times D_m) = D_m \left[T_g + \frac{T_1 + T_2}{3} \right] \tag{3.21}$$

If the sections are connected in series, total reactance is S times that of one section,

$$\%X = 2.48 \times 10^{-5} f \frac{NI}{H_{eq}(V/N)} \frac{1}{S} \sum_{k=1}^{n} ATD \tag{3.22}$$

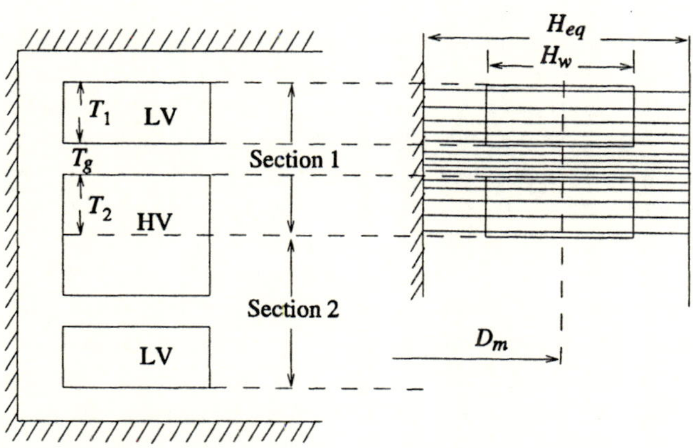

Figure 3.4 Sandwich winding

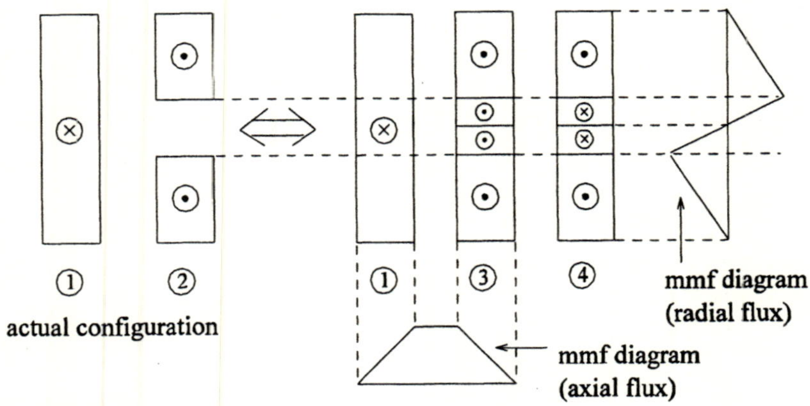

Figure 3.5 Unequal AT/m distribution

Similarly, if sections are connected in parallel, the formula can be derived by taking number of turns in one section as N with current as I/S.

3.1.3 Concentric windings with non-uniform distribution of ampere turns

Generally, on account of exclusion of tap winding turns at various tap positions, we get different ampere-turn/height (AT/m) for LV and HV windings. This results in a higher amount of radial flux at tapped out sections. When taps are in the main body of a winding (no separate tap winding), it is preferable to put taps symmetrically in the middle or at the ends to minimize the radial flux. If taps are provided only at one end, the arrangement causes an appreciable asymmetry and higher radial component of flux resulting into higher eddy losses and axial short circuit forces. For different values of AT/m along the height of LV and HV windings, the reactance can be calculated by resolving the AT distribution as shown in figure 3.5. The effect of gap in the winding 2 can be accounted by replacing it with the windings 3 and 4. The winding 3 has same AT/m distribution as that of the winding 1, and the winding 4 has AT/m distribution such that the addition of ampere-turns of the windings 3 and 4 along the height gives the same ampere-turns as that of the winding 2. The total reactance is the sum of two reactances; reactance between the windings 1 and 3 calculated by equation 3.16 and reactance of the winding 4 calculated by equation 3.22 (for sections connected in series).

Since equation 3.22 always gives a finite positive value, a non-uniform AT distribution (unequal AT/m of LV and HV windings) always results into higher reactance. The increase in reactance can be indirectly explained by stating that the effective height of windings in equation 3.16 is reduced if we take the average of heights of the two windings. For example, if the tapped out section in one of the windings is 5% of the total height at the tap position corresponding to the rated voltage, the average height is reduced by 2.5%, giving the increase in reactance of 2.5% as compared to the case of uniform AT/m distribution.

3.2 Different Approaches for Reactance Calculation

The first approach for reactance calculation is based on the fundamental definition of inductance in which inductance is defined as the ratio of total flux linkages to a current which they link

$$L = \frac{\psi}{I} = \frac{N\phi}{I} \tag{3.23}$$

and this approach has been used in Section 3.1 for finding the inductance and reactance (equations 3.13 and 3.14).

In the second approach, use is made of an equivalent definition of inductance from the energy point of view,

$$L = \frac{2\,W_m}{I^2} \tag{3.24}$$

where W_m is energy in the magnetic field produced by a current I flowing in a closed path. Now, we will see that the use of equation 3.24 leads us to the same formula of reactance as given by equation 3.16.

Energy per unit volume in the magnetic field in air, with linear magnetic characteristics ($H = B/\mu_0$), when the flux density is increased from 0 to B, is

$$w = \int_0^B H\,dB = \int_0^B \frac{B}{\mu_0}\,dB = \frac{B^2}{2\mu_0} \tag{3.25}$$

Hence, the differential energy dW_x for a cylindrical ring of height H_{eq}, thickness dx and diameter ($ID + 2x$) is

$$dW_x = \frac{B_x^2}{2\mu_0} \times (volume\ of\ cylindrical\ ring) = \frac{B_x^2}{2\mu_0}\pi\,(ID + 2x)\,H_{eq}\,dx \tag{3.26}$$

Now the value of B_x can be substituted from equation 3.4 for the simple case of flux tube with the conditions of $a = 0$ and $b = 1$ (with reference to figure 3.3).

$$\therefore dW_x = \frac{\mu_0\pi(NI)^2 x^2 (ID + 2x)dx}{2\,R^2 H_{eq}} \tag{3.27}$$

For the winding configuration of figure 3.2, the total energy stored in LV winding (with the term *R* replaced by the radial depth T_1 of the LV winding) is

$$W_1 = \frac{\mu_0 \pi (NI)^2}{2T_1^2 H_{eq}} \int_0^{T_1} x^2 (ID + 2x) dx = \frac{\mu_0 \pi (NI)^2}{2H_{eq}} \frac{1}{3} \left\{ ID + \frac{3T_1}{2} \right\} T_1 \tag{3.28}$$

As seen in Section 3.1.1, the term in the brackets can be approximated as mean diameter (D_1) of the LV winding,

$$\therefore W_1 = \frac{\mu_0 \pi (NI)^2}{2 H_{eq}} \frac{1}{3} D_1 T_1 \tag{3.29}$$

Similarly, the energy in HV winding can be calculated as

$$W_2 = \frac{\mu_0 \pi (NI)^2}{2H_{eq}} \frac{1}{3} D_2 T_2 \tag{3.30}$$

Since flux density is constant in the gap between the windings, energy in it can be directly calculated as

$$W_g = \frac{B_g^2}{2\mu_0} \times (volume\ of\ cylindrical\ gap) = \frac{1}{2\mu_0} \left[\frac{\mu_0 (NI)}{H_{eq}} \right]^2 \pi D_g T_g H_{eq} \tag{3.31}$$

$$\therefore W_g = \frac{\mu_0 \pi (NI)^2}{2H_{eq}} D_g T_g \tag{3.32}$$

Substituting the values of energies from equations 3.29, 3.30 and 3.32 in equation 3.24,

$$L = \frac{2 W_m}{I^2} = \frac{2(W_1 + W_2 + W_g)}{I^2} = \frac{\mu_0 \pi N^2}{H_{eq}} \left[\frac{1}{3}(T_1 D_1 + T_2 D_2) + T_g D_g \right] \tag{3.33}$$

If the term in the brackets is substituted by $\sum ATD$ as per equation 3.17, we see that equation 3.33 derived for the leakage inductance from the energy view point is the same as equation 3.13 calculated from the definition of flux linkages per ampere.

In yet another approach, when numerical methods like Finite Element Method are used, solution of the field is generally obtained in terms of magnetic vector potential, and the inductance is obtained as

$$L = \frac{1}{I^2} \int_{vol} \mathbf{A} \cdot \mathbf{J}\, dv \tag{3.34}$$

where $\mathbf{A}$ is magnetic vector potential and $\mathbf{J}$ is current density vector. Equation 3.34 can be derived [2] from equation 3.24,

$$L = \frac{2\,W_m}{I^2} = \frac{1}{I^2}\int_{vol} \mathbf{B}\cdot\mathbf{H}\,dv = \frac{1}{I^2}\int_{vol} \mathbf{A}\cdot\mathbf{J}\,dv \tag{3.35}$$

The leakage reactance between two windings of a transformer can also be calculated by the equation,

$$X_{12} = X_1 + X_2 - 2\,M_{12} \tag{3.36}$$

where X_1 and X_2 are the self reactances of the windings and M_{12} is the mutual reactance between them. It is difficult to calculate or accurately test the self and mutual reactances which depend on saturation effects. Also, since the values of $(X_1 + X_2)$ and $2M_{12}$ are nearly equal and are very high as compared to the leakage reactance X_{12}, it is very difficult to calculate accurately the value of leakage reactance as per equation 3.36. Hence, it is always easier to calculate the leakage reactance of a transformer directly without using formulae involving self and mutual reactances. Therefore, for finding the effective leakage reactance of a system of windings, the total power of the system is expressed in terms of leakage impedances instead of self and mutual impedances. Consider a system of windings 1, 2, ----, n, with leakage impedances Z_{jk} between pairs of windings j and k. For a negligible magnetizing current (as compared to the rated currents in the windings) the total power can be expressed as [3]

$$P + jQ = -\frac{1}{2}\sum_{k=1}^{n}\sum_{j=1}^{n} \overline{Z}_{jk}\,\overline{I}_j\,\overline{I}_k^{*} \tag{3.37}$$

where $\overline{I}_k^{*}$ is the complex conjugate of $\overline{I}_k$. The resistances can be neglected in comparison with much larger reactances. When current vectors of windings are parallel (in phase or phase-opposition), the expression for Q (which is given by the imaginary part of above equation) becomes

$$Q = \mathrm{Im}\left[-\frac{1}{2}\sum_{k=1}^{n}\sum_{j=1}^{n} \overline{Z}_{jk}\,\overline{I}_j\,\overline{I}_k^{*}\right] = -\frac{1}{2}\sum_{k=1}^{n}\sum_{j=1}^{n} X_{jk}\,I_j\,I_k \tag{3.38}$$

Equation 3.38 gives the total reactive volt-amperes consumed by all the leakage reactances of the system of windings. The effective or equivalent leakage reactance of the system of n windings, referred to source (primary) winding with current I_p is given by

$$X_{eff} = \frac{Q}{I_p^2} \text{ ohms} \tag{3.39}$$

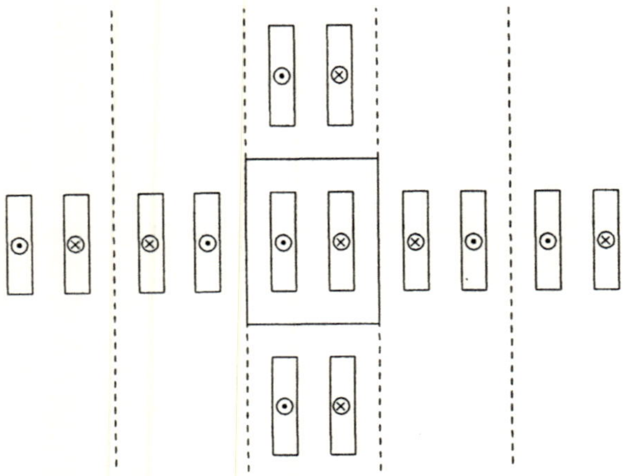

Figure 3.6 Method of images

If X_{jk} and currents are expressed in per-unit in equation 3.38, the value of Q (calculated with rated current flowing in the primary winding) gives directly the per-unit reactance of a transformer with n windings. Use of this reactive KVA approach is illustrated in Sections 3.6 and 3.7.

3.3 Two-Dimensional Analytical Methods

The classical method described in Section 3.1 has certain limitations. The effect of core is not taken into account. It is also tedious to take into account axial gaps in windings and asymmetries in ampere-turn distribution. Some of the more commonly used analytical methods, in which these difficulties are overcome, are now described. The leakage reactance calculation by more accurate numerical methods (e.g., Finite Element Method) is described in Section 3.4.

3.3.1 Method of images

When computers were not available, many attempts were made to devise accurate methods of calculating axial and radial components of the leakage field, and subsequently the reactance. One popular approach was to use simple Biot-Savart's law with the effect of iron core taken into account by method of images. The method basically works in Cartesian (x-y) coordinate system in which windings are represented by straight coils (assumed to be of infinite dimension along the z axis perpendicular to plane of the paper) placed at an appropriate distance from a plane surface bounding a semi-infinite mass of infinite permeability. The effect of iron is represented by images of coils as far behind the surface as the coils are in the front. Parallel planes have to be added to get accurate results as shown in figure 3.6, giving an arrangement of infinite number of images in all four directions [4]. The idea is that all these coils give the same value of leakage field at any point as that with the original geometry of two windings enclosed in an iron boundary. A new plane (mirror) can be added one at a time till the difference between the results is less than the admissible value of error; generally the first three or four images are sufficient. Biot-Savart's law is then applied to this arrangement of currents, which is devoid of magnetic mass (iron), to find the value of field at any point.

3.3.2 Roth's method

The method of field analysis by double Fourier series originally proposed by Roth was extended in [5] to calculate the leakage reactance for irregular distribution of windings. The advantage of this method is that it is applicable to uniform as well as non-uniform ampere-turn distributions of windings. The arrangement of windings in the core window may be entirely arbitrary but divisible into rectangular blocks, each block having a uniform current density within itself.

In this method, the core window is considered as π radians wide and π radians long, regardless of its absolute dimensions. The ampere-turn density distribution as well as the flux distribution is conceived to be consisting of components which vary harmonically along both the x and y axes. The method uses a similar principle to that of the method of images; for every harmonic the maximum occurs at fictitious planes about which mirroring is done to simulate the effect of iron boundary. Reactive volt-amperes (I^2X) are calculated in terms of these current harmonics for a depth of unit dimension in the z direction. The total volt-amperes are estimated by multiplying the obtained value by mean perimeter. The per-unit value of reactance is calculated by dividing I^2X by base volt-amperes. For a reasonable accuracy, the number of space harmonics for double Fourier series should be at least equal to 20 when the ampere-turn distribution is identical in the LV and HV windings [6]. The accuracy is higher with the increase in number of space harmonics. Figure 3.7 shows plots of radial flux density along the height of a transformer winding having uniform ampere-turn distribution in LV and HV windings. As the number of harmonics is increased, the variation of radial flux density becomes smooth, indicating the higher accuracy of field computations.

3.3.3 Rabin's method

If the effect of winding curvature is required to be taken into account in the Roth's formulation, the method becomes complicated, and in that case Rabin's method is more suitable [4, 7]. It solves the following Poisson's equation in polar co-ordinates,

$$\nabla^2 \mathbf{A} = -\mu \, \mathbf{J} \tag{3.40}$$

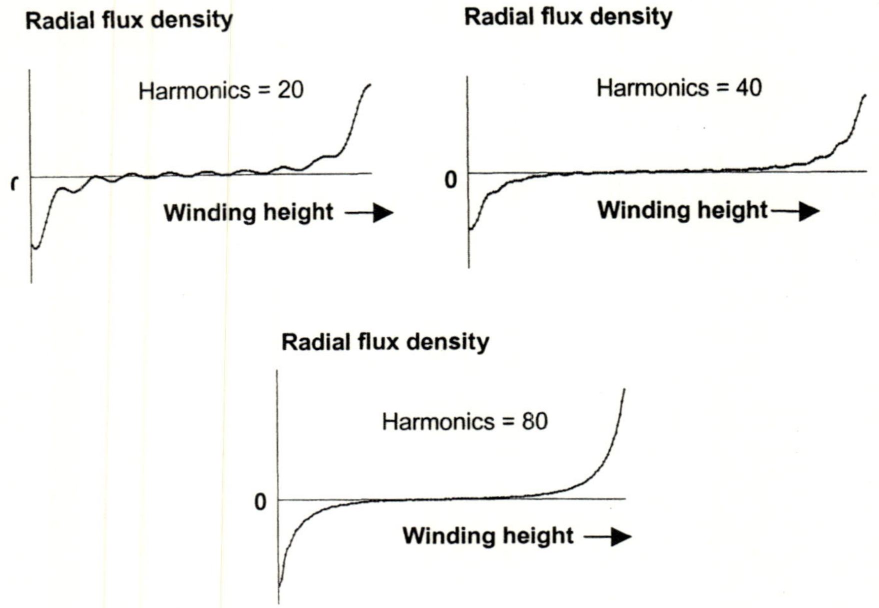

Figure 3.7 Radial flux density with increasing number of space harmonics

where **A** is magnetic vector potential and **J** is current density having only the angular component. Therefore, in circular co-ordinates the equation becomes

$$\frac{\partial^2 A_\theta}{\partial r^2} + \frac{1}{r}\frac{\partial A_\theta}{\partial r} - \frac{A_\theta}{r^2} + \frac{\partial^2 A_\theta}{\partial z^2} = -\mu \; J_\theta \tag{3.41}$$

In this method, the current density is assumed to depend only on the axial position and hence can be represented by a single Fourier series with coefficients which are Bessel and Struve functions. For reasonable accuracy, the number of space harmonics should be about 70 [6].

3.4 Numerical Method for Reactance Calculation

Finite Element Method (FEM) is the most commonly used numerical method for reactance calculation of non-standard winding configurations and asymmetrical / non-uniform ampere-turn distributions, which cannot be easily and accurately handled by the classical method given in Section 3.1. The FEM analysis can be more accurate than the analytical methods described in Section 3.3. User-friendly commercial FEM software packages are now available. Two-dimensional FEM analysis can be integrated into routine design calculations. The main advantage of FEM is that any complex geometry can be analyzed since the FEM formulation depends only on the class of problem and is independent of its geometry. It can also take into account material discontinuities easily. The FEM formulation makes use of the fact that Poisson's partial-differential equation is satisfied when total magnetic energy function is a minimum [8, 9]. The problem geometry is divided into small elements. Within each element, the flux density is assumed constant so that the magnetic vector potential varies linearly within each element. For better accuracy, the vector potential is assumed to vary as a polynomial of a degree higher than one. The elements are generally of triangular or tetrahedral shape. Windings are modeled as rectangular blocks. If ampere-turn distribution is not uniform (different ampere-turn densities), the windings are divided into suitable sections so that the ampere-turn distribution in each section is uniform. A typical configuration of LV and HV windings in a transformer window is shown in figure 3.8. The main steps of analysis are now outlined below:

1. ***Creation of geometry***: The geometry shown in figure 3.8 is quite simple. In case of complex 2-D or 3-D geometries, many commercial FEM programs allow importing of figures drawn in drafting packages, which makes it easier and less time consuming to create a geometry. The geometry has to be always bounded by a boundary like *abcda* shown in the figure. The two-dimensional problems can be solved in either Cartesian or Axisymmetric coordinate systems. Since a transformer is a three-dimensional electromagnetic structure, both the systems are approximate but sufficiently accurate for magnetostatic problems such as reactance estimation. In Axisymmetric (r-z) coordinate system, line *ab* represents the axis (center-line) of the core and hence the horizontal distance between lines *ab* and *ef* equals half the core diameter.

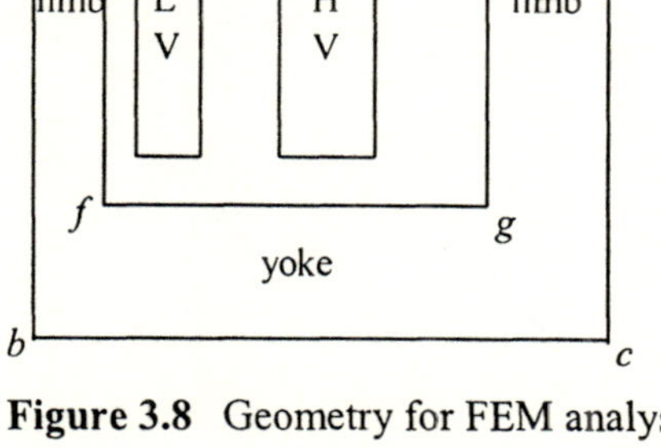

Figure 3.8 Geometry for FEM analysis

2. ***Meshing***: This step involves division of geometry into small elements. For most accurate results, the element size should be as small as possible if flux density is assumed constant in it. Thus, logically the element (mesh) size should be smaller only in the regions where there is an appreciable variation in values of flux density. Such an intelligent meshing reduces the number of elements and computation time. An inexperienced person may not always know the regions where the solution is changing appreciably; hence one can start with a very coarse mesh, get a solution, and then refine the mesh in the regions where the solution is changing rapidly. Ideally, one has to go on refining the mesh till there is no appreciable change in the value of solution (flux density in this case) at any point in the geometry. For the geometry of figure 3.8, the radial component of flux density changes appreciably at the winding ends, necessitating the use of finer mesh in these regions as shown in figure 3.9.

3. ***Material properties***: Core is defined with relative permeability (μ_r) of some tens of thousands. It really does not matter whether we define it as 10000 or 50000 because almost all the energy is stored in the non-magnetic regions ($\mu_r = 1$) outside the core. While estimating the leakage reactance, ampere-turns of LV and HV are assumed to be exactly equal and opposite (magnetizing ampere-turns are neglected), and hence there is no mutual component of flux in the core (there is no flux contour in the core enclosing both the windings). Other parts, including windings, are defined with μ_r of 1. Here, the conductivity of winding material is not defined since the effect of eddy currents in winding conductors on the leakage field is usually neglected in reactance calculations (the problem is solved as a magnetostatic problem). Individual conductor / turn may have to be modeled for estimation of circulating currents in parallel strands of a winding, which is a subject of discussion in Chapter 4.

4. ***Source definition***: In this step, the ampere-turn density for each winding / section (ampere-turns divided by cross-sectional area) is defined.

Figure 3.9 FEM mesh

5. ***Boundary conditions***: There are two types of boundary conditions, viz. Dirichlet and Neumann. The boundary conditions in which potential is prescribed are called as Dirichlet conditions. In the present case, Dirichlet condition is defined for the boundary *abcda* (flux lines are parallel to this boundary) with the value of magnetic vector potential taken as zero for convenience. It should be noted that a contour of equal values of magnetic vector potential is a flux line. The boundary conditions on which the normal derivative of potential is prescribed are called as Neumann conditions. The flux lines cross orthogonally (at 90° angle) at these boundaries. A boundary on which the Dirichlet condition is not defined, the Neumann condition gets automatically specified. If the core is not modeled, no magnetic vector potential should be defined on the boundary *efgh* (iron-air boundary). The flux lines then impinge on this boundary orthogonally, which is in line with the valid assumption that the core is infinitely permeable. But in the absence of core, one reference potential should be defined in the whole geometry (usually at a point in the gap between windings along their center-line).

6. ***Solution***: Matrix representation of each element, formation of global coefficient matrix and imposition of boundary conditions are done in this step (commercial FEM software does these things internally). Solution of resulting simultaneous algebraic equations is subsequently obtained. Solution proceeds broadly in the following way:

- approximation of magnetic vector potential A within each element in a standardized fashion. For example in Cartesian coordinate system,

$$A = a + b\,x + c\,y \tag{3.42}$$

- the constants a, b, c can be expressed in terms of values of A at the nodes of an element. The above expression then gives A over the entire element as linear interpolation between the nodal values
- potential distributions in various elements are inter-related so as to constrain the potential to be continuous across inter-element boundaries
- minimization of energy then determines the values of A at the nodes

7. ***Post-processing***: Leakage field plot (like in figure 3.1) can be obtained and studied. The total stored energy is calculated as per equation

$$W_m = \frac{1}{2}\int_{vol} \mathbf{A}\cdot\mathbf{J}\,dv \tag{3.43}$$

If the problem is solved in Cartesian coordinate system, energy obtained is per unit length in the z direction. In order to obtain the total energy, the value of energy for each section of the geometry is multiplied by the corresponding mean diameters. Finally, the leakage inductance can be calculated by equation 3.24.

Example 3.1

The relevant dimensions (in mm) of 31.5 MVA, 132/33 kV, 50 Hz, Yd1 transformer are indicated in figure 3.10. The value of volts/turn is 76.21. The transformer is having -0% to +10% taps on HV winding. It is having linear type of on-load tap changer; there are 10% tapping turns placed symmetrically in the middle of HV winding giving a total voltage variation of 10%. It is required to calculate the leakage reactance of the transformer at the nominal tap position (corresponding to HV voltage of 132 kV) by the classical method and FEM analysis.

Solution:

We will calculate the leakage reactance by method given in Section 3.1.3 as well as by FEM analysis.

1. Classical method

At the nominal tap position, TAP winding has zero ampere-turns since all its turns are cut out of the circuit. This results into unequal AT/m distribution between LV and HV windings along their height.

$$\text{HV current} = \frac{31.5\times 10^6}{\sqrt{3}\times 132\times 10^3} = 137.78 \text{ A}$$

$$\text{HV turns} = \frac{132\times 10^3/\sqrt{3}}{76.21} = 1000 \text{ and LV turns} = \frac{33\times 10^3}{76.21} = 433$$

The HV winding is replaced by a winding (HV1) with the uniformly distributed ampere-turns (1000 turns distributed uniformly along the height of 1260 mm) and a second winding (HV2) having ampere-turns distribution such that the superimposition of ampere-turns of both these windings gives the ampere-turn distribution of the original HV winding.

We will first calculate reactance between LV and HV1 windings by using the formulation given in Section 3.1.

$$T_1 = 7.0 \text{ cm}, \ T_2 = 5.0 \text{ cm}, \ T_3 = 10.0 \text{ cm}, \ H_w = 126.0 \text{ cm}$$

Equations 3.18 and 3.1 give

$$K_R = 0.944 \quad \text{and} \quad H_{eq} = H_w/K_R = 126/0.944 = 133.4 \text{ cm}$$

The term $\sum ATD$ is calculated as per equation 3.17,

$$\sum ATD = \frac{1}{3}\times 7.0\times 67.6 + 5.0\times 79.6 + \frac{1}{3}\times 10.0\times 94.6 = 871.1 \text{ cm}^2$$

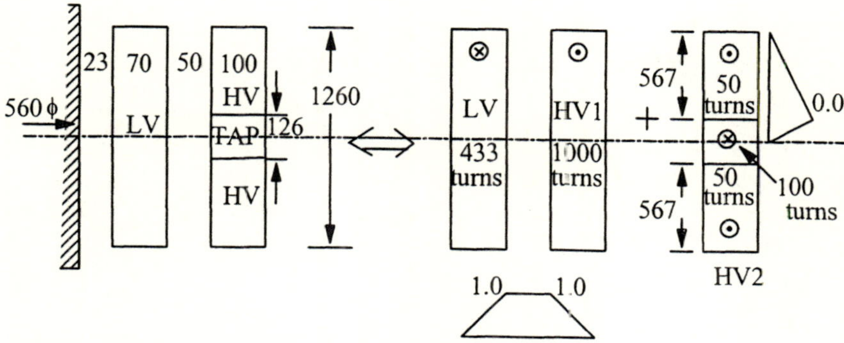

Figure 3.10 Details of transformer of Example 3.1

Leakage reactance can be calculated from equation 3.16 as

$$X_{\text{LV\_HV1}} = 2.48 \times 10^{-5} \times 50 \times \frac{(137.78 \times 1000)}{133.4 \times 76.21} \times 871.1 = 14.64\%$$

The winding HV2 is made up of two sections. The ampere-turn diagram for the top section is shown in figure 3.10. The section has two windings, each having ampere-turns of 0.05 per-unit $[=(50 \times 137.78)/(1000 \times 137.78)]$. For this section,

$$T_1 = 56.7 \text{ cm}, T_g = 0.0 \text{ cm}, T_2 = 6.3 \text{ cm}, H_w = 10.0 \text{ cm}$$

It is to be noted here that the winding height is actually the dimension in radial direction, which is equal to 10.0 cm. Equations 3.18 and 3.1 give

$$K_R = 0.213 \text{ and } H_{eq} = H_w / K_R = 10/0.213 = 47.0 \text{ cm}$$

The term ATD is calculated for each part as per equation 3.12 with the corresponding values of a and b, and the mean diameter of HV winding,

$$\sum ATD = \frac{56.7}{3} \times 0.05^2 \times 94.6 + 0.0 \times 0.05^2 \times 94.6 + \frac{6.3}{3} \times 0.05^2 \times 94.6 = 5.0$$

The leakage reactance of the section can be calculated from equation 3.16 as

$$X_S = 2.48 \times 10^{-5} \times 50 \times \frac{(137.78 \times 1000)}{47.0 \times 76.21} \times 5.0 = 0.24\%$$

The HV2 winding comprises of two such sections connected in series. Hence, the total reactance contributed by HV2 is two times the reactance of one section as explained in Section 3.1.2.

$$\therefore X_{\text{HV2}} = 2 \times 0.24 = 0.48\%$$

Therefore, the total reactance is,

$$X = X_{\mathrm{LV\_HV1}} + X_{\mathrm{HV2}} = 14.64 + 0.48 = 15.12\%$$

2. FEM analysis

The analysis is done as per the steps outlined in Section 3.4. The winding to yoke distance is 130 mm for this transformer. The stored energy in different parts of geometry as given by the FEM analysis is:

LV	:	438 J
HV	:	773 J
HV center gap	:	87 J
Portion of whole geometry excluding LV and HV windings:		1205 J

The energy stored in the core is negligible. The total energy is 2503 J. Using equation 3.24, the leakage inductance can be found as

$$L = \frac{2 \times 2503}{(137.78)^2} = 0.264 \text{ H}$$

The value of base impedance is

$$Z_b = \frac{\mathrm{kV}^2}{\mathrm{MVA}} = \frac{132^2}{31.5} = 553.1\ \Omega$$

$$\therefore X = \frac{2\pi \times 50 \times L}{Z_b} \times 100 = \frac{2\pi \times 50 \times 0.264}{553.1} \times 100 = 15.0\%$$

Thus, the values of leakage reactance given by the classical method and FEM analysis are quite close.

Example 3.2

Calculate the leakage reactance of a transformer having 10800 ampere-turns in each of the LV and HV windings. The rated voltage of LV is 415 volts and current is 300 A. The two windings are sandwiched into 4 sections as shown in figure 3.11. The relevant dimensions (in mm) are given in the figure. The value of volts/turn is 11.527. The mean diameter of windings is 470 mm.

Solution:

The leakage reactance will be calculated by the method given in Section 3.1.2 and FEM analysis.

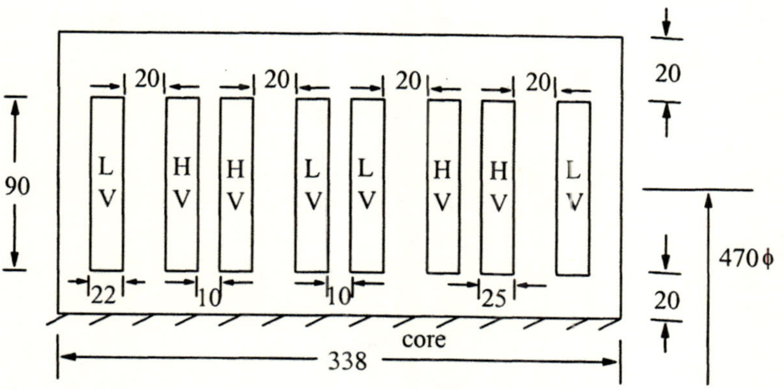

Figure 3.11 Details of transformer with sandwiched windings

1. Classical method

The whole configuration consists of four sections, each having 1/4th part of both the LV and HV windings. For any one section,

$$T_1 = 2.2 \text{ cm}, \ T_g = 2.0 \text{ cm}, \ T_2 = 2.5 \text{ cm}, \ H_w = 9.0 \text{ cm}$$

Equations 3.18 and 3.1 give

$$K_R = 0.767 \text{ and } H_{eq} = H_w / K_R = 9/0.767 = 11.7 \text{ cm}$$

The term $\sum ATD$ is (as per equation 3.17)

$$\sum ATD = \frac{1}{3} \times 2.2 \times 47.0 + 2.0 \times 47.0 + \frac{1}{3} \times 2.5 \times 47.0 = 167.6$$

The leakage reactance between LV and HV windings can be calculated from equation 3.22 with number of sections as $S = 4$,

$$\%X = 2.48 \times 10^{-5} \times 50 \times \frac{10800}{11.7 \times 11.527} \times \frac{1}{4} \times 167.6 = 4.16\%$$

2. FEM analysis

The full geometry as given in figure 3.11 is modeled and the analysis is performed as per the steps outlined in Section 3.4. The stored energy in the different parts of the geometry is:

LV (all 4 parts)	:	1.44 J
HV (all 4 parts)	:	1.66 J
Portion of whole geometry excluding LV and HV windings:		5.26 J

The total energy is 8.36 J. Using equation 3.24, the leakage inductance can be found to be

$$L = \frac{2 \times 8.36}{(300)^2} = 0.000186 \text{ H}$$

Base impedance is

$$Z_b = \frac{Rated\ voltage}{Rated\ current} = \frac{415}{300} = 1.38\ \Omega$$

$$\therefore X = \frac{2\pi \times 50 \times L}{Z_b} \times 100 = \frac{2\pi \times 50 \times 0.000186}{1.38} \times 100 = 4.23\%$$

3.5 Impedance Characteristics of Three-Winding Transformer

A three-winding (three-circuit) transformer is generally required when actual loads or auxiliary loads (reactive power compensating devices such as shunt reactors or condensers) are required to be supplied at a voltage different from that of either primary or secondary voltage. An unloaded tertiary winding is also used just for the stabilizing purpose (which is discussed in Section 3.8). The phenomena related to leakage field (efficiency, regulation, parallel operation and short circuit currents) of a multi-circuit transformer cannot be analyzed in the same way as that for a two-winding transformer. Each winding is interlinked with the leakage fields of other windings, and hence a load current in one winding affects voltages in other windings, sometimes in a surprising way. For example, a lagging load on one winding may increase the voltage of other windings due to negative leakage reactance (capacitive reactance).

The leakage reactance characteristics of a three-winding transformer can be represented by the equivalent circuit method in which it is assumed that each circuit has an individual leakage reactance. When the magnetizing current is neglected (which is quite justified in the calculations related to leakage fields) and if all the quantities are expressed in per-unit or percentage notation, magnetically interlinked circuits of a three-winding transformer can be represented by electrically interlinked circuits as shown in figure 3.12. The equivalent circuit can be either star or mesh network. The star equivalent circuit is more commonly used and is discussed here.

The percentage leakage reactances between pairs of windings can be expressed in terms of their individual percentage leakage reactances (all expressed on common volt-amperes base) as

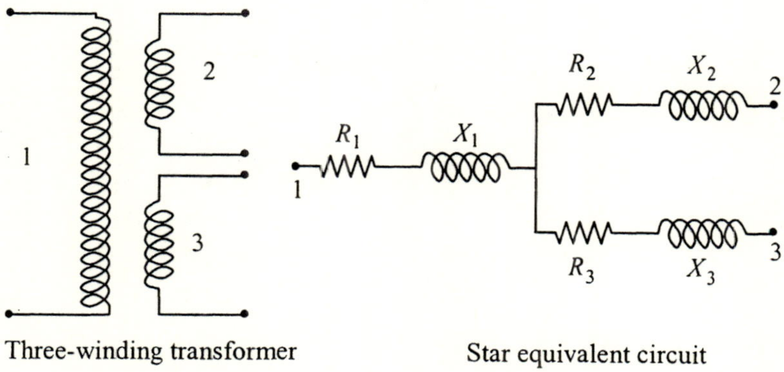

Figure 3.12 Representation of a three-winding transformer

$$X_{12} = X_1 + X_2 \tag{3.44}$$

$$X_{23} = X_2 + X_3 \tag{3.45}$$

$$X_{31} = X_1 + X_3 \tag{3.46}$$

It follows from the above three equations that the individual reactances in the star equivalent circuit are given by

$$X_1 = \frac{1}{2}\left[X_{12} + X_{13} - X_{23}\right] \tag{3.47}$$

$$X_2 = \frac{1}{2}\left[X_{12} + X_{23} - X_{13}\right] \tag{3.48}$$

$$X_3 = \frac{1}{2}\left[X_{13} + X_{23} - X_{12}\right] \tag{3.49}$$

A rigorous derivation for above three equations and evolution of the star equivalent circuit is given in [10].

Similarly, percentage resistances can be derived as

$$R_1 = \frac{1}{2}\left[R_{12} + R_{13} - R_{23}\right] \tag{3.50}$$

$$R_2 = \frac{1}{2}\left[R_{12} + R_{23} - R_{13}\right] \tag{3.51}$$

$$R_3 = \frac{1}{2}\left[R_{13} + R_{23} - R_{12}\right] \tag{3.52}$$

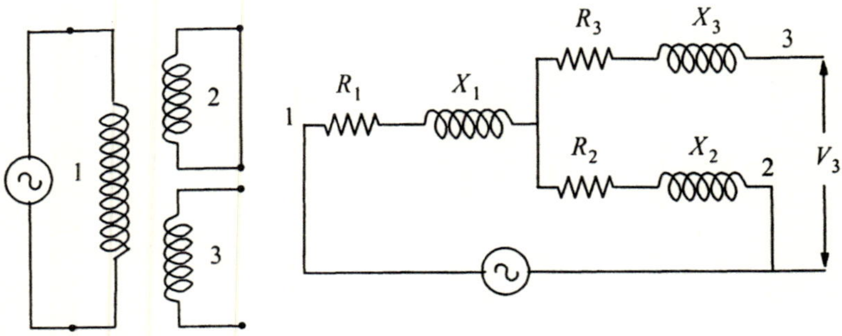

Figure 3.13 Mutual effect in star equivalent network

It is to be noted that these percentage resistances represent the total load loss (DC resistance I^2R loss in windings, eddy loss in windings and stray losses in structural parts).

The leakage reactances in the star equivalent network are basically the mutual load reactances between different circuits. For example, the reactance X_1 in figure 3.12 is the common or mutual reactance to loads in circuits 2 and 3. A current flowing from circuit 1 to either 2 or 3, produces drop in R_1 and X_1, and hence affects voltages of circuits 2 and 3. When a voltage is applied to winding 1 with winding 2 short-circuited as shown in figure 3.13, the voltage across open-circuited winding 3 is equal to the voltage drop across the leakage impedance, Z_2, of circuit 2.

As said earlier, the individual leakage reactance of a winding may be negative. The total leakage reactance between a pair of windings cannot be negative but depending upon how the leakage field of one interlinks with the other, the mutual effect between circuits may be negative when a load current flows [11]. Negative impedances are virtual values, and they reproduce faithfully the terminal characteristics of transformers and cannot be necessarily applied to internal windings. Similarly, a negative resistance may appear in the star equivalent network of an autotransformer with tertiary or of a high efficiency transformer having stray losses quite high as compared to winding ohmic losses (e.g., when a lower value of current density is used for windings).

In Chapter 1, we have seen how the regulation of a two-winding transformer is calculated. Calculation of voltage regulation of a three-winding transformer is explained with the help of following example.

Example 3.3

Find the regulation between terminals of a three-winding transformer, when the load on IV winding is 70 MVA at power factor of 0.8 lagging and the load on LV winding is 30 MVA at power factor of 0.6 lagging. The transformer data is:

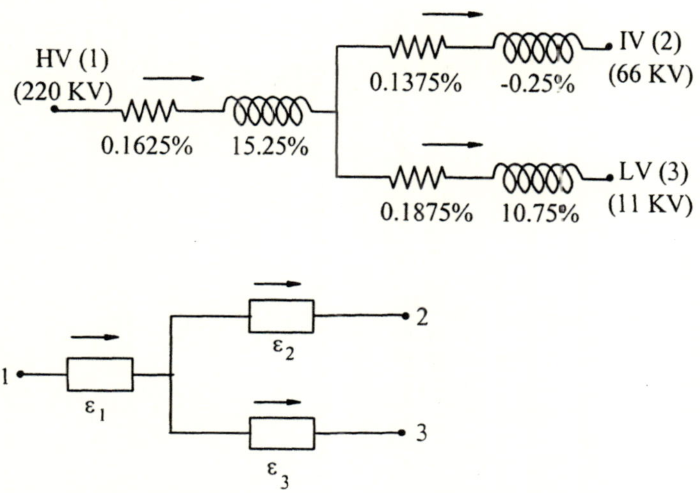

Figure 3.14 Star-equivalent circuit and regulation of Example 3.3

Rating: 100/100/30 MVA, 220/66/11 kV
Results of load loss (short circuit) test referred to 100 MVA base:

HV-IV : $R_{1-2} = 0.30\%$, $X_{1-2} = 15.0\%$
HV-LV : $R_{1-3} = 0.35\%$, $X_{1-3} = 26.0\%$
IV-LV : $R_{2-3} = 0.325\%$, $X_{2-3} = 10.5\%$

Solution:

The star equivalent circuit derived using equations 3.47 to 3.52 is shown in figure 3.14. It is to be noted that although HV, IV and LV windings are rated for different MVA values, for finding the equivalent circuit, we have to work on a common MVA base (in this case it is 100 MVA).

The IV winding is loaded to 70 MVA; let the constant K_2 denote the ratio of actual load to the base MVA,

$$K_2 = 70/100 = 0.7$$

Similarly, for LV winding which is loaded to 30 MVA,

$$K_3 = 30/100 = 0.3$$

The regulations for circuits 2 and 3 are calculated using equation 1.65,

$$\varepsilon_2 = K_2(R_2 \cos\theta_2 + X_2 \sin\theta_2) = 0.7(0.1375 \times 0.8 + (-0.25) \times 0.6) = -0.03\%$$
$$\varepsilon_3 = K_3(R_3 \cos\theta_3 + X_3 \sin\theta_3) = 0.3(0.1875 \times 0.6 + 10.75 \times 0.8) = 2.61\%$$

The effective power factor ($\cos\theta_1$) and the effective load taken as a fraction of base MVA (constant K_1) for the primary (220 kV) circuit are found by solving the following two equations:

$$K_1 \cos\theta_1 = K_2 \cos\theta_2 + K_3 \cos\theta_3 = 0.7\times0.8+0.3\times0.6=0.74 \tag{3.53}$$

$$K_1 \sin\theta_1 = K_2 \sin\theta_2 + K_3 \sin\theta_3 = 0.7\times0.6+0.3\times0.8=0.66 \tag{3.54}$$

Solving these two equations we get

$$K_1 = 0.99,\quad \cos\theta_1 = 0.75,\quad \sin\theta_1 = 0.67$$

Therefore, the primary circuit regulation is

$$\varepsilon_1 = K_1(R_1 \cos\theta_1 + X_1 \sin\theta_1) = 0.99(0.1625\times0.75+15.25\times0.67) = 10.23\%$$

Now, the regulation between terminals can be calculated as

$$\varepsilon_{1-2} = \varepsilon_1 + \varepsilon_2 = 10.23 + (-0.03) = 10.2\%$$

$$\varepsilon_{1-3} = \varepsilon_1 + \varepsilon_3 = 10.23 + 2.61 = 12.84\%$$

$$\varepsilon_{2-3} = -\varepsilon_2 + \varepsilon_3 = 0.03 + 2.61 = 2.64\%$$

While calculating ε_{2-3}, negative sign is taken for ε_2 because the drop in voltage is in direction opposite to the direction of flow of current.

When there are more than three windings (circuits), transformers cannot be in general represented by a pure star or mesh equivalent circuit. The equivalent circuit should have $n(n-1)/2$ independent impedance links, where n is the number of circuits. A four-winding transformer has six independent links with four terminal points. The procedures for determination of the values of equivalent network impedances for transformers with four or more independent circuits are given in [1, 12, 13].

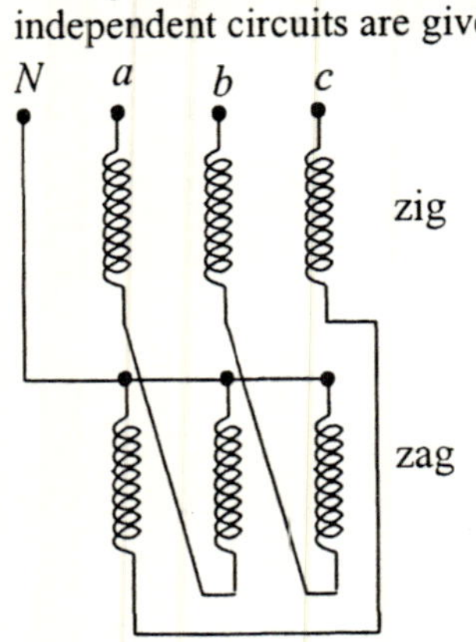

Figure 3.15 Zigzag transformer

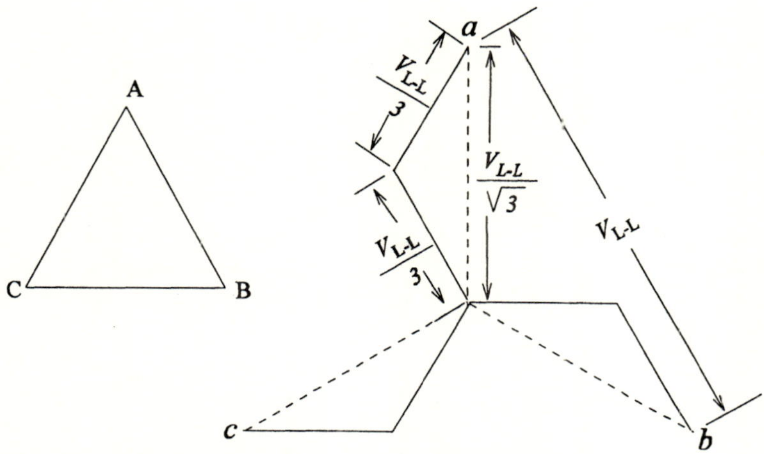

Figure 3.16 Vector diagram of delta-zigzag transformer

3.6 Reactance Calculation for Zigzag Transformer

A connection diagram of zigzag transformer, in which windings on different limbs (in a 3-phase 3-limb construction) are cross-connected, is shown in figure 3.15. It is termed as zigzag or interconnected star winding because the zig winding of one phase is connected in series with the zag winding of one of the other two phases. The vector diagram for a delta-zigzag transformer is shown in figure 3.16. The interconnection of windings of different phases introduces 30° (or 150°) phase-shift between zig (or zag) winding and the corresponding line-to-neutral voltage ($V_{L-N} = V_{L-L}/\sqrt{3}$). The zig and zag windings have 15.47% more turns as compared to conventional windings to get the same magnitude of phase / line voltages. Hence, the zigzag transformer is costlier than the conventional transformer, but its use is essential in some cases. Its main characteristics and uses are as follows:

1) It can be used as an earthing transformer in a delta connected system or an ungrounded star connected system, wherein neutral is not available for earthing / grounding. Consider a zigzag transformer connected to a delta connected source as shown in figure 3.17. The zigzag transformer provides neutral for the grounding purpose. For a single line-to-ground fault, zero-sequence current flows in the ground circuit allowing the protection system to act. The voltages of other two healthy line terminals are maintained at their respective line-to-neutral voltage levels. In absence of the grounded neutral, voltages of healthy phases would increase to line-to-line voltage level, stressing the insulation of connected equipment. Thus, zigzag earthing transformer not only helps in protection but also reduces the voltage stresses under asymmetrical fault conditions.

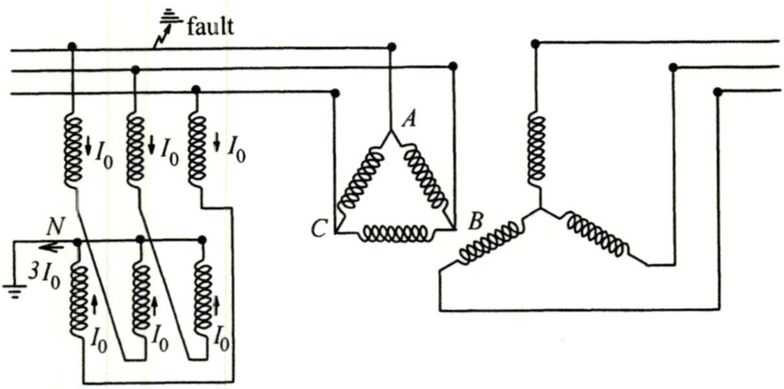

Figure 3.17 Zigzag earthing transformer

2) A zigzag transformer has a specific advantage when used in an application involving power electronic converters. The DC magnetization, as a result of asymmetry in firing angles, gets cancelled in each limb due to opposite directions of DC magnetizing component of currents flowing in the two windings on the same limb. Similarly, DC magnetization inherent in a 3-pulse mid-point rectifier connection is eliminated if the secondary is a zigzag winding [14].

3) The earthing transformer offers a low impedance path to the zero-sequence currents under fault conditions because the only magnetic flux which results from zero-sequence currents is the leakage flux around each winding section (the flux in a limb is zero due to opposite directions of equal currents flowing in the two windings wound on it as shown in figure 3.17). Owing to a very small impedance of the earthing transformer under fault conditions, it may be necessary to limit the value of fault current by connecting a resistor of suitable value between the neutral and earth. Under normal operating conditions, only a small exciting current circulates in the windings of the earthing transformer.

4) Third harmonic voltage components present in zig and zag windings get cancelled in the lines.

The reactance of a transformer consisting of zigzag and star (or delta) windings can be calculated from reactive volt-amperes. Consider a star-zigzag transformer shown in figure 3.18. The directions of currents I_a, I_b and I_c are shown in the figure as per convenience of three-phase system (all currents flowing towards neutral). The corresponding vector diagram is also shown in the figure.

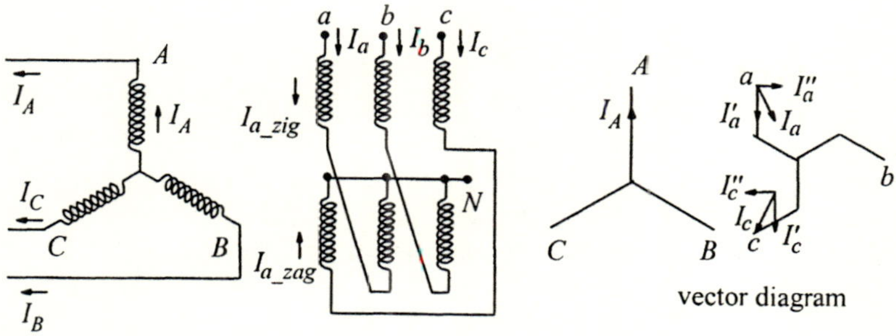

Figure 3.18 Reactance calculation of a transformer having zigzag winding

All the currents are resolved into two mutually perpendicular sets of components. Since the vector sum of all ampere-turns is zero on each limb, sum of all components indicated by prime is zero and sum of all components indicated by double prime is also zero. The phase A current of the star connected winding (I_A) is taken as the reference vector, and all the other currents are resolved in the directions along and perpendicular to this vector. Since I_A is taken as the reference vector, in p.u. notations $I_A^{'}$ and $I_A^{''}$ have values of 1 p.u. and 0 p.u. respectively. Vectors of zigzag winding currents of phases a and c are shown in the vector diagram. Current I_{a_zig} is in phase with I_a and current I_{a_zag} is in phase with I_c. The zig and zag windings have 0.577 p.u. ampere-turns (1.1547 $\times$ 0.5). Also, the currents I_a and I_c are at an angle of 30° with respect to the reference vector. The direction of current I_c (I_{a_zag}) has been reversed in the vector diagram so that at any instant the corresponding ampere-turns of the zag winding oppose that of the star winding. Thus, we can write

$$I_A^{'} = 1 \tag{3.55}$$

$$I_A^{''} = 0 \tag{3.56}$$

$$(I_{a\_zig})^{'} = (I_{a\_zag})^{'} = -0.577 \cos 30° = -0.5 \tag{3.57}$$

$$(I_{a\_zig})^{''} = -(I_{a\_zag})^{''} = 0.577 \sin 30° = 0.289 \tag{3.58}$$

The equations 3.55 to 3.58 satisfy the following two equations as required by the condition that the vector sum of all ampere-turns on the limb corresponding to phase A is zero,

$$I_A^{'} + (I_{a\_zig})^{'} + (I_{a\_zag})^{'} = 0 \tag{3.59}$$

$$I_A'' + (I_{a\_zig})'' + (I_{a\_zag})'' = 0 \tag{3.60}$$

The expression for Q can be generalized using equation 3.37 for the case where the current vectors of windings are not parallel (not in phase or phase-opposition) by resolving them into two mutually perpendicular sets. The resistances are neglected since they are much smaller than the reactances. The value of Q is calculated separately for each set and the two values are added algebraically,

$$Q = -\frac{1}{2}\sum_{k=1}^{n}\sum_{j=1}^{n} X_{jk}(I_j)'(I_k)' - \frac{1}{2}\sum_{k=1}^{n}\sum_{j=1}^{n} X_{jk}(I_j)''(I_k)'' \tag{3.61}$$

The value of Q for the phase A having three windings, viz. star, zig, and zag, which are now denoted by numbers 1, 2, and 3, respectively, is

$$\begin{aligned} Q = &-\frac{1}{2}\left\{2X_{12}\, I_1' I_2' + 2X_{13}\, I_1' I_3' + 2X_{23}\, I_2' I_3'\right\} \\ &-\frac{1}{2}\left\{2X_{12}\, I_1'' I_2'' + 2X_{13}\, I_1'' I_3'' + 2X_{23}\, I_2'' I_3''\right\} \end{aligned} \tag{3.62}$$

where X_{12}, X_{13}, and X_{23} are the per-unit leakage reactances between the corresponding windings. Substituting the values of all currents from equations 3.55 to 3.58, and remembering that $I_1 = I_A, I_2 = I_{a\_zig}$ and $I_3 = I_{a\_zag}$,

$$\begin{aligned} Q = &-\{X_{12}\times 1\times(-0.5) + X_{13}\times 1\times(-0.5) + X_{23}\times(-0.5)\times(-0.5)\} \\ &-\{0 + 0 + X_{23}\times 0.289\times(-0.289)\} \end{aligned} \tag{3.63}$$

Since the currents and reactances are expressed in per-unit notation, the value of Q directly gives the per-unit reactance of star-zigzag windings.

$$\therefore Q = X_{\text{star\_zigzag}} = \frac{1}{2}[X_{12} + X_{13}] - \frac{1}{6}X_{23} \tag{3.64}$$

Thus, the leakage reactance of a transformer with the star (or delta) and zigzag connected windings can be computed easily, once the leakage reactances between the pairs of windings are calculated on a common volt-amperes base.

Example 3.4

Find the leakage reactance of 31.5 MVA, 132/11 kV, 50 Hz, star/zigzag transformer whose various relevant dimensions (in mm) are indicated in figure 3.19. Value of volts/turn is 76.39.

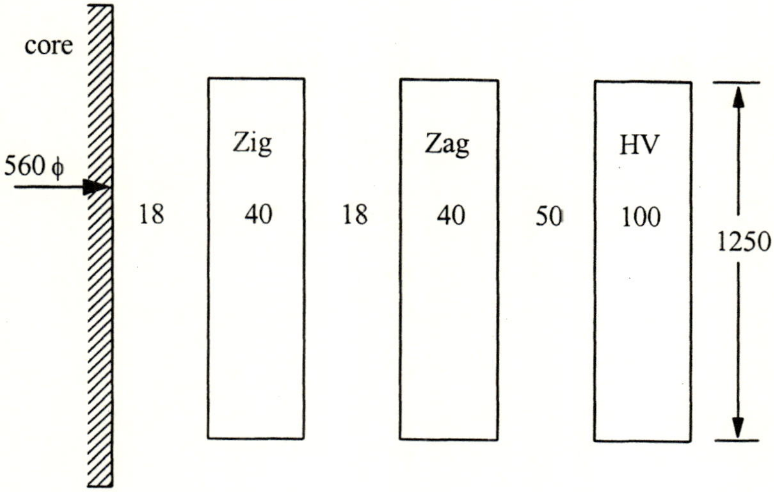

Figure 3.19 Dimensions of star-zigzag transformer

Solution:

Let us first calculate the leakage reactances between pairs of windings, viz. zig – zag, zig – HV and zag – HV by the procedure given in Section 3.1.1 for the concentric windings.

Reactance between zig and zag:

$$T_1 = 4.0 \text{ cm}, \; T_g = 1.8 \text{ cm}, \; T_2 = 4.0 \text{ cm}, \; H_w = 125.0 \text{ cm}$$

Equations 3.18 and 3.1 give: $K_R = 0.975$ and $H_{eq} = 125/0.975 = 128.2$ cm

Equation 3.17 gives

$$\sum ATD = \frac{1}{3} \times 4.0 \times 63.6 + 1.8 \times 69.4 + \frac{1}{3} \times 4.0 \times 75.2 = 310 \text{ cm}^2$$

All the three reactance values need to be calculated on a common MVA base. The value of base MVA is taken as 31.5 MVA. The corresponding current and turns of HV side are used in the reactance formula.

$$\text{HV current} = \frac{31.5 \times 10^6}{\sqrt{3} \times 132 \times 10^3} = 137.78 \text{ A}$$

$$\text{HV turns} = \frac{132 \times 10^3 / \sqrt{3}}{76.39} = 998$$

The leakage reactance is calculated from equation 3.16 as

$$X_{\text{zig\_zag}} = 2.48 \times 10^{-5} \times 50 \times \frac{(137.78 \times 998)}{128.2 \times 76.39} \times 310 = 5.4\%$$

Similarly, other two reactances can be calculated:

Reactance between zig and HV:

$$T_1 = 4 \text{ cm},\ T_2 = 10.8 \text{ cm},\ T_3 = 10 \text{ cm}$$

$$\therefore K_R = 0.937,\ H_{eq} = 133.4 \text{ cm},\ \sum ATD = 1262.2 \text{ and } X_{\text{zig\_HV}} = 21.1\%$$

Reactance between zag and HV:

$$T_1 = 4 \text{ cm},\ T_2 = 5 \text{ cm},\ T_3 = 10 \text{ cm}$$

$$\therefore K_R = 0.95,\ H_{eq} = 131.4 \text{ cm},\ \sum ATD = 851.9 \text{ and } X_{\text{zag\_HV}} = 14.5\%$$

By using equation 3.64, we get reactance between star and zigzag windings as

$$X_{\text{star\_zigzag}} = \frac{1}{2}\left[X_{\text{HV\_zig}} + X_{\text{HV\_zag}}\right] - \frac{1}{6} X_{\text{zig\_zag}}$$

$$= \frac{1}{2} \times [21.1 + 14.5] - \frac{1}{6} \times 5.4 = 16.9\%$$

3.7 Zero-Sequence Reactance Estimation

The method of symmetrical components is commonly used in power system analysis. Unlike rotating machines, for a static apparatus like a transformer, positive-sequence and negative-sequence impedances ($\cong$ reactances) are equal. Under symmetrical loading conditions, only positive-sequence reactances need to be considered. In case of asymmetrical loading / disturbances or single-phase faults, the system response is largely decided by the zero-sequence reactances of the network. It is comparatively easy to understand and calculate positive-sequence reactances. The zero-sequence reactance of a transformer may differ considerably from its positive-sequence reactance depending upon the type of magnetic circuit and winding connections.

In order to measure the zero-sequence reactance, a voltage is applied between the shorted line terminals of a star connected winding and neutral as shown in figure 3.20. Since the zero-sequence currents cannot exist in the line terminals connected to a delta winding, its zero-sequence reactance is infinitely large, except for a large capacitive reactance. With reference to the test arrangement of figure 3.20, the zero-sequence reactance ($\cong$ impedance) of a star connected winding with the grounded neutral is calculated as

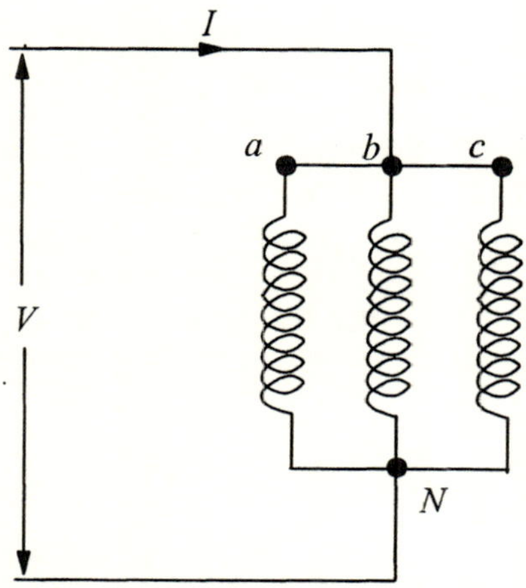

Figure 3.20 Measurement of zero-sequence impedance

$$X_0 = \frac{V}{(I/3)} = 3\frac{V}{I} \tag{3.65}$$

There are basically two types of zero-sequence reactances for a winding:

1. Open circuit (magnetizing) reactance (with terminals of all other windings kept open-circuited)
2. Short circuit (leakage) reactance (with terminals of only one other winding short-circuited)

The significance of these two types of zero-sequence reactances and methods for their estimation are now discussed.

3.7.1 Open circuit zero-sequence reactance without delta connected winding

A. Three-phase three-limb transformers

For a three-phase three-limb core, since fluxes in three limbs are in the same direction, they have to return through a path outside the core. In this case, tank acts as an equivalent delta winding, and the magnetizing zero-sequence reactance is the reactance between the tank and excited winding. Since the gap / area between the excited winding and tank is much larger than the gap / area between the windings, the magnetizing zero-sequence reactance is considerably higher than the positive-sequence leakage reactance. The tank influences the zero-sequence reactance in the following way. It provides a comparatively lower reluctance return path (as compared to air) to the zero-sequence flux, which has the effect of increasing the reactance; on the other hand the tank enclosing three phases acts as a short-circuited winding reducing the reactance. The latter effect is more dominant and hence the zero-sequence reactance inside a tank is appreciably less than that without it [15].

The reactance between the excited winding and tank can be calculated by the methodology given in Section 3.1.1 with the tank represented as an equivalent (fictitious) winding with zero radial depth (contribution of tank to the ampere-turn diagram area can be neglected due to its small thickness). The distance between the excited winding and tank can be taken as average equivalent distance which gives approximately the same area of space between them. Thus, in this case the zero-sequence reactance can be calculated by using equation 3.16 as

$$\%X_0 = 2.48\times 10^{-5} f \frac{(Ampere \times Turns)}{H_{eq}\times (Volts/Turn)} \times (\frac{1}{3}\times T_w \times D_w + T_g \times D_g) \tag{3.66}$$

where

T_w = radial depth of excited winding in cm

D_w = mean diameter of excited winding in cm

T_g = gap between excited winding and tank in cm

D_g = mean diameter of gap between excited winding and tank in cm

A sufficient voltage should be applied during the test so that the tank, which is generally made of mild steel material, is in saturated condition (due to flux passing through it). If the test is done at a voltage which does not saturate the tank, its higher permeability will increase the reactance. Hence, the reactance calculated as per equation 3.66 has to be suitably corrected by an empirical factor which can be derived based on the measurements done at various voltage levels. The correction factor depends on the applied voltage as shown in figure 3.21 [16] and it is a function of permeability value. The curve has the same nature as that of permeability versus magnetic field intensity graph derived by measurements [17] on a typical grade of mild steel material (figure 3.22).

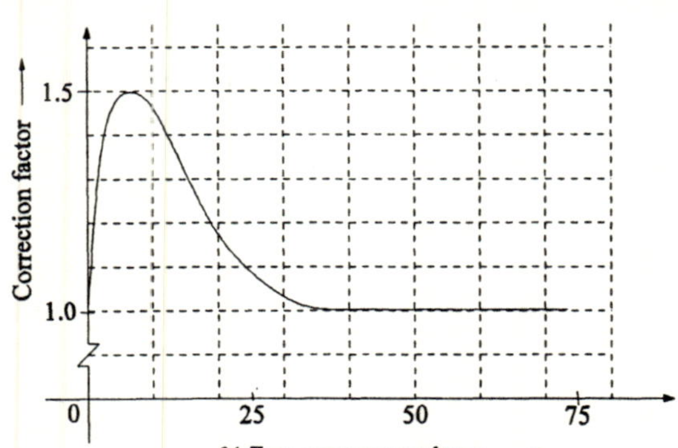

Figure 3.21 Correction factor for zero-sequence reactance [16]

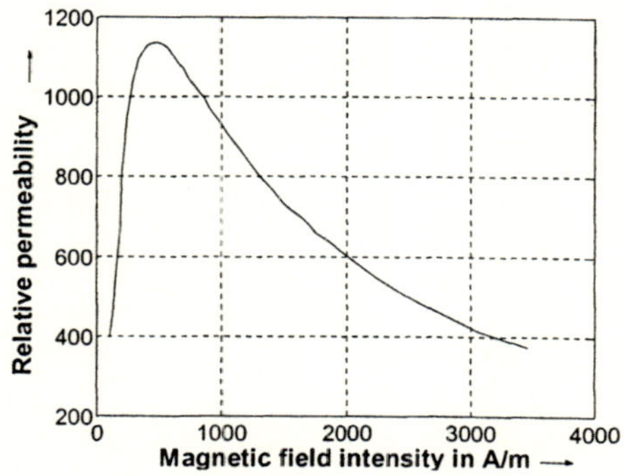

Figure 3.22 Variation of μ_r with field intensity in mild steel material [17]

B. Three-phase five-limb and single-phase three-limb transformers

In a three-phase five-limb core, the zero-sequence flux has a return path through the end yokes and end limbs and hence the zero-sequence magnetizing reactance has the same value as the corresponding positive-sequence magnetizing reactance (quite a high value) unless the voltage applied is such that yokes and end limbs saturate. For an applied value of zero-sequence voltage, close to the rated voltage, yokes and end limbs will get completely saturated (area of yokes and end limbs are too less to carry the zero-sequence flux of all the three phases at the rated voltage) giving a lower value of magnetizing zero-sequence reactance close to that of a transformer with three-phase three-limb core.

For single-phase three-limb cores, since the zero-sequence flux has a return path through end limbs, the zero-sequence magnetizing reactance is equal to the positive-sequence magnetizing reactance (~ infinite value). Thus, for a three-phase bank of single-phase transformers, the zero-sequence reactance for any type connection is usually equal to the positive-sequence reactance [18].

3.7.2 Open circuit zero-sequence reactance with delta connected winding

With one delta connected secondary winding, a current flows in the delta winding. Thus, the transformer acts as if it is short-circuited regardless of whether the winding is loaded or not. Let us again consider following two cases:

A. Three-phase three-limb transformers

In this case, when zero-sequence voltage is applied to a star connected winding, currents flow in the delta connected winding and the tank (which acts as an equivalent delta winding). In order to estimate the current division between these two delta windings and the value of zero-sequence reactance, we will use the general method of finding reactance for a composite system of windings explained in the Section 3.2. Let the star connected primary winding, delta connected secondary winding and the equivalent delta winding (tank) be represented by notations 1, 2 and 3 respectively. Using equation 3.38, for the present case of 3 windings (one star and two delta connected windings), the per-unit value of zero-sequence reactance is

$$Q = X_0 = -\frac{1}{2}\left[2X_{12}I_1(-I_2) + 2X_{23}(-I_2)(-I_3) + 2X_{13}I_1(-I_3)\right] \qquad (3.67)$$

The currents in windings 2 and 3 are assigned with negative sign as their direction is opposite to that in the primary winding. Since the magnetizing current can be neglected,

$$I_1 = I_2 + I_3 \qquad (3.68)$$

If the magnitude of applied zero-sequence voltage is such that the rated current flows in the star connected primary winding (I_1 has a value of 1 per-unit), then

$$I_3 = 1 - I_2 \qquad (3.69)$$

$$\therefore Q = X_{12}\,I_2 - X_{23}\,I_2\,(1 - I_2) + X_{13}\,(1 - I_2) \qquad (3.70)$$

The currents get distributed in the windings in such a way that the total energy is minimized. Hence, differentiating Q with respect to current I_2 and equating it to zero, we get

$$\frac{dQ}{dI_2} = X_{12} - X_{23} + 2X_{23}I_2 - X_{13} = 0 \qquad (3.71)$$

$$\therefore I_2 = \frac{X_{13} + X_{23} - X_{12}}{2X_{23}} \qquad (3.72)$$

The above value of I_2 when substituted in equation 3.69 gives the tank current as

$$I_3 = \frac{X_{23} + X_{12} - X_{13}}{2\,X_{23}} \qquad (3.73)$$

The delta connected secondary winding can be either inner or outer winding.

If delta connected winding (2) is outer winding:

$$X_{13} \cong X_{12} + X_{23} \qquad (3.74)$$

Substituting the value of X_{13} in equations 3.72 and 3.73, we get

$$I_2 \cong 1 \quad \text{and} \quad I_3 \cong 0 \qquad (3.75)$$

Substituting these values of I_2 and I_3 in equation 3.70, the value of zero-sequence reactance can be obtained as

$$X_0 = Q \cong X_{12} \tag{3.76}$$

Thus, with the outer delta winding, the zero-sequence reactance is approximately equal to the positive-sequence leakage reactance (X_{12}); this is an expected result since the outer delta winding acts effectively as a shield and no current flows in the tank (the outer delta winding almost entirely shields the tank).

If delta connected winding (2) is inner winding:

$$X_{23} \cong X_{21} + X_{13} = X_{12} + X_{13} \tag{3.77}$$

Substituting this value of X_{23} in equations 3.72 and 3.73, we get

$$I_2 \cong \frac{X_{13}}{X_{23}} \quad \text{and} \quad I_3 \cong \frac{X_{12}}{X_{23}} \tag{3.78}$$

Substituting these values of I_2 and I_3 in equation 3.70,

$$X_0 = Q \cong \frac{X_{13}}{X_{23}} X_{12} \tag{3.79}$$

Thus, with an inner delta winding, the zero-sequence reactance is always less than the positive-sequence leakage reactance (X_{12}). This is due to the fact that in this case $X_{23} > X_{13}$ (outer star connected winding 1 is closer to the tank).

It is to be noted that equations 3.74 and 3.77 are approximately valid; for accurate calculations, expressions for I_2 and I_3 given by equations 3.72 and 3.73 should be directly substituted in equation 3.70.

B. *Three-phase five-limb and single-phase three-limb transformers*

For a transformer with three-phase five-limb core, the value of zero-sequence reactance is equal to that of positive-sequence leakage reactance between the windings until the applied voltage saturates the yokes and end limbs. At such a high applied voltage, it acts as a three-limb transformer, and the zero-sequence reactance can be calculated accordingly.

For a single-phase three-limb core, the zero-sequence reactance is equal to the positive-sequence leakage reactance between star and delta connected windings, since current can flow in the closed delta (as if short-circuited) and there is a path available for flux in the magnetic circuit.

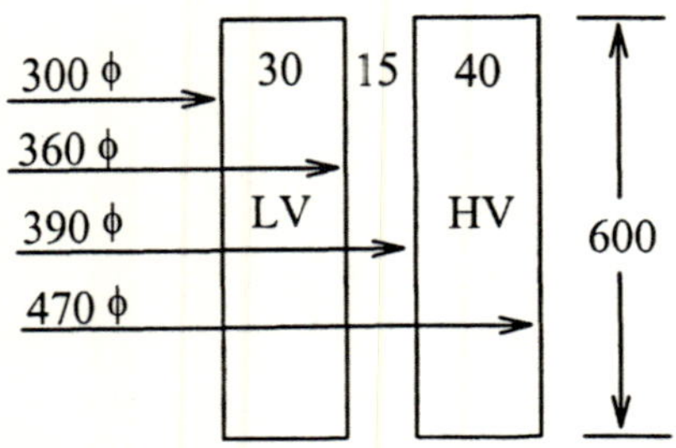

Figure 3.23 Dimensions of 2 MVA transformer

Example 3.5

Calculate the positive-sequence and zero-sequence reactances of a 2 MVA, 11/0.433 kV, 50 Hz, Dyn11 transformer whose various relevant dimensions in mm are indicated in figure 3.23. The value of volts/turn is 15.625.

Solution:

Positive-sequence leakage reactance is calculated by the procedure given in Section 3.1.1 for concentric windings.

The Rogowski factor is calculated by equation 3.18 as

$$K_R = 1 - \frac{1 - e^{-\pi 60.0/(3.0+1.5+4.0)}}{\pi 60.0/(3.0+1.5+4.0)} = 0.955$$

Equivalent winding height as per equation 3.1 is

$$H_{eq} = H_w / K_R = 60/0.955 = 62.8$$

The term $\sum ATD$ is calculated as per equation 3.17,

$$\sum ATD = \frac{1}{3} \times 3.0 \times 33.0 + 1.5 \times 37.5 + \frac{1}{3} \times 4.0 \times 43.0 = 146.6$$

For calculating the reactance, either LV or HV ampere-turns are taken (values of which are equal since the magnetizing ampere-turns are neglected).

$$\text{LV current} = \frac{2 \times 10^6}{\sqrt{3} \times 433} = 2666.67\ \text{A} \ \text{ and LV turns} = \frac{433/\sqrt{3}}{15.625} = 16$$

The positive-sequence leakage reactance is given by equation 3.16,

$$\%X_p = 2.48 \times 10^{-5} \times 50 \times \frac{(2666.67 \times 16)}{62.8 \times 15.625} \times 146.6 = 7.9\%$$

Since the delta connected HV winding is the outer winding, the zero-sequence reactance of star connected winding is approximately equal to the positive-sequence leakage reactance as explained in Section 3.7.2. However, during the actual test, one usually gets the value of zero-sequence reactance higher than the positive-sequence leakage reactance by an amount corresponding to the voltage drop in the neutral bar. The reactance of a neutral bar of rectangular dimensions ($a \times b$) is given by the expression [19, 20]:

$$X_n = 2\pi f \times 0.002 L_b \left\{ \log_e \frac{2L_b}{D_s} - 1 + \frac{D_s}{L_b} \right\} \times 10^{-6} \tag{3.80}$$

where

L_b = length of bus-bar in cm

D_s = geometric mean distance from itself = $0.2235 \times (a+b)$ cm

If the neutral bus-bar dimensions are: a = 5 cm and b = 0.6 cm, with a length of 50 cm,

$$D_s = 0.2235 \times (5 + 0.6) = 1.2516$$

$$\therefore X_n = 2\pi \times 50 \times 0.002 \times 50 \times \left\{ \log_e \frac{2 \times 50}{1.2516} - 1 + \frac{1.2516}{50} \right\} \times 10^{-6}$$

$$= 107 \times 10^{-6} \Omega$$

The base impedance on LV side is

$$Z_b = \frac{(0.433)^2}{2} = 0.0937$$

Since the current flowing in the neutral bar is 3 times that in the phase, the neutral bar contributes 3 times the value of X_n in the zero-sequence reactance.

$$\therefore \% \; X_n = \frac{3 \times 107 \times 10^{-6}}{0.0937} \times 100 = 0.34 \;\%$$

Hence, the measured zero-sequence reactance will test close to

$$(X_0)_{actual} = 7.9 + 0.34 = 8.24 \;\%$$

3.7.3 Short circuit zero-sequence reactance

Zero-sequence reactance under short circuit is applicable, for example, when a star connected secondary winding is short-circuited.

A. *Three-phase three-limb transformers*

The procedure for calculation of short circuit zero-sequence reactance of a three-phase three-limb transformer is now explained with the help of an example.

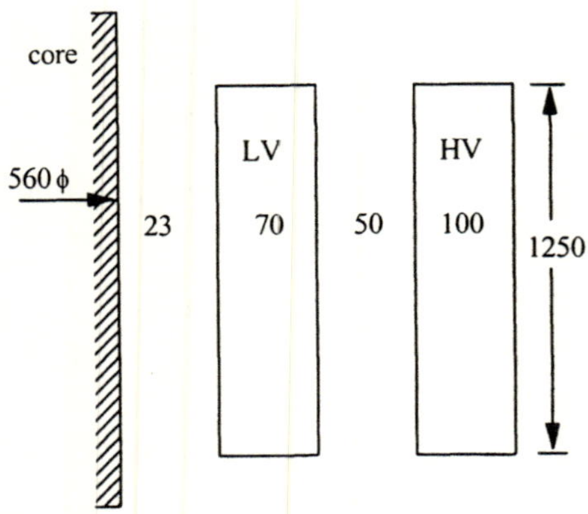

Figure 3.24 Dimensions of transformer in Example 3.6

Example 3.6

The relevant dimensions (in mm) are given in figure 3.24 for a 31.5 MVA, 132/33 kV, 50 Hz, YNyn transformer. The volts/turn is 83.93. Calculate the zero-sequence reactance of LV and HV windings and the parameters of the zero-sequence network.

Solution:

The value of positive-sequence leakage reactance can be calculated in line with previous examples as 12.16%. Let the inner 33 kV winding and outer 132 kV winding be denoted by numbers 1 and 2 respectively.

$$X_p = (X_p)_{12} = (X_p)_{21} = 12.16\%$$

The open circuit zero-sequence reactance of LV and HV windings can be calculated by the procedure given in Section 3.7.1 (and equation 3.66) with the average LV to tank and HV to tank distances of 400 mm and 250 mm (for this transformer) respectively.

$$T_1 = 7 \text{ cm}, \ T_g = 40 \text{ cm}, \ T_2 \approx 0, \ H_w = 125 \text{ cm}, \ K_R = 0.88, \ H_{eq} = 142 \text{ cm}$$

$$\text{HV current} = 137.78 \text{ A}$$

$$\text{HV turns} = \frac{132 \times 10^3 / \sqrt{3}}{83.93} = 908$$

$$(X_z)_{1\_oc} = \frac{2.48 \times 10^{-5} \times 50 \times (137.78 \times 908)}{142 \times 83.93} \times \left(\frac{1}{3} \times 7.0 \times 67.6 + 40 \times 114.6 \right)$$

$$= 61.73\%$$

Similarly,

$$(X_z)_{2\_oc} = \frac{2.48\times10^{-5}\times50\times(137.78\times908)}{137.2\times83.93}\times\left(\frac{1}{3}\times10\times94.6+25\times129.6\right)$$

$$= 47.89\%$$

The zero-sequence reactance of the inner LV winding with the short-circuited outer HV winding is same as the positive-sequence leakage reactance (if the reactance contributed by neutral bar is neglected) ,

$$(X_z)_{12\_sc} = X_p = 12.16\%$$

Zero-sequence reactance of HV winding with short-circuited LV winding is given as per equation 3.79,

$$(X_z)_{21\_sc} = X_p \times \frac{(X_z)_{2\_oc}}{(X_z)_{1\_oc}} = 12.16\times\frac{47.89}{61.73} = 9.44\%$$

The zero-sequence network [18] of this two winding transformer is shown in figure 3.25 which satisfies all the calculated zero-sequence reactance values, viz.

$(X_z)_{1\_oc}$, $(X_z)_{2\_oc}$, $(X_z)_{12\_sc}$ and $(X_z)_{21\_sc}$

For example, the zero-sequence reactance with HV as the excited winding and LV as the short-circuited winding is

$$(X_z)_{21\_sc} = -0.835 + (13.0 // 48.73) = -0.835 + \frac{13.0\times48.73}{13.0+48.73} = 9.43\%$$

which matches the value calculated previously.

In a similar way, the zero-sequence reactance of a three-winding three-phase three-limb transformer can be estimated as shown by the following example.

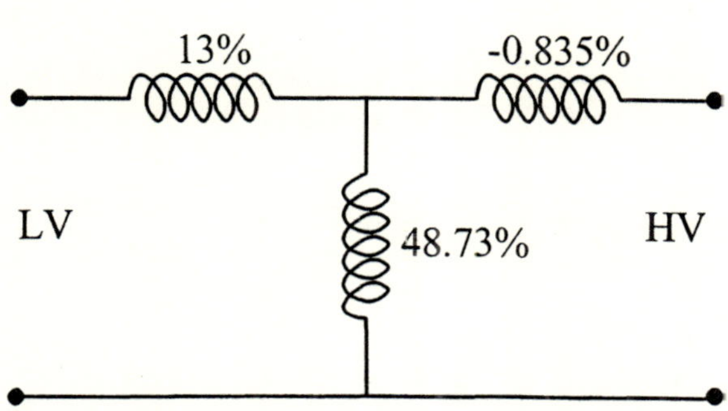

Figure 3.25 Zero-sequence network of two-winding transformer

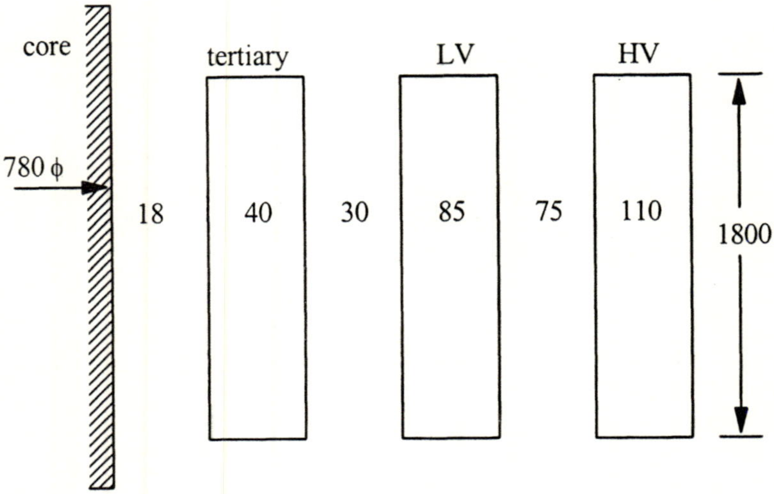

Figure 3.26 Dimensions of the 100 MVA transformer in Example 3.7

Example 3.7

The relevant dimensions (in mm) of 100 MVA, 220/66/11 kV, 50 Hz, YNynd1 transformer are given in figure 3.26. The volts/turn is 160. Calculate the zero-sequence reactance of HV winding with LV short-circuited (and tertiary delta closed).

Solution:

Let tertiary (11 kV), LV (66 kV) and HV (220 kV) be denoted by numbers 1, 2, and 3 respectively. The values of positive-sequence leakage reactances for three pairs of windings are calculated as:

$$(X_p)_{12} = 6.0\%, \quad (X_p)_{23} = 14.6\%, \quad (X_p)_{13} = 22.64\%$$

Open circuit zero-sequence reactances of tertiary, LV and HV windings with average distance between HV and tank of 250 mm are calculated as

$$(X_z)_{1\_oc} = 64.81\%$$

$$(X_z)_{2\_oc} = 57.71\%$$

$$(X_z)_{3\_oc} = 40.93\%$$

The various zero-sequence reactances between pairs of windings can be calculated as per equation 3.79 as

1. Zero-sequence voltage applied to HV, with LV open-circuited and tertiary delta closed

$$(X_z)_{31\_sc} = \frac{(X_z)_{3\_oc}}{(X_z)_{1\_oc}} \times (X_p)_{31} = \frac{40.93}{64.81} \times 22.64 = 14.3\%$$

2. Zero-sequence voltage applied to LV, with HV open-circuited and tertiary delta closed

$$(X_z)_{21\_sc} = \frac{(X_z)_{2\_oc}}{(X_z)_{1\_oc}} \times (X_p)_{21} = \frac{57.71}{64.81} \times 6.0 = 5.34\%$$

3. Zero-sequence voltage applied to HV, tertiary delta open and LV short-circuited

$$(X_z)_{32\_sc} = \frac{(X_z)_{3\_oc}}{(X_z)_{2\_oc}} \times (X_p)_{32} = \frac{40.93}{57.71} \times 14.6 = 10.36\%$$

Individual zero-sequence reactance of windings can be calculated by using equations given in Section 3.5 as

$$(X_z)_1 = \frac{1}{2} \times \left[(X_z)_{21\_sc} + (X_z)_{31\_sc} - (X_z)_{32\_sc}\right] = 4.64\%$$

$$(X_z)_2 = \frac{1}{2} \times \left[(X_z)_{21\_sc} + (X_z)_{32\_sc} - (X_z)_{31\_sc}\right] = 0.7\%$$

$$(X_z)_3 = \frac{1}{2} \times \left[(X_z)_{31\_sc} + (X_z)_{32\_sc} - (X_z)_{21\_sc}\right] = 9.66\%$$

The zero-sequence star equivalent network of the three-winding transformer is shown in figure 3.27. The zero-sequence reactance of HV with LV short-circuited (and tertiary delta closed) can be found as

$$(X_z)_{3\_21} = 9.66 + (0.7 // 4.64) = 10.27\%$$

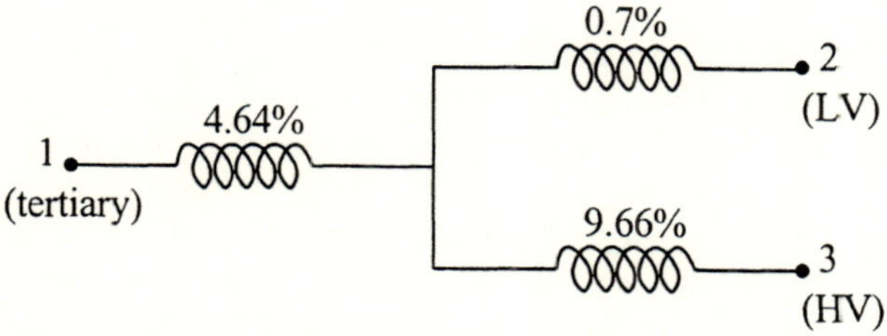

Figure 3.27 Zero-sequence star equivalent circuit of three-winding transformer

Here, the system of four windings (tertiary, LV, HV and tank) is converted into an equivalent three-winding system by taking into account the effect of the tank while calculating short circuit zero-sequence reactance between any two of the three windings. Hence, the calculation of $(X_z)_{3\_21}$ by the star equivalent circuit of figure 3.27 is an approximate way. Actually the problem needs to be solved with an additional equivalent delta winding.

Let us now calculate the reactance by more accurate and logical method in which the tank is treated as the 4th winding. The accurate value of zero-sequence reactance can be calculated by the procedure given in Section 3.2 (i.e., reactive kVA approach). Let I_1, I_2 and I_4 be the currents flowing through the tertiary, LV and equivalent delta (tank) windings respectively. The current flowing through the HV winding (I_3) is 1 per-unit and we know that

$$I_3 = 1 = I_1 + I_2 + I_4$$

The expression for Q is

$$Q = -\frac{1}{2}\begin{pmatrix} 2(X_p)_{12}(-I_1)(-I_2) + 2(X_p)_{13}(-I_1)I_3 + 2(X_p)_{14}(-I_1)(-I_4) \\ +2(X_p)_{23}(-I_2)I_3 + 2(X_p)_{24}(-I_2)(-I_4) + 2(X_p)_{34}I_3(-I_4) \end{pmatrix}$$

The reactance between the tank and any other winding has already been calculated (e.g., $(X_p)_{14} = (X_z)_{1\_oc} = 64.81\,\%$). By putting the values of all reactances and using $I_1 = 1 - I_2 - I_4$, the above expression becomes

$$Q = \left(22.64 - 14.04\,I_2 - 46.52\,I_4 + 13.1\,I_2 I_4 + 6 I_2^2 + 64.81\,I_4^2\right)$$

Differentiating the above expression with respect to I_2 and I_4, and equating it to zero, we get two simultaneous equations. These two equations are solved to get

$$I_2 = 0.8747 \text{ and } I_4 = 0.2705$$

By putting the values of I_2 and I_4 in the expression for Q, we get directly the zero-sequence reactance of HV with LV short-circuited (and tertiary delta closed) as $(X_z)_{3\_21} = 10.21\%$, which is close to that calculated by the approximate method using the star equivalent circuit (i.e., 10.27%).

The methods described for calculation of zero-sequence reactance can give reasonably accurate results and should be refined by empirical correction factors based on results of tests conducted on a number of transformers. For more accurate calculations, numerical methods like Finite Element Method can be used [21, 22] in which the effect of the level of tank saturation on the zero-sequence reactance can be exactly simulated. The tank material can be modeled in FEM formulation by defining its conductivity and non-linear B-H characteristics. In power transformers, tank shunts of CRGO material are commonly put on the inner tank wall to reduce stray losses due to the leakage field. These shunts provide a low reluctance path to the zero-sequence flux reducing the effect of the tank. As explained earlier, the tank acts as an equivalent delta winding reducing the zero-sequence reactance. The placement of magnetic shunts on the tank wall has the effect of increasing the zero-sequence reactance. In such cases, FEM analysis is essential for correct estimation of the reactance.

B. Three-phase five-limb and single-phase three-limb transformers

Since short circuit reactances are much smaller than open circuit reactances, zero-sequence voltage applied to circulate rated currents is usually much smaller than the rated voltage. Hence, in case of three-phase five-limb cores, yokes and end limbs (which provide the path for zero-sequence flux) do not saturate. Therefore, there will not be any currents in tank since all the flux is contained within the core. Hence, the short circuit zero-sequence reactance is equal to the positive-sequence leakage reactance in three-phase five-limb transformers.

In a single-phase three-limb transformer, yokes and end limbs provide a path for zero-sequence flux and hence the zero-sequence leakage reactance is equal to the positive-sequence leakage reactance.

The same inferences can be arrived at by analysis of the zero-sequence network (see figure 3.25). The shunt branch reactance is very high (~ infinity) due to low reluctance path of end limbs in single-phase three-limb and three-phase five-limb transformers (the shunt branch now represents the equivalent magnetizing reactance in place of the reactance representing equivalent short-circuited tank winding). This makes the zero-sequence reactance equal to the corresponding positive-sequence short circuit (leakage) reactance with HV excited and LV short-circuited (or with LV excited and HV short-circuited).

3.8 Stabilizing Tertiary Winding

As mentioned earlier, in addition to primary and secondary windings (both of which are star or auto-star connected), transformers are sometimes provided with tertiary winding. It can be used for the following purposes:

1) Static capacitors or synchronous condensers can be connected to the tertiary winding for the injection of reactive power into a system for maintaining voltage within certain limits.

2) Auxiliary equipment in a substation can be supplied at a voltage which is different and lower than that of primary and secondary windings.

3) Three windings may be required for interconnecting three transmission lines at three different voltages.

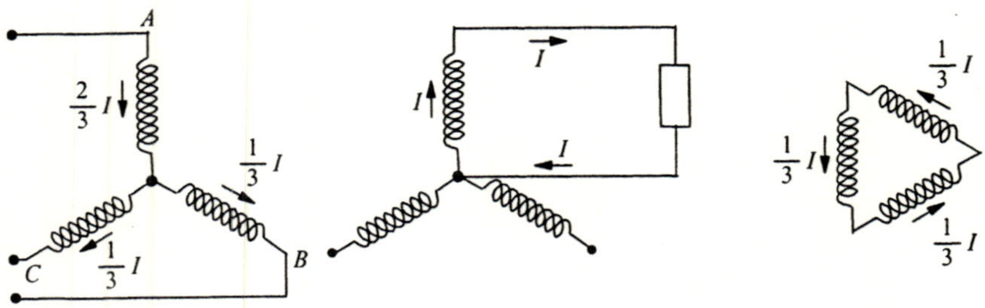

Figure 3.28 Unbalanced load

In all the above cases, the tertiary winding is loaded. In some applications, delta connected tertiary winding is not loaded, in which case it is called as stabilizing winding. The functions of the stabilizing winding are:

1) Third harmonic magnetizing currents flow in closed delta, making induced voltages and core flux almost sinusoidal (refer to Section 2.8 of Chapter 2).

2) It stabilizes the neutral point; the zero-sequence impedance is lower and an unbalanced load can be taken without undue unbalancing of phase voltages. When an unloaded tertiary winding is provided to stabilize the neutral under asymmetrical loading conditions, the currents flow in such a way that there is ampere-turn balance between the three windings as shown for the case of single-phase loading in figure 3.28. The load in each phase of tertiary is equal to one-third of the single-phase (unbalanced) load. Hence, the rating of tertiary windings is usually one-third that of the main windings.

3) It can prevent interference in telephone lines caused by third harmonic currents and voltages in the lines and earth circuits.

In the previous section, we have seen that the zero-sequence characteristics of three-phase three-limb and three-phase five-limb / bank of single-phase transformers are different. The magnetic circuit of three-phase three-limb type can be considered as open circuit offering a high reluctance to the zero-sequence flux, and hence a lower zero-sequence reactance is obtained. On the other hand, for three-phase five-limb / bank of single-phase transformers, the magnetic circuit can be visualized as a closed one giving a very high value of the zero-sequence reactance. If a delta connected stabilizing winding is added to these three types of transformers (three-phase three-limb transformers, a bank of single-phase three-limb transformers, and three-phase five-limb transformers), the difference between zero-sequence reactance characteristics of three-phase three-limb transformers and that of the other two types diminishes. As long as one delta connected winding is present, it makes a very little difference whether it is an effectively open or closed magnetic circuit from the zero-sequence reactance point of view.

A frequently asked question is whether stabilizing windings can be dispensed with for three-phase three-limb star-star (or auto-star) connected transformers with grounded neutrals. This is because as seen in the previous section, the high reluctance path and presence of tank (as an equivalent one turn delta connected winding) together give a lower zero-sequence reactance compared to the positive-sequence reactance, and thus to some extent the effect of stabilizing winding is achieved. Whether or not the stabilizing winding can be omitted depends mainly on whether zero-sequence and third harmonic characteristics are compatible with the system into which the transformer is going to be installed. If these two characteristics are not adversely affected in absence of the stabilizing winding, it may be omitted [23]. Developments in power systems have led to more balanced loads. Also if the telephone-interference problem due to harmonic currents is within limits and if the zero-sequence currents during asymmetrical fault conditions are large enough to be easily detected, the provision of stabilizing winding in three-phase three-limb transformers should be critically reviewed. This is because, since the stabilizing winding is generally unloaded, its conductor dimensions tend to be designed smaller. Such a winding becomes quite weak and vulnerable under asymmetrical fault conditions, which is a subject of discussion in Chapter 6.

Consequences of omitting the stabilizing winding in bank of single-phase transformers / three-phase five-limb transformers are significant. The zero-sequence reactance will be higher, and if the disadvantages of high zero-sequence reactance are not tolerable, the stabilizing winding cannot be omitted. For very large transformers, a low voltage stabilizing winding with its terminal brought out helps in carrying out tests such as no-load loss test at the manufacturer's works. In the absence of this winding, it may not be possible to do the no-load loss test if the manufacturer does not have a high voltage source or a suitable step-up transformer. Alternatively, a star connected auxiliary winding can be provided for testing purposes, whose terminals can be buried inside the tank after testing.

References

1. Blume, L. F., Boyajian, A., Camilli, G., Lennox, T. C., Minneci, S., and Montsinger, V. M. *Transformer engineering*, John Wiley and Sons, New York, and Chapman and Hall, London, 1951.
2. Hayt, W. H. *Engineering electromagnetics*, McGraw-Hill Book Company, Singapore, 1989, pp. 298-301.
3. Garin, A. N. and Paluev, K. K. Transformer circuit impedance calculations, *AIEE Transactions - Electrical Engineering*, June 1936, pp. 717-729.
4. Waters, M. *The short circuit strength of power transformers*, Macdonald, London, 1966, pp. 24-25, p. 53.

5. Boyajian, A. Leakage reactance of irregular distributions of transformer windings by method of double Fourier series, *AIEE Transactions - Power Apparatus and Systems,* Vol. 73, Pt. III-B, 1954, pp. 1078-1086.
6. Sollergren, B. Calculation of short circuit forces in transformers, *Electra*, Report no. 67, 1979, pp. 29-75.
7. Rabins, L. Transformer reactance calculations with digital computers, *AIEE Transactions - Communications and Electronics,* Vol. 75, Pt. I, 1956, pp. 261-267.
8. Silvester, P. P. and Ferrari, R. L. *Finite elements for electrical engineers*, Cambridge University Press, New York, 1990.
9. Andersen, O. W. Transformer leakage flux program based on Finite Element Method, *IEEE Transactions on Power Apparatus and Systems*, Vol. PAS-92, 1973, pp. 682-689.
10. Rothe, P. S. *An introduction to power system analysis*, John Wiley and Sons, New York, 1953, pp. 45-50.
11. Boyajian, A. Theory of three-circuit transformers, *AIEE Transactions*, February 1924, pp. 508-528.
12. Starr, F. M. An equivalent circuit for the four-winding transformer, *General Electric Review*, Vol. 36, No. 3, March 1933, pp. 150-152.
13. Aicher, L. C. A useful equivalent circuit for a five-winding transformer, *AIEE Transactions – Electrical Engineering*, Vol. 62, February 1943, pp. 66-70.
14. Schaefer, J. *Rectifier circuits: theory and design*, John Wiley and Sons, New York, 1965, pp. 12-19.
15. Christoffel, M. Zero-sequence reactances of transformers and reactors, *The Brown Boveri Review*, Vol. 52, No. 11/12, November/December 1965, pp. 837-842.
16. Clarke E. *Circuit analysis of AC power systems*, Vol. II, John Wiley and Sons, New York, Chapman and Hall, London, 1957, p. 153.
17. Jha, S. K. *Evaluation and mitigation of stray losses due to high current leads in transformers*, M. Tech Dissertation, Department of Electrical Engineering, IIT-Bombay, India, 1995.
18. Garin, A. N. Zero phase sequence characteristics of transformers, *General Electric Review*, Vol. 43, No. 3, March 1940, pp. 131-136.
19. Copper Development Association, *Copper for busbars*, Publication No. 22, January 1996, p. 53.
20. Schurig, O. R. Engineering calculation of inductance and reactance for rectangular bar conductors, *General Electric Review*, Vol. 36, No. 5, May 1933, pp. 228-231.
21. Allcock, R., Holland, S., and Haydock, L. Calculation of zero phase sequence impedance for power transformers using numerical methods, *IEEE Transactions on Magnetics*, Vol. 31, No. 3, May 1995, pp. 2048-2051.

22. Ngnegueu, T., Mailhot, M., Munar, A., and Sacotte, M. Zero phase sequence impedance and tank heating model for three-phase three-leg core type power transformers coupling magnetic field and electric circuit equations in a Finite Element software, *IEEE Transactions on Magnetics*, Vol. 31, No. 3, May 1995, pp. 2068-2071.
23. Cogbill, B. A. Are stabilizing windings necessary in all Y-connected transformers, *AIEE Transactions – Power Apparatus and Systems*, Vol. 78, Pt. 3, 1959, pp. 963-970.

4

Eddy Currents and Winding Stray Losses

The load loss of a transformer consists of losses due to ohmic resistance of windings (I^2R losses) and some additional losses. These additional losses are generally known as stray losses, which occur due to leakage field of windings and field of high current carrying leads / bus-bars. The stray losses in the windings are further classified as eddy loss and circulating current loss. The other stray losses occur in structural steel parts. There is always some amount of leakage field in all types of transformers, and in large power transformers (limited in size due to transport and space restrictions) the stray field strength increases with growing rating much faster than in smaller transformers. The stray flux impinging on conducting parts (winding conductors and structural components) gives rise to eddy currents in them. The stray losses in windings can be substantially high in large transformers if conductor dimensions and transposition methods are not chosen properly.

Today's designer faces challenges like higher loss capitalization and optimum performance requirements. In addition, there could be constraints on dimensions and weight of the transformer which is to be designed. If the designer lowers current density to reduce the DC resistance copper loss (I^2R loss), the eddy loss in windings increases due to increase in conductor dimensions. Hence, the winding conductor is usually subdivided with a proper transposition method to minimize the stray losses in windings.

In order to accurately estimate and control the stray losses in windings and structural parts, in-depth understanding of the fundamentals of eddy currents starting from basics of electromagnetic fields is desirable. The fundamentals are described in first few sections of this chapter. The eddy loss and circulating current loss in windings are analyzed in subsequent sections. Methods for evaluation and control of these two losses are also described. Remaining components of stray losses, mostly the losses in structural components, are dealt with in Chapter 5.

4.1 Field Equations

The differential forms of Maxwell's equations, valid for static as well as time dependent fields and also valid for free space as well as material bodies are:

$$\nabla \times \mathbf{E} = -\frac{\partial \mathbf{B}}{\partial t} \tag{4.1}$$

$$\nabla \times \mathbf{H} = \mathbf{J} + \frac{\partial \mathbf{D}}{\partial t} \tag{4.2}$$

$$\nabla \cdot \mathbf{B} = 0 \tag{4.3}$$

$$\nabla \cdot \mathbf{D} = \rho \tag{4.4}$$

where **H** = magnetic field strength (A/m)
E = electric field strength (V/m)
B = flux density (wb/m^2)
J = current density (A/m^2)
D = electric flux density (C/m^2)
ρ = volume charge density (C/m^3)

There are three constitutive relations,

$$\mathbf{J} = \sigma\,\mathbf{E} \tag{4.5}$$

$$\mathbf{B} = \mu\,\mathbf{H} \tag{4.6}$$

$$\mathbf{D} = \varepsilon\,\mathbf{E} \tag{4.7}$$

where μ = permeability of material (henrys/m)
ε = permittivity of material (farads/m)
σ = conductivity (mhos/m)

The ratio of the conduction current density (**J**) to the displacement current density ($\partial \mathbf{D}/\partial t$) is given by the ratio $\sigma/(j\omega\varepsilon)$, which is very high even for a poor metallic conductor at very high frequencies (where ω is frequency in rad/sec). Since our analysis is for the (smaller) power frequency, the displacement current density is neglected for the analysis of eddy currents in conducting parts in transformers (copper, aluminum, steel, etc.). Hence, equation 4.2 gets simplified to

$$\nabla \times \mathbf{H} = \mathbf{J} \tag{4.8}$$

The principle of conservation of charge gives the point form of the continuity equation,

$$\nabla \cdot \mathbf{J} = -\frac{\partial \rho}{\partial t} \tag{4.9}$$

In the absence of free electric charges in the present analysis of eddy currents in a conductor we get

$$\nabla \cdot \mathbf{J} = 0 \tag{4.10}$$

To get the solution, the first-order differential equations 4.1 and 4.8 involving both **H** and **E** are combined to give a second-order equation in **H** or **E** as follows.

Taking curl of both sides of equation 4.8 and using equation 4.5 we get

$$\nabla \times \nabla \times \mathbf{H} = \nabla \times \sigma \mathbf{E}$$

For a constant value of conductivity (σ), using vector algebra the equation can be simplified as

$$\nabla (\nabla \cdot \mathbf{H}) - \nabla^2 \mathbf{H} = \sigma \nabla \times \mathbf{E} \tag{4.11}$$

Using equation 4.6, for linear magnetic characteristics (constant μ) equation 4.3 can be rewritten as

$$\nabla \cdot \mathbf{B} = \nabla \cdot \mu \mathbf{H} = \mu \nabla \cdot \mathbf{H} = 0 \tag{4.12}$$

which gives

$$\nabla \cdot \mathbf{H} = 0 \tag{4.13}$$

Using equations 4.1 and 4.13, equation 4.11 gets simplified to

$$\nabla^2 \mathbf{H} = \sigma \frac{\partial \mathbf{B}}{\partial t} \tag{4.14}$$

or

$$\nabla^2 \mathbf{H} - \mu \sigma \frac{\partial \mathbf{H}}{\partial t} = 0 \tag{4.15}$$

Equation 4.15 is a well-known diffusion equation. Now, in the frequency domain, equation 4.1 can be written as follows:

$$\nabla \times \mathbf{E} = -j\omega \mathbf{B} \tag{4.16}$$

In above equation, term $j\omega$ appears because the partial derivative of a sinusoidal field quantity with respect to time is equivalent to multiplying the corresponding phasor by $j\omega$. Using equation 4.6 we get

$$\nabla \times \mathbf{E} = -j\omega\mu\mathbf{H} \tag{4.17}$$

Taking curl of both sides of the equation,

$$\nabla \times \nabla \times \mathbf{E} = -j\omega\mu\nabla \times \mathbf{H} \tag{4.18}$$

Using equation 4.8 we get

$$\nabla \times \nabla \times \mathbf{E} = -j\omega\mu\mathbf{J} \tag{4.19}$$

Following the steps similar to those used for arriving at the diffusion equation 4.15 and using the fact that $\nabla \cdot \mathbf{E} = (1/\varepsilon)\nabla \cdot \mathbf{D} = \rho/\varepsilon = 0$ (since no free electric charges are present) we get

$$\nabla \times \nabla \times \mathbf{E} = -\nabla^2\mathbf{E} = -j\omega\mu\mathbf{J} \tag{4.20}$$

Substituting the value of **J** from equation 4.5,

$$\nabla^2\mathbf{E} = j\omega\mu\sigma\mathbf{E} \tag{4.21}$$

Now, let us assume that the vector field **E** has component only along the x axis.

$$\therefore \nabla^2 E_x = j\omega\mu\sigma E_x \tag{4.22}$$

The expansion of the operator ∇ leads to the second-order partial differential equation,

$$\frac{\partial^2 E_x}{\partial x^2} + \frac{\partial^2 E_x}{\partial y^2} + \frac{\partial^2 E_x}{\partial z^2} = j\omega\mu\sigma E_x \tag{4.23}$$

Suppose, if we further assume that E_x is a function of z only (does not vary with x and y), then equation 4.23 reduces to the ordinary differential equation

$$\frac{d^2 E_x}{dz^2} = j\omega\mu\sigma E_x \tag{4.24}$$

We can write the solution of equation 4.24 as

$$E_x = E_{xp}e^{-\gamma z} + E_{xp}e^{+\gamma z} \tag{4.25}$$

where E_{xp} is the amplitude factor and γ is the propagation constant, which can be given in terms of the attenuation constant α and phase constant β as

$$\gamma = \alpha + j\beta \tag{4.26}$$

Substituting the value of E_x from equation 4.25 in equation 4.24 we get

$$E_{xp}\gamma^2 e^{-\gamma z} + E_{xp}\gamma^2 e^{+\gamma z} = j\omega\mu\sigma\,(E_{xp}e^{-\gamma z} + E_{xp}e^{+\gamma z}) \tag{4.27}$$

which gives

$$\gamma^2 = j\omega\mu\sigma \tag{4.28}$$

$$\therefore \gamma = \pm\sqrt{j\omega\mu\sigma} \tag{4.29}$$

If the field E_x is incident on a surface of a conductor at $z = 0$ and gets attenuated inside the conductor ($z > 0$), then only the plus sign has to be taken for γ (which is consistent for the case considered).

$$\therefore \gamma = \sqrt{j\omega\mu\sigma} = \sqrt{\omega\mu\sigma} \times \sqrt{1\angle 90^\circ} = \sqrt{\omega\mu\sigma} \times 1\angle 45^\circ \tag{4.30}$$

$$\therefore \gamma = (1+j)\sqrt{\frac{\omega\mu\sigma}{2}} \tag{4.31}$$

Substituting $\omega = 2\pi f$ we get

$$\gamma = (1+j)\sqrt{\pi f\mu\sigma} \tag{4.32}$$

Hence,

$$\alpha = \beta = \sqrt{\pi f\mu\sigma} \tag{4.33}$$

The electric field intensity (having a component only along the x axis and traveling / penetrating inside the conductor in $+z$ direction) expressed in the complex exponential notation in equation 4.25 becomes

$$E_x = E_{xp}\, e^{-\gamma z} \tag{4.34}$$

which in time domain can be written as

$$E_x = E_{xp}\, e^{-\alpha z}\cos(\omega t - \beta z) \tag{4.35}$$

Substituting the values of α and β from equation 4.33 we get

$$E_x = E_{xp}e^{-z\sqrt{\pi f\mu\sigma}}\cos(\omega t - z\sqrt{\pi f\mu\sigma}) \tag{4.36}$$

The conductor surface is represented by $z = 0$. Let $z > 0$ and $z < 0$ represent the regions corresponding to the conductor and perfect loss-free dielectric medium respectively. Thus, the source field at the surface which establishes fields within the conductor is given by

$$(E_x)_{z=0} = E_{xp} \cos \omega t$$

Making use of equation 4.5, which says that the current density within a conductor is directly related to the electrical field intensity, we can write

$$J_x = \sigma E_x = \sigma E_{xp} e^{-z\sqrt{\pi f \mu \sigma}} \cos(\omega t - z\sqrt{\pi f \mu \sigma}) \tag{4.37}$$

Equations 4.36 and 4.37 tell us that away from the source at the surface and with penetration into the conductor there is an exponential decrease in the electric field intensity and (conduction) current density. At a distance of penetration $z = 1/\sqrt{\pi f \mu \sigma}$, the exponential factor becomes $e^{-1}(=0.368)$, indicating that the value of field (at this distance) reduces to 36.8% of that at the surface. This distance is called as the skin depth or depth of penetration δ,

$$\delta = \frac{1}{\sqrt{\pi f \mu \sigma}} = \sqrt{\frac{2}{\omega \mu \sigma}} \tag{4.38}$$

All the fields at the surface of a good conductor decay rapidly as they penetrate few skin depths into the conductor. Comparing equations 4.33 and 4.38, we get the relationship,

$$\delta = \frac{1}{\alpha} = \frac{1}{\beta} \quad \text{and} \quad \gamma = \alpha + j\beta = \frac{1}{\delta} + \frac{j}{\delta} = (1+j) \times \frac{1}{\delta} \tag{4.39}$$

The depth of penetration or skin depth is a very important parameter in describing the behavior of a conductor subjected to electromagnetic fields. The conductivity of copper conductor at 75°C (temperature at which load loss of a transformer is usually calculated and guaranteed) is 4.74×10^7 mhos/m. Copper being a non-magnetic material, its relative permeability is 1. Hence, the depth of penetration of copper at the power frequency of 50 Hz is

$$\delta_{Cu} = \frac{1}{\sqrt{\pi \times 50 \times \mu_0 \mu_r \times \sigma}} = \frac{1}{\sqrt{\pi \times 50 \times 4\pi \times 10^{-7} \times 1 \times 4.74 \times 10^7}} = 0.0103\,\text{m}$$

or 10.3 mm. The corresponding value at 60 Hz is 9.4 mm. For aluminum, whose conductivity is approximately 61% of that of copper, the skin depth at 50 Hz is 13.2 mm. Most of the structural elements inside a transformer are made of either mild steel or stainless steel material. For a typical grade of mild steel (MS) material with relative permeability of 100 (assuming that it is saturated) and conductivity of 7×10^6 mho/m, the skin depth is $\delta_{MS} = 2.69$ mm at 50 Hz. A non-magnetic stainless steel is commonly used for structural components in the vicinity of the field due to high currents. For a typical grade of stainless steel (SS) material with relative permeability of 1 (non-magnetic) and conductivity of 1.136×10^6 mho/m, the skin depth is $\delta_{SS} = 66.78$ mm at 50 Hz.

4.2 Poynting Vector

Poynting's theorem is the expression of the law of conservation of energy applied to electromagnetic fields. When the displacement current is neglected, as in the previous section, Poynting's theorem can be mathematically expressed as [1, 2]

$$-\oint_s (\mathbf{E}\times\mathbf{H})\cdot\mathbf{n}\,ds = \int_v \mathbf{H}\cdot\frac{\partial\mathbf{B}}{\partial t}dv + \int_v \mathbf{J}\cdot\mathbf{E}\,dv \tag{4.40}$$

where v is the volume enclosed by the surface s and $\mathbf{n}$ is the unit vector normal to the surface directed outwards. Using equation 4.5, the above equation can be modified as,

$$-\oint_s (\mathbf{E}\times\mathbf{H})\cdot\mathbf{n}\,ds = \int_v \mathbf{H}\cdot\frac{\partial\mathbf{B}}{\partial t}dv + \frac{1}{\sigma}\int_v |\mathbf{J}|^2\,dv \tag{4.41}$$

This is a simpler form of Poynting's theorem which states that the net inflow of power is equal to the sum of the power absorbed by the magnetic field and the ohmic loss. The Poynting vector is given by the vector product,

$$\mathbf{P} = \mathbf{E}\times\mathbf{H} \tag{4.42}$$

which expresses the instantaneous density of power flow at a point.

Now, with $\mathbf{E}$ having only the x component which varies as a function of z only, equation 4.17 becomes

$$\frac{\partial E_x}{\partial z} = -j\omega\mu H_y, \qquad H_x = 0, \quad H_z = 0 \tag{4.43}$$

Substituting the value of E_x from equation 4.34 and rearranging we get

$$H_y = \frac{1}{-j\omega\mu}\times\frac{\partial}{\partial z}(E_{xp}e^{-\gamma z}) = \frac{\gamma E_x}{j\omega\mu} \tag{4.44}$$

The ratio of E_x to H_y is defined as the intrinsic impedance,

$$\eta = \frac{E_x}{H_y} = \frac{j\omega\mu}{\gamma} \tag{4.45}$$

Substituting the value of γ from equation 4.30 we get

$$\eta = \frac{j\omega\mu}{\sqrt{j\omega\mu\sigma}} = \sqrt{\frac{j\omega\mu}{\sigma}} \tag{4.46}$$

Using equation 4.38, the above equation can be rewritten as

$$\eta = \sqrt{j}\left[\frac{1}{\delta}\sqrt{\frac{2}{\sigma}}\right]\frac{1}{\sqrt{\sigma}} = \frac{\sqrt{1\angle 90^\circ}\times\sqrt{2}}{\sigma\,\delta} = \frac{\sqrt{2}\angle 45^\circ}{\sigma\,\delta} \tag{4.47}$$

Now, equation 4.36 can be rewritten in terms of skin depth as

$$E_x = E_{xp}e^{-z/\delta}\cos(\omega t - z/\delta) \tag{4.48}$$

Using equations 4.45 and 4.47, H_y can be expressed as

$$H_y = \frac{\sigma\delta}{\sqrt{2}}E_{xp}e^{-z/\delta}\cos(\omega t - \frac{z}{\delta} - \frac{\pi}{4}) \tag{4.49}$$

Since **E** is in the *x* direction and **H** is in the *y* direction, the Poynting vector, which is a cross product of **E** and **H** as per equation 4.42, is in the z direction.

$$\therefore P_z = E_x\,H_y = \frac{\sigma\delta}{\sqrt{2}}\,E_{xp}^2\,e^{-2z/\delta}\cos(\omega t - z/\delta)\,\cos(\omega t - \frac{z}{\delta} - \frac{\pi}{4}) \tag{4.50}$$

Using the identity $\cos A\cos B = 1/2\left[\cos(A+B) + \cos(A-B)\right]$, the above equation simplifies to

$$P_z = \frac{\sigma\delta}{\sqrt{2}}E_{xp}^2 e^{-2z/\delta}\left[\frac{1}{2}\left\{\cos(2\omega t - \frac{2z}{\delta} - \frac{\pi}{4}) + \cos\frac{\pi}{4}\right\}\right] \tag{4.51}$$

The time average Poynting vector is then given by

$$(P_z)_{avg} = \frac{1}{2}\frac{\sigma\delta}{\sqrt{2}}E_{xp}^2 e^{-2z/\delta}\cos\frac{\pi}{4} = \frac{\sigma\delta}{4}E_{xp}^2 e^{-2z/\delta} \tag{4.52}$$

Thus, it can be observed that at a distance of one skin depth ($z = \delta$), the power density is only 0.135 ($=e^{-2}$) times its value that at the surface. This is very important fact for the analysis of eddy currents and losses in structural components of transformers. If the eddy losses in the tank of a transformer due to incident leakage field emanating from windings are being analyzed by using FEM analysis, then there should be at least two or three elements in one skin depth for getting accurate results.

Let us now consider a conductor with field E_{xp} and the corresponding current density J_{xp} at the surface as shown in figure 4.1. The fields have the value of 1 p.u. at the surface. The total power loss in height (length) *h* and width *b* is given by the value of power crossing the conductor surface [2] within the area ($h \times b$),

$$P=\int_S (P_z)_{avg}\,dS=\int_0^b\int_0^h\left[\frac{\sigma\delta}{4}E_{xp}^2e^{-2z/\delta}\right]_{z=0}dxdy=\frac{\sigma\delta bh}{4}E_{xp}^2=\frac{\delta bh}{4\sigma}J_{xp}^2 \tag{4.53}$$

The total current in the conductor is found out by integrating the current density over the infinite depth of the conductor. Using equations 4.34 and 4.39 we get

$$\begin{aligned}I&=\int_0^\infty\int_0^b J_x\,dy\,dz=\int_0^\infty\int_0^b\frac{E_x}{\sigma}\,dy\,dz=\int_0^\infty\int_0^b\frac{E_{xp}\,e^{-\gamma z}}{\sigma}\,dy\,dz\\&=\int_0^\infty\int_0^b J_{xp}\,e^{\frac{-(1+j)z}{\delta}}\,dy\,dz=\frac{J_{xp}b\delta}{1+j}=\frac{J_{xp}b\delta}{\sqrt{2}\angle 45^\circ}\end{aligned} \tag{4.54}$$

If this total current is assumed to be uniformly distributed in one skin depth, the uniform current density can be expressed in the time domain as

$$J_u=\frac{J_{xp}}{\sqrt{2}}\cos\left(\omega t-\frac{\pi}{4}\right) \tag{4.55}$$

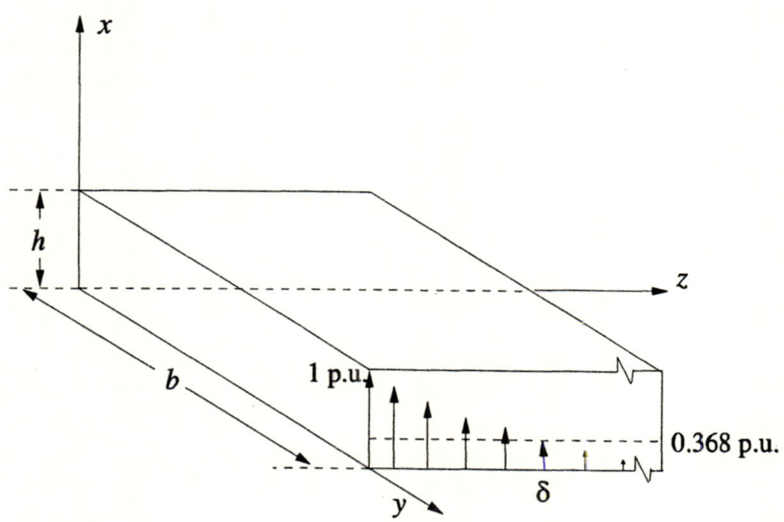

Figure 4.1 Penetration of field inside a conductor

The total ohmic power loss is given by

$$P = \frac{1}{\sigma}\int_v J_u^2\,\mathrm{d}v = \frac{1}{\sigma}\left(\frac{J_{xp}^2}{2}\cos^2\left(\omega t - \frac{\pi}{4}\right)\right)bh\delta \tag{4.56}$$

The average value of power can be found out as

$$P_{avg} = \frac{bh\delta}{2\sigma}J_{xp}^2 avg\left(\cos^2\left(\omega t - \frac{\pi}{4}\right)\right) = \frac{bh\delta}{2\sigma}J_{xp}^2 avg\left(\frac{1}{2}\left\{1 + \cos 2\left(\omega t - \frac{\pi}{4}\right)\right\}\right) \tag{4.57}$$

Since the average value of a cosine term over integral number of periods is zero we get

$$P_{avg} = \frac{\delta bh}{4\sigma}J_{xp}^2 \tag{4.58}$$

which is the same as equation 4.53. Hence, the average power loss in a conductor may be computed by assuming that the total current is uniformly distributed in one skin depth. This is a very important result, which is made use of in calculation of eddy current losses in conductors by numerical methods. When a numerical method such as Finite Element Method (FEM) is used for estimation of stray losses in the tank (made of mild steel) of a transformer, it is important to have element size less than the skin depth of the tank material as explained earlier. With the other transformer dimensions in meters, it is difficult to have very small elements inside the tank thickness. Hence, it is convenient to use analytical results to simplify the numerical analysis. For example in [3], equation 4.58 is used for estimation of tank losses by 3-D FEM analysis. The method assumes uniform current density in the skin depth allowing the use of relatively larger element sizes.

The above-mentioned problem of modeling and analysis of skin depths can also be taken care by using the concept of surface impedance. The intrinsic impedance can be rewritten from equation 4.46 as

$$\eta = R + jX = \sqrt{\frac{j\omega\mu}{\sigma}} = 1\angle 45^\circ\sqrt{\frac{\omega\mu}{\sigma}} = \sqrt{\frac{\omega\mu}{2\sigma}} + j\sqrt{\frac{\omega\mu}{2\sigma}} \tag{4.59}$$

The real part of the impedance, termed as surface resistance, is given by

$$R = \sqrt{\frac{\omega\mu}{2\sigma}} \tag{4.60}$$

After calculating the r.m.s. value of the tangential component of the magnetic field intensity (H_{rms}) at the surface of the tank or any other structural component in the transformer by either numerical or analytical method, the specific loss per unit surface area can be calculated by the expression [4, 5]

$$P = R H_{rms}^2 = \sqrt{\frac{\omega\mu}{2\sigma}}\, H_{rms}^2 \tag{4.61}$$

Thus, the total losses in the transformer tank can be determined by integrating the specific loss on its internal surface.

4.3 Eddy Current and Hysteresis Losses

All the analysis done previously assumed linear material (*B-H*) characteristics meaning that the permeability (μ) is constant. The material used for structural components in transformers is usually magnetic steel (mild steel), which is a ferromagnetic material having a much larger value of relative permeability (μ_r) as compared to the free space (for which μ_r =1). The material has non-linear *B-H* characteristics and the permeability itself is a function of *H*. Moreover, the characteristics also exhibit hysteresis property. Equation 4.6 ($\mathbf{B} = \mu \mathbf{H}$) has to be suitably modified to reflect the non-linear characteristics and hysteresis behavior. Hysteresis introduces a time phase difference between **B** and **H**; **B** lags **H** by an angle (θ) known as the hysteresis angle. One of the ways in which the characteristics can be mathematically expressed is by complex or elliptical permeability,

$$\mu_h = \mu e^{-j\theta} \tag{4.62}$$

In this formulation, where harmonics introduced by saturation are ignored, the hysteresis loop becomes an ellipse with the major axis making an angle of θ with the *H* axis as shown in figure 4.2. The significance of complex permeability is that a functional relationship between **B** and **H** is now realized in which the permeability is made independent of **H** resulting into a linear system [6]. Let us now find an expression for the eddy current and hysteresis loss for an infinite half-space shown in figure 4.3.

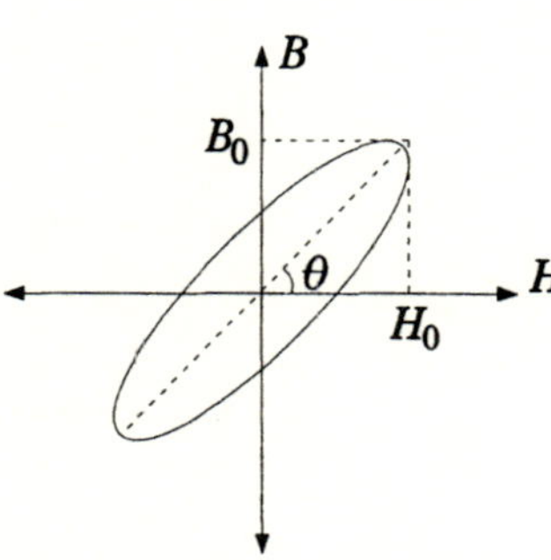

Figure 4.2 Elliptic hysteresis loop

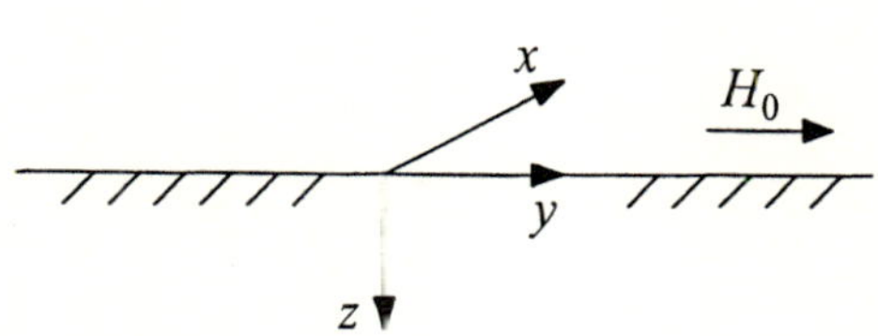

Figure 4.3 Infinite half space

The infinite half-space is an extension of the geometry shown in figure 4.1 in the sense that the region of the material under consideration extends from $-\infty$ to $+\infty$ in the x and y directions, and from 0 to ∞ in the z direction. Similar to Section 4.1, we assume that **E** and **H** vectors have components in only the x and y directions respectively, and that they are function of z only. The diffusion equation 4.15 can be rewritten for this case with the complex permeability as

$$\frac{d^2 H_y}{dz^2} = j\omega(\mu e^{-j\theta})\sigma H_y \tag{4.63}$$

A solution satisfying boundary conditions,

$$H_y = H_0 \text{ at } z = 0 \text{ and } H_y = 0 \text{ at } z = \infty \tag{4.64}$$

is given by

$$H_y = H_0 e^{-kz} \tag{4.65}$$

where constant k is

$$k = \sqrt{j\omega\sigma\mu e^{-j\theta}} = (1+j)\sqrt{\frac{\omega\mu\sigma}{2}}\; e^{-j\theta/2} = \frac{\alpha + j\beta}{\delta} \tag{4.66}$$

and

$$\alpha = \cos(\theta/2) + \sin(\theta/2) \text{ and } \beta = \cos(\theta/2) - \sin(\theta/2) \tag{4.67}$$

$$\therefore H_y = H_0\, e^{-(\alpha + j\beta)z/\delta} \tag{4.68}$$

Using equation 4.8 and the fact that $H_x = H_z = 0$ we get

$$J_x = -\frac{d}{dz}(H_y) = H_0\,\frac{\alpha + j\beta}{\delta}\, e^{-(\alpha+j\beta)z/\delta} \tag{4.69}$$

The time average density of eddy and hysteresis losses can be found by computing the real part of the complex Poynting vector evaluated at the surface [1],

$$P_{e+h} = \frac{1}{2} Re\left[\mathbf{E} \times \mathbf{H}^*\right]_{z=0} \tag{4.70}$$

Now,

$$\mathbf{E} = \left(J_x/\sigma\right)_{z=0} \bar{i} = \frac{H_0}{\sigma\delta}(\alpha + j\beta)\bar{i} \tag{4.71}$$

and

$$\mathbf{H} = H_0 \bar{j} \tag{4.72}$$

$$\therefore P_{e+h} = \frac{1}{2}\mathrm{Re}\left[\frac{H_0}{\sigma\delta}(\alpha + j\beta)H_0 \bar{k}\right] = \frac{1}{2}\left[\frac{H_0^2}{\sigma\delta}\right]\alpha \tag{4.73}$$

In the absence of hysteresis ($\theta = 0$), $\alpha = \beta = 1$ as per equation 4.67. Hence, the eddy loss per unit surface area is given by

$$P_e = \frac{1}{2}\left[\frac{H_0^2}{\sigma\delta}\right] \tag{4.74}$$

Substituting the expression for skin depth from equation 4.38 and using the r.m.s. value of magnetic field intensity (H_{rms}) at the surface we get

$$P_e = \frac{1}{2}\left[\frac{2H_{rms}^2}{\sigma}\right]\sqrt{\frac{\omega\mu\sigma}{2}} = \sqrt{\frac{\omega\mu}{2\sigma}}\; H_{rms}{}^2 \tag{4.75}$$

which is same as equation 4.61, as it should be in the absence of hysteresis (for linear *B-H* characteristics).

4.4 Effect of Saturation

In a transformer, the structural components (mostly made from magnetic steel) are subjected to the leakage field and / or high current field. The incident field gets predominantly concentrated in the skin depth (1 to 3 mm) near the surface. Hence, the structural components may be in a state of saturation depending upon the magnitude of the incident field. The eddy current losses predicted by the calculations based on a constant relative permeability are found to be smaller than the actual experimental values. Thus, although the magnetic saturation is part of same physical phenomenon as the hysteresis effect, it is considerably more important in its effect on the eddy current losses. The step function magnetization curve, as shown in figure 4.4 (a), is the simplest way of taking the saturation into account for an analytical solution of eddy current problems. It can be expressed by an equation,

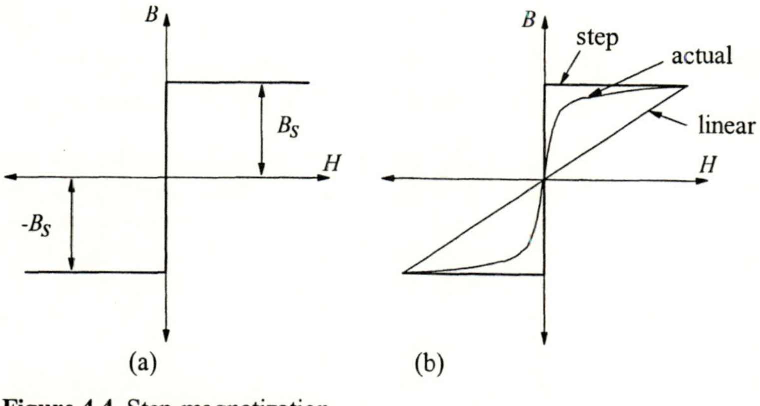

Figure 4.4 Step-magnetization

$$\mathbf{B} = (\text{sign of } \mathbf{H})\, B_s \tag{4.76}$$

where B_s is the saturation flux density. The magnetic field intensity **H** at the surface is sinusoidally varying with time $(= H_0 \sin \omega t)$. The extreme depth to which the field penetrates and beyond which there is no field is called as depth of penetration δ_s. This depth of penetration has a different connotation as compared to that with constant or linear permeability. In this case, the depth of penetration is simply the maximum depth the field will penetrate at the end of each half period. The depth of penetration for a thick plate (thickness much larger than the depth of penetration so that it can be considered as infinite half space) is given by [1, 7, 8]

$$\delta_s = \sqrt{\frac{2H_0}{\omega \sigma B_s}} \tag{4.77}$$

It can be observed that for this non-linear case with step magnetization characteristics, the linear permeability in equation 4.38 gets replaced by the ratio B_s/H_0. Further, the equation for average power per unit area can be derived as

$$P_e = \frac{8}{3\pi} \frac{H_0^2}{\sigma \delta_s} = 1.7 \frac{H_0^2}{2\sigma \delta_s} \tag{4.78}$$

Comparing this with equation 4.74, it can be noted that if we put $\mu = B_s/H_0$, δ_s will be equal to δ and in that case the loss in the saturated material is 70% higher than the loss in the material having linear *B-H* characteristics. Practically, the actual *B-H* curve is in between the linear and step characteristics, as shown in figure 4.4 (b). In [7], it is pointed out that as we penetrate inside into the material, each succeeding inner layer is magnetized by a progressively smaller number of exciting ampere-turns because of shielding effect of eddy currents in the region between the outermost surface and the layer under consideration. In step-magnetization characteristics, the flux density has the same magnitude irrespective of the magnitude of mmf. Due to this departure of the step curve response from the actual response, the value of B_s is replaced by $0.75 \times B_s$. From equations 4.77 and 4.78, it is clear that $P_e \propto \sqrt{B_s}$, and hence the constant 1.7 in equation 4.78 would reduce to $1.7 \times \sqrt{0.75}$, i.e., 1.47. As per Rosenberg's theory, the constant is 1.33 [7]. Hence, in the simplified analytical formulations, linear characteristics are assumed after taking into account the non-linearity by the linearization coefficient in the range of 1.3 to 1.5. For example, a coefficient of 1.4 is used in [9] for the calculation of losses in tank and other structural components in transformers.

After having seen in details the fundamentals of eddy currents, we will now analyze eddy current and circulating current losses in windings in the following sections. Analysis of stray losses in structural components, viz. tank, frames, flitch plates, high current terminations, etc., is covered in Chapter 5.

4.5 Eddy Loss in Transformer Winding

4.5.1 Expression for eddy loss

Theory of eddy currents explained in the previous sections will be useful while deriving the expression for the eddy loss in windings. The losses in a transformer winding due to an alternating current are usually more than that due to direct current of the same effective (r.m.s.) value. There are two different approaches of analyzing this increase in losses. In the first approach, we assume that the load current in the winding is uniformly distributed in the conductor cross section (similar to the direct current) and, in addition to the load current, there exist eddy currents which produce extra losses. Alternatively, one can calculate losses due to the combined action of the load current and eddy currents. The former method is more suitable for the estimation of eddy loss in winding conductors, in which eddy loss due to the leakage field (produced by the load current) is calculated separately and then added to the DC I^2R loss. The latter method is preferred for calculating circulating current losses, in which the resultant current in each conductor is calculated first, followed by the calculation of losses (which give the total of DC I^2R loss and circulating current loss). We will first analyze eddy losses in windings in this section; the circulating current losses are dealt with in the next section.

Consider a winding conductor, as shown in figure 4.5, which is placed in an alternating magnetic field along the *y* direction having the peak amplitude of H_0. The conductor can be assumed to be infinitely long in the *x* direction. The current density J_x and magnetic field intensity H_y are assumed as functions of *z* only. Rewriting the (diffusion) equation 4.15 for the sinusoidal variation of the field quantity and noting that the winding conductor, either copper or aluminum, has constant permeability (linear *B-H* characteristics),

$$\frac{d^2 H_y}{dz^2} = j\omega\mu\sigma H_y \tag{4.79}$$

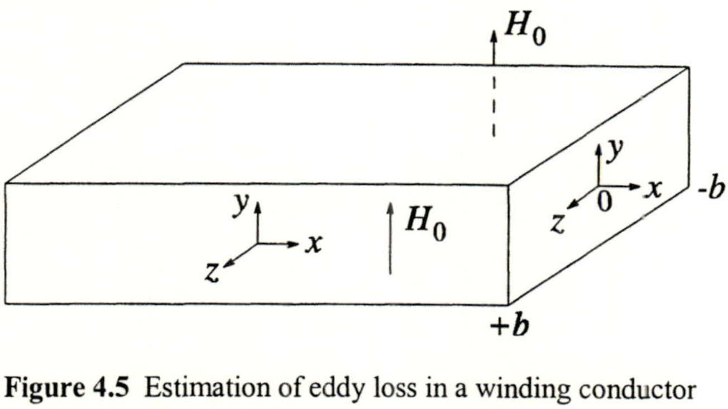

Figure 4.5 Estimation of eddy loss in a winding conductor

A solution satisfying this equation is

$$H_y = C_1 e^{\gamma z} + C_2 e^{-\gamma z} \tag{4.80}$$

where γ is defined by equation 4.32. In comparison with equation 4.65, equation 4.80 has two terms indicating waves traveling in both +z and –z directions (which is consistent with figure 4.5). The incident fields on both the surfaces, having peak amplitude of H_0, penetrate inside the conductor along the z axis in opposite directions (it should be noted that equation 4.80 is also a general solution of equation 4.63, in which case $C_1 = 0$ and $C_2 = H_0$ for the boundary conditions specified by equation 4.64). For the present case, the boundary conditions are

$$H_y = H_0 \text{ at } z = +b \text{ and } H_y = H_0 \text{ at } z = -b \tag{4.81}$$

Using these boundary conditions, we can get the expression for the constants as

$$C_1 = C_2 = \frac{H_0 e^{\gamma b} - H_0 e^{-\gamma b}}{e^{2\gamma b} - e^{-2\gamma b}} = \frac{H_0}{e^{\gamma b} + e^{-\gamma b}} \tag{4.82}$$

Putting these values of constants in equation 4.80 we get

$$H_y = \frac{H_0 (e^{\gamma z} + e^{-\gamma z})}{e^{\gamma b} + e^{-\gamma b}} = H_0 \frac{\cosh \gamma z}{\cosh \gamma b} \tag{4.83}$$

Using equation 4.8 and the fact that $H_x = H_z = 0$, the current density is

$$J_x = -\frac{dH_y}{dz} = -\gamma H_0 \frac{\sinh \gamma z}{\cosh \gamma b} \tag{4.84}$$

The loss produced per unit surface area (of the x-y plane) of the conductor in terms of the peak value of current density $\left(|J_x|\right)$ is given by

$$P_e = \frac{1}{2\sigma} \int_{-b}^{+b} |J_x|^2 dz \tag{4.85}$$

Now, using equation 4.39 we get

$$|J_x| = \left| -\gamma H_0 \frac{\sinh \gamma z}{\cosh \gamma b} \right| = H_0 \left| -\frac{1+j}{\delta} \frac{\sinh(1+j)z/\delta}{\cosh(1+j)b/\delta} \right|$$

$$= H_0 \left| -\frac{1+j}{\delta} \times \frac{\sinh(z/\delta)\cos(z/\delta) + j\cosh(z/\delta)\sin(z/\delta)}{\cosh(b/\delta)\cos(b/\delta) + j\sinh(b/\delta)\sin(b/\delta)} \right|$$

$$=\frac{\sqrt{2}H_0}{\delta}\left[\frac{(\sinh(z/\delta)\cos(z/\delta))^2+(\cosh(z/\delta)\sin(z/\delta))^2}{(\cosh(b/\delta)\cos(b/\delta))^2+(\sinh(b/\delta)\sin(b/\delta))^2}\right]^{1/2}$$

$$=\frac{\sqrt{2}H_0}{\delta}\left[\frac{\left(e^{z/\delta}-e^{-z/\delta}\right)^2\cos^2(z/\delta)+\left(e^{z/\delta}+e^{-z/\delta}\right)^2\sin^2(z/\delta)}{\left(e^{b/\delta}+e^{-b/\delta}\right)^2\cos^2(b/\delta)+\left(e^{b/\delta}-e^{-b/\delta}\right)^2\sin^2(b/\delta)}\right]^{1/2}$$

$$=\frac{\sqrt{2}H_0}{\delta}\left[\frac{\left(e^{2z/\delta}+e^{-2z/\delta}\right)-2\left(\cos^2(z/\delta)-\sin^2(z/\delta)\right)}{\left(e^{2b/\delta}+e^{-2b/\delta}\right)+2\left(\cos^2(b/\delta)-\sin^2(b/\delta)\right)}\right]^{1/2}$$

$$=\frac{\sqrt{2}H_0}{\delta}\left[\frac{\cosh(2z/\delta)-\cos(2z/\delta)}{\cosh(2b/\delta)+\cos(2b/\delta)}\right]^{1/2} \tag{4.86}$$

Substituting this magnitude of current density in equation 4.85 we get

$$P_e=\frac{1}{2\sigma}\frac{2H_0^2}{\delta^2}\int_{-b}^{+b}\frac{\cosh(2z/\delta)-\cos(2z/\delta)}{\cosh(2b/\delta)+\cos(2b/\delta)}\,dz \tag{4.87}$$

$$\therefore P_e=\frac{H_0^2}{\sigma\delta}\left[\frac{\sinh(2b/\delta)-\sin(2b/\delta)}{\cosh(2b/\delta)+\cos(2b/\delta)}\right] \tag{4.88}$$

or

$$P_e=\frac{H_0^2}{\sigma\delta}\left[\frac{e^{\xi}-e^{-\xi}-2\sin\xi}{e^{\xi}+e^{-\xi}+2\cos\xi}\right] \tag{4.89}$$

where $\xi = 2b/\delta$.

When $2b \gg \delta$, i.e., $\xi \gg 1$, equation 4.89 can be simplified to

$$P_e \approx \frac{H_0^2}{\sigma\delta} \tag{4.90}$$

Equation 4.90 gives the value of eddy loss per unit surface area of a conductor with its dimension, perpendicular to the applied field, much greater than the depth of penetration. Such a case, with the field applied on both the surfaces of the conductor, is equivalent to two infinite half spaces. Therefore, the total eddy loss given by equation 4.90 is two times that of the infinite half space given by equation 4.74. For such thick conductors / plates (winding made of copper bars, structural component made of magnetic steel having sufficiently large thickness, etc.), the resultant current distribution is greatly influenced and limited by the effect its own field and the currents are said to be inductance limited (currents are confined to the surface layers).

Now, let us analyze the case when dimension (thickness) of the conductor is quite small as compared to its depth of penetration, which is usually the case for rectangular paper insulated conductors used in transformers. For $2b << \delta$, i.e., $\xi << 1$, equation 4.89 can be simplified to

$$P_e = \frac{H_0^2}{\sigma\delta}\left[\frac{\left(1+\xi+\frac{\xi^2}{2!}+\frac{\xi^3}{3!}\cdots\right)-\left(1-\xi+\frac{\xi^2}{2!}-\frac{\xi^3}{3!}\cdots\right)-2\left(\xi-\frac{\xi^3}{3!}\cdots\right)}{\left(1+\xi+\frac{\xi^2}{2!}+\frac{\xi^3}{3!}\cdots\right)+\left(1-\xi+\frac{\xi^2}{2!}-\frac{\xi^3}{3!}\cdots\right)+2\left(1-\frac{\xi^2}{2!}\cdots\right)}\right]$$

$$= \frac{H_0^2}{\sigma\delta}\left[\frac{\left(4\frac{\xi^3}{3!}+\cdots\right)}{4+4\frac{\xi^4}{4!}+\cdots}\right] \tag{4.91}$$

Neglecting higher order terms and substituting the expression of δ from equation 4.38 we get

$$P_e = \frac{H_0^2}{\sigma\delta}\frac{\xi^3}{6} = \frac{H_0^2}{\sigma\delta}\frac{8b^3}{6\delta^3} = \frac{H_0^2}{\sigma}\frac{8b^3\omega^2\mu^2\sigma^2}{24} = \frac{1}{3}(\mu H_0)^2\sigma\omega^2 b^3 \tag{4.92}$$

Now, if the thickness of the winding conductor is *t*, then substituting $b = t/2$ in equation 4.92 we get

$$P_e = \frac{1}{3}B_0^2\sigma\omega^2\left(\frac{t}{2}\right)^3 = \frac{\omega^2 B_0^2 t^3 \sigma}{24} \tag{4.93}$$

It is more convenient to find an expression for the mean eddy loss per unit volume (since the volume of the conductor in the winding is usually known). Hence, dividing by t and finally substituting resistivity (ρ) in place of conductivity, we get the expression for the eddy loss in the winding conductor per unit volume as

$$P_E = \frac{\omega^2 B_0^2 t^2}{24\rho} \tag{4.94}$$

In case of thin conductors, the eddy currents are restricted by the lack of space or high resistivity and are said to be resistance limited. In other words, since the field of the eddy currents is negligible for thin conductors, the behavior is resistance limited. Equation 4.94 matches exactly with that derived in [10] by ignoring the magnetic field produced by the eddy currents. These currents are 90° out of phase with the load current (uniformly distributed current which produces the leakage field and is also responsible for DC I^2R loss in windings) flowing in the conductor. The eddy currents are shown to be lagging by 90° with respect to the load (source) current for a thin circular conductor in the later part of this section. The total current flowing in the conductor can be visualized to be a vector sum of the eddy current (I_{eddy}) and load current (I_{load}), having the magnitude of $\sqrt{(I_{eddy})^2 + (I_{load})^2}$, because these two current components are 90° out of phase in a thin conductor. This is a very important and convenient result because it means that the I^2R losses due to load current and eddy current losses can be calculated separately and then added later for thin conductors.

Equation 4.94 is very well-known and useful formula for calculation of eddy losses in windings. If we assume that the leakage field in windings is in axial direction only, then we can calculate the mean value of eddy loss in the whole winding by using the equations of Section 3.1.1. The axial leakage field for an inner winding (with a radial depth of R and height of H_W) varies linearly from inside diameter to outside diameter as shown in figure 4.6. The thickness of the conductor, which is its dimension perpendicular to the axial field, is usually quite small. Hence, the same value of flux density (B_0) can be assumed along both its vertical surfaces (along width w). The position of the conductor changes along the radial depth as the turns are wound. Hence, in order to calculate the mean value of the eddy loss of the whole winding, we have to first calculate the mean value of B_0^2. The r.m.s. value of ampere turns are linearly changing from 0 at the inside diameter (ID) to NI at the outside diameter (OD). The peak value of flux density at a distance x from the inside diameter is

$$B_x = \frac{\sqrt{2}\,\mu_0}{H_W}\left[\left(\frac{x}{R}\right)NI\right] \tag{4.95}$$

The mean flux density value, which gives the same overall loss, is given by

$$(B_0^2)_{mean} = \frac{1}{R}\int_0^R \left[\frac{\sqrt{2}\,\mu_0 NI}{H_W}\right]^2 \frac{x^2}{R^2}\,dx \tag{4.96}$$

Simplifying we get

$$(B_0^2)_{mean} = \frac{(B_{gp}^2)}{3} \tag{4.97}$$

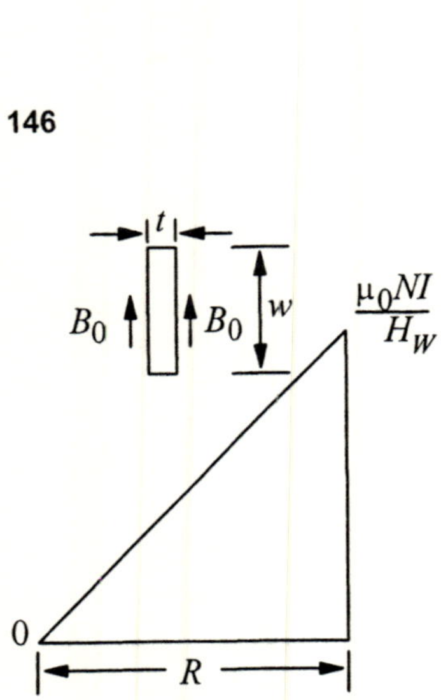

Figure 4.6 Leakage flux density in winding

where B_{gp} is the peak value of flux density in the LV-HV gap,

$$B_{gp} = \frac{\sqrt{2}\,\mu_0 NI}{H_W} \tag{4.98}$$

Hence, using equations 4.97 and 4.94, the mean eddy loss per unit volume of the winding due to the axial leakage field is expressed as

$$(P_E)_{mean} = \frac{\omega^2 t^2}{24\rho}\,\frac{B_{gp}^2}{3} \tag{4.99}$$

If we are interested in finding the mean value of eddy loss in a section of a winding in which ampere-turns are changing from $a(NI)$ at ID to $b(NI)$ at OD, where NI are rated r.m.s. ampere-turns, the mean value of B_0^2 can be easily found out by using the procedure similar to that given in Section 3.1.1 as

$$(B_0^2)_{mean} = \frac{(B_{gp}^2)}{3}\left[a^2 + ab + b^2\right] \tag{4.100}$$

and the mean eddy loss per unit volume in the section is

$$(P_E)_{mean} = \frac{\omega^2 t^2}{24\rho}\,\frac{(B_{gp}^2)}{3}\left[a^2 + ab + b^2\right] \tag{4.101}$$

Equation 4.101 tells us that for a winding consisting of a number of layers, the mean eddy loss of the layer adjacent to LV-HV gap is higher than that of others. For example, in the case of a 2-layer winding,

$$(B_0^2)_{mean} = \frac{(B_{gp}^2)}{3}\left[\left(\frac{1}{2}\right)^2 + \frac{1}{2}\times 1 + 1^2\right] = 1.75\frac{(B_{gp}^2)}{3} \quad \text{(for second layer)}$$

indicating that the mean eddy loss in the second layer close to the gap is 1.75 times the mean eddy loss for the entire winding. Similarly, for a 4-layer winding,

$$(B_0^2)_{mean} = \frac{(B_{gp}^2)}{3}\left[\left(\frac{3}{4}\right)^2 + \frac{3}{4}\times 1 + 1^2\right] = 2.31\frac{(B_{gp}^2)}{3} \quad \text{(for fourth layer)}$$

giving the mean eddy loss in the 4th layer as 2.31 times the mean eddy loss for the entire winding. Hence, it is always advisable to calculate the total loss ($I^2R + eddy$) in each layer separately and estimate the temperature rise of each layer. Such a calculation procedure helps designers to take countermeasures to eliminate high temperature rise in windings. Also, the temperatures measured by fiber-optic sensors (if installed) will be closer to the calculated values when such a calculation procedure is adopted.

Eddy loss calculated by equation 4.99 is approximate since it assumes the leakage field entirely in the axial direction. As seen in Chapter 3, there exists a radial component of the leakage field at winding ends and in winding zones where ampere-turns per unit height are different for LV and HV windings. For small distribution transformers, the error introduced by neglecting the radial field may not be appreciable, and equation 4.99 is generally used with some empirical correction factor applied to the total calculated stray loss value. Analytical / numerical methods, described in Chapter 3, need to be used for the correct estimation of the radial field. The amount of efforts required for getting the accurate eddy loss value may not get justified for very small distribution transformers. For medium and large power transformers, however, the eddy loss due to the radial field has to be estimated and the same can be found out by using equation 4.94, for which the dimension of the conductor perpendicular to the radial field is its width w. Hence, the eddy losses per unit volume due to axial (B_y) and radial (B_x) components of leakage field are

$$(P_E)_{axial} = \frac{\omega^2 B_y^2 t^2}{24\rho} \tag{4.102}$$

$$(P_E)_{radial} = \frac{\omega^2 B_x^2 w^2}{24\rho} \tag{4.103}$$

Thus, the leakage field incident on a winding conductor (see figure 4.7) is resolved into two components, viz. B_y and B_x, and losses due to these two components are calculated separately by equations 4.102 and 4.103 and then added. This is permitted because the eddy currents associated with these two perpendicular components do not overlap (since the angle between them is 90°).

We have assumed that the conductor dimension is very much less than the depth of penetration while deriving equation 4.94. This is particularly true in the case of loss due to the axial field. The conductor thickness used in transformers mostly falls in the range of 2 to 3.5 mm, which is considerably less than the depth of penetrations of copper and aluminum which are 10.3 mm and 13.2 mm respectively at 50 Hz. The conductor width is usually closer to the value of depth of penetration. If the conductor width is equal to the depth of penetration ($w = 2b = \delta$), equation 4.89 becomes

$$(P_e)_{2b=\delta} = \frac{H_0^2}{\sigma\delta}\left[\frac{e^1 - e^{-1} - 2\sin 1}{e^1 + e^{-1} + 2\cos 1}\right] = \frac{1}{6.24}\frac{H_0^2}{\sigma\delta} \quad \cdots\cdots \textit{from exact equation}$$

Comparing this value with that given by equation 4.92,

$$(P_e)_{2b=\delta} = \frac{H_0^2}{\sigma\delta}\frac{\xi^3}{6} = \frac{1}{6}\frac{H_0^2}{\sigma\delta} \quad \cdots\cdots \textit{from approximate equation}$$

the error of just 4% is obtained, which is quite acceptable. Hence, it can be concluded that the eddy loss due to the radial field can also be calculated with a reasonable accuracy from equation 4.103 for the conductor widths comparable to the depth of penetration.

For thin circular conductors of radius of R, if the ratio R/δ is small, we can neglect the magnetic field of eddy currents. If the total current in the conductor is $I\cos\omega t$, the uniform current density is given by

$$J_0 = \frac{I}{\pi R^2}\cos\omega t \tag{4.104}$$

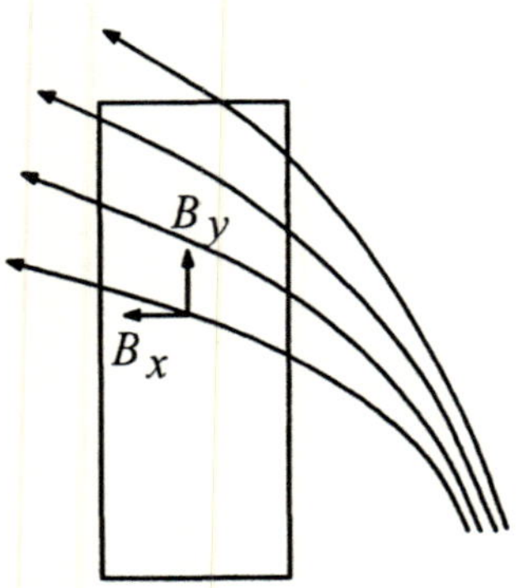

Figure 4.7 Winding conductor in a leakage field

and the eddy current density at any radius r inside the conductor is [8]

$$J_e = \frac{I}{2\pi\delta^2}\left[\frac{1}{2} - \frac{r^2}{R^2}\right]\sin\omega t \tag{4.105}$$

Thus, it can be observed from equations 4.104 and 4.105 that for thin conductors (resistance-limited behavior), the eddy currents lag the exciting current (the current which produces the field responsible for eddy currents) by 90°. Contrary to this, for thick conductors (thickness or radius much larger than the depth of penetration), the eddy currents lag the exciting current by 180° (inductance-limited behavior in which the currents are confined to the surface layers).

The power loss per unit length of the thin circular conductor can be found out by using equations 4.104 and 4.105 as (J_0 and J_e are 90° apart, square of their sum is sum of their squares)

$$P_1 = \int_0^R \left[\frac{|J_0|^2 + |J_e|^2}{2\sigma}\right] 2\pi r\, dr = \frac{I^2}{2\pi R^2 \sigma} + \frac{I^2 R^2}{96\pi\sigma\delta^4} \tag{4.106}$$

The power loss per unit length due to the exciting current alone is

$$P_2 = \left(\frac{I}{\sqrt{2}}\right)^2 \times \frac{1}{\sigma} \times \frac{1}{\pi R^2} \tag{4.107}$$

Therefore, the ratio of effective AC resistance to DC resistance of a thin circular conductor can be deduced from equations 4.106 and 4.107 as

$$\frac{R_{AC}}{R_{DC}} = \frac{P_1}{P_2} = 1 + \frac{1}{48}\left(\frac{R}{\delta}\right)^4 \quad \cdots \textit{for thin circular conductors} \tag{4.108}$$

For thick circular conductors ($R >> \delta$), the effective resistance is that of the annular ring of diameter $2R$ and thickness δ, since all the current can be assumed to be concentrated in one depth of penetration as seen in Section 4.2. Hence, the effective AC resistance per unit length is

$$R_{AC} = 1/(2\pi R\delta\sigma) \tag{4.109}$$

and

$$\frac{R_{AC}}{R_{DC}} = \frac{R}{2\delta} \quad \cdots \textit{for thick circular conductors} \tag{4.110}$$

4.5.2 Methods of estimation

As said earlier, the axial and radial components of a field can be estimated by analytical or numerical methods. Accurate estimation of eddy loss due to the radial leakage field by means of empirical formulae is not possible. The analytical methods [11, 12] and two-dimensional FEM [13, 14] can be used to calculate the eddy loss due to axial and radial leakage fields. It is assumed that the eddy currents do not have influence on the leakage field (the case of thin conductors). The FEM analysis is quite commonly used for the eddy loss calculations. The winding is divided into many sections. For each section the corresponding ampere-turn density is defined. The value of conductivity is not defined for these sections. The values of B_y and B_x for each conductor can be obtained from the FEM solution, and then the axial and radial components of the eddy loss are calculated for each conductor by using equations 4.102 and 4.103 respectively. The B_y and B_x values are assumed to be constant over a single conductor and equal to the value at the center of the conductor. If the cylindrical coordinate system is used, B_y and B_x components are replaced by B_z and B_r components respectively. The total eddy loss for each winding is calculated by integrating the loss components of all its conductors.

Sometimes a very quick but reasonably accurate calculation of eddy loss is required. At the tender design stage, an optimization program may have to work hundreds of designs to arrive at the optimum design. In such cases, expressions for the eddy loss in windings for their simple configurations can be found out using multiple regression method in conjunction with *Orthogonal Array Design of Experiments* technique [15]. With the quantum improvement in the speed of computational tools, it is now possible to integrate the accurate analytical / numerical methods in the main design optimization program.

For cylindrical windings in core-type transformers, the two-dimensional methods give sufficiently accurate eddy loss values. For getting most accurate results, three-dimensional magnetic field calculations have also been used [16, 17, 18]. Once the three-dimensional field solution is obtained, the three components of the flux density (B_x, B_y and B_z) are resolved into two components, viz. the axial and radial components, which enables the use of equations 4.102 and 4.103 for the eddy loss evaluation.

For small distribution transformers with LV winding having crossmatic conductor (thick rectangular bar conductor), each and every turn of LV winding has to be modeled (with the value of conductivity defined) in FEM analysis. This is because the thickness of the bar conductor is usually comparable to or sometimes more than the depth of penetration and its width is usually more than 5 times the depth of penetration. With such a conductor having large dimensions, a significant modification of the leakage field occurs due to the eddy currents, which cannot be neglected in the calculations.

The problem of accurate estimation of winding eddy loss seems to be quite resolved by method such as 2-D FEM. The analysis of winding eddy loss by 3-D FEM analysis is the most accurate one, but the computational efforts involved should be compared with the improvement obtained in the accuracy.

4.5.3 Optimization of losses and elimination of winding hot spots

In order to reduce the DC resistance (I^2R) loss, if the designer increases the conductor dimensions, the eddy loss in windings increases. Hence, optimization of the total of I^2R and eddy loss should be done.

The knowledge of flux density distribution in a winding helps in choosing proper dimensions of conductors. This is particularly important for a winding with tappings within its body, in which the high value of radial flux density can cause excessive loss and temperature rise. For the minimization of radial flux, balancing of ampere-turns per unit height of LV and HV windings should be done (for various sections along their height) at the average tap position. The winding can be designed with different conductor dimensions in the tap zone to minimize the risk of hot spots. Guidelines are given in [19] for choosing the conductor width for eliminating hot spots in windings. For 50 Hz frequency, the maximum width that can be used is usually in the range of 12 to 14 mm, whereas for 60 Hz it is of the order of 10 to 12 mm. This guideline is useful in the absence of detailed analysis which involves calculation of temperature rise in the part of the winding where a hot spot is expected. For calculating the temperature rise of a disk / turn, its I^2R loss and eddy loss should be added. An idle winding between LV and HV windings links the high gap flux resulting in higher eddy loss. Hence, its conductor dimensions should be properly decided.

In gapped core shunt reactors, there is considerable flux fringing between limb packets (separated by non-magnetic gap), resulting in an appreciable radial flux causing excessive losses in the reactor winding if the distance between the reactor winding and core is small or if the conductor width is large.

One of the most logical ways of reducing the eddy loss of a winding is to sub-divide winding conductors into a number of parallel conductors. If a conductor having thickness t is sub-divided into 2 insulated parallel conductors of thickness $t/2$, the eddy loss due to axial leakage field reduces by a factor of $1/4$ (refer to equation 4.102). In actual practice, from the short circuit withstand considerations there is a limitation imposed on the minimum thickness that can be used. Also, if the width to thickness ratio of a rectangular conductor is more than about 6, there is difficulty in winding it. The sub-division of the conductor also impairs the winding space factor in the radial direction. This is because each individual parallel conductor in a turn has to be insulated increasing the total insulation thickness in the radial direction. In order to improve the space factor, sometimes a bunch conductor is used in which usually two or three parallel conductors are bunched in a common paper covering. The advantage is that the individual conductor needs to be insulated with a lower paper insulation thickness because of the outermost common paper covering. A single bunch conductor is also easier to wind, since no crossovers are required at ID and OD of the winding. In contrast to this, for example in the case of two parallel conductors, the two conductors are usually crossed over at ID and OD of each disk for the ease of winding. Three rectangular strip conductors and the corresponding bunch conductor are shown in figure 4.8. For a strip conductor of radial thickness 2 mm with a radial paper covering of 0.5 mm, the total radial dimension of three strip conductors is 9 mm, whereas for a bunch conductor with the individual radial insulation of 0.175 mm and the common outermost radial insulation of 0.325 mm, the total radial dimension is 7.7 mm ($=3\times(2+2\times0.175)+2\times0.325$) giving 15% improvement in the space factor.

For a continuously transposed cable (CTC) conductor, in which a number of small rectangular conductors are transposed at regular intervals, it is possible to use quite a small thickness (as low as 1.2 mm) and width (as low as 3.8 mm), resulting into considerably less amount of eddy loss. A cost-benefit analysis has to be done to check the advantage gained by the reduction of losses versus the higher cost of the CTC conductor. For transformer specifications with higher value of load loss capitalization (dollars per kW of load loss), the CTC conductor may give a lower total capitalized cost (material cost plus loss cost) particularly in large transformers. The low voltage winding in large generator transformers is usually designed with the CTC conductor to minimize the eddy loss and improve productivity. The CTC conductor is shown in figure 4.9.

One more way of reducing the eddy loss is by use of two conductors with different widths along the winding height. At the winding ends, there may be appreciable eddy loss due to the fact that radial flux is incident perpendicularly on the wider conductor dimension (width). This loss can be decreased by reducing the conductor width for few disks at the ends based on the detailed FEM analysis and increasing the width for remaining length of the winding where the leakage field is predominantly axial, so that I^2R losses are maintained approximately constant. Such an arrangement can give reduction in the winding loss ($I^2R+eddy$) by about 1 to 2% in large transformers.

In another method, end disks can have less turns per disk (one or two turns less as compared to the other part of winding) in HV winding having a large number of turns per disk. A conductor with higher thickness and lower width can be used for these disks so that I^2R loss is kept approximately constant and eddy losses are reduced. Here, the assumption is that the increase in axial eddy loss due to higher thickness is more than compensated by the decrease in radial eddy loss due to lower width (which means that the radial field is contributing more to the eddy loss as compared to the axial field).

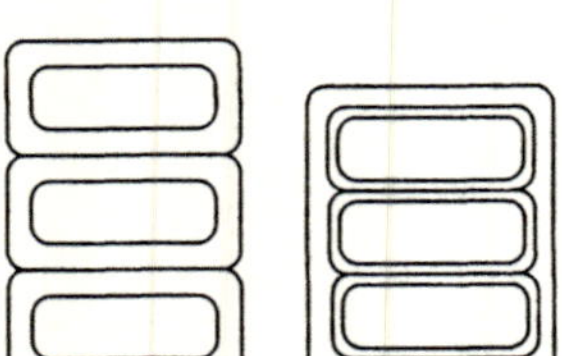

Figure 4.8 Strip and bunch conductor

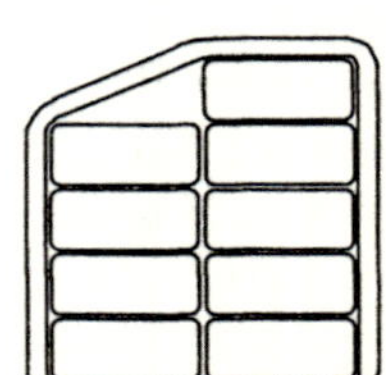

Figure 4.9 CTC conductor

In both these methods, ampere-turns of HV and LV windings per unit height should be equal (for various sections along their height); otherwise there will be an extra radial flux and corresponding losses, nullifying the advantages gained.

Some manufacturers use the combination of strip and CTC conductors to minimize winding losses. The CTC conductor is used for a few top and bottom disks minimizing the eddy loss, whereas the remaining disks are made out of the strip conductor. This technique requires establishment of a proper method for making the joint of the strip and CTC conductors.

4.5.4 Eddy loss in foil windings

In small transformers, foil windings (made up of thin sheets of copper or aluminum) are quite popular because of the simplicity obtained in the winding operations. The metal foil and the insulation layer can be wound simultaneously on a special machine and the whole process can be easily mechanized. In foil windings, eddy loss due to axial leakage field is insignificant because of very small thickness of the foil. On the contrary, the radial field results into higher eddy loss at the ends of the foil windings. The eddy loss in a foil winding can be evaluated by analytical or numerical methods. The current density has been obtained as a solution of integral equation (Fredholm type) for a two winding transformer in [20]. For specific transformer dimensions given in the paper, the coefficient of additional loss is 1.046, i.e., the eddy loss is 4.6% of the DC I^2R loss. In another approach [21], a boundary-value field problem is solved for magnetic vector potential in cylindrical coordinate system using modified Bessel and Struve functions. For symmetrically placed LV (foil) and HV windings, the additional loss factor reported is 5.8%. For an asymmetrical arrangement of windings this loss increases.

In [22], the foil winding is assumed as a vertical section of an infinitely wide and deep conducting plate, which is assumed symmetrically penetrated on both sides by plane electromagnetic waves. The geometry is given in figure 4.10, where h is the height of the foil winding and B_x denotes the value of flux density at winding ends. The induced eddy current density (with the vertical dimensions specified with respect to the middle of the winding height) is [22]

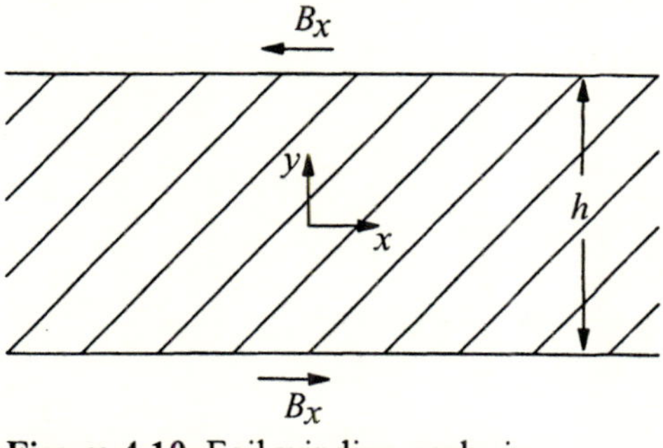

Figure 4.10 Foil winding analysis

$$J_e = \frac{B_x}{\mu_0 \delta} e^{\left[-\left(\frac{h}{2}-y\right)/\delta\right]} \left[\cos\left(\frac{\pi}{4} - \frac{\frac{h}{2}-y}{\delta} \right) + j \sin\left(\frac{\pi}{4} - \frac{\frac{h}{2}-y}{\delta} \right) \right] \tag{4.111}$$

The eddy loss in terms of the peak value of current density is given as

$$P_e = \frac{1}{2\sigma} \int_v \left|J_e\right|^2 dv \tag{4.112}$$

It is clear from equations 4.111 and 4.112 that

$$P_e \propto \left|J_e\right|^2 \propto e^{\left[-2\left(\frac{h}{2}-y\right)/\delta\right]} \tag{4.113}$$

Hence, the maximum value of eddy loss occurs at $y = h/2$ (at two ends of the winding) since at this height the exponential term has the maximum value of 1,

$$e^{\left[-2\left(\frac{h}{2}-y\right)/\delta\right]} = 1$$

whereas its value is just 0.0067 at $y = (h/2) - 2.5\delta$, which means that more than 99% of the induced eddy current loss occurs in just 2.5δ depth from the winding ends. For an aluminum foil winding, whose skin depth is 13.2 mm at 50 Hz, almost all the eddy loss is concentrated in 33 mm from the winding ends. The foil winding eddy current density distribution calculated by equation 4.111 for a typical transformer is shown in figure 4.11.

Thus, the current density at the ends can be about 1.5 to 2 times the uniform current density, resulting into local heating of 2.25 to 4 times that in the middle portion of the winding. Hence, the temperature rise at the foil winding ends should be carefully assessed.

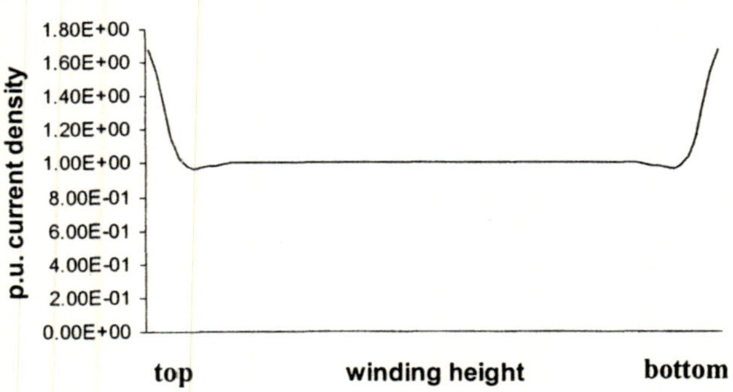

Figure 4.11 Eddy current distribution in a foil winding

Usually, there is no temperature rise problem in a foil winding since the thermal conductivity of aluminum / copper is quite good, and the winding edges are well exposed to the cooling medium. If a transformer with foil winding is supplying a power electronic load, there is increase in the heating effect at winding ends on account of harmonics [23].

With the development in computational facilities, the numerical tools such as FEM can easily analyze the foil winding eddy problem without any simplifications done in analytical formulations. It is actually a three-dimensional problem. The losses in a foil winding calculated in the core-window cross section are lower than that calculated in the region outside the window. The radial flux at winding ends is usually more in the region outside the window, and hence the additional loss factor may be of the order of 10 to 15%. The position of foil winding affects its eddy loss significantly. If the foil winding is the outer winding, the radial flux density at ends reduces resulting in reduction of eddy loss [24].

4.6 Circulating Current Loss in Transformer Windings

We have seen that the leakage field in a transformer is responsible for extra losses in windings in addition to the DC I^2R loss. If the windings have a single solid conductor then eddy currents flow causing the eddy loss, and no circulating current loss appears. In the windings with large current ratings, conductor has to be sub-divided into a number of parallel conductors to reduce the eddy loss to an acceptable / optimum value. Without proper transposition, the leakage flux linked by these parallel conductors is different, inducing different voltages in them. Since the conductors are paralleled at the ends, there will be net induced voltages in the loops formed by these conductors resulting into circulating currents. These circulating currents flow back and forth within the parallel conductors, and thus do not flow outside the windings. The appearing value of circulating current loss adds to the other stray loss components resulting into higher load loss.

The circulating current loss, if not controlled, may cause hot spots in the windings of large power transformers. The circulating current loss depends very much on the physical locations of the conductors in the leakage field. This loss can be avoided if the location of each conductor in the layer and from layer to layer is such that all conductors are linked with an equal amount of leakage flux. Such a winding is said to have ideal / complete transposition in which each conductor occupies the position of every other conductor. This can be achieved by doing suitable transpositions between the parallel conductors. In the absence of transpositions, the advantage gained through the eddy loss reduction (by sub-division of conductors) will be lost due to the circulating current loss. Inaccurate or inadequate transpositions introduce circulating currents in transformer windings, which can be calculated by either analytical or numerical methods.

4.6.1 Analytical methods

Analytical formulae are given in [25] to calculate the circulating current loss for various types of windings. In this formulation, the ampere-turn diagram is assumed to have trapezoidal form. The leakage flux density (B) at the inside of the conductor and in the paper covering is calculated. Integrating this value of B over the thickness, the flux/cm is obtained. The voltage gradient is then calculated as the sum of the resistance drops of load current and stray currents (eddy plus circulating currents), and the reactive drop induced by the flux. The circulating current component is calculated assuming that it is uniform in the whole cross section of the conductor. The effect of circulating currents on the leakage field has been neglected in this approximate method. Hence, it gives higher value of circulating current losses than the actual value.

It is shown in the paper that circulating current loss for an untransposed layer winding is the same as the eddy loss in an equivalent one solid conductor. In other words, if a winding conductor of thickness t is split into two conductors of thickness $t/2$, the circulating current loss for the untransposed case of these two conductors is same as the eddy loss in the conductor of thickness t. Hence, if a winding conductor is sub-divided into parallel conductors which are untransposed, the total stray loss in the winding (eddy and circulating losses) will theoretically remain unchanged.

In order that each conductor occupies the position of every other conductor, various transposition schemes are used depending upon the type of winding and its conductor. Two such schemes are shown in figure 4.12 for a layer winding with four parallel conductors. In both the cases, all the conductors are occupying all the four positions and hence they are linked with the same value of leakage field (assuming that the field is entirely in the axial direction) resulting in no circulating currents.

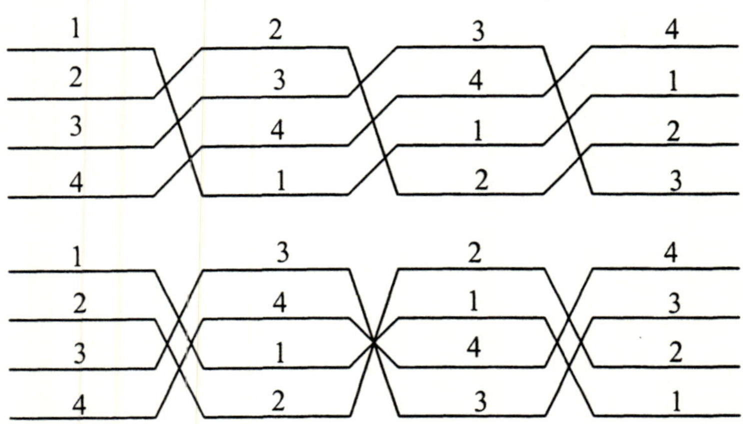

Figure 4.12 Transposition schemes in a layer winding

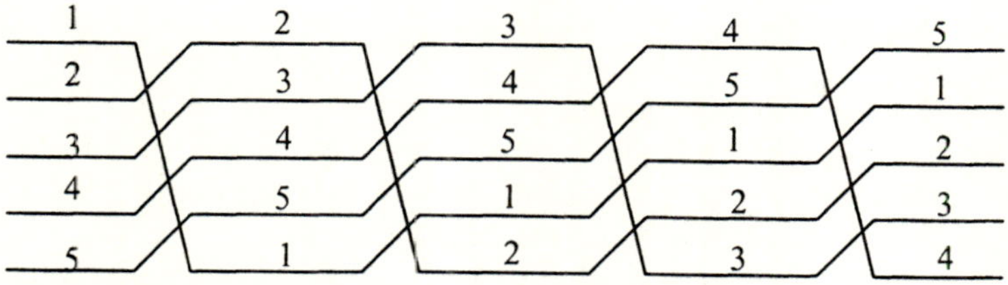

Figure 4.13 Transposition scheme for 5 parallel conductors

In general, for n parallel conductors one ideally needs $n-1$ transpositions equally spaced along the winding height. For example, in the case of 5 parallel conductors, there should be 4 transpositions as shown in figure 4.13. At every transposition, the top conductor is brought to the bottom position. A thin rectangular copper strip of nearly the same cross-sectional area (as that of the conductor) can be used to join the two cut parts of the conductor by a brazing operation. The conductor can be brought to the bottom position without the cutting and brazing operations by providing a slanting support made of an insulation material.

There is one more scheme [26, p. 147 of 25] which can be used in small transformers in which only 3 transpositions are done in place of $n-1$ transpositions in order to reduce the manufacturing time. A typical scheme is shown in figure 4.14 for 12 parallel conductors. In this scheme, parallel conductors are divided into four groups for the middle transposition, and hence the number of parallel conductors has to be multiple of 4.

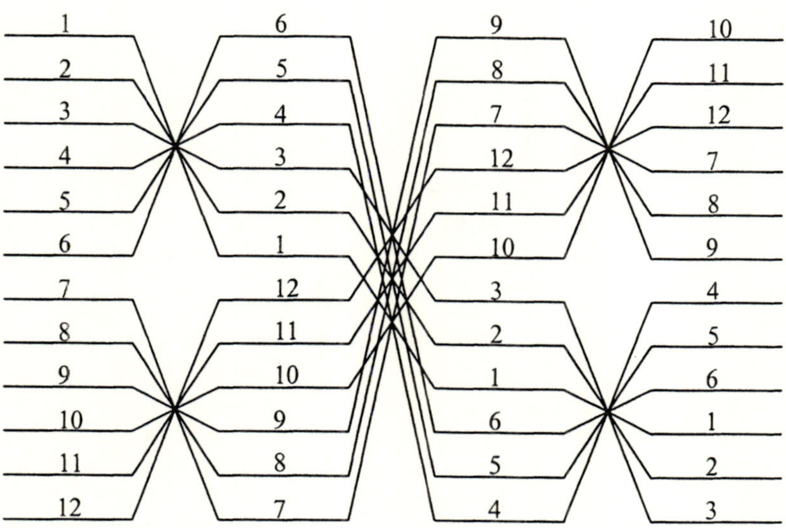

Figure 4.14 Transposition scheme for layer winding with 12 parallel conductors

This is not the ideal transposition scheme since none of the conductors is occupying the position of every other conductor. But each conductor occupies such 4 positions that the average value of the field linked by a conductor is not much different than that of the others, when only the axial field is considered. The transposition scheme may be used with caution in small transformers when the radial field at ends and within body of windings is insignificant.

In the discussion of paper [25], Rabins and Cogbill have given a very useful and quick method to check the correctness of any transposition scheme. As per this method, for each conductor a sum

$$s = \sum \frac{N_p}{N} p(p-1)$$

is calculated where p is a particular position and N_p is number of turns (out of total turns N) in position p. Let us apply the method to check the transposition scheme given in figure 4.14. Conductor 1 in figure 4.14 occupies positions 1, 6, 9 and 10 for which $s_1 = 0.25 \times (1 \times 0 + 6 \times 5 + 9 \times 8 + 10 \times 9) = 48$. Similarly, the value of s for remaining conductors can be found out. Here, it is assumed that the conductor occupies each of the 4 positions for 25% of the winding height. If this is not true or if a transposition is missed by one turn or one disk by mistake during manufacturing, the corresponding number of turns in each position should be taken. The closeness of s values for all the parallel conductors is a good indicator of the correctness of the transposition scheme. The average of s values for all conductors is calculated (s_{avg}). The circulating current (P_{cc}) loss is calculated in terms of the winding eddy loss (P_{eddy}) as

$$P_{cc} = P_{eddy} \left(\frac{45}{5n^2 - 1} \right) \left(\frac{t + t_i}{t} \right)^2 \left[\frac{1}{n} \sum_{1}^{n} (s_n - s_{avg})^2 \right] \tag{4.114}$$

where t and t_i are individual conductor thickness and insulation thickness between adjacent conductors respectively. For example, if the values of t and t_i are 2.3 mm and 0.5 mm respectively for the winding shown in figure 4.14, $P_{cc} = 0.0206\, P_{eddy}$, which is negligible. This indicates that the transposition scheme given in figure 4.14, although not ideal, is sufficiently good to reduce the circulating current loss to a very low value. It should be noted that the calculation assumed the axial leakage field only. Hence, the method may be used for small transformers only, when the radial field in them is low. The method will not be effective if there is an appreciable radial field in any part of the winding. Also, at the middle transposition, the conductors have to be properly supported while they are being transposed, which otherwise may be vulnerable when subjected to short circuit forces.

Now, let us analyze continuous disk windings having a number of turns in the radial direction with each turn having two or more parallel conductors. Usually, the number of parallel conductors in a continuous disk winding is less than that in a layer winding. Hence, the problem of circulating current loss is less ominous in continuous disk windings since the radial depth of one turn (consisting of two or more parallel conductors) is smaller. A continuous disk winding having N_d turns/disk and n parallel conductors per turn, with crossovers between conductors at inside and outside diameters of every disk, is equivalent to a multi-layer winding with N_d layers with each layer having n parallel conductors transposed at its center (middle height position).

A continuous disk winding with 8 parallel conductors and 3 turns per disk is shown in figure 4.15. Kaul's formula [25] for estimation of circulating current loss for such a continuous winding with n parallel conductors is

$$P_{cc} = 2.4\times 10^{6}\times\frac{1}{180}(kR)^{4}\left[1-\frac{5}{n^{2}}+\frac{4}{n^{4}}\right]\times(I^{2}R\ Loss) \tag{4.115}$$

where

R = radial depth of insulated turn = $n(t+t_i)$

$$k = 4.8\times 10^{-4}\times\sqrt{\frac{h_c}{h_w}\frac{t}{(t+t_i)}\frac{f}{\rho}} \tag{4.116}$$

ρ = resistivity of conductor material (copper or aluminum)

h_c = total length (height) of copper or aluminum in the axial direction

h_w = length (height) of winding in the axial direction

Ratios h_c/h_w and $t/(t+t_i)$ define winding space factors in the axial and radial directions respectively. Let us calculate the circulating current loss for the winding shown in figure 4.15 with the following parameters:

$f = 50$ Hz

$t = 2.4$ mm, $t_i = 0.7$ mm

$h_c/h_w = 0.8$

$\rho = 0.0211\times 10^{-6}$ ohm-m for a copper conductor

$$\therefore R = 8\times(2.4+0.7)\times 10^{-3} = 0.0248\ \text{m}$$

and

$$k = 0.48\times 10^{-4}\times\sqrt{0.8\times\frac{2.4}{2.4+0.7}\times\frac{50}{0.0211\times 10^{-6}}} = 1.839$$

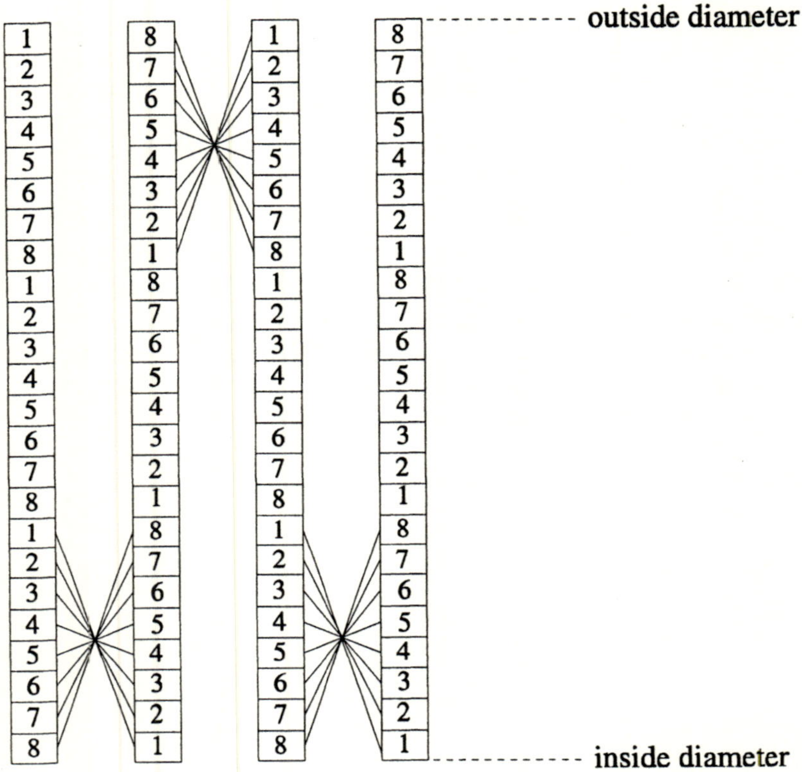

Figure 4.15 Continuous disk winding

The circulating current loss in the winding is

$$P_{cc} = 2.4 \times 10^{6} \times \frac{1}{180} (1.839 \times 0.0248)^{4} \left[1 - \frac{5}{8^{2}} + \frac{4}{8^{4}} \right] \times (I^{2}R\ Loss)$$

$$P_{cc} = 0.053 \times (I^{2}R\ Loss)$$

For I^2R loss of 130 kW for this winding (for all 3-phases), total circulating current loss is about 6.9 kW, which is an appreciable value. This circulating current loss can be almost eliminated by doing the group transpositions at appropriate places similar to that for a layer winding shown in figure 4.14.

It should be noted that in the analytical methods discussed till now, only the axial leakage field is considered and the effect of the radial leakage field is neglected. Due to this assumption the calculation is quite easy, and the extra loss on account of circulating currents can be found very quickly while the transformer design is in progress. Although these methods may not give the accurate loss value, designers can use them to check the correctness of the transposition method used and arrive at the best possible transposition scheme.

4.6.2 Method based on multi-winding transformer theory

This method is based on circuit equations of a multi-winding transformer. Let us consider the layer winding shown in figure 4.14. The LV winding, consisting of 12 parallel conductors, can be considered as made of 12 windings. Hence, with HV winding, it becomes a system of 13 windings. Following circuit equations can be written for this system of 13 windings in terms of per-unit short circuit (leakage) impedances between the pairs of windings [27],

$$V_{13}-V_1=-I_1Z_{13-1}-I_2\frac{Z_{13-1}+Z_{13-2}-Z_{1-2}}{2}-I_3\frac{Z_{13-1}+Z_{13-3}-Z_{1-3}}{2}-\cdots\cdots-I_{12}\frac{Z_{13-1}+Z_{13-12}-Z_{1-12}}{2} \tag{4.117}$$

$$V_{13}-V_2=-I_1\frac{Z_{13-2}+Z_{13-1}-Z_{1-2}}{2}-I_2Z_{13-2}-I_3\frac{Z_{13-2}+Z_{13-3}-Z_{2-3}}{2}-\cdots\cdots-I_{12}\frac{Z_{13-2}+Z_{13-12}-Z_{2-12}}{2} \tag{4.118}$$

Similarly, remaining 10 equations for $V_{13}-V_3,\ V_{13}-V_4,\cdots\cdots,V_{13}-V_{12}$ can be written. If the load loss test is performed by applying the impedance voltage to HV winding (number 13) with all 12 LV windings short-circuited, then

$$I_1+I_2+I_3+\cdots\cdots+I_{13}=0 \tag{4.119}$$

and

$$V_1=V_2=V_3=\cdots\cdots=V_{12}=0 \tag{4.120}$$

The leakage impedance between the pairs of windings can be calculated by using the conventional formulae (of Section 3.1.1) considering the four distinct positions occupied by each of these 12 LV windings along their height. The per-unit leakage impedance ($R + jX$) is calculated with the transformer MVA as the base. A more accurate estimation of leakage impedance can be made by using FEM analysis [28]. This combined method, in which FEM is used (for calculation of impedances) in conjunction with the multi-winding theory, is quite accurate method in two dimensions. The following system of 12 equations can be easily solved to give the currents flowing in 12 parallel conductors:

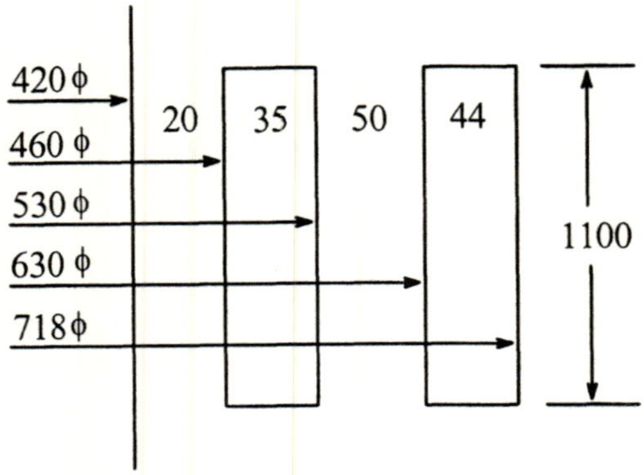

Figure 4.16 Details of 10 MVA transformer

$$[I]_{12\times1} = [Z]^{-1}_{12\times12} \times [V]_{12\times1} \tag{4.121}$$

Let us now calculate the circulating current loss for a transformer with LV winding having 12 parallel conductors. The dimensions of windings are shown in figure 4.16. The transformer specification and relevant design details are:

10 MVA, 33/6.9 kV, star/star, 50 Hz Transformer
Volts/turn = 46.87, Z = 7.34%
LV conductor: thickness = 2.3 mm, insulation between conductors = 0.5 mm, area of one conductor = 22.684×10^{-6} m^2
Total conductor area = 12 × area of one conductor = 272.21×10^{-6} m^2
Mean diameter of LV winding = 0.495 m
Turns in LV winding = 85
LV winding volume for 3 phases $= 272.21\times10^{-6} \times (\pi\times0.495\times85\times3)$
$= 0.1079$ m^3

Currents calculated by the method are given in table 4.1. The currents in parallel conductors are almost equal. The total loss (I^2R + circulating current loss) calculated by the above method based on the multi-winding transformer theory is 21.882 kW (for all 3 phases) out of which I^2R loss is 21.87 kW. Thus, the circulating current loss is 12 watts, which is just 0.055% of the I^2R loss, suggesting that the transposition scheme, although not ideal, is good enough to minimize the circulating current loss. Hence, it can be used for small rating transformers with low radial fields in windings as mentioned earlier. We had calculated for the same transformer, the circulating current loss by method of Rabins and Cogbill as

$$P_{cc} = 0.0206\ P_{eddy}$$

Table 4.1 Current distribution in parallel strands

Conductor	1	2	3	4	5	6
Current (A)	68.97	69.0	69.08	69.08	69.0	68.97
Conductor	7	8	9	10	11	12
Current (A)	69.08	69.0	68.97	68.97	69.0	69.08

The total eddy loss in LV winding can be found out using equation 4.99.

$$P_{eddy} = \frac{\omega^2 t^2}{24\rho} \frac{B_{gp}^2}{3} \times volume$$

$$B_{gp} = \sqrt{2}\,\frac{\mu_0 NI}{H_w} = \sqrt{2}\,\frac{4\pi \times 10^{-7} \times 85 \times 836.74}{1.1} = 0.1149 \text{ T}$$

$$\therefore P_{eddy} = \frac{(2\pi \times 50)^2 (0.0023)^2}{24 \times 0.0211 \times 10^{-6}} \frac{0.1149^2}{3} \times 0.1079 = 489.55 \text{ W}$$

Hence,

$$P_{cc} = 0.0206 \times 489.55 = 10 \text{ W}$$

The above answer matches closely with that calculated from the method based on the theory of multi-winding transformer (12 W).

4.6.3 FEM analysis

As mentioned earlier, the analytical methods which consider only the axial leakage field are approximate. The effect of radial fields at winding ends or at tap breaks on the circulating currents cannot be analyzed by these methods. Also, the effect of circulating currents on the main leakage field is neglected. These effects are dominant in large power transformers as well as low voltage high current transformers (e.g., furnace transformers). Hence, for more accurate estimation of the circulating current loss and for arriving at the correct transposition scheme, a numerical method such as FEM is used [28, 29].

In [29], it has been concluded that in case of windings without any transpositions or with inadequate transpositions, the analytical formulations give much higher and inaccurate circulating current losses. This is particularly true when the number of parallel conductors is more. This is due to the fact that in such cases, the effect of field due to large circulating currents significantly affects the leakage field, and this effect is neglected in the analytical formulations.

For FEM analysis, a time-harmonic formulation is used in which each and every conductor needs to be modeled. It requires quite a good amount of modeling efforts. The conductors are defined as stranded conductors meaning that the eddy currents in them are neglected. Hence, the loss given by FEM analysis is a total of the I^2R loss and circulating current loss. Let us take a case where the loss is required to be found in LV winding having a number of parallel conductors. The impedance voltage is applied to HV winding (which is modeled as one or very few rectangular block/s with the definition of ampere-turn density for the block/s). In [30], the circulating currents and the corresponding losses have been calculated for a continuous disk winding with 12 parallel conductors by using the 2-D FEM analysis and the analytical method of multi-winding transformers. The transposition scheme was not the ideal one for this winding. The currents calculated by both the methods are quite close, which are compared in figure 4.17. The circulating current losses calculated by the two methods are also close. When the transposition scheme was changed to the ideal one, in which each of the 12 parallel conductors occupied the position of every other conductor, the load loss of the transformer reduced by an amount corresponding to the calculated value of the circulating current loss, thus verifying the results of the calculation by the two different methods.

In [17, 18], 3-D leakage field distribution is obtained by 3-D FEM analysis. The voltages that circulate the currents between parallel conductors are found out by integrating the flux density components over three dimensions taking into account the transpositions between parallel conductors. Based on these voltages and conductor resistances, circulating currents in the loops are found out from the simple network solution of the circuit shown in figure 4.18.

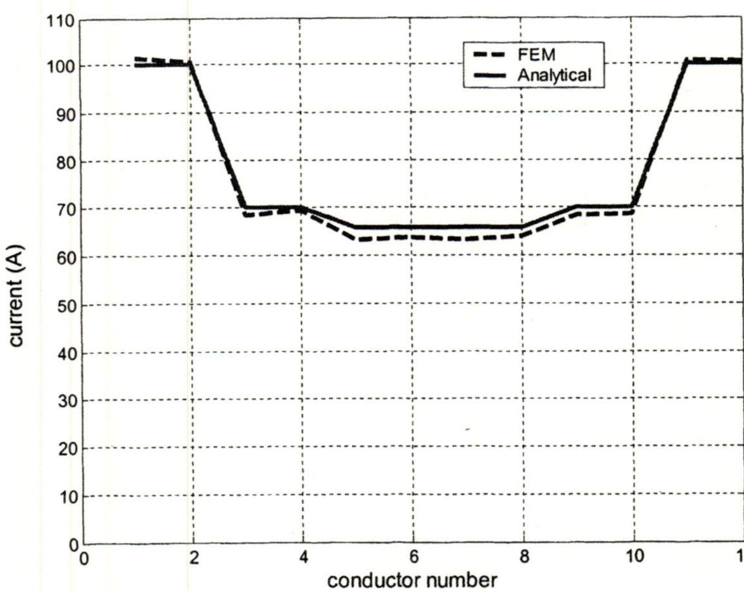

Figure 4.17 Circulating currents in continuous disk winding

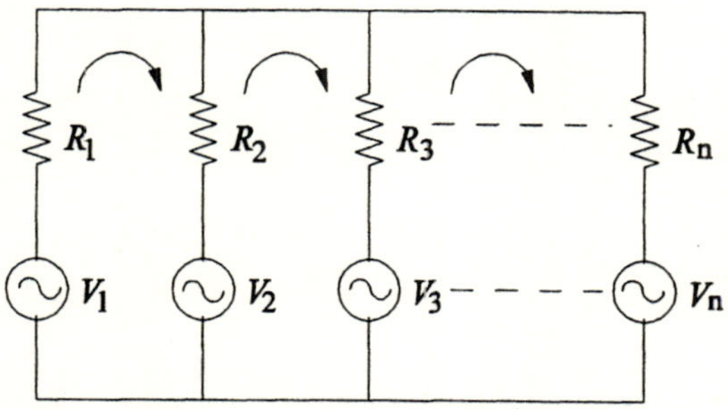

Figure 4.18 Equivalent circuit for calculation of circulating currents

The path of circulating currents is assumed to be resistive because the inductance between parallel conductors is much smaller on account of adequate transpositions, small spacing between conductors and one-turn path of circulating current. It is reported in [17] that a significant reduction in circulating current losses can be achieved by deciding the correct locations of transpositions based on such 3-D field calculations.

Till now, we considered only radially placed parallel conductors. There can be axially placed parallel conductors also. If the radial field is symmetrical at both the ends and if there are two parallel conductors placed axially as shown in figure 4.19, the net leakage field enclosed by the loop formed by the two conductors is zero (since the radial field is in opposite directions at the top and bottom ends of the winding). Hence, there will not be any circulating currents on account of the radial field, obviating any need for the transpositions in axial direction for these two conductors. If the radial field is significantly different at the top and bottom ends, the axial transpositions (exchange of axial positions) may be required at the pre-determined locations (based on FEM analysis). In such cases, the approximate analytical formulations (which consider only the axial field) cannot be used.

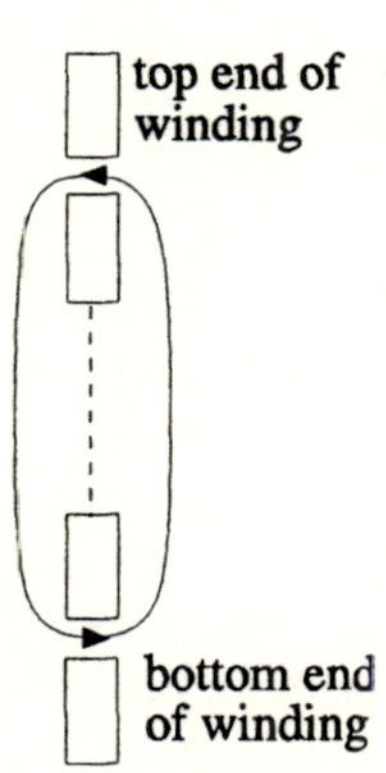

Figure 4.19 Axially placed parallel conductors in a winding

References

1. Stoll, R. L. *The analysis of eddy currents*, Clarendon Press, Oxford, 1974.
2. Hayt, W. H. *Engineering electromagnetics*, McGraw-Hill Book Company, Singapore, 1989, pp. 336-363.
3. Dexin, X., Yunqiu, T., and Zihong, X. FEM analysis of 3-D eddy current field in power transformer, *IEEE Transactions on Magnetics*, Vol. 23, No. 5, May 1987, pp. 3786-3788.
4. Holland, S. A., O'Connell, and G. P., Haydock, L. Calculating stray losses in power transformers using surface impedance with finite elements, *IEEE Transactions on Magnetics*, Vol. 28, No. 2, March 1992, pp. 1355-1358.
5. Berezovski, A. A. The calculation of stray losses in transformer tanks, *Elektrichestvo,* No. 9, 1966, pp. 1-7.
6. Subbarao, V. *Eddy currents in linear and nonlinear media*, Omega Scientific Publishers, New Delhi, 1991.
7. Agarwal, P. D. Eddy current losses in solid and laminated iron, *AIEE Transactions,* Vol. 78, May 1959, pp. 169-181.
8. Hammond, P. *Applied electromagnetism,* Pergamon Press, Oxford, 1971.
9. Turowski, J., Turowski, M., and Kopec, M. Method of three-dimensional network solution of leakage field of three-phase transformers, *IEEE Transactions on Magnetics*, Vol. 26, No. 5, September 1990, pp. 2911-2919.
10. Bean, R. L., Chackan, N., Moore, H. R., and Wentz, E. C. *Transformers for the electric power industry*, McGraw-Hill Book Company, New York, 1959, pp. 396-402.
11. Boyajian, A. Leakage reactance of irregular distributions of transformer windings by method of double Fourier series, *AIEE Transactions - Power Apparatus and Systems,* Vol. 73, Pt. III-B, 1954, pp. 1078-1086.
12. Rabins, L. Transformer reactance calculations with digital computers, *AIEE Transactions - Communications and Electronics,* Vol. 75, Pt. I, 1956, pp. 261-267.
13. Andersen, O. W. Transformer leakage flux program based on finite element method, *IEEE Transactions on Power Apparatus and Systems*, Vol. PAS- 92, 1973, pp. 682-689.
14. Kulkarni, S. V., Gulwadi, G. S., Ramachandran, R., and Bhatia, S. Accurate estimation of eddy loss in transformer windings by using FEM analysis, *International Conference on Transformers, TRAFOTECH-94,* Bangalore, India, January 1994, pp. I7-I10.
15. Kulkarni, S. V. and Ramachandrudu, B. Orthogonal array design of experiments for analysis of key performance parameter of transformer, *First World Congress on Total Quality Management,* Organized by Royal Statistical Society and Sheffield Hallam University, Sheffield, U.K., 10-12 April 1995, pp. 201-204.

16. McWhirter, J. H., Oravec, J. J., and Haack, R. Computation of magnetostatic fields in three dimensions based on Fredholm integral equations, *IEEE Transactions on Magnetics*, Vol. MAG-18, No. 2, March 1982, pp. 373-378.
17. Girgis, R. S., Yannucci, D. A., and Templeton, J. B. Performance parameters of power transformers using 3D magnetic field calculations, *IEEE Transactions on Power Apparatus and Systems*, Vol. PAS-103, No. 9, September 1984, pp. 2708-2713.
18. Girgis, R. S., Scott, D. J., Yannucci, D. A., and Templeton, J. B. Calculation of winding losses in shell form transformers for improved accuracy and reliability – Part I: Calculation procedure and program description, *IEEE Transactions on Power Delivery,* Vol. PWRD-2, No. 2, April 1987, pp. 398-410.
19. Kozlowski, M. and Turowski, J. Stray losses and local overheating hazard in transformers, *CIGRE 1972*, Paper No. 12-10.
20. Mullineux, N., Reed, J. R., and Whyte, I. J. Current distribution in sheet and foil wound transformers, *Proceedings IEE,* Vol. 116, No. 1, January 1969, pp. 127-129.
21. El-Missiry, M. M. Current distribution and leakage impedance of various types of foil-wound transformers, *Proceedings IEE*, Vol. 125, No. 10, October 1978, pp. 987-992.
22. Turowski, J. Additional losses in foil and bar wound transformers, *IEEE PES Winter Meeting and Symposium,* New York, January 25-30, 1976, Paper No. A-76-151-1.
23. Genon, A., Legros, W., Adriaens, J. P., and Nicolet, A. Computation of extra joule losses in power transformers, *ICEM 88*, pp. 577-581.
24. Ram, B. S. Loss and current distribution in foil windings of transformers, *Proceedings IEE - Generation, Transmission and Distribution*, Vol. 145, No. 6, November 1998, pp. 709-716.
25. Kaul, H. J. Stray current losses in stranded windings of transformers, *AIEE Transactions,* June 1957, pp. 137-149.
26. de Buda, R. G. Winding transpositions for electrical apparatus, U. S. Patent No. 2,710,380, June, 1955.
27. Rothe, F. S. *An introduction to power system analysis,* John Wiley, New York, 1953, pp. 45-50.
28. Isaka, S., Tokumasu, T., and Kondo, K. Finite element analysis of eddy currents in transformer parallel conductors, *IEEE Transactions on Power Apparatus and Systems,* Vol. PAS-104, No. 10, October 1985, pp. 2731-2737.
29. Koppikar, D. A., Kulkarni, S. V., Ghosh, G., Ainapure, S. M., and Bhavsar, J. S. Circulating current loss in transformer windings, *Proceedings IEE - Science, Measurement and Technology*, Vol. 145, No. 4, July 1998, pp. 136-140.

30. Kulkarni, S. V., Palanivasagam, S., and Yaksh, M. Calculation of circulating current loss in transformer windings using FE analysis, *ANSYS Users Conference*, Pittsburgh, USA, August 27-30, 2000.

5

Stray Losses in Structural Components

The previous chapter covered the theory and fundamentals of eddy currents. It also covered in detail, the estimation and reduction of stray losses in windings, viz., eddy loss and circulating current loss. This chapter covers estimation of remaining stray losses, which predominantly consist of stray losses in structural components. Various countermeasures required for the reduction of these stray losses and elimination of hot spots are discussed.

The stray loss problem becomes increasingly important with growing transformer ratings. Ratings of generator transformers and interconnecting auto-transformers are steadily increasing over last few decades. Stray losses of such large units can be appreciably high, which can result in higher temperature rise, affecting their life. This problem is particularly severe in the case of large auto-transformers, where actual impedance on equivalent two-winding rating is higher giving a very high value of stray leakage field. In the case of large generator transformers and furnace transformers, stray loss due to high current carrying leads can become excessive, causing hot spots. To become competitive in the global marketplace it is necessary to optimize material cost, which usually leads to reduction in overall size of the transformer as a result of reduction in electrical and magnetic clearances. This has the effect of further increasing stray losses if effective shielding measures are not implemented. Size of a large power transformer is also limited by transportation constraints. Hence, the magnitude of stray field incident on the structural parts increases much faster with growing rating of transformers. It is very important for a transformer designer to know and estimate accurately all the stray loss components because each kW of load loss may be capitalized by users from US$750 to US$2500. In large transformers, a reduction of stray loss by even 3 to 5 kW can give a competitive advantage.

Stray losses in structural components may form a large part (> 20%) of the total load loss if not evaluated and controlled properly. A major part of stray losses occurs in structural parts with a large area (e.g., tank). Due to inadequate shielding of these parts, stray losses may increase the load loss of the transformer substantially, impairing its efficiency. It is important to note that the stray loss in some clamping elements with smaller area (e.g., flitch plate) is lower, but the incident field on them can be quite high leading to unacceptable local high temperature rise seriously affecting the life of the transformer.

Till 1980, a lot of work was done in the area of stray loss evaluation by analytical methods. These methods have certain limitations and cannot be applied to complex geometries. With the fast development of numerical methods such as Finite Element Method (FEM), calculation of eddy loss in various metallic components of the transformer is now easier and less complicated. Some of the complex 3-D problems when solved by using 2-D formulations (with major approximations) lead to significant inaccuracies. Developments of commercial 3-D FEM software packages since 1990 have enabled designers to simulate the complex electromagnetic structure of transformers for control of stray loss and elimination of hot spots. However, FEM analysis may require considerable amount of time and efforts. Hence, wherever possible, a transformer designer would prefer fast analysis with sufficient accuracy so as to enable him to decide on various countermeasures for stray loss reduction. It may be preferable, for regular design use, to calculate some of the stray loss components by analytical / hybrid (analytically numerical) methods or by some formulae derived on the basis of one-time detailed analysis. Thus, the method of calculation of stray losses should be judiciously selected; wherever possible, the designer should be given equations / curves or analytical computer programs providing a quick and reasonably accurate calculation.

Computation of stray losses is not a simple task because the transformer is a highly asymmetrical and three-dimensional structure. The computation is complicated by

- magnetic non-linearity
- difficulty in quick and accurate computation of stray field and its effects
- inability in isolating exact stray loss components from tested load loss values
- limitations of experimental verification methods for large power transformers

Stray losses in various clamping structures (frame, flitch plate, etc.) and the tank due to the leakage field emanating from windings and due to the field of high current carrying leads are discussed in this chapter. The methods used for estimation of these losses are compared. The effectiveness of various methods used for stray loss control is discussed. Some interesting phenomena observed during three-phase and single-phase load loss tests are also reported.

5.1 Factors Influencing Stray Losses

With the increase in ratings of transformers, the proportion of stray losses in the load loss may increase significantly. These losses in structural components may exceed the stray losses in windings in large power transformers (especially autotransformers). A major portion of these stray losses occurs in structural components with a large area (e.g., tank) and core clamping elements (e.g., frames). The high magnitude of stray flux usually does not permit designers to disregard the non-linear magnetic characteristics of steel elements. Stray losses in structural steel components depend in a very complicated manner on the parameters such as the magnitude of stray flux, frequency, resistivity, type of excitation, etc.

In the absence of hysteresis and non-linearity of magnetic characteristics, the expression for the eddy loss per unit surface area of a plate, subjected to (on one of its surfaces) a magnetic field of r.m.s. value (H_{rms}), has been derived in Chapter 4 as

$$P_e = \sqrt{\frac{\omega\mu}{2\sigma}}\, H_{rms}^2 \tag{5.1}$$

Hence, the total power loss in a steel plate with a permeability μ_s can be given in terms of the peak value of the field (H_0) as

$$P = \iint_S \sqrt{\frac{\omega\mu_s}{2\sigma}}\, \frac{H_0^2}{2}\, ds \qquad \text{..... linear characteristics} \tag{5.2}$$

This equation assumes a constant permeability. It is necessary to take into account the non-linear magnetic saturation effect in structural steel parts because their surfaces are often saturated due to the skin effect. Non-linearity of magnetic characteristics can be taken into account by a linearization coefficient as explained in Section 4.4. Thus, the total power loss with the consideration of non-linear characteristics can be given by

$$P = a_l \iint_S \sqrt{\frac{\omega\mu_s}{2\sigma}}\, \frac{H_0^2}{2}\, ds \qquad \text{..... non-linear characteristics} \tag{5.3}$$

The term a_l in the above equation is the linearization coefficient. Equation 5.3 is applicable to a simple geometry of a plate excited by a tangential field on one of its sides. It assumes that the plate thickness is sufficiently larger than the depth of penetration (skin depth) so that it becomes a case of infinite half space. For magnetic steel, as discussed in Section 4.4, the linearization coefficient has been taken as 1.4 in [1]. For a non-magnetic steel, the value of the coefficient is 1 (i.e., a_l=1).

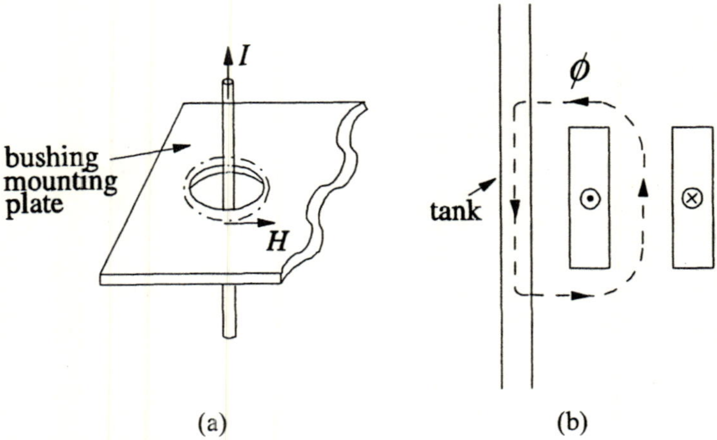

Figure 5.1 Types of excitation

5.1.1 Type of surface excitation

In transformers, there are predominantly two kinds of surface excitation as shown in figure 5.1. In case (a), the incident field is tangential (e.g., bushing mounting plate). In this case, the incident tangential field is directly proportional to the source current since the strength of the magnetic field (*H*) on the plate surface can be determined approximately by the principle of superposition [2]. In case (b), for estimation of stray losses in the tank due to a leakage field incident on it, only the normal (radial) component of the incident field (ϕ) can be considered as proportional to the source current. The relationship between the source current and the tangential field component is much more complicated. In many analytical formulations, the loss is calculated based on the tangential components (two orthogonal components in the plane of plate), which need to be evaluated from the normal component of the incident field with the help of Maxwell's equations. The estimated values of these two tangential field components can be used to find the resultant tangential component and thereafter the tank loss as per equation 5.3.

Let us use the theory of eddy currents described in Chapter 4 to analyze the effect of different types of excitation on the stray loss magnitude and distribution. Consider a structural component as shown in figure 5.2 (similar to that of a winding conductor of figure 4.5) which is placed in an alternating magnetic field in the y direction having peak amplitudes of H_1 and H_2 at its two surfaces. The structural component can be assumed to be infinitely long in the x direction. Further, it can be assumed that the current density J_x and magnetic field intensity H_y are functions of z only. Proceeding in a way similar to that in Section 4.3 and assuming that the structural component has linear magnetic characteristics, the diffusion equation is given by

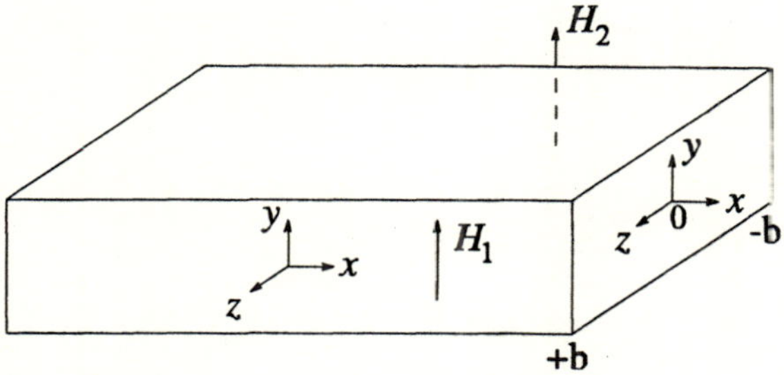

Figure 5.2 Stray loss in a structural component

$$\frac{d^2H_y}{dz^2} = j\omega\mu\sigma H_y \tag{5.4}$$

The solution of this equation is

$$H_y = C_1 e^{\gamma z} + C_2 e^{-\gamma z} \tag{5.5}$$

where γ is propagation constant given by equation 4.39, viz. $\gamma = (1+j)/\delta$, δ being the depth of penetration or skin depth. Now, for the present case the boundary conditions are

$$H_y = H_1 \text{ at } z = +b \text{ and } H_y = H_2 \text{ at } z = -b \tag{5.6}$$

Using these boundary conditions, we can get expressions for the constants as

$$C_1 = \frac{H_1 e^{\gamma b} - H_2 e^{-\gamma b}}{e^{2\gamma b} - e^{-2\gamma b}} \text{ and } C_2 = -\frac{H_1 e^{-\gamma b} - H_2 e^{\gamma b}}{e^{2\gamma b} - e^{-2\gamma b}} \tag{5.7}$$

Substituting these values of constants back into equation 5.5 we get

$$H_y = \left[\frac{H_1 e^{\gamma b} - H_2 e^{-\gamma b}}{e^{2\gamma b} - e^{-2\gamma b}}\right] e^{\gamma z} - \left[\frac{H_1 e^{-\gamma b} - H_2 e^{\gamma b}}{e^{2\gamma b} - e^{-2\gamma b}}\right] e^{-\gamma z} \tag{5.8}$$

Since $\nabla \times \mathbf{H} = \mathbf{J}$ and $\mathbf{J} = \sigma\mathbf{E}$, and only the y component of **H** and x component of **J** are non-zero we get

$$E_x = -\frac{1}{\sigma}\frac{dH_y}{dz} \tag{5.9}$$

$$\therefore\ E_x = -\frac{\gamma}{\sigma}\left\{\left[\frac{H_1 e^{\gamma b} - H_2 e^{-\gamma b}}{e^{2\gamma b} - e^{-2\gamma b}}\right] e^{\gamma z} + \left[\frac{H_1 e^{-\gamma b} - H_2 e^{\gamma b}}{e^{2\gamma b} - e^{-2\gamma b}}\right] e^{-\gamma z}\right\} \tag{5.10}$$

In terms of complex vectors, the (time average) power flow per unit area of the plate (in the x-y plane) can be calculated with the help of Poynting's theorem [3]:

$$P + jQ = (1/2)\left(E_x H_y^*\right)_{z=b} - (1/2)\left(E_x H_y^*\right)_{z=-b} \tag{5.11}$$

Substituting the values of H_y and E_x from equations 5.8 and 5.10, the value of eddy loss per unit area of the plate can be calculated. Figure 5.3 shows the plot of the normalized value of eddy loss, $P/(H^2/2\sigma\delta)$, versus the normalised plate thickness ($2b/\delta$) for three different cases of the tangential surface excitation.

Case 1 ($H_1 = H$ and $H_2 = 0$): As expected, the eddy loss for this case decreases with the increase in plate thickness until the thickness becomes 1 to 2 times the skin depth. This situation resembles the case in a transformer when a current carrying conductor is placed parallel to a conducting plate (mild steel tank / pocket). For this case (see figure 5.3), the normalised active power approaches unity as the thickness and hence the ratio $2b/\delta$ increases. This is because it becomes a case similar to an infinite half space, where the power loss equals $H^2/(2\sigma\delta)$. It is to be remembered that H, H_1 and H_2 denote peak values.

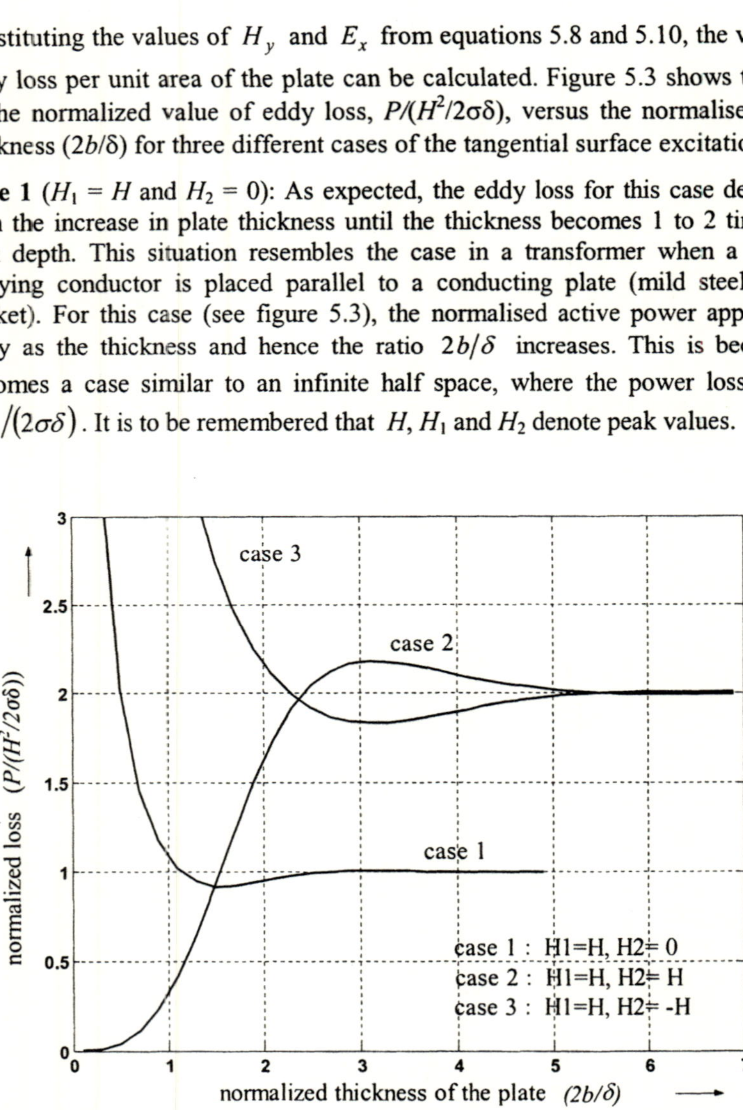

Figure 5.3 Eddy Loss in a structural plate for different surface excitations

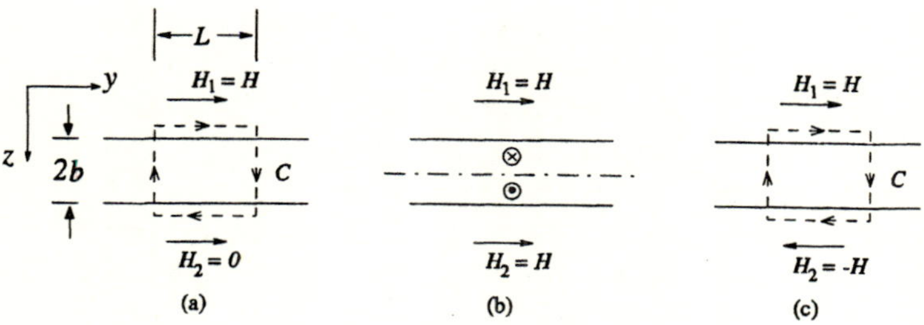

Figure 5.4 Explanation for curves in Figure 5.3

The plot also shows that the active power loss is very high for a thin plate. A qualitative explanation for this phenomenon can be given with reference to figure 5.4 (a). Consider a contour C shown in the figure. By applying Ampere's circuital law on the contour we obtain

$$\oint_C \mathbf{H} \cdot \mathbf{dL} = \int_S \mathbf{J} \cdot \mathbf{ds} \tag{5.12}$$

Noting that **H** is only in the y direction with $H_1 = H$ and $H_2 = 0$, the equation simplifies to

$$H\,L = I$$

As the thickness $2b$ decreases, the same amount of current passes through a smaller cross section of the plate and thus through a higher resistance, resulting in more loss.

Case 2 ($H_1 = H_2 = H$): Here, the eddy loss increases with the increase in the plate thickness. This situation arises in lead terminations / bushing mounting plates, where a current passes through holes in the metallic plates. In this case, as the thickness increases, normalized active power loss approaches the value of 2 because, for $2b/\delta >> 1$, the problem reduces to that of two infinite half-spaces, each excited by the peak value of field (H) on their surfaces. Therefore, the total loss adds up to 2 per-unit. As the thickness decreases, the active power loss decreases in contrast with Case 1. As shown in figure 5.4 (b), the currents in two halves of the plate are in opposite directions (as forced by the boundary conditions of H_1 and H_2). For a sufficiently small thickness, the effects of these two currents tend to cancel each other reducing the loss to zero.

Case 3 ($H_1 = -H_2 = H$): Here, the eddy loss decreases with the increase in thickness. For very high thickness (much greater than the skin depth), the loss approaches the value corresponding to two infinite half-spaces, i.e., $H^2/(\sigma\delta)$. As the thickness decreases, the power loss approaches very high values. For the representation given in figure 5.4 (c), an explanation similar to that for Case 1 can be given. The application of Ampere's circuital law gives double the value of current (i.e., $2\,H\,L = I$) as compared to Case 1. Hence, as the thickness ($2b$) decreases, the current has to pass through a smaller cross section of the plate and thus through a higher resistance causing more loss.

In the previous three cases, it is assumed that the incident magnetic field intensity is tangential to the surface of a structural component (e.g., bushing mounting plate). If the field is incident radially, the behavior of stray loss is different. Based on a number of 2-D FEM simulations involving a configuration in which the leakage field from the windings is radially incident on a structural component (e.g., tank or flitch plate), the typical curves are presented in figure 5.5. The figure gives the variation of loss in a structural component as the thickness is increased, for three different types of material: magnetic steel, non-magnetic steel and aluminum. The curves are similar to those given in [4] wherein a general formulation is given for the estimation of losses in a structural component for any kind of spatial distribution of the incident magnetic field.

Let us now analyse the graphs of three different types of materials given in figure 5.5.

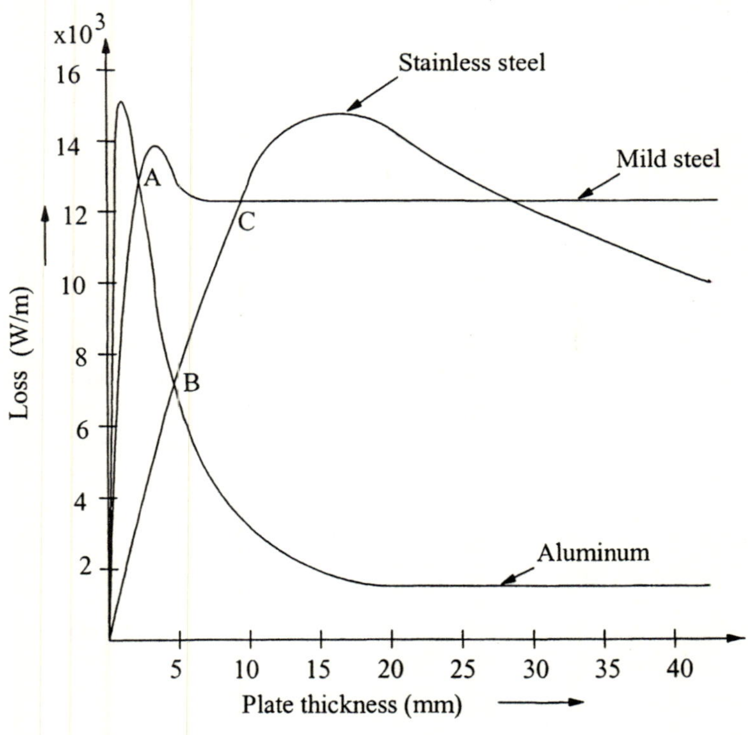

Figure 5.5 Loss in different materials for radial excitation

1) **Magnetic steel**: One can assume that the magnetic steel plate is saturated due to its small skin depth. Hence, the value of relative permeability corresponding to the saturation condition is taken ($\mu_r = 100$). With $\sigma = 7 \times 10^6$ mho/m, we get the value of skin depth as 2.69 mm at 50 Hz. It can be seen from the graph that the power loss value reaches a maximum in about two skin depths and thereafter remains constant. This behavior is in line with the theory of eddy currents and skin depth elaborated in Chapter 4. Since eddy currents and losses are concentrated at the surface only, increasing the plate thickness beyond few skin depths does not change the effective resistance offered to the eddy currents and hence the loss remains constant (at a value which is governed by equation 4.74).

2) **Aluminum**: In case of aluminum with $\mu_r = 1$ and $\sigma = 29 \times 10^6$ mho/m, the skin depth at 50 Hz is 13.2 mm. It can be observed from the graph that the loss first increases with thickness and then reduces. The phenomenon can be analyzed qualitatively from the supply end as an equivalent resistive-inductive circuit. For small thickness (thin plates), it becomes a case of resistance-limited behavior (as discussed in Section 4.5.1) and the effective resistance is larger compared to the inductance. Hence, the equivalent circuit behaves as a predominantly resistive circuit, for which the loss can be given as $P = (V^2/R)$, where V is the supply voltage. An increase of the thickness of the aluminum plate leads to a decrease of resistance, due to the increased cross section available for the eddy-currents, and hence the loss increases. This is reflected in a near-linear increase in losses with the increase of plate thickness.

Upon further increase of the plate thickness, the resistance continues to decrease while the inductance gradually increases, and the circuit behavior changes gradually from that of a purely resistive one to that of a series $R-L$ circuit. The power loss undergoes a peak, and starts to decrease as the circuit becomes more inductive. Finally, when the thickness is near or beyond the skin depth, the field and eddy currents are almost entirely governed by the inductive effects (inductance-limited behavior). The field does not penetrate any further when the plate thickness is increased. The equivalent resistance and inductance of the circuit become independent of the increase in the plate thickness. The power loss also approaches a constant value as the thickness increases significantly more than the skin depth making it a case of infinite half space. Since the product ($\sigma \cdot \delta$) is much higher for the aluminum plate than that for the mild steel plate, the constant (minimum) value of loss for the former is much lower (the loss is inversely proportional to the product ($\sigma \cdot \delta$) as per equation 4.74). The curves of aluminum and mild steel intersect at about 3 mm (point A).

3) **Non-magnetic stainless steel**: For the non-magnetic steel plate, the behavior is similar to that of the aluminum plate, both being non-magnetic materials. The curve is more flat as compared to aluminum as the skin depth of stainless steel is quite high. For a typical grade of stainless steel material with relative permeability

of 1 and conductivity of 1.136×10^6 mho/m, the skin depth is 66.78 mm at 50 Hz. Another difference is that as the thickness is increased, loss approaches a constant value higher than the aluminum plate but lower than the magnetic steel plate since the product $(\sigma\cdot\delta)$ for stainless steel lies between that of mild steel and aluminum. The intersection point (B) of the curves for stainless steel and aluminum occurs at about 5 mm and the intersection point (C) of the curves for stainless steel and mild steel occurs at about 10 mm. The location of intersection points depends on the configuration being analyzed and the nature of the incident field.

With the increase in the plate thickness, the values of losses in the mild steel (MS), aluminum (AL) and stainless steel (SS) plates stabilize to 12.2 kW/m, 1.5 kW/m and 5.7 kW/m respectively for particular values of currents in the windings. For large thickness, it becomes a case of infinite half space and the three loss values should actually be in proportion to $(1/\sigma\,\delta)$ for the same value of tangential component of magnetic field intensity (H_0) on the surface of the plate (as per equation 4.74). The magnitude and nature of eddy currents induced in these three types of plates are different, which makes the value of H_0 different for these cases. Also, the value of H_0 is not constant along the surface (as observed from the FEM analysis). Hence, the losses in the three materials are not in the exact proportion of their corresponding ratios $(1/\sigma\,\delta)$. Nevertheless, the expected trend is there; the losses follow the relationship $(loss)_{MS} > (loss)_{SS} > (loss)_{AL}$ since $(1/\sigma\,\delta)_{MS} > (1/\sigma\,\delta)_{SS} > (1/\sigma\,\delta)_{AL}$.

A few general conclusions can be drawn based on the above discussion:

1) When a plate made of non-magnetic and highly conductive material (aluminum or copper) is used in the vicinity of field due to high currents or leakage field from windings, it should have thickness at least comparable to its skin depth (13.2 mm for aluminum and 10.3 mm for copper at 50 Hz) to reduce the loss in it to a low value. For the field due to a high current, the minimum value of loss is obtained for a thickness of [5],

$$t_{\min} = \frac{\pi}{2}\,\delta \tag{5.13}$$

For aluminum (with $\delta = 13.2$ mm), we get the value of $t_{\min}$ as 20.7 mm at 50 Hz. The ratio $t_{\min}/\delta$ corresponding to the minimum loss value is 1.57. This agrees with the graph of figure 5.3 corresponding to Case 1 (assuming that tangential field value $H_2 \cong 0$, which is a reasonable assumption for a thickness 50% more than δ), in which the minimum loss is obtained for the normalized thickness of 1.57. For the case of radial incident field also (figure 5.5), the loss reaches a minimum value at the thickness of about 20 mm. For $t < (0.5\times\delta)$, the loss becomes substantial and may lead to overheating of the plate. Hence, if aluminum or copper is used as an electromagnetic (eddy current) shield, then it should have sufficient thickness to eliminate its overheating and minimize the stray loss in the structural component (shielded by it). Sufficient thickness of a shield ensures that its effective resistance is close to the minimum value.

2) Since the skin depth of mild steel (2.69 mm) is usually much less than the thickness required from mechanical design considerations, one may not be able to change its thickness to control the eddy loss. Hence, either magnetic shunts (made of low reluctance steel material) or electromagnetic shields (aluminum or copper) are used to minimize the stray losses in structural components made of mild steel material in medium and large transformers.

3) From figure 5.5, it is clear that the loss in a stainless steel plate is less than a mild steel plate for lower values of thickness. Hence, when a structural component is made of stainless steel, its thickness should be as small as possible (permitted from mechanical design considerations) in order to get a lower loss value. Thus, if a mild steel flitch plate is replaced by a stainless steel one, the stray loss in it is lower only if its thickness is about 10 mm or lower.

5.1.2 Effect of load, temperature and frequency

Generally, it is expected that the load loss test is conducted at the rated current. For large power transformers the tested load loss value at a lower current when extrapolated to the rated condition in the square proportion of currents may result in a value less than the actual one. This is because the stray losses in structural components, which form an appreciable part of the total load loss in large power transformers, may increase more than the square proportion. With the increase in winding currents and leakage field values, saturation effects in the (mild steel) material used for structural components increase. If magnetic or electromagnetic shield is not adequately designed, it becomes less effective at higher currents increasing stray losses. The exponent of current for stray losses may even be of the order of 2.3 to 2.5 instead of 2 in such cases [2]. Hence, depending upon the proportion of stray losses in the total load loss, the latter will be higher than that extrapolated with the exponent of 2. Hence, it is preferable to do the load loss test at the rated current for large transformers. If the test plant is having some limitation, the test can be done at a current less than the rated value subject to the agreement between user and manufacturer.

It should be noted that equation 5.2 or 5.3 can be used for a plate excited by a tangential field on one side, the plate thickness being sufficiently larger than the skin depth so that it becomes a case of infinite half space. By using an analytical approximation for the magnetization curves of a commonly used mild steel material, equation 5.2 or 5.3 for the power loss per unit surface area in a massive steel element subjected to a tangential field of H_0 at the surface, can be rewritten in terms of the source current I as [6]

$$P_e \propto I^{1.6} \times f^{0.5} \times \sigma^{-0.5} \tag{5.14}$$

The above equation is valid when H_0 is proportional to I, which is true for example in the case of bushing mounting plates. The current exponent of 1.5 is reported in [4] for the loss in bushing mounting plates.

For stray losses in magnetic steel plates subjected to the field of high current carrying bars (leads), the exponent of current is slightly less than 2. The exponent is a function of distance between the bar and the plate [7] ($=1.975-0.154\log_{10} h$, where h is distance from center of bar to plate surface in inches). For aluminum plates, the current exponent can be taken as 2 [7, 8].

The power loss per unit area for an incident flux ϕ, which is radial in nature (incident normally on the plate), is given by [6]

$$P_e \propto \phi^{2.8} \times f^{1.9} \times \sigma^{0.9} \tag{5.15}$$

Equation 5.15 is applicable to the case of tank plate subjected to the stray leakage field emanating from windings. The inter-winding gap flux is proportional to the current in windings. It has been reported [9] that in the case of tank plate being penetrated by a part of stray (leakage) field originating from windings, the relation between this radial field and the winding current is

$$\phi \propto I^{\kappa} \tag{5.16}$$

where κ is in the range of 0.8 to 0.9. Hence, equation 5.15 can be re-written in terms of current as

$$P_e \propto I^{\eta} \times f^{1.9} \times \sigma^{0.9} \tag{5.17}$$

where η, the exponent of current, is in the range of 2.2 to 2.5, which is in line with the value of 2.3 given in [2]. Hence, some stray loss components increase with the load current having an exponent greater than 2. Since these losses generally do not form the major part of load losses (if adequate shielding is done) and other stray losses vary with the current exponent of 2 or less than 2, the load loss dependence on the current is not much different than the square proportion. This is particularly true when the load loss test is done at or below the rated currents. Under overloading conditions, however, the load loss may increase with the current having an exponent higher than 2.

Losses due to high current field (e.g., in bushing mounting plate) vary in direct proportion of $\sqrt{f\rho}$ as per equation 5.14. From equation 5.17, it is clear that the stray losses in tank vary in almost the inverse proportion of resistivity and square proportion of frequency. Since the eddy losses in windings are also inversely proportional to resistivity (see equation 4.94), the total stray losses may be assumed to vary in the inverse proportion of resistivity (because the winding eddy losses and tank stray losses form the major part of stray losses for most of the transformers). For simplicity in calculations, they are assumed to be varying in the inverse proportion of the resistivity of winding conductor. Thus, the total stray losses in transformers can be related to resistivity as

$$P_{stray} \propto \frac{1}{\rho} \quad or \quad P_{stray} \propto \sigma \tag{5.18}$$

Since metals have positive temperature coefficient of resistance (resistivity increases with temperature), the stray losses can be taken to vary in the inverse proportion of temperature. If the load loss is guaranteed at 75°C, the stray loss component of the measured load loss at temperature t_m is converted to 75°C by the formula (when the copper conductor is used in windings)

$$P_{stray\_75°C} = P_{stray\_t_m} \times \frac{235 + t_m}{235 + 75} \tag{5.19}$$

For aluminum conductor the constant 235 is replaced by 225. Contrary to stray losses, the DC I^2R loss in windings varies in direction proportion of resistivity and hence the temperature. Therefore, for the copper conductor,

$$P_{I^2R\_75°C} = P_{I^2R\_t_m} \times \frac{235 + 75}{235 + t_m} \tag{5.20}$$

The I^2R loss at t_m (obtained by converting the value of I^2R loss corresponding to DC resistance test done at temperature t_r to temperature t_m) is subtracted from the measured value of load loss at t_m to calculate $P_{stray\_t_m}$. In order to calculate I^2R loss at t_m, the average winding temperature should be accurately determined. This is done by taking the average of top and bottom cooler temperatures. A substantial error may occur if, after oil processing and filtration cycles at about 50 to 60°C, a sufficient time is not provided for oil to settle down to a lower temperature (close to ambient temperature). In such a case, the temperature of windings may be quite different than the average of top and bottom cooler temperatures. Hence, it is preferable to wait till the oil temperature settles as close to the ambient temperature or till the difference between the top and bottom oil temperatures is small enough (the difference should not exceed 5°C as per ANSI Standard C57.12.90-1993) for accurate measurements. For the forced oil cooling system, a pump may be used to mix the oil to minimize the difference between top and bottom oil temperatures.

Regarding the effect of frequency variation on the total stray losses, it can be said that since the eddy loss in windings is proportional to the square of frequency, the stray loss in tank is proportional to frequency with an exponent less than 2 as per equation 5.17, and the stray loss due to the field of high current varies with $f^{0.5}$ as per equation 5.14, the total stray loss varies with frequency with an exponent x, whose value depends on the proportion of these losses in the total stray loss.

$$\therefore P_{stray} \propto f^{x} \tag{5.21}$$

If the winding eddy loss and stray losses in all structural components are treated separately, the winding eddy loss (stray loss in windings) is taken to be varying with frequency in the square proportion, whereas remaining stray losses can be assumed to vary with frequency having an exponent close to 1. According to IEC 61378 Part-1, 1997, *Transformers for industrial applications*, the winding eddy losses are assumed to depend on frequency with the exponent of 2, whereas stray losses in structural parts are assumed to vary with frequency with the exponent of 0.8. The frequency conversion factors for various stray loss components are reported and analyzed in [10].

For a transformer subjected to a non-sinusoidal duty, at higher frequencies the skin depth is lower than the thickness of the winding conductor. Hence, the relationship given by equation 4.90 or 5.1 is more valid (frequency exponent of 0.5) instead of that given by equation 4.94 (frequency exponent of 2 when thickness is less than the skin depth). Therefore, at higher frequencies the frequency exponent for the winding eddy loss reduces from 2 to a lower value [11].

5.2 Overview of Methods for Stray Loss Estimation [12]

After having seen basic theory of stray loss in structural components, we will now take a look at how methods of computation of stray losses have evolved from approximate 2-D analytical methods to present day advanced 3-D numerical methods.

5.2.1 Two-dimensional methods

A method is given in [13] for estimating leakage field, in which any kind of current density distribution can be resolved into space harmonics by a double Fourier series. The leakage field distribution obtained in the core window by this method can be used to calculate the approximate value of losses in flitch plate and first step of the core (in addition to eddy loss in windings). A two-dimensional Axisymmetric finite element formulation based on magnetic vector potential is used in [14] to obtain the tank losses. A computer program based on 2-D FEM formulation for skin effect and eddy current problems is presented in [15]. The formulation is suitable for both Cartesian and Axisymmetric 2-D problems. In [16], analogy between magnetic field equations for 2-D Cartesian and Axisymmetric problems is presented, and usefulness of this analogy for numerical calculations has been elaborated. The relation between finite element and finite difference methods is also clarified. Results of measurement of flux densities and eddy currents on a 150 MVA experimental transformer are reported. In [17], a 2-D finite element formulation based on magnetic vector potential is presented, which takes into account the varying distance between the winding and tank (due to 3-D geometry) by a correction factor. The 2-D FEM is used to get a static magnetic field solution in [18], and losses in tank are calculated by analytical formulae. The paper has reported test results of tank losses with magnetic and eddy current shielding. The geometric parameters affecting tank losses are explained through graphs. The need is emphasized in [19] for analyzing the stray losses as a complete system and not on an individual component basis. For example, placement of magnetic shunts on tank surface has the effect of reducing stray losses in clamping elements of the core since the leakage field gets more oriented towards the tank. The magnetic tank shunts also increase the radial field at the ends of outer winding and may increase the winding eddy loss if the width of its conductor is high enough to compensate the reduction in eddy loss due to reduced axial field at the ends. A number of 2-D FEM simulations are done to understand the effect of tank shields (magnetic / eddy current) on the other stray loss components (winding, flitch plate, frame and core edge losses). The simulations have shown that the effectiveness of magnetic shunts is quite dependent on the permeability of material indicating that the magnetic shunts should have adequate thickness so that their permeability does not reduce due to saturation.

In this era of 3-D calculations, 2-D methods are preferred for routine calculations of stray losses. These 2-D methods can be integrated into transformer design optimization programs which need reasonably accurate determination of stray losses.

5.2.2 Three-dimensional analytical formulations

A quasi 3-D formulation is given in [20], which obtains the radial flux density distribution on the tank wall by using method of images. The calculation of flux density does not consider the effect of the tank eddy currents on the incident field. This assumption is made to simplify the analytical formulation. From this radially incident peak value of the flux density (say in the z direction), the tangential components of magnetic field intensity (H_x and H_y) are calculated from Maxwell's equations. The resultant peak value ($H_0 = \sqrt{H_x^2 + H_y^2}$) is used to calculate the power loss per unit area with the assumption of step-magnetization characteristics (similar to the theory given in Section 4.4). The total losses in the tank are calculated by integration carried over the entire area. The method given in [21] calculates 3-D magnetic flux density distribution on the tank wall using a 2-D solution for one phase of a three-phase transformer.

These analytical methods may not get easily applied to complicated tank shapes and for finding the effects of tank shielding accurately. For such cases 3-D numerical methods are commonly used.

5.2.3 Three-dimensional numerical methods

Advent of high speed and large memory computers has made possible the application of numerical methods such as FEM, Finite Difference Method,

Boundary Element Method, etc., for the calculation of 3-D fields inside a transformer and accurate estimation of stray losses in structural components.

Boundary Element Method (BEM) is more suitable for open boundary problems involving structural parts of non-magnetic stainless steel, where it is difficult to determine the boundary conditions [22, 23]. For such open boundary conditions, some researchers have used [24] Integral Equation Method (IEM). In order to make the grid (mesh) generation easier, IEM with surface impedance modeling is proposed in [25]. Improved $\mathbf{T}-\Omega$ (electric vector potential – magnetic scalar potential) formulation is used in [26], wherein the total problem region is divided into source, non-conductive and conductive regions simplifying computational efforts.

An overview of methods for eddy current analysis is presented in [27]. The paper compares methods based on differential formulation (analytical, finite difference method, reluctance network method), integral formulation (volume integral, boundary element method), and variational methods (weighted residual, FEM) on attributes such as accuracy, ease of use, practicality and flexibility. The advantages of BEM for transient and open boundary problems are enumerated in the paper. There is continuous ongoing development in 3-D numerical formulations (which started gaining importance in 1980s) to improve their modeling capabilities and accuracy for the analysis of eddy currents.

After having seen the different approaches for the calculation of stray losses, we will now discuss in detail each stray loss component and its control.

5.3 Core Edge Loss

Core edge loss is the stray loss occurring due to flux impinging normally (radially) on core laminations. The amount and path of leakage field in the core depends on the relative reluctances of the alternative magnetic circuits. Load conditions of the transformer also have significant influence; the phase angle between the leakage field and magnetizing field decides the loading of the magnetic circuit and the total core losses during operation at site. During factory tests, the leakage flux path in the core depends largely on whether the inner or outer winding is short-circuited as explained in Section 5.12.2. The incident leakage flux density on the limb and clamping elements is quite appreciable in case of generator transformers due to relative closeness of the limb from the inter-winding gap as compared to autotransformers. Hence, there are more possibilities of hot spots being generated in these parts in generator transformers. However, the stray loss magnitude may be of the same order in generator transformers and autotransformers due to more leakage field in autotransformers on equivalent two winding basis.

In large transformers, the radially incident flux may cause considerable eddy currents to flow in the core laminations resulting in local hot spots. The flux penetration phenomenon is quite different in a laminated core structure as compared to a solid one. In a solid block of finite dimensions, the eddy currents tending to concentrate at the edges can complete their path through the side faces, and the field is confined to the surface (skin depth) in all faces. In the laminated case, there is restriction to the flow of eddy currents and the field penetrates much deeper as compared to the solid case. The leakage flux penetration into the laminated core poses an anisotropic and three-dimensional non-linear field problem. The problem is formulated in terms of electric vector potential and magnetic scalar potential (**T** - Ω formulation) in [28]. The solution is expressed in the form of three different characteristic modes, two associated with the core surfaces and the third describing the flux penetration into the interior. All the three modes are represented in a network model by complex impedances, and then the current distribution and losses are derived from the solution of the network. The core discontinuities (holes) are accounted by change of appropriate impedances. Thus, the method provides a means of studying effects of core steps, holes, ducts and discontinuities (due to lapped joints). The network has to be modified with any change in the geometry or type of excitation. The formulation in the paper has been verified on two experimental models of a core [29, 30]. Approximate formulae for finding the loss and temperature rise of a core due to an incident field are also given. The effect of type of flitch plate (magnetic or non-magnetic) on the core edge loss is also explained. A non-magnetic (stainless steel) flitch plate increases the core edge loss since it allows (due to its higher skin depth) the flux to penetrate through it to impinge on the laminations. Hence, although the use of non-magnetic flitch plate may reduce the loss in it (assuming that its thickness is sufficiently small as explained in Section 5.1.1), the core edge loss is generally increased.

The first step of the core is usually slit into two or three parts to reduce the core edge loss in large transformers. If the stack height of the first step of the core is less than about 12 mm, slitting may have to be done for the next step also. The use of a laminated flitch plate for large generator transformers and autotransformers is preferable since it also acts as a magnetic shunt (as described in Section 5.5).

The evaluation of exact stray loss in the core poses a challenge to transformer designers. With the developments in 3-D FEM formulations with features of anisotropic modeling (of permeability and conductivity), the computational difficulties can be overcome now.

5.4 Stray Loss in Frames

Frames (also called as yoke beams), serving to clamp yokes and support windings, are in vicinity of stray magnetic field of windings. Due to their large surface area and efficient cooling, hot spots seldom develop in them. The stray loss in frames has been calculated by Finite Difference Method and an analytical method in [31]. The loss in frames made up of mild steel, aluminum and non-magnetic steel are compared. It has been shown that the losses in frame and tank have mutual effect on each other. Non-magnetic steel is not recommended as a material for frames. It is expensive, difficult to machine and stray losses will be lower only if its thickness is sufficiently small. A quick and reasonably accurate calculation of the frame loss can be done by using 3-D Reluctance Network Method (RNM) [32]. The numerical methods such as FEM are also commonly used.

The loss in frames due to leakage field can be reduced by either aluminum shielding or by use of non-metallic platforms for supporting the windings. In distribution transformers, the stray loss in the tank may not be much since the value of leakage field is low. But the loss in frames due to currents in low voltage leads running parallel to them can be significant. For example, the current of a star connected LV winding of a 2 MVA, 11/0.433 kV transformer is 2666.67 A, which can result into stray loss in the frames of the order 1 kW (which is substantial for a 2 MVA transformer). Non-metallic frames can be used (after thorough assessment of their short circuit withstand capability) for eliminating the stray loss.

Another way of minimizing this loss is by having go and return arrangement of LV winding leads passing close to the frame. These two leads can be either firmly supported from the frame or they can pass through a hole made in the frame as shown in figure 5.6. The net field responsible for eddy current losses in the metallic frame is negligible as the two currents are in opposite directions.

A single lead may be allowed to pass through a hole in the frame with a non-magnetic insert (e.g., stainless steel material with high resistivity) as shown in figure 5.7 upto a certain value of current.

In power transformers, sometimes a frame of non-magnetic material (stainless steel) is used. As explained in Section 5.1.1, its thickness should be as small as mechanically possible; otherwise its loss may exceed the corresponding value for frame made of (magnetic) mild steel material.

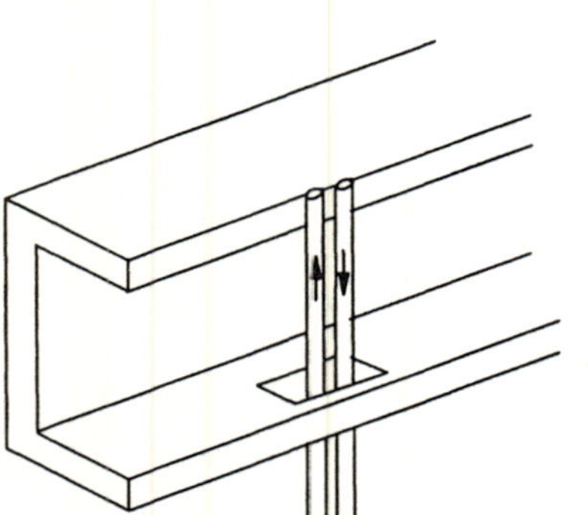

Figure 5.6 Go and return arrangement

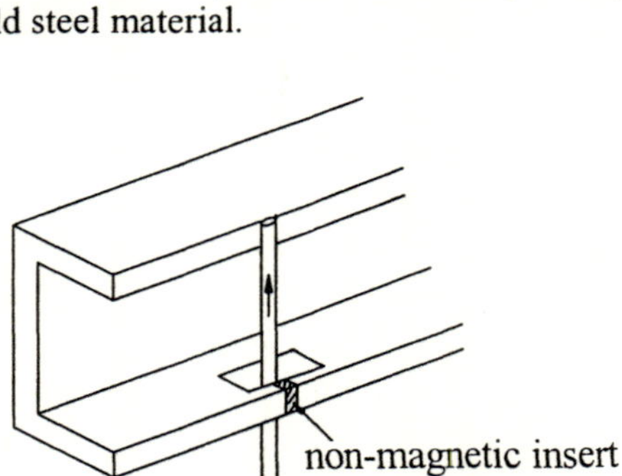

Figure 5.7 Non-magnetic insert

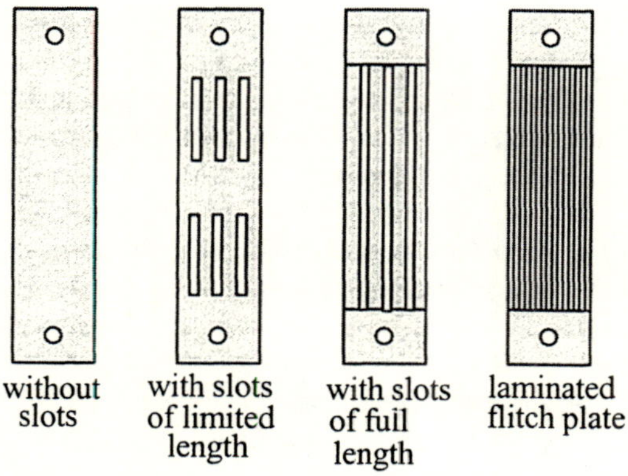

Figure 5.8 Types of flitch plates

5.5 Stray Loss in Flitch Plates

Stray flux departing radially through the inner surface of windings hits fittings such as flitch plates mounted on the core. On the surface of the flitch plate (lying on the outermost core-step of limbs for holding core laminations together vertically), the stray flux density may be much higher than that on the tank. Hence, although the losses occurring in a flitch plate may not form a significant part of the total load loss of a transformer, the local temperature rise can be much higher due to high value of incident flux density and poorer cooling conditions. The loss density may attain levels that may lead to a hazardous local temperature rise if the material and type of flitch plate are not selected properly. The higher temperature rise can cause deterioration of insulation in the vicinity of flitch plate, thereby seriously affecting the transformer life.

There are a variety of flitch plate designs being used in power transformers as shown in figure 5.8. For small transformers, mild steel flitch plate without any slots is generally used because the incident field is not large enough to cause hot spots. As the incident field increases in larger transformers, a plate with slots at the top and bottom ends can be used (where the incident leakage field is higher). Sometimes, flitch plates are provided with slots in the part corresponding to the tap zone in taps-in-body designs. These slots of limited length may be adequate if the incident field on the flitch plates is not high. Fully slotted plates are even better, but they are weak mechanically, and their manufacturing process is a bit more complicated. The plates can be made of non-magnetic stainless steel having high resistivity only if their thickness is small as explained in Section 5.1.1. When the incident leakage field on the flitch plate is very high, as in large generator transformers, the best option would be to use a laminated flitch plate. It consists of a stack of CRGO laminations, which are usually held together by epoxy molding to make the assembly mechanically strong. The top and bottom ends of laminations are welded to solid (non-magnetic) steel pads which are then locked to the frames. A laminated flitch plate not only minimizes its own eddy loss but it also acts as a magnetic shunt reducing the loss in the first step of the core.

The literature available on the analysis of flitch plate loss is quite scarce. An approximate but practical method for calculation of the loss and temperature rise of a flitch plate is given in [4], which makes certain approximations based on the experimental data given in [33]. The field strength at the inner edge of LV winding is assumed to vary periodically with a sinusoidal distribution in the space along the height of the winding, and the non-sinusoidal nature is accounted by multiplying the loss by a factor. The eddy current reaction is neglected in this analytical formulation. For a fully slotted flitch plate, the formulation is modified by considering that the plate is split into distinct parts. A more accurate 2-D / 3-D FEM analysis is reported in [34], in which many limitations of analytical formulations are overcome. The paper describes details of statistical analysis, *orthogonal array design of experiments*, used in conjunction with 2-D FEM for quantifying the effect of various factors influencing the flitch plate loss. This Section contains results of authors' paper [34] © 1999 IEEE. Reprinted, with permission, from *IEEE Transactions on Power Delivery*, Vol. 14, No. 3, July 1999, pp. 996-1001. The dependence of flitch plate loss on the axial length of windings, core-LV gap, winding to yoke clearance and LV-HV gap is observed to be high. The flitch plate loss varies almost linearly with LV-HV gap. A quadratic surface derived by multiple regression analysis can be used by designers for a quick but approximate estimation of the flitch plate loss. The loss value obtained can be used to decide type (with slots / without slots) and material (magnetic mild steel / non-magnetic stainless steel) of the flitch plate to control its loss and avoid hot spots. The effectiveness of number and length of slots in reducing losses can be ascertained accurately by 3-D field calculations. In the paper, in-depth analysis of eddy current paths has been reported for slotted mild steel and stainless steel flitch plates, having dimensions of 1535 mm × 200 mm × 12 mm, used in a single-phase 33 MVA, 220/132/11 kV autotransformer.

For this analysis, a mild steel (MS) flitch plate with μ_r = 1000 and σ = 4×10^6 mho/m has been studied. The corresponding skin depth is 1.1 mm at 50 Hz. The results obtained are summarized in table 5.1. The loss values shown are for one fourth of the complete plate.

Table 5.1 Loss in MS flitch plate

Case number	Description	Loss in watts
1	No slots	120
2	1 slot throughout	92
3	3 slots throughout	45
4	7 slots throughout	32
5	1 slot of 400mm length	100
6	3 slots of 400mm length	52
7	7 slots of 400mm length	45

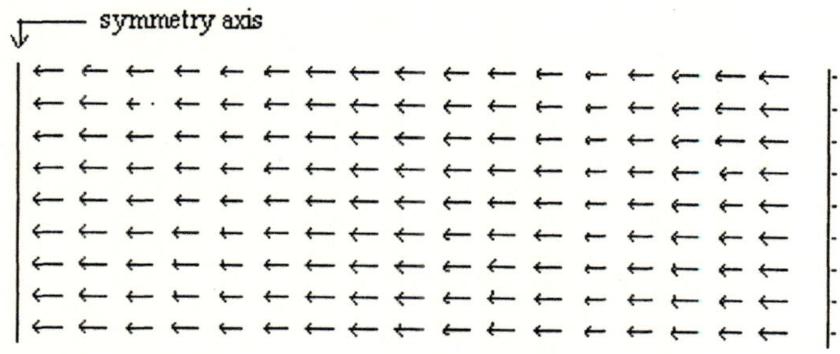

Figure 5.9 Eddy currents in MS plate with no slots

The loss for the '7 slots throughout' case is approximately 4 times less than that of the 'no slots' case. Theoretically, the loss is proportional to the square of width, hence for *n* slots, the loss should reduce approximately by a factor of $(n+1)$, i.e., 8 (if a plate width of $3w$ is divided by 2 slots into 3 plates of width w, then loss will theoretically reduce by a factor of $(3w)^2$ divided by $3w^2$, i.e., 3). The reason for this discrepancy can be explained as follows. The pattern of eddy currents is complex in a mild steel material. Eddy loss in it has two components, viz. loss due to radial incident field, and the other due to axial field (the incident radial flux changes its direction immediately once it penetrates inside the plate due to very small skin depth). This phenomenon is evident from the eddy current pattern in the plate cross section, taken at 0.5 mm from the surface facing the windings (figure 5.9 and figure 5.10). There is hardly any change in the eddy current pattern in this cross section after the introduction of slots. The direction of eddy currents suggests the predominance of axial field at 0.5 mm from the surface. Hence, there are eddy current loops in the thickness of the plate as shown in figure 5.11. These are the reasons for the ineffectiveness of slots in the MS plate, which is responsible for the fact that the reduction of losses is not by a factor of 8.

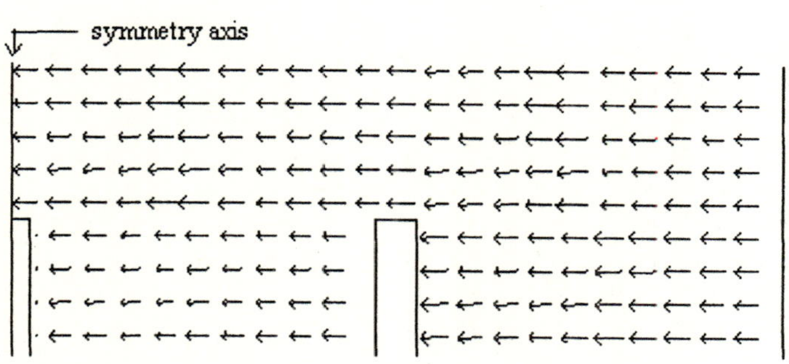

Figure 5.10 Eddy currents in MS plate with 3 slots

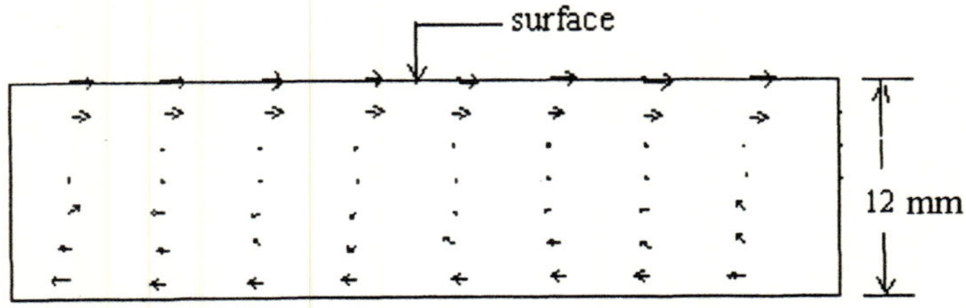

Figure 5.11 Eddy currents across thickness of MS plate with 3 slots

For a non-magnetic stainless steel (SS) flitch plate ($\mu_r = 1$, $\sigma = 1.13 \times 10^6$ mho/m), due to its large penetration depth (67 mm at 50 Hz), the incident field penetrates through it and hits the core laminations. This phenomenon is evident from the eddy current pattern at the plate cross section taken at 0.5 mm from the surface (figures 5.12 and 5.13). There is an appreciable distortion in the eddy current pattern after the introduction of slots.

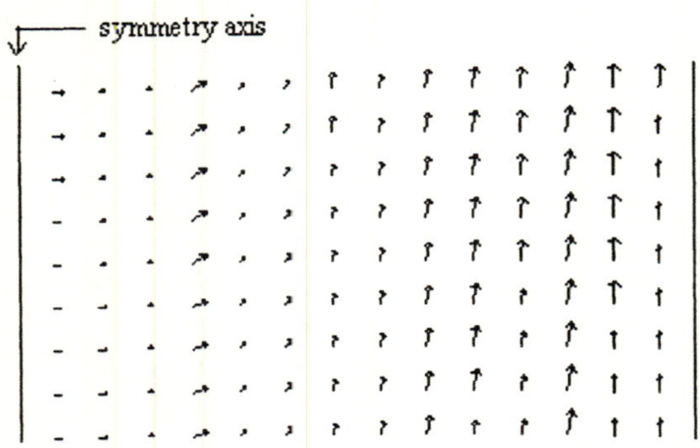

Figure 5.12 Eddy currents in SS plate with no slots

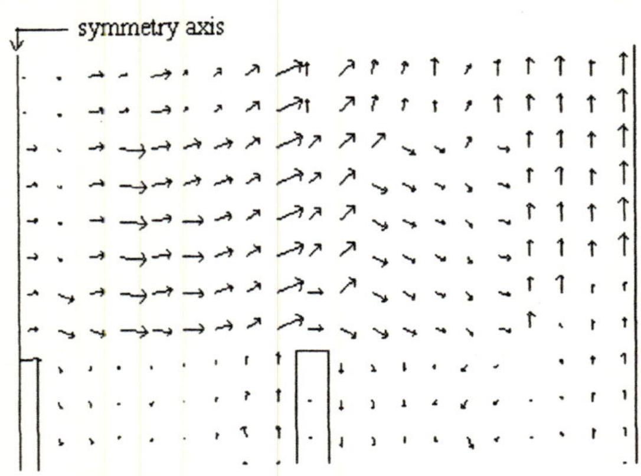

Figure 5.13 Eddy currents in SS plate with 3 slots

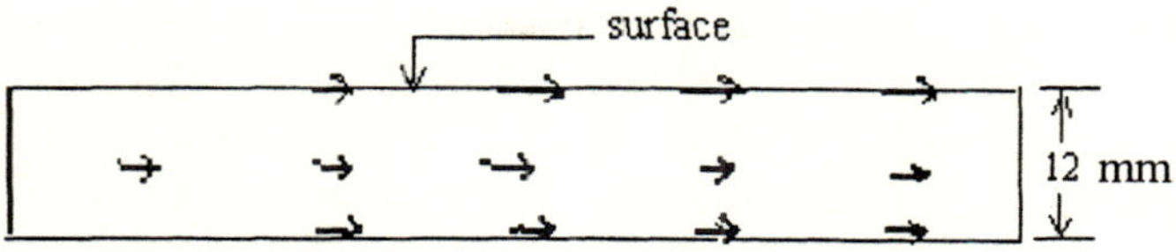

Figure 5.14 Eddy currents across thickness in SS plate with 3 slots

The direction of eddy currents indicates the predominance of radial field at the cross section, 0.5 mm from the surface. There are no eddy current loops in thickness of the plate (see figure 5.14). These are the reasons for the effectiveness of slots in the SS plate. The eddy current loops are parallel to the surface (on which the flux in incident) indicating that the eddy loss in the SS plate is predominantly due to the radial field. Hence, the slots in the SS plate are more effective as compared to the MS plate. This means that the loss should reduce approximately by a factor of (n+1). From the first two results given in table 5.2, we see that the reduction in the loss is more (12 times) than expected (8 times). This may be due to fact that each slot is 5 mm wide causing a further reduction in the loss due to the reduced area of conduction.

Due to higher resistivity of SS, the losses in the SS plate are lower than the MS plate. If results from tables 5.1 and 5.2 are compared for the 'no slots' case, it can be seen that the SS plate loss is not significantly lower than the MS plate loss for 12 mm thickness. For a higher thickness, the loss in the SS plate may exceed the loss in the MS plate, which is in line with the graphs in figure 5.5. It shows that in order to get lower losses with SS material, its thickness should be as small as possible with due considerations to mechanical design requirements. With the SS plate, shielding effect is not available. Hence, although losses in the flitch plate are reduced with SS material, the stray loss in the first step of the core may increase substantially if it is not split. Therefore, thicker flitch plates with a low incident flux density should be of MS material.

A laminated flitch plate (consisting of M4 grade CRGO laminations) has also been analyzed through 3-D FEM analysis by taking anisotropy into account. The direction along the flitch plate length is defined as soft direction and other two directions are defined as hard directions. The loss value obtained for the laminated flitch plate is just 2.5 watts, which is quite lower than the SS plate. Hence, laminated flitch plates are generally used for large power transformers, particularly generator transformers, where the incident flux density is quite high.

Table 5.2 Losses in SS flitch plate

Case Number	Description	Loss in watts
1	No slots	98
2	7 slots throughout	8
3	7 slots 400 mm long	11
4	3 slots 400 mm long	17

The eddy loss distribution obtained by 3-D FEM electromagnetic analysis is used for estimation of the temperature rise of the flitch plate by 3-D FEM thermal analysis [34, 35]. The heat generation rates (watts/m^3) for various zones of the flitch plate are obtained from the 3-D FEM electromagnetic analysis. The computed temperatures have been found to be in good agreement with that obtained by measurements. Thus, the method of combined 3-D electromagnetic FEM analysis and thermal FEM analysis can be used for the analysis of eddy loss and temperature rise of a flitch plate. Nowadays, commercial FEM software packages are available having multi-physics capability. Hence, the temperature rise can be found more easily without manual interface between the electromagnetic FEM analysis and thermal FEM analysis.

5.6 Stray Loss in Tank

The tank stray loss forms a major part of the total stray loss in large power transformers. Stray flux departing radially from the outer surface of winding gives rise to eddy current losses in transformer tank walls. Though the stray flux density in the tank wall is low, the tank loss may be high due to its large area. Hot spots seldom develop in the tank, since the heat is carried away by the oil. A good thermal conductivity of the tank material also helps to mitigate hot spots. The stray loss in tank is controlled by magnetic / eddy current shields.

Methods for estimation of tank loss have evolved from approximate analytical methods to present day more accurate three-dimensional numerical methods. The radial incident flux density at various points on the tank is found in [36] by neglecting the effect of eddy currents on the incident field. It is assumed that the ampere-turns of windings are concentrated at the longitudinal center of each winding as a current sheet, and the field at any point on the tank is calculated by superimposition of the fields due to all windings. The tank loss is calculated using the estimated value of the radial field at each point. The analytical formulation in [37] determines the field in air without the presence of tank, from the construction of the transformer and the currents in windings. Based on this field and the coefficient of transmission, the tangential component of the magnetic field strength on the inner surface of the tank is determined. The specific power loss at a point is then calculated by using the value of active surface resistance of the tank material. The total losses are determined by summing the specific losses on the surface of the tank. The analytical method, presented in [38], takes into account the hysteresis and non-linearity by using complex permeability. A current sheet, the sum of trigonometric functions in between the core and tank (both treated as infinite half space), represents mmf of windings. The calculated value of the radial component of the flux density at the tank surface is corrected by a coefficient accounting for the influence of eddy currents. The tank loss is found by Poynting's vector. The method can be applied for a specific tank shape only. The effect of magnetic / eddy current shields on the tank wall is not accounted in the method. The analytical approach in [39] expresses the incident flux density (obtained by any method) on the tank in terms of double Fourier series. Subsequently, after getting the field and eddy current distribution within the tank plate, the loss is evaluated by using volume integral. The results are verified by an experimental set-up in which a semi-circular electromagnet is used to simulate the radial incident field on the tank plate.

Thus, since the 1960s the research reported for calculation of tank loss has been mainly concentrating on various analytical methods involving intricate formulations, which approximate the three-dimensional transformer geometry to simplify the calculations. Transformer designers prefer fast interactive design with sufficient accuracy to enable them to decide the method for reducing tank stray losses. Reluctance Network Method [1] can fulfill the requirements of very fast estimation and control of the tank stray loss. It is based on a three-dimensional network of reluctances. The reluctances are calculated from various geometrical dimensions and electrical parameters of the transformer. There are two kinds of elements: magnetic resistances for non-conductive areas and magnetic impedances for conductive parts. The first ones are calculated purely from the geometrical dimensions of the elements, whereas the latter ones take into account analytically the skin effect, eddy current reactions with phase shift, non-linear permeability inside solid metals, and the effect of eddy current shields (if placed on the tank wall). Hence, the method is a hybrid method in which the analytical approach is used (for the portion of the geometry involving eddy currents) in conjunction with the numerical formulation.

The equivalent reluctance of the solid iron can be determined with the help of the theory of eddy currents explained in Chapter 4. For a magnetic field applied on the surface of solid iron in the y direction, and assuming that it is function of z only (figure 5.15), the amplitude of flux per unit length in the x direction is

$$\phi_0 = \mu \int_0^\infty H_y \, dz \tag{5.22}$$

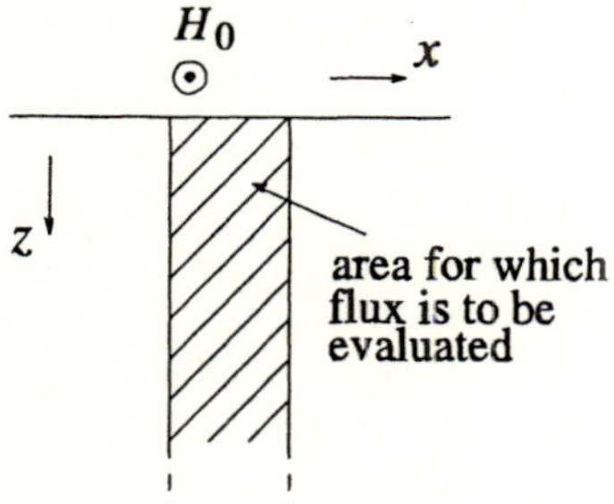

Figure 5.15 Equivalent reluctance for tank

In line with Section 4.3, for the peak value of the magnetic field intensity (H_0) at the surface, we can write from equation 4.68 for linear *B-H* characteristics $(\alpha = \beta = 1)$,

$$\phi_0 = \mu \int_0^{\infty} H_0 e^{-(1+j)z/\delta} dz \tag{5.23}$$

$$\therefore \phi_0 = \mu H_0 \frac{\delta}{1+j} \tag{5.24}$$

If $\Re$ is the equivalent reluctance per unit length at the surface of the tank in the *y* direction, then

$$H_0 \times 1 = \Re \times \phi_0 \tag{5.25}$$

From equations 5.24 and 5.25 we get

$$\Re = \frac{1}{\mu\delta} + j\frac{1}{\mu\delta} \quad \cdots\cdots \textit{ linear characteristics} \tag{5.26}$$

The above equation gives the value of complex reluctance per unit length in the *y* and *x* directions (per unit surface area) along the surface of tank (with the tank thickness along the *z* direction) for linear material characteristics. It is assumed that the tank thickness is more than 3 times the skin depth. For non-linear characteristics of magnetic steel, semi-empirical correction factors are used [32]

$$\Re \cong 0.52 \times \frac{1}{\mu\delta} + j0.86 \times \frac{1}{\mu\delta} \quad \cdots\cdots \textit{ non-linear characteristics} \tag{5.27}$$

The proper values of these elements (real and complex reluctances), corresponding to the transformer parameters and frequency of the supplied voltage, are placed into a network scheme along with the voltage sources which model magnetomotive forces in the windings. All the network elements are expressed in per unit (relative) values referred to the data of the inter-winding gap. The power losses are calculated from the surface field strength as per equation 5.3, with a semi-empirical linearization coefficient of 1.4 for the solid steel. The three-dimensional Reluctance Network Method for tank loss estimation has been verified [40, 41] on various ratings of power transformers from 31.5 MVA upto 315 MVA. The method is based on the assumption of four-quarter symmetric structure of a three-phase transformer. Later on the method has been further improved [42] to take into account various deformations of the symmetric model. Due to the increased number of non-standard reluctance elements, they are first expressed with the help of analytical formulae and then introduced into the matrix equation of the entire 3-D model.

As compared to analytical and semi-analytical methods, numerical methods can give more accurate results but higher computational efforts are required. Numerical methods can be combined with analytical formulations to reduce the computational efforts. A 3-D FEM analysis of eddy current problems presented in [43] uses complex magnetic vector potential. Eddy current losses in steel materials are computed by combining the numerical method with the analytical formulation because of the discretization problem due to very thin skin depth of the magnetic steels. A power transformer has dimensions of few meters, whereas skin depths are in millimeters resulting into errors due to a poor aspect ratio of elements. A method of modeling tank wall and other fittings with surface elements is outlined in [44], which obviates the need of complex layers of thin elements to account for the skin effect. The basis of formulation has been explained in Section 4.2. The surface impedance element modeling approach helps designers to calculate the tank loss efficiently and accurately.

The presence of the tank has some influence on the stray loss in other structural components (frame, flitch plate, etc.), which depends on the relative closeness of the tank from windings as compared to the core. If C and T are the distances of the core and tank from the inter-winding gap-center as shown in figure 5.16, the fringing of the leakage flux at winding ends towards the core is independent of the presence or absence of the tank with magnetic shunts if [45]

$$\frac{T}{C} \geq \frac{\pi}{(8C/h)} - 1 \tag{5.28}$$

In this case, the stray loss in the tank can be isolated by doing the load loss test with and without it. If the tank is lined with aluminum or copper (eddy current) shields, its effect on stray losses in other structural components is more pronounced. A much higher distance between the tank and outside winding is required [45] to make the tank's influence negligible, which is governed by the relation $T \geq h$.

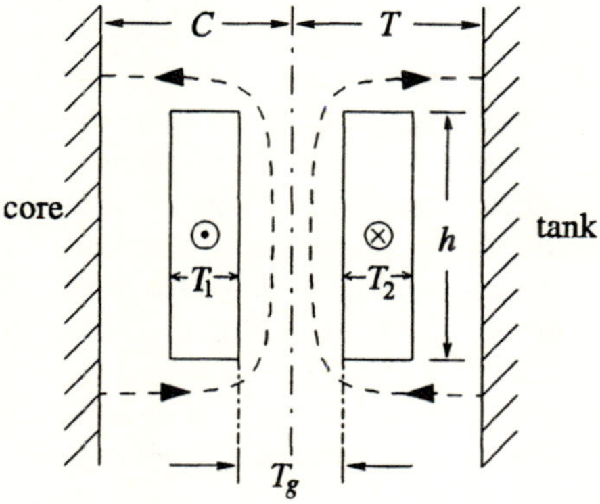

Figure 5.16 Effect of tank on other stray losses

Another important aspect related to the electromagnetic field calculations of a tank wall is the analysis of temperature rise of the bolted joint between the tank and cover. The currents induced in the tank and cover due to leakage and high current fields, are forced to complete their path through flange bolts. The bolts are overheated if the induced currents flowing in the tank and cover are large. The flow of these induced currents through the bolts can be avoided by completely isolating them from the tank and cover. This results in a number of problems. First, the induced currents may concentrate in the larger cross-sectional area of flanges causing local overheating, which leads to deterioration of gaskets over a period of time. Second, due to the bad electromagnetic contact between the tank and cover, there is an increase in magnetic voltage drop (magnetomotive force), leading to a greater magnetic field strength on the bolt surface, which may give rise to excessive local eddy current losses in the magnetic steel bolts. Also, the condition that the tank and cover should be at the same electrical (ground) potential is not satisfied. This phenomenon of overheating hazard in the case of bad electromagnetic contact has been investigated in [46] by representing the bolted joint by an equivalent reluctance. The local eddy currents in these bolts may cause dangerous hot spots, damaging the gaskets / sealing between the flanges. Hence, the better option would be to connect the two parts by metallic strips (links) made of high conductivity materials like aluminum or copper to maintain both the parts at the same electrical potential and to provide a low resistance alternative path to the induced currents in the tank and cover. The electrical connection through the metallic bolts cannot be relied upon (either due to presence of paint or bad contact resistance). The number of connecting links depends on the amount of leakage field / high current field. For small transformers with a low field at the tank surface, two connecting links may be enough.

Currents flowing in these connecting links can be hundreds of amperes in large power transformers (say, 300 MVA rating). These currents are high particularly in a bell type of tank construction (described in Chapter 10), in which curb (flanged) joint is at the position of the bottom yoke. A typical leakage field plot is shown in figure 5.17. The construction is not suitable for providing an effective shielding arrangement. This is because if a vertical magnetic shunt is placed with its bottom end at point B, its length will be lower than the ideal one. An additional small vertical shunt placed between points C and D will result into a higher temperature on the tank surface at the locations near these points (since the flux will leave the shunt and enter the tank at these locations). Hence, it is always preferable to have a magnetic shunt in one piece between the point at which the flux enters the tank to the point where it leaves the tank. Therefore, it is advisable to have the curb joint as low as possible (in the bell construction) or as high as possible (in the conventional construction with the curb joint at top). If it is not possible, a complex arrangement of shunts as shown in figure 5.18 may have to be tried. The shunt S2, which can be fixed to shunt S1, overlaps on the shunts S1 and S3 shielding the curb joint.

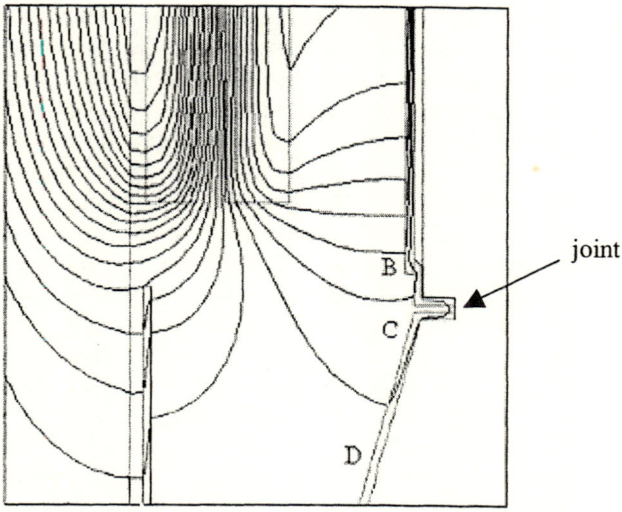

Figure 5.17 Leakage field plot for bell tank

When high current leads pass nearby the curb joint of the tank, excessive heating of bolts may occur resulting in deterioration of gaskets. In such cases, either adequate number and size of external links connecting the tank and cover should be used (for shunting the currents) or the magnitudes of currents should be controlled by careful positioning of the leads. The arrangement of figure 5.18 is not useful in the case of field due to high current leads.

5.7 Stray Loss in Bushing Mounting Plates

As the transformer rating increases, the current in its low voltage side as well as the high voltage side increases. This leads to increase in the eddy currents in the structures that surround the bushings (e.g., bushing mounting plate).

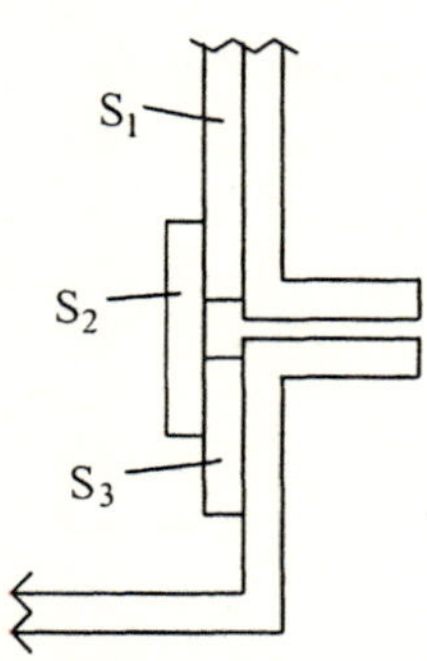

Figure 5.18 Shielding of curb joint

The bushing mounting plates are made of mild steel (MS) or stainless steel (SS) material. As the rating increases, the eddy current loss and the related heating effects increase. The loss and temperature rise of the MS plate (being a magnetic material) are more, and hence for higher ratings SS plates are used. But compared to the MS material, the SS material is expensive; hence instead of using a SS plate, an MS plate with SS inserts is used up to a certain current.

An experimental analysis of eddy current phenomenon in the structure that surrounds the high current bushings of a large capacity transformer is presented in [47]. A basic model with a conducting current of 20 kA is constructed to investigate the eddy current phenomenon. An eddy current probe is used for the direct measurement of the magnitude and phase of eddy currents. Two-dimensional formulation is used to estimate the eddy current patterns. The formulation is based on few approximations, and the experimentally measured data is used for the calculations. The eddy current losses in high current (10 to 20 kA) terminations are analyzed in [48]. Experimental investigations have been carried out on actual physical models with different geometric dimensions and types of shielding. An analytical method is presented in [32, 49] to determine and prevent hot spots in bushing mounting plates. The instantaneous field intensity at any point on the plate surface is calculated as a vector sum of field intensities due to currents in the conductors of three-phases. Equation 5.3 is used to calculate the loss if the plate thickness is sufficiently larger than the skin depth (the calculated loss value is multiplied by 2 since the plate is excited on its both surfaces). If the thickness is smaller than the skin depth, a correction factor based on figure 5.3 can be used corresponding to Case 2 (excitation of the plate by the same value of field on both surfaces).

A formula for calculating the permissible r.m.s. value of bushing current is given in [49], over which there is a possibility of excessive overheating,

$$I_{max} = 2460 \times d \times \left[\sqrt{1 + 3.9 \times \left(1 + \frac{2.4 \times \sqrt{t}}{d} \right)} - 1 \right] \tag{5.29}$$

where d is the center-to-center distance between phases as shown in figure 5.19 and t is plate thickness; both the dimensions are in meters. This formula has been derived for a typical grade of (magnetic) structural steel having conductivity of about 7×10^6 mho/m at 20°C. For a plate of thickness 6 mm with a distance of 150 mm between the phases, the maximum permitted current is about 780 amperes. As the current rating increases, non-magnetic inserts made of high resistivity SS material are used as shown in figure 5.20. Depending upon the width of inserts, which can be in the range of 20 mm to 50 mm, the current rating can be increased substantially. For still higher currents, a mounting plate of SS material is used.

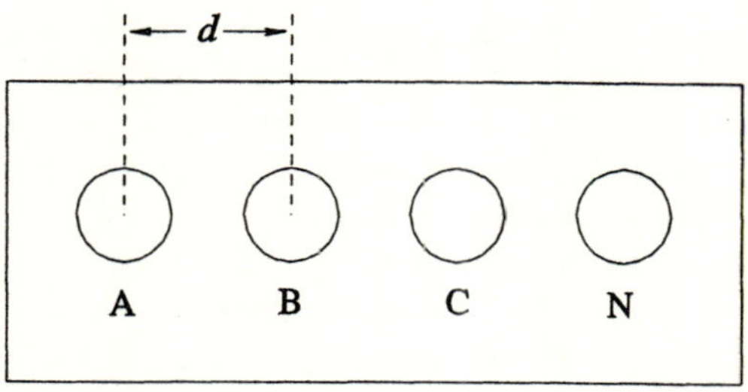

Figure 5.19 Bushing mounting plate

Different shapes of non-magnetic inserts have been analyzed using 3-D FEM formulation in [50] to reduce the tank wall loss in small pad-mounted transformers. In these transformers LV leads are usually terminated on the tank wall. The results of the FEM simulation have been verified on a transformer in which T shaped SS inserts were used.

In [35], the results of the analytical formulation given in [32, 49] are compared with that of 3-D FEM analysis and experimental measurements. The loss occurring in a bushing mounting plate is calculated indirectly from the measured values of initial temperature rise and steady-state temperature rise. These methods, in which indirect verification of loss is done, are described in Section 5.10. These methods are useful because it is very difficult to verify the calculation by conventional experimental measurements. This is due to the fact that the loss occurring in the plate cannot be exactly isolated from the tested load loss value of the transformer. If the conditions are simulated by using a low-voltage high-current source in a laboratory, the loss measurement is very difficult (if not impossible) at a low voltage (of few volts).

5.8 Evaluation of Stray Loss Due to High Current Leads

In furnace transformers and large generator transformers, the stray loss due to induced eddy currents in structural components in the vicinity of high current leads can become substantial. It could lead to hot spots if adequate magnetic clearances are not provided or shielding measures are not taken.

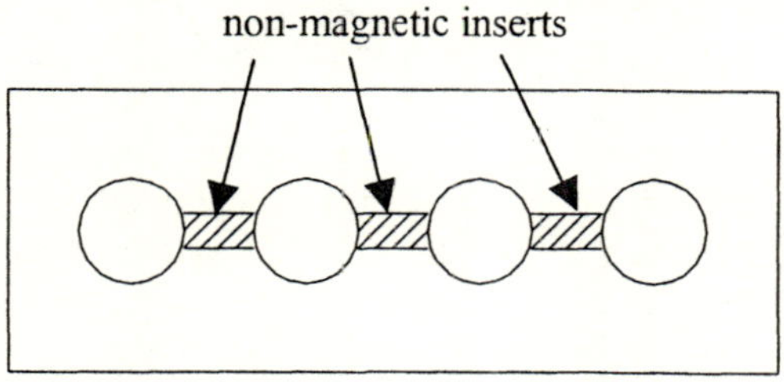

Figure 5.20 Bushing mounting plate with non-magnetic inserts

Evaluation of stray loss due to a high current field has improved gradually from 2-D analytical methods to the present day 3-D numerical methods. The method discussed in [51] evaluates eddy losses in a semi-infinite solid nearby a filament current. Practical cases, however, involve plates of finite thickness, and hence a modification of the solution is required. An empirical formula is presented in [7] for eddy loss calculation in steel plates based on measurements on an experimental set-up. The field pattern and eddy current losses due to current carrying strip bus bars are evaluated in [5] for an aluminum sheet. The current distribution is expressed in terms of an infinite number of sinusoidal distributions with the help of Fourier integral, and then the field due to any current distribution is obtained by superimposition of fields due to sinusoidal components. In [52] also, the solution method first transforms the original current distribution into sinusoidal current sheets. The method is therefore quite general and can be used for different current distributions (single-phase current and three-phase current) and different shapes of conductors (filament, rectangular conductor, etc.). A laboratory experiment has been conducted to verify the theoretical results. A method for analyzing the electromagnetic field in a system that comprises of parallel current carrying bars placed above a steel wall is presented in [53]. The current density within the cross section of bars is computed using the integral equation technique.

All these methods, due to some assumptions and approximations, are useful for simplified 2-D geometries and cannot be applied to complex 3-D structures. With the quantum improvement in 3-D FEM software capabilities, 3-D problems can be easily simulated and analyzed. The analysis of eddy current patterns in the LV lead termination of a furnace transformer is done in [52] by using 3-D FEM analysis. The total eddy loss estimated by using time-harmonic 3-D FEM formulation is found to be in close agreement with that observed during the testing of the transformer. The arrow plot of eddy currents in the front plate of the termination is shown in figure 5.21 (reproduced from [52] with permission from IEE). The three rectangular holes in the plate carry three-phase currents of 56 kA magnitude. The plot shows the eddy currents at the instant when R-phase current is maximum. There is concentration of eddy currents at locations A, B, C, D, E and F. Hence, hot spots developed at these locations during the testing of the furnace transformer. Therefore, for transformers having very high currents on the LV side, magnetic clearances and material of termination structures have to be judiciously selected. In most of the cases, it requires 3-D analysis in the absence of data of previous proven designs.

It should be noted that time-harmonic 2-D and 3-D formulations find the time-harmonic magnetic field in and around current-carrying conductors, in the presence of isotropic materials that may be conducting, magnetic, or both. Theoretically, the time-harmonic analysis is not possible for non-linear materials because sinusoidally-varying sources do not give rise to sinusoidally-varying fields. The basis of the time-harmonic analysis is as follows [54].

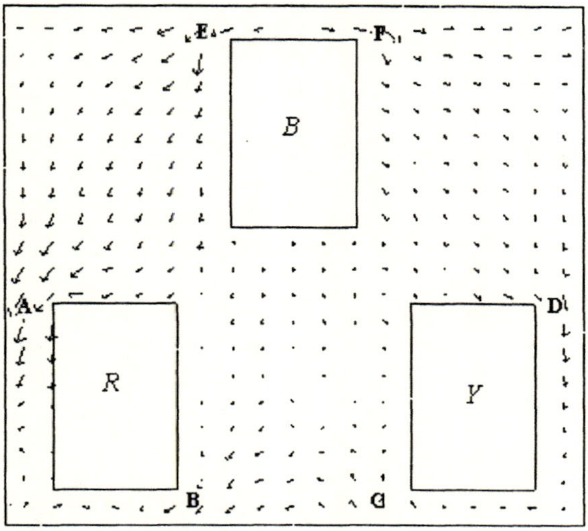

Figure 5.21 Eddy currents in the front plate of the furnace transformer [52]

In the time-harmonic formulations, the sources and fields are assumed to be time-harmonic at one specified frequency; complex phasors are used to represent them. For the eddy current analysis involving non-linear materials, a transient formulation may have to be used. In the time harmonic formulation based on the magnetic vector potential **A**, the solution of the following complex diffusion equation is found.

$$\nabla^2 \mathbf{A} = \mu(-\sigma \nabla V - j\omega\, \sigma\, \mathbf{A}) \tag{5.30}$$

In 2-D problems, **A** is assumed in the z direction only, and is a complex quantity. The term ($-\sigma \nabla V$) represents the source current density vector $\mathbf{J}_0$ and is refereed to as DC distribution of current density (applicable to the three-phase conductors of figure 5.21 when eddy currents in them are not considered). The other term $(-j\omega\sigma\mathbf{A})$ represents the induced current density (applicable to the conducting structural components in which eddy currents are induced). Equating real parts and imaginary parts on both sides of the above equation we get

$$\nabla^2 \operatorname{Re}[\mathbf{A}] - \mu\omega\sigma \operatorname{Im}[\mathbf{A}] = -\mu\sigma \operatorname{Re}[\nabla V] \tag{5.31}$$

$$\nabla^2 \operatorname{Im}[\mathbf{A}] + \mu\omega\sigma \operatorname{Re}[\mathbf{A}] = -\mu\sigma \operatorname{Im}[\nabla V] \tag{5.32}$$

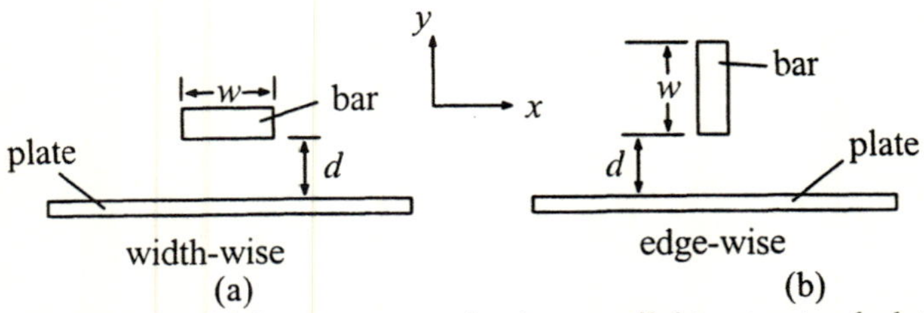

Figure 5.22 High current carrying bar parallel to structural plate

The solution of equation 5.31, Re[**A**], corresponds to the instant when Re[$\mathbf{J}_0$] is at peak with Im[$\mathbf{J}_0$] as zero. Whereas, the solution of equation 5.32, Im[**A**], corresponds to the instant when Re[$\mathbf{J}_0$] is zero with Im[$\mathbf{J}_0$] at peak value. Obviously at any other time instant the solution is

$$\mathbf{A}(t)=\mathrm{Re}[\mathbf{A}]\ \cos\omega t+\mathrm{Im}[\mathbf{A}]\ \sin\omega t \tag{5.33}$$

Thus, the time-harmonic (frequency-domain) analysis technique is computationally quite efficient since the data required to be stored is just twice that of the static solution. For a time-domain analysis of eddy current problems involving non-linear material and / or non-sinusoidal excitations, the solution has to be computed at each sufficiently small time-step requiring a large computational effort and time.

After having seen some of the methods reported for analyzing stray loss due to a field of high currents, let us see some practical aspects of design. When a high current bar runs parallel to a structural plate (e.g., tank), there are two possible configurations as shown in figure 5.22. The configuration in which the edge (thickness) of the bar faces the plate (Case b: edge-wise) is better than the other one in which the width of the bar (w) faces the plate (Case a: width-wise). General guidelines and rules given below for designing magnetic clearances are based on the results of a number of time-harmonic 2-D FEM simulations (with mild steel plate having $\sigma=7\times10^6$ mho/m and $\mu_r=100$). The simulations are done with the bar current of 4 kA (r.m.s. value).

1) The edge-wise configuration gives a loss advantage of about 15 to 20% over the width-wise configuration. About the same amount of advantage is reported in references [22, 4]. For a given clearance d between the bar and plate, the edge-wise configuration always results into a lower loss. Sometimes the space restrictions may not allow the edge-wise placement. The loss values obtained by 2-D FEM simulations are somewhat higher than that reported in [4] (based on experimental measurements done on 1 m $\times$ 2 m plate). The difference could be because of the fact that the actual loss is different due to the finite length of plate in the z direction and the associated end effects (2-D FEM calculates loss per unit length in the z direction assuming that the plate has infinite length in the z direction). Moreover, a constant value of permeability is assumed in 2-D FEM analysis, which is not the case in reality.

Table 5.3 Loss in plate for a distance of 120 mm from the center of bar

Configuration	Distance of plate from center-line of bar (mm)	Width of bar (mm)	Loss in plate (watts/m)
Edge-wise	120	60	586.3
Edge-wise	120	80	587.8
Edge-wise	120	100	589.8
Edge-wise	120	120	592.3
Width-wise	120	120	579.8

2) It is generally observed that the loss in the plate (per meter length in the *z* direction) for the edge-wise and width-wise configurations are approximately equal for a given distance between the center-line of the bar and the plate surface (refer to table 5.3). The plate thickness and width are 10 mm and 1000 mm respectively, whereas the bar thickness is 20 mm.

3) Theoretically, the plate loss varies in the square proportion of the bar current for linear material characteristics. An exponent of current slightly less than 2 is reported in [7] as explained in Section 5.1.2.

4) It can be seen from figure 5.23 that the loss varies as the frequency is varied with an exponent in the range of 0.7 to 0.8. The effect of frequency has to be considered in the case of excitation currents with a high harmonic content. An exponent of 0.8 is specified for frequency in IEC 61378 - Part 1, as explained in Section 5.1.2.

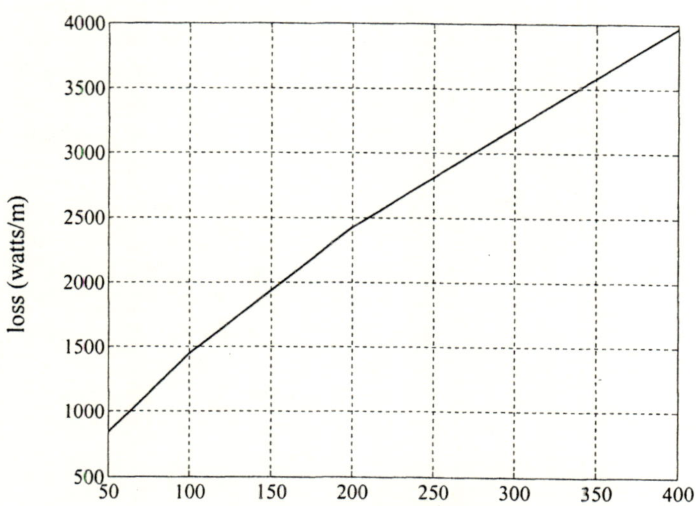

Figure 5.23 Effect of frequency variation

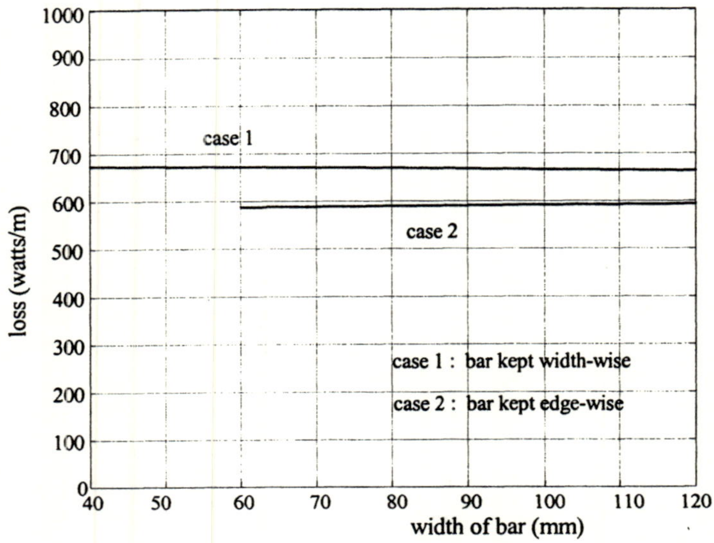

Figure 5.24 Effect of change in bar width on plate loss

5) In width-wise case, when width of the bar is changed keeping the current flowing through it and its distance (*d* in figure 5.22) from the plate constant, there is no noticeable change in the plate loss as shown in figure 5.24 (although the loss in the bar changes). The plate loss is basically affected by the magnitude of current and the clearance. But with the increase in bar width (for the same current), the loss density and temperature rise in the plate reduce. Hence, for a given current with more width, the plate loss remains almost constant and one can have lower clearance for the same temperature rise. The loss in the edge-wise case is lower than that in the width-wise case as explained in point 1.

6) Losses occurring in the plate are substantially influenced by the variation of its material conductivity as shown in figure 5.25 for a non-magnetic material. For a very low conductivity value (e.g., stainless steel material), the plate thickness is less than the skin depth. It becomes a case of resistance-limited behavior, and the loss increases with the increase in conductivity (this is in agreement with the discussion in Section 4.5.1 and equation 4.93).

As the conductivity increases (e.g., aluminum material), the induced eddy currents in the plate become substantial reducing the incident field and losses in the plate. It gradually displays the inductance-limited behavior as the skin depth becomes quite thin as compared to the plate thickness.

The behavior is similar to that explained for aluminum material in Section 5.1.1 with reference to figure 5.5.

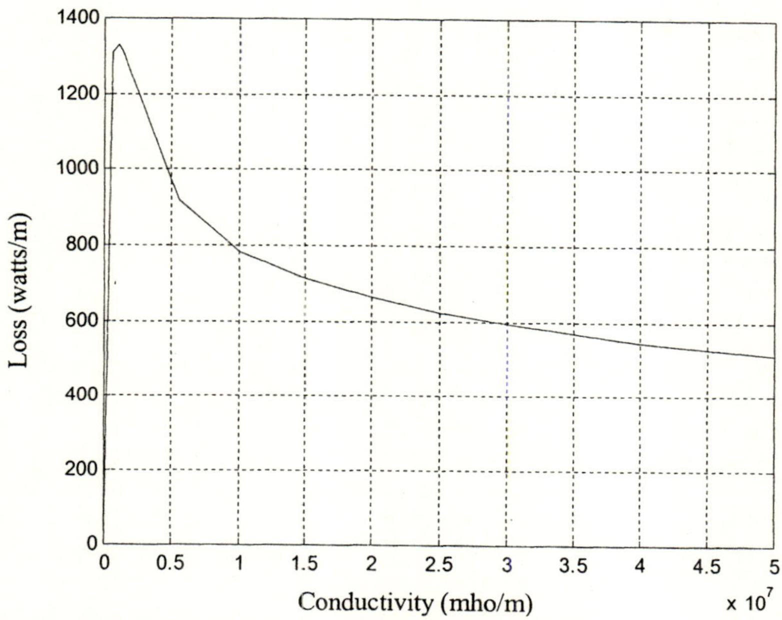

Figure 5.25 Effect of change in conductivity of plate

7) For two bars carrying currents in opposite directions (go and return), there are three possible configurations as shown in figure 5.26. The calculated loss in Case (c) is the least and that in Case (a) is the highest. In all the three cases, the loss increases with the increase in the distance of separation (s) between the two bars. Hence, the distance of separation should be minimum as permitted by mechanical, thermal and electrical considerations.

For all the configurations of the bar and plate discussed in this section, the manufacturers usually conduct a number of experiments on various configurations and measure rise in temperatures. Based on these measurements, the clearances are standardized for various currents and configurations to limit the plate temperature rise to a specific permitted value.

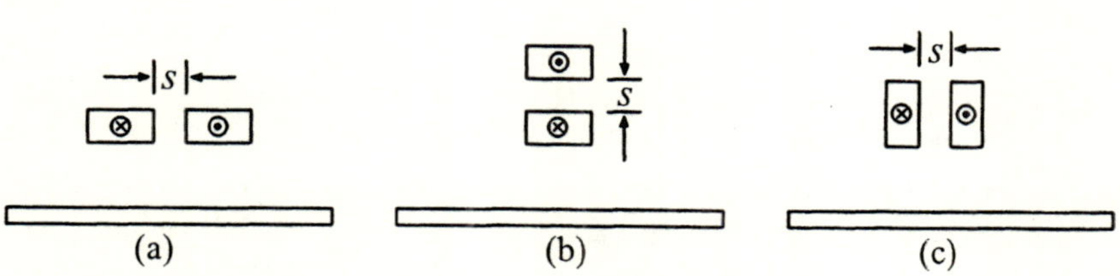

Figure 5.26 Different configurations of bars carrying go and return currents

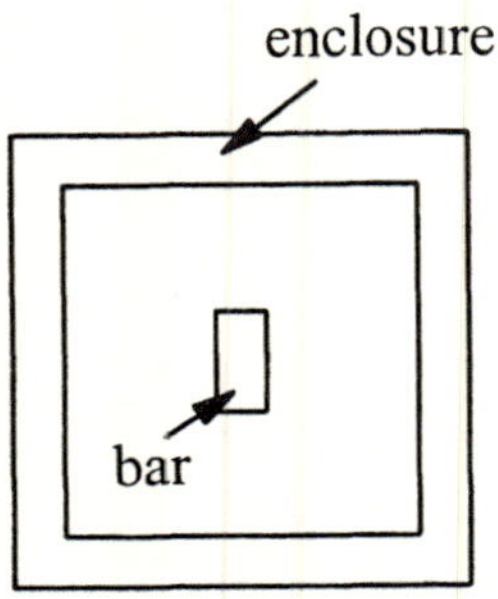

Figure 5.27 High current bar in a metallic enclosure

When a high current bar passes through a metallic enclosure / bus-duct (see figure 5.27), the eddy loss in the enclosure is a function of its perimeter or its distance from the bar. The higher the perimeter or distance, the higher is the current that can be allowed to pass through the enclosure. The loss can be controlled by making the enclosure of a non-magnetic material having high resistivity or by use of eddy current shields (aluminum / copper shields). If three-phase bars are passing through it, the eddy loss can be reduced substantially if the bars are placed symmetrically inside the enclosure and close to each other (to make the resultant field small).

When magnetic steel is shielded by aluminum or copper shield, eddy currents induced in the shield are forced out to its edges and may cause local overheating at the edges or in structural magnetic parts close to the edges. Hence, the shield should of sufficient width and should be made in one piece (without discontinuity).

5.9 Measures for Stray Loss Control

The stray loss in a structural component is reduced by a number of ways:

- use of laminated material
- use of high resistivity material
- reduction of flux density in the component by use of material with lower permeability
- reduction of flux density in the component by provision of a parallel magnetic path having low reluctance and loss
- reduction of flux density in the component by diverting / repelling the incident flux by use of a shielding plate having high conductivity

These are general guidelines, and the application of these measures depends on the configuration and type of excitation as discussed in Section 5.1. The measures for stray loss control of structural components, except tank, have been discussed in previous sections. In this section, methods for reducing stray losses in tank are elaborated.

Three methods are commonly used for reducing stray losses in tanks. The first method uses yoke shunts which collect the leakage field coming out of the windings so that there is very little flux external to the core and windings. Secondly, magnetic shunts can be provided on the tank so that they carry most of the leakage flux. Thirdly, the tank may be lined with aluminum or copper plates (shields); the eddy currents in these plates tend to shield the tank from most of the radial incident flux.

Shielding measures are not usually adopted for small distribution transformers and the cost-benefit analysis is done to take the decision regarding the use of these measures [55]. The measures for stray loss control are discussed with reference to large power transformers in [56, 57, 58]. The methods for reducing the loss in various structural components due to high current and leakage fields are described briefly in [56]. An approximate formula is given for the calculation of incident radial flux density on the tank wall in [57] and the effectiveness of magnetic shunts is also investigated. The permissible values of the tangential component of the magnetic field strength on various constructional elements are given in [59] to eliminate local overheating hazards. The principles of design (type and thickness) of tank shields are also explained.

Guidelines and useful curves are given in [60] to prevent overheating hazards in windings, first step of core, flitch plates, frames and tank. The experimental results of a 37 MVA transformer are presented in [61] for various proportions of tank shielding by magnetic shunts. The effect of positioning of inter-phase connections of high current carrying leads on the tank loss has been explained. The effects of tank and its shields on the magnetic field and stray loss in windings are evaluated and compared with the corresponding measurements carried out on a 150 MVA experimental transformer in [62]. A number of findings are reported, viz. flux density distribution varies along the circumference of the winding when placed in tank with shields, flux density distribution in a winding may be calculated with a single-phase representation regardless of the tank and tank shield conditions, winding eddy loss in the absence of tank is almost the same as that with tank without any shield, winding eddy loss increases when placed in a tank with magnetic shields (shunts) and reduces when placed in a tank with aluminum shields, and circulating current loss does not depend much on the tank and shielding conditions.

Some geometrical factors have a significant influence on the tested stray losses. A small difference in heights of LV and HV windings can affect various stray loss components in different ways. For example, if LV winding is taller by 1% and is placed symmetrically with respect to HV winding height, the losses in the core clamping structures reduce, whereas the losses in the tank increase. This is because the leakage field at winding ends is now more oriented towards the tank. Some manufacturers may purposely adopt this design to reduce the axial short circuit forces in LV winding (which is generally weaker if it is a layer winding). On the contrary, if HV winding is taller, the stray losses in the core, frames and flitch plates increase, and those in the tank reduce. This shows that a manufacturing variation in geometrical dimensions may have significant effect on the leakage field pattern and stray losses, and this is one of the reasons why the tested load loss values of transformers, manufactured with an identical design, sometimes differ noticeably.

5.9.1 Magnetic shielding

The magnetic shunts are more effective in controlling stray losses as compared to the non-magnetic (eddy current) shields. They offer a low reluctance path to the leakage flux constraining its path in a predetermined fashion. In the case of eddy current shields, the flux repelled by them may find a path through nearby structural components negating the advantages of shielding. An ideal magnetic shunt (infinite permeability) has no magnetic voltage drop across its length. The magnetic shunts are basically useful to shield structural components from the leakage field. They are not used for shielding against the field of high currents. If magnetic shunts are of adequate thickness and are made of CRGO laminations with lower watts/kg characteristics, the losses in them are almost negligible. Usually, left over pieces of core laminations (from original rolls) are used to make a magnetic shunt. The height of magnetic shunts should be higher than the height of windings. The optimum height can be decided based on the leakage field pattern obtained from the FEM analysis. The shunts can be placed on the tank from the level of top yoke-center to the level of bottom yoke-center in the absence of any detailed analysis. For a 360 MVA, 500 kV transformer, it is suggested in [26] that the height of shunts should be more than that of windings by one meter (500 mm at each end).

There are basically two types of magnetic shunts, viz. width-wise and edge-wise shunts. The width-wise shunts (more commonly used) are placed on the tank as shown in figure 5.28. The width of shunts should be as small as possible to reduce entry losses at their top and bottom portions where the leakage field impinges on them radially. As the shunt width reduces, the number of shunts increases leading to a higher manufacturing time. Hence, each manufacturer arrives at an optimum width of shunts. It can be as high as 300 mm and as small as 50 mm. The gaps between shunts have to be kept as minimum as mechanically possible (some space is required for their fixing arrangement on the tank) for effective shielding. By taking advantage of the fact that the incident radial flux enters the shunt (mostly in its end portions) at different points and that the (axial) flux density in the shunt (corresponding to the collected flux) is maximum at the center, a magnetic shunt of varying length can also be used as shown in figure 5.29. Most of the incident flux can enter the shunts through their thin edges reducing the entry loss. The saving of shunt material (laminations) obtained has to be compared with the extra manufacturing time required. An optimum design of a similar tank shield arrangement is reported in [63] by using the *orthogonal array design of experiments* technique in conjunction with 3-D FEM.

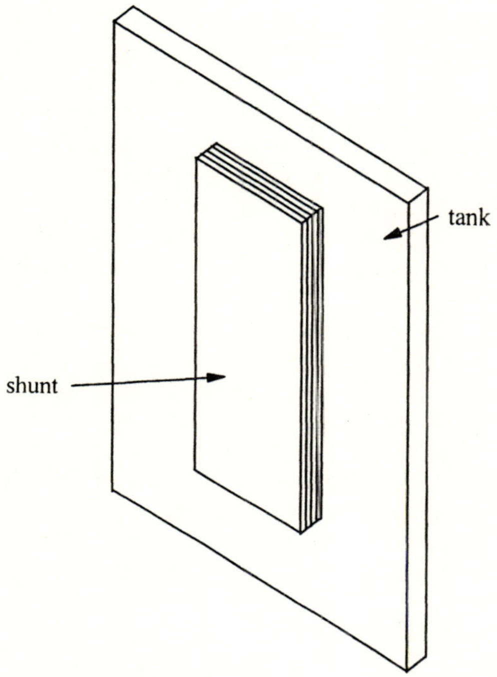

Figure 5.28 Width-wise shunt

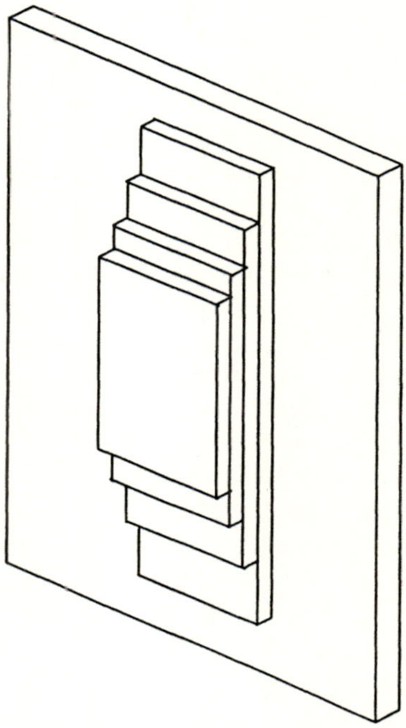

Figure 5.29 Optimum width-wise shunt

A practical formula for calculation of thickness of width-wise tank shunt is given in [6, 32] with reference to figure 5.16:

$$ST_{min} = \frac{C}{C+T} \frac{\sqrt{2}\,\mu_0\, N\, I\left[\dfrac{T_1+T_2}{2}+T_g\right]}{h\,B} \text{ meters} \tag{5.34}$$

where B, the flux density value to be limited in shunts, can be taken as 1.7 Tesla for CRGO material and 1.4 Tesla for CRNGO material. Since the effective permeability of the width-wise shunt is less due to inter-laminar non-magnetic gaps, it is preferable to take values of B about 20% lower than the above values (i.e., 1.4 Tesla for CRGO and 1.15 Tesla for CRNGO). Under overloading conditions, shunts may get saturated and become ineffective; hence a design with a lower value of flux density also helps under overloading conditions.

The other type of magnetic shunt, edge-wise shunt, is better than width-wise shunt because the flux is incident on the thickness (edge) of laminations resulting in negligible eddy loss in them. A typical edge-wise shunt is shown in figure 5.30. The effective permeability of laminations as seen by the incident flux is much higher for this shunt as compared to the width-wise shunt since the flux does not encounter any non-magnetic gaps once it enters the shunt. In the width-wise shunt, due to non-magnetic gaps (however small they be), the effective permeability at the entry point reduces making it less effective as compared to the edge-wise shunt. The flux distribution at the entry point is quite complicated. The presence of inter-laminar non-magnetic gap reduces the average permeability in the direction normal to the laminations to a low value, hence the flux tends to stay within a particular lamination until it saturates. The flux finds its way through the next lamination when the earlier lamination saturates and so on. Thus, it can be seen that the effectiveness of the width-wise shunt is less as compared to the edge-wise shunt.

The manufacturing process of edge-wise shunts is quite elaborate. In one of the forms, a set of laminations are epoxy moulded (like that of laminated flitch plates). In another design, it can be made into a wound form. The loss advantage with the edge-wise shunts has to be assessed vis-à-vis their higher cost and manufacturing time as compared to the width-wise shunts. The performances of these two types of shunts are compared in [64] by finding their effective anisotropic permeability. A substantial reduction in tank stray losses is reported in [26] by the use of edge-wise shunts. It is preferable to experimentally check the quantum of stray loss reduction before standardizing the use of edge-wise shunts.

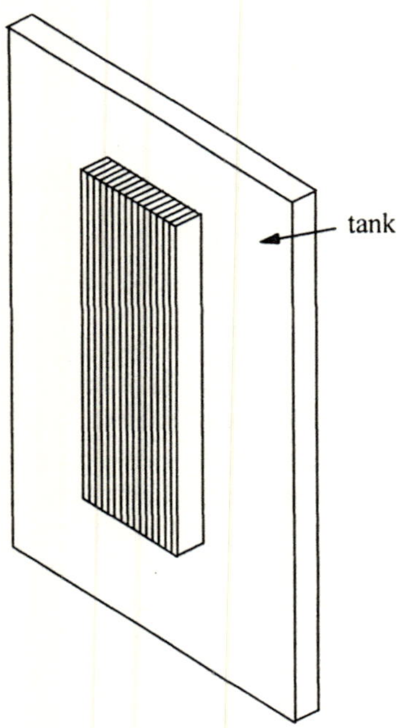

Figure 5.30 Edge-wise shunt

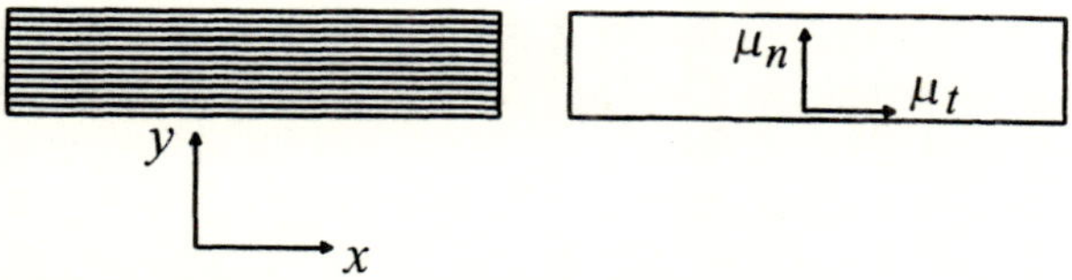

Figure 5.31 Anisotropic permeability

For a stack of laminations subjected to a magnetic field, the directional effects of material anisotropy and lamination stacking factor have to be taken into account. Because of the obvious difficulties of treating an individual lamination separately in any kind of formulation, a stack of laminations can be represented by a solid anisotropic block. For a stack of laminations (figure 5.31) with a stacking (space) factor of k, subjected to a field in the y direction, the effective relative permeability of the equivalent solid anisotropic block across the laminations is [44, 65]

$$\mu_n = \frac{\mu_y}{k(1-\mu_y)+\mu_y} \cong \frac{1}{1-k} \tag{5.35}$$

and tangential to the laminations is

$$\mu_t = k(\mu_x - 1) + 1 \cong k\mu_x \tag{5.36}$$

where μ_x and μ_y are the relative permeabilities of the lamination material in the x and y directions respectively (μ_x, μ_y >>1). The reduced effective permeability across the laminations (μ_n) gives the correct representation of much deeper penetration of the flux in the stack of laminations. This tends to make the flux density distribution more uniform as compared to the inaccurate isotropic modeling, where the flux concentrates only in the surface layers giving a highly non-uniform flux density distribution. A 3-D FEM formulation, which takes electric and magnetic anisotropies into account, is reported in [66]. The electric conductivity and magnetic permeability are represented by tensor quantities in the Cartesian system of coordinates.

Yoke shunts are another form of magnetic shunts (flux collectors), which are placed parallel to the yoke at the top and bottom ends of the windings. These shunts can be quite effective since the fluxes coming out from the three phases can add up to zero in them. The yoke shunts provide an excellent means of guiding the leakage field safely back to the core minimizing stray losses in the tank and other structural components. They tend to make the leakage field in the windings axial, minimizing the winding eddy loss due to the radial field at the winding ends. The analysis and design of yoke shunts in large power transformers must deal with a full three-dimensional solution of the leakage field because of the complicated geometries and anisotropic materials used. The effects of a yoke shunt are studied in [67] assuming that the shunt is connected directly to the yoke. The work is extended in [68] to include the effects of a small gap between the shunt and yoke on the leakage field distribution. All the surfaces of magnetic circuit components are considered as magnetic equipotentials (infinitely permeable and non-saturated). The equipotentials for all the surfaces except that of shunt are zero. The shunt floats at a potential, which is a function of the gap between the shunt and yoke. The leakage flux transferred to the yoke decreases as the shunt is spaced away from it. Hence, the gap between the shunt and yoke must be kept sufficiently small for the effective control of the leakage field. In further work [69], the field in the gap and yoke is analyzed in detail. The laminated iron (yoke) is treated as a solid anisotropic block with effective permeabilities in three directions calculated as per the method illustrated in [65]. The references [67, 68, 69] give useful practical guidelines for the yoke shunt design. In [70], the Reluctance Network Method is used to study the effectiveness of yoke shunts in controlling stray losses. The effect of gap length between the windings and yoke shunt on the stray losses in the tank and other structural components is reported. Yoke shunts can be conveniently used for three-phase five-limb and single-phase three-limb constructions, where the transfer of flux can be easily achieved through the yoke steps on the either side of windings. For the three-phase three-limb construction, the collection of flux by the yoke from the outer phases is not straight-forward and a special transfer arrangement may be needed. Hence, yoke shunts are usually not used alone in the three-phase three-limb construction. They are aided by either eddy current shields or magnetic shunts on the tank.

The merits and demerits of magnetic shunts are explained in [71]. The main disadvantage of magnetic shunts is that they cannot be used on the tank surfaces of irregular shapes. The losses measured under various combinations of shielding (yoke shunts, magnetic shunts and eddy current shields) arrangements are reported in the paper. Some manufacturers use wound steel pressure ring on the top of the windings, which not only acts as a clamping ring (for mechanical stability during short circuits) but it also reduces the stray losses in structural components. The steel ring provides a low reluctance path for the leakage field coming out of the windings and diverts it into the yoke away from the structural components. Thus, the axial component of the leakage field increases and the radial component reduces affecting the winding eddy loss. The increase or decrease in the winding eddy loss is decided by the conductor dimensions and winding configurations.

A combination of horizontal and vertical magnetic shunts can also be used on the tank as shown in figure 5.32. The vertical shunts are placed in front of three phases, while the horizontal shunts are placed at the level of yokes. A significant part of the incident leakage field from three phases gets effectively cancelled in the horizontal shunts if it is properly collected by them. The effectiveness of this arrangement can be enhanced by putting eddy current shields on the short sides of the tank. The leakage flux repelled from eddy current shields is collected by horizontal shunts aiding the cancellation effect.

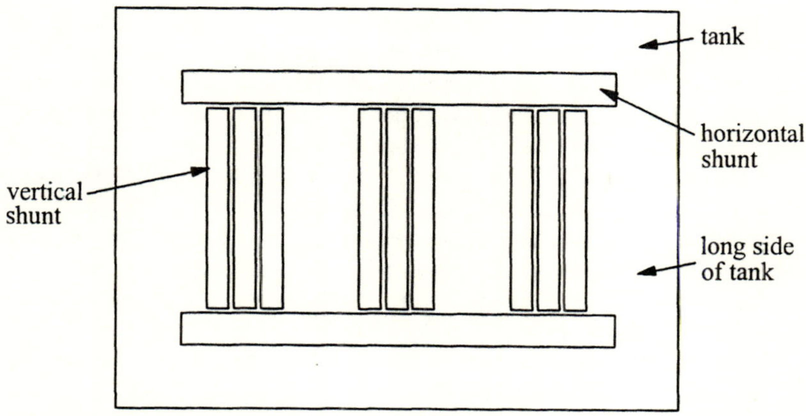

Figure 5.32 Combination of vertical and horizontal magnetic shunts

Usually, the tank shapes are not so conducive for the placement of magnetic shunts and eddy current shields in such an ideal manner.

5.9.2 Eddy current shielding

Aluminum or copper shields are used for shielding structural components from the high current and leakage fields. Eddy currents induced in them repel the incident field reducing the losses in structural components. As discussed in Section 5.1, the thickness of these shields should be adequate for their effectiveness and for reducing the loss in shields themselves. In most of the cases, the loss in the structural component and eddy current shield is more than that of the structural component and magnetic shunt. However, the eddy current shields have the advantage that they can be fitted on odd shapes of the tank unlike magnetic shunts. The weight of the eddy current shield is also usually lower than the magnetic shunt. For shielding a tank from the high current field, the eddy current shields are better than the magnetic shunts. This is because there are gaps between magnetic shunts reducing their effectiveness as shields.

An analytical formulation is given in [72] for calculating loss in the eddy current shield and the tank shielded by it. The paper has used a two-dimensional approximation and has first outlined the method of calculation for eddy loss of a tank, shielded by an aluminum shield, due to a line current. The method is then extended to transformer windings, wherein the windings are replaced by an infinite array of line currents by using the theory of images. The eddy current loss in the shields used in air core reactors is evaluated by the image method using Fourier-Bessel integral in [73]. For the finite dimensions of shields, 2-D approximations and end effects make the analytical formulations inaccurate and such problems can be simulated by 3-D numerical techniques.

The stray loss control by this flux rejection technique is suitable for structural components of odd shapes. The components required to make the eddy current shielding arrangement are of simpler construction and the shields can be suitably formed to protect the areas having complex shapes. The disadvantage of this method is that there are losses produced in the shield itself and these must be accurately evaluated. The shield dimensions have to be properly designed and adequate cooling needs to be provided to limit its temperature rise. Secondly, the diverted flux from the shield may cause overheating in the nearby unprotected structural parts. Hence, the design and positioning of the eddy current shields have to be done more carefully as compared to the magnetic shunts. The shield should of sufficient width as explained earlier in Section 5.8.

A combination of eddy current shields (on the tank) and yoke shunts can be used. This arrangement makes the leakage field predominantly axial (which gets collected by the yoke shunts) minimizing stray losses in the tank and other structural components.

5.10 Methods for Experimental Verification

Conventional search coil / Hall effect probe measurements have been used by many researchers for verifying the calculated values of flux densities or current densities in the structural components. There are some other indirect methods of predicting the eddy loss in the structural components based on their temperature rise. One method uses the measured steady-state temperature rise, while the other uses the initial temperature rise. These methods are described below.

5.10.1 Steady-state temperature rise

The structural component (e.g., bushing mounting plate), wherein eddy current losses need to be calculated, is allowed to reach a steady-state temperature rise. The component is assumed as a vertical plate in the method described below. The oil film temperature, which is the average of measured values of the plate and oil temperatures, is then calculated. If the test is performed in a laboratory in the air medium, then the air film temperature is found out as the average of measured values of the plate and ambient air temperature. The properties of oil / air film (ρ, k, υ, Pr, and β) are then obtained [4] where

$\rho =$ Density in kg m^{-3}

$k =$ Thermal conductivity in W m^{-1} $^{o}C^{-1}$

$\upsilon =$ Kinematic viscosity in m^2 s^{-1}

Pr = Prandtl number

$\beta =$ Coefficient of thermal cubic expansion in $^{o}C^{-1}$

The Rayleigh number is given by

$$Ra = \frac{g \beta \Delta\theta l^3}{\upsilon^2} \Pr \tag{5.37}$$

where

g = Acceleration due to gravity in m s^{-2}

l = Vertical length of plate in m

$\Delta\theta$ = Temperature difference between plate and oil / air in °C

The Nusselt number is then calculated [74, 75] from one of the following two equations:

$$Nu = 0.68 + \frac{0.67 Ra^{1/4}}{\left[1 + (0.492/\Pr)^{9/16}\right]^{4/9}} \qquad for\ Ra < 10^9 \tag{5.38}$$

$$Nu^{1/2} = 0.825 + \frac{0.387 Ra^{1/6}}{\left[1 + (0.492/\Pr)^{9/16}\right]^{8/27}} \qquad for\ 10^{-1} < Ra < 10^{12} \tag{5.39}$$

The heat transfer coefficient (h) is calculated as

$$h = \frac{Nu \times k}{l} \quad \text{W m}^{-2}\ {}^\circ\text{C}^{-1} \tag{5.40}$$

Finally, the loss in the plate is given by

$$P = h \times A \times \Delta\theta \tag{5.41}$$

where A is area of the convection surface in m^2. The estimation of losses in bushing mounting plate from the measured steady-state temperature rise and its comparison with that calculated by the analytical method and 3-D FEM analysis are reported in [35].

5.10.2 Initial temperature rise

The eddy loss in the plate can also be calculated from the initial temperature rise measured in the first few seconds of the application of current. The power developed in a unit volume inside a solid can be expressed in terms of temperature θ as [6]

$$P_E = c\rho \frac{\partial\theta}{\partial t} - \left\{ k_x \frac{\partial^2\theta}{\partial x^2} + k_y \frac{\partial^2\theta}{\partial y^2} + k_z \frac{\partial^2\theta}{\partial z^2} \right\} \tag{5.42}$$

The terms k_x, k_y and k_z are the thermal conductivities in x, y and z directions respectively, and c is specific heat (J kg^{-1} ^{0}C^{-1}). If the plate temperature is measured sufficiently rapidly after switching on the current (at $t = 0$) so that the temperature of the body may be considered uniform, the term inside the bracket in the equation reduces to zero giving

$$P_E = c\,\rho \left(\frac{\partial \theta}{\partial t} \right)_{t=0} \tag{5.43}$$

The term $(\partial\theta/\partial t)_{t=0}$ represents the gradient of the initial part of the heating curve. Thus, one can find power losses occurring in a structural component by a fast (say in the first 30 seconds) and accurate temperature rise measurement by using thermocouples. The loss in a bushing mounting plate has been calculated from the initial temperature rise measurements in [35], which agrees well with that calculated from the steady-state temperature rise and 3-D FEM analysis. The loss in frames is estimated from the initial temperature rise measurements in [31].

The method works well for thin structural plates (5 to 6 mm thickness). When the thickness increases, an appreciable error may be introduced due to faster heat transfer from the hot surface layer (where almost the entire loss is taking place) to the inner colder layers. The improved thermometric method proposed in [76] takes into account the heat removal from the surface to the surroundings and to the interior of the structural plate by a correction factor (which is a function of plate thickness). The method sets up equations of heat flow and electromagnetic wave propagation in the material, which take into account the non-uniform distribution of heat sources, the variation of permeability with the field strength and the excitation modes of eddy currents. These equations are then solved numerically using a computer. An example of the analysis of the tank wall loss of a transformer is given in which the direct measurement of the surface field strength is compared with that derived from the thermometric method.

5.11 Estimation of Stray Losses in Overexcitation Condition

The overvoltages appearing across the transformer terminals are classified according to their duration. The specifications regarding overvoltages are usually provided by users as: 110 or 115% continuous, 125% for 1 minute, 140% for 5 seconds, and 150% for 1 second. Temporary overvoltages can occur due to the disconnection of a load at the remote end of a long transmission line, ferroresonance, etc. The ferroresonant overvoltages initiated by the energization processes last through several cycles or even few seconds depending on the decay rate of the transformer inrush current. Apart from all these, there can be a voltage variation due to load fluctuations. An overexcitation leading to the overfluxing condition in a transformer causes additional losses due to the core saturation. The resulting spill-over flux from the core can cause intense local losses in conducting parts leading to hot spots. This stray flux cuts the winding conductors and other metallic structural parts and causes eddy currents to flow in them. Unlaminated structural parts can be overheated rapidly by such eddy currents, and the condition of adjacent insulation may deteriorate. To avoid these eddy current losses and hot spots, the structural parts have to be properly designed. The effects of overexcitation are elaborated in [77]. Typical loss curves in windings and structural parts as a function of overexcitation are given. The harmonic analysis of excitation current at various overexcitation levels is also given.

Finding a suitable analytical model for the purpose of studying overfluxing phenomenon is a difficult task. Good working models exist for transformers in steady-state. There are several methods, which try to find a suitable model for transient-state analysis. The models presented in [78, 79] analyze the ferroresonance phenomena in distribution transformers connected via a long cable. A model based on EMTP (Electro-Magnetic Transient Program) is reported in [80] for simulating a transformer under the condition of out-of-phase synchronization. A model for a three-phase, two-winding and five-limb transformer with its supply cables is presented in [81]. In [82], an approach for studying the behavior of a transformer under overfluxing conditions is presented, which is developed by suitable modifications to the model presented in [81]. The lumped parameter approach is used for representing limb and yoke reluctances. The part of flux fringing out of core, and not linked to the windings, is lumped into one parameter called air-flux. This air-flux path is located between the upper and lower yokes. To take into account the saturation effect of the limbs, modification in the basic model has been done by incorporating one additional path parallel to each limb. The analysis showed that the yoke reinforcement can help in controlling the path of the stray flux under overfluxing conditions.

The exact simulation of overexcitation of a transformer under the saturated core condition poses a real challenge to researchers. In [83], an attempt has been made to analyze the transformer performance under the overfluxing conditions by using 3-D FEM transient formulation. In this work, a 2 MVA, 11/0.433 kV transformer has been simulated under the 10% continuous overfluxing condition. The temperature rise of frame is calculated by thermal analysis which uses the loss values obtained in the electromagnetic analysis. The flux density distribution (arrow plot) is shown in figure 5.33 (at the instant when R phase voltage is at the positive peak, and Y and B phase voltages are at half the negative peak value).

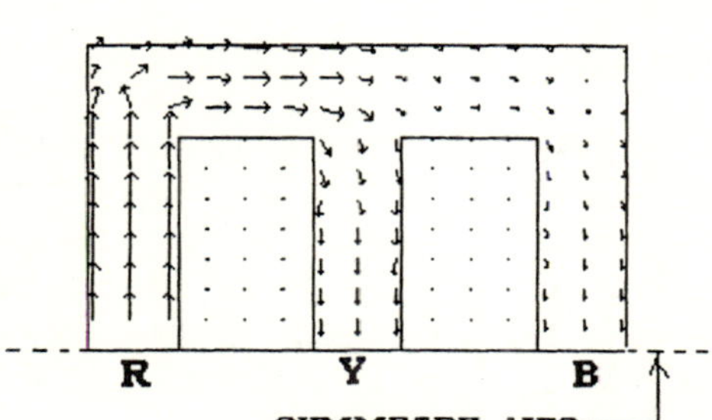

Figure 5.33 Flux density distribution during overfluxing condition

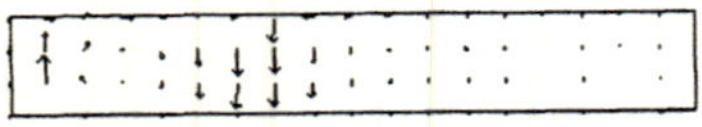

Figure 5.34 Eddy current distribution in frame

Some amount of flux in the yoke spills over and hits the frame part facing the yoke above the R phase winding, which is confirmed by the plot of distribution of eddy currents in the frame as shown in figure 5.34. The effect of yoke reinforcement on the overfluxing performance has also been analyzed. The frame loss with yoke reinforcement is considerably less than that without it. The temperature rise of the frame is calculated by using the method described in Section 5.10.1. The solution is obtained by an iterative process as follows. Some initial value of temperature rise is assumed. The oil film temperature (the average of frame and oil temperatures) is calculated. The properties of oil, at this temperature, are used to calculate the Rayleigh and Nusselt numbers. The heat transfer coefficient is then calculated by using equation 5.40, and subsequently the temperature rise ($\Delta\theta$) is calculated by equation 5.41. The procedure is repeated till $\Delta\theta$ converges. Thus, the performance of a transformer under the overfluxing conditions can be assessed by calculating temperature rise of various structural components by this procedure. The method described is based on some approximations. The effect of eddy currents in the core on the field is neglected. Also, the core is modeled as an isotropic material. The core needs to be modeled more exactly using anisotropic properties of permeability and conductivity in *x*, *y* and *z* directions. An appreciable amount of flux flows into the air paths, only during the portion of the cycle when the core gets saturated. Therefore, the waveform of the air-flux is of pulse-like form. This type of waveform has a high harmonic content increasing the winding eddy loss considerably, which needs to be calculated accurately. The simulation of overvoltage condition, say 125% for 1 minute or 140% for 5 seconds, as specified in technical requirements by users, may necessitate the use of a coupled circuit-field formulation.

5.12 Load Loss Measurement

The dependence of stray losses on various factors has been discussed in Section 5.1. The stray losses are quite sensitive to the magnitude of load current, temperature and frequency. In a very competitive market scenario, as it exists today, designers may be forced to keep a very small margin between the guaranteed and calculated values of the load loss in order to optimize the material cost. The penalty for every kW exceeded is quite high, and it is important to understand the effects of various design, manufacturing and test conditions on the stray losses (and the measured load loss).

The accuracy of measurement of the load loss in power transformers is an important issue. With the increase in loss capitalization rates, many of the large power transformers are being designed with lower losses (due to improved design and manufacturing techniques and / or higher material content used). This has the effect of improving further the efficiency of already quite efficient transformers (as compared to most of the other electrical equipment) lowering their power factor. At very low power factors, a small phase angle error in current or voltage transformers can cause a large error in the measurement of the real power [58, 84, 85]. During the load loss measurement of large power transformers, where a power factor of 0.015 is not uncommon, a phase angle error of 1 minute results in an error of 1.9% in the measured load loss. As the power factor reduces, the error increases.

Sometimes at sites, the impedance measurement test is done (with the available low voltage source) as one of the investigative tests. Care should be taken that the voltage should not be too low, which otherwise results in the core operation near the origin of *B-H* curve, where the permeability can be quite low decreasing the magnetizing reactance [86]. In Chapter 1, we have seen that the shunt branch in the transformer equivalent circuit is neglected while estimating the leakage reactance from the short circuit test. If the test is done at a site with a very low voltage, the effect of magnetizing reactance cannot be neglected and may have some noticeable influence on the leakage reactance measurement.

The losses occurring at the site under the rated voltage and current conditions can be noticeably different than that measured during the open circuit (no-load loss) and short circuit (load loss) tests. This is because the amount of leakage flux completing its path through the core (and causing extra stray losses in it) depends on the load power factor. At different time instants as the flux density in the core varies from zero to the peak value, the core permeability also varies. The load power factor decides the phase angle between the main (mutual) flux and stray flux, and hence the path of leakage flux [61, 87] and the magnitude of stray losses in the core. For example, when the load is inductive, the stray flux (in phase with the load current) lags the terminal voltage by 90°. The main flux in the core also lags the terminal voltage by approximately 90°. Hence, the main flux and leakage flux are almost in phase. When the main flux is at its maximum value, the core permeability is low and the leakage flux finds an alternate path reducing the core stray losses.

We will now discuss two typical phenomena observed during the load loss measurement.

5.12.1 Half-turn effect

In single-phase transformers with unwound end limbs, (e.g., single-phase three-limb transformer), the tested load loss value can be higher than the calculated value due to extra losses occurring in the core on account of the half-turn effect.

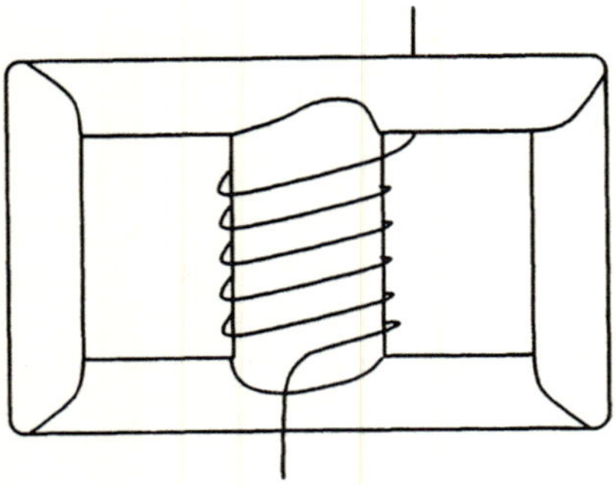

Figure 5.35 Half-turn effect in single-phase three-limb transformer

This phenomenon is particularly observed in autotransformers, in which high voltage (HV) and intermediate voltage (IV) bushings are fitted on the tank cover on the opposite sides of the core. In this case, either common winding (IV) and / or series winding (HV) of the autotransformer will contain a half-turn if the lead crosses the core to the opposite side for the termination. In figure 5.35, one such arrangement resulting into a half-turn in a winding is shown. In double wound transformers (non-auto) also, the half-turn effect is present if the line and neutral terminals are not on the same side of the core for any of the two windings. The half-turn effect becomes clearly evident during the ratio test (done phase-by-phase) in three-phase three-limb transformers. For an arrangement of three phases R, Y and B from left to right with turns of the windings wound in the clockwise direction (looking from top), if neutral and line leads are on opposite sides, R phase does not have the half-turn effect as the last turn while going on the opposite side does not cross the core window. Hence, the turns ratio measured on R phase will be less than the other two phases. Thus, if there are physically, say 500½ turns in the winding, the R phase turns ratio will correspond to 500 turns, the Y phase turns ratio will correspond to 500½ turns (the last Y phase turn links only half the core flux while crossing), and the B phase turns ratio will correspond to 501 turns (the last B phase turn links the full core flux while crossing). During the three-phase load loss measurement in a three-phase three-limb transformer, the net flux due to these unbalances in three phases has to pass through yokes and air-path since the limbs will not allow this flux in them due to a short-circuited winding on them. Since the air path has extremely high reluctance, a negligible flux would be set up in the path (yokes plus air-path) adding no extra loss during the load loss measurement.

In a single-phase three-limb transformer, a low reluctance magnetic path consisting of yokes and end limbs is available for the half ampere-turn (half turn multiplied by the corresponding current) to set up an appreciable flux in this closed path as shown in figure 5.36.

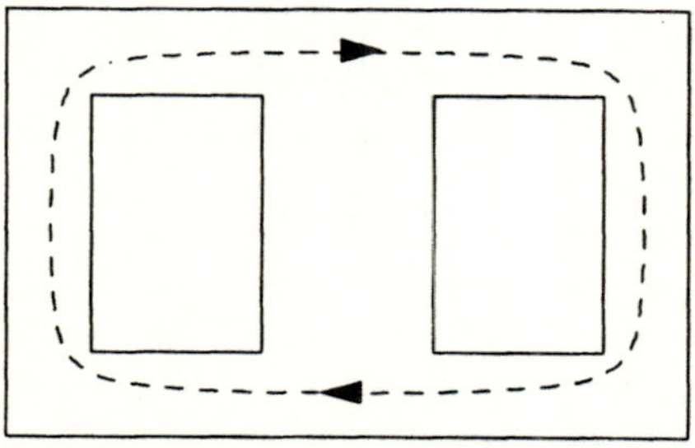

Figure 5.36 Circulating flux due to half-turn effect

Let us calculate the extra losses on account of this circulating flux for a 66.67 MVA, $(220/\sqrt{3})/(132/\sqrt{3})/11$ kV single-phase autotransformer having a half-turn in the HV series winding with 334 turns. The IV common winding has 500 turns, giving a total of 834 turns in IV common and HV series windings (corresponding to the HV voltage of $220/\sqrt{3}$ kV). The no-load current in today's very efficient transformers with the availability of ever improving core materials, can be as low as 0.1% of the rated current. The rated current of the HV winding is 524.9 A giving a no-load current of 0.525 A, which is required to set up the rated flux density of, say, 1.7 T. The magnetizing ampere-turns are therefore 438 (= 0.525 × 834) to set up 1.7 T in the entire core. For a given flux density, less ampere-turns will be required for the path consisting of yokes and end limbs as compared to the entire core. Hence, ampere-turns of 262.45 (= 524.9/2), corresponding to the half-turn effect during the load loss measurement (with rated current flowing in the windings), will set up a flux density (in the path of yokes and end limbs) of about rated value resulting in an extra core loss. This approximate calculation shows that the core loss value of the order of rated no-load loss of the transformer occurs and gets additionally measured during the load loss test when the half-turn effect is present.

During investigative tests, phase-by-phase load loss measurements are sometimes done on three-phase three-limb transformers. If one of the windings is having the half-turn effect, the value of load loss for the middle phase Y (with windings of R and B phases open-circuited) is quite high as compared to that of R and B phases. This is because it becomes a case of single-phase three-limb transformer with extra core losses occurring in the limbs of R and B phases and yokes due to the half-turn effect. The voltage induced in the open-circuited windings of R and B phases can be quite high depending upon the value of flux density induced in R and B phase limbs due to the half-turn flux [88].

In case of three-phase five-limb transformers, due to presence of end yokes and end limbs, a resultant flux under the combined action of the half-turn effects of three phases flows in the magnetic path formed by yokes and end limbs. In this path the three fluxes (corresponding to three phases) tend to cancel each other being displaced in phase by 120°, resulting in no extra core loss during the three-phase load loss measurement under balanced load conditions. Under appreciable unbalanced load conditions (e.g., at site), the effects of three-phase fluxes will not cancel and there will be extra core loss, the value of which will depend on the amount of unbalance between three phases.

If countermeasures are not taken, the half-turn effect will result into an unacceptable value of load loss during the factory test and excessive overfluxing / temperature rise of the core in service. The half-turn effect, if present, can be eliminated in the case of single-phase three-limb transformers by winding a few compensating turns on the end limbs. The direction of winding these turns should be same on both the end limbs as shown in figure 5.37 so that the main (mutual) useful flux induces a voltage in them in the same direction and hence they can be paralleled (connection of A_1 to B_1 and A_2 to B_2). The net voltage in the loop $A_1\,A_2\,B_2\,B_1\,A_1$ is zero resulting in zero value of circulating current, and hence the main flux flows unhindered. On the contrary, the flux due to the half-turn effect links these two windings on the two end limbs in opposite directions. This induces voltages in opposite directions in these two compensating windings causing a circulating current to flow (as shown by the arrows in figure 5.37), which opposes its cause, viz. the flux due to the half-turn effect. Thus, the half-turn effect gets nullified and there is no extra core loss during the load loss measurement, except for the small copper loss in the compensating turns due to the circulating current. The current in the compensating turns is equal to $I_1/(2n)$, where I_1 is the current in the winding having the half-turn effect during the load loss measurement and n is the number of compensating turns on each end limb. The cross section of the turns should be designed to carry this much amount of current.

Another way of avoiding the half-turn effect in autotransformers with HV and IV terminals on opposite sides of the core is to take the terminal of the winding with half-turn to the opposite side through space between the core and tank without crossing the core window. This alternative may become quite costly because extra space and material content are required (since adequate insulation clearances have to be provided for routing high voltage leads).

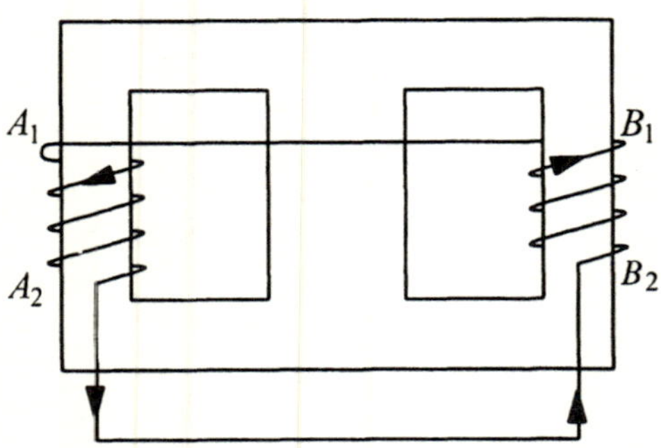

Figure 5.37 Compensating turns

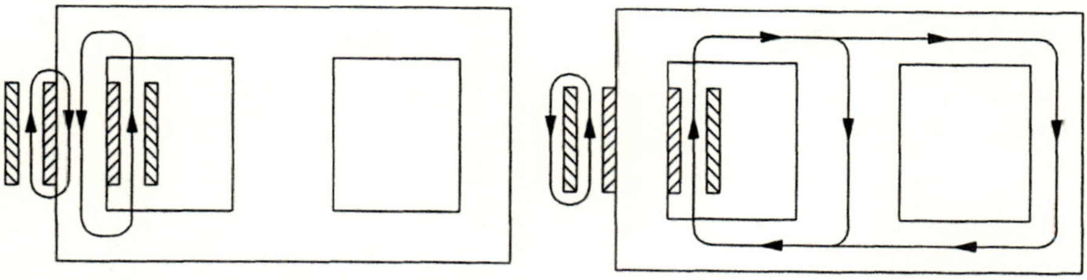

(a) Outer winding short-circuited (b) Inner winding short-circuited

Figure 5.38 Single-phase load loss measurement on three-phase transformer

It is to be noted that the half-turn effect in three-phase five-limb transformers, which becomes significant under appreciable unbalanced load conditions, cannot be eliminated by the arrangement of compensating turns shown in figure 5.37 since the main fluxes in the two end limbs are not in phase.

5.12.2 Single-phase load loss measurement on a three-phase transformer

In three-phase transformers with a delta connected winding, one peculiar phenomenon is observed during the investigative test of single-phase load loss measurement. The path of leakage flux for two cases is shown in figure 5.38. The path is influenced by the fact that the resultant flux linkages of a short-circuited winding are nearly zero.

When the inner winding is excited and the outer winding is short-circuited, the path of leakage field predominantly consists of the leakage channel (gap between two windings) and the core limb of the same phase. On the contrary, when the outer winding is excited and the inner winding is short-circuited, there is no flux in the core limb of the phase under test in order to satisfy the zero resultant flux linkage condition for the inner winding. Thus, with the inner winding short-circuited, the path of the leakage flux predominantly consists of the leakage channel, yokes, limbs of other two phases and tank.

Let us consider a case of the delta connected inner short-circuited winding with the impedance voltage applied to the outer star connected winding. When the leakage flux tries to complete its path through the limbs of other two phases, the voltages are induced in these phases on the delta side, which results in a circulating current in the closed delta. This current produces a counteracting flux forcing the initial flux out of the core, which is forced to complete its path through yoke, air-path and tank causing extra stray losses in tank and I^2R loss in the delta winding. There is also a reduction in the leakage reactance. Hence, the extra losses measured or the lower value of leakage reactance in such a case should not be mistaken as some defect in the transformer. These extra losses during the single-phase load loss measurements can be avoided by doing single-phase load loss measurement by exciting the inner delta winding and short-circuiting the outer winding.

References

1. Turowski, J., Turowski, M., and Kopec, M. Method of three-dimensional network solution of leakage field of three-phase transformers, *IEEE Transactions on Magnetics*, Vol. 26, No. 5, September 1990, pp. 2911-2919.
2. Jezierski, E. and Turowski, J. Dependence of stray losses on current and temperature, *CIGRE 1964*, Report 102, pp. 1-13.
3. Stoll, R. L. *The analysis of eddy currents*, Clarendon Press, Oxford, 1974.
4. Karsai, K., Kerenyi, D., and Kiss, L. *Large power transformers*, Elsevier Publication, Amsterdam, 1987.
5. Jain, M. P. and Ray, L. M. Field pattern and associated losses in aluminum sheet in presence of strip bus bars, *IEEE Transactions on Power Apparatus and Systems*, Vol. PAS-89, No. 7 September/October 1970, pp.1525 –1539.
6. Sykulski, J. K. *Computational magnetics*, Chapman and Hall, London, 1995.
7. Deuring, W. G. Induced losses in steel plates in the presence of an alternating current, *AIEE Transactions*, June 1957, pp. 166-173.
8. Allan, D. J., Mullineux, N., and Reed, J. R. Some effects of eddy currents in aluminum transformer tanks, *Proceedings IEE,* Vol. 120, No. 6, June 1973, pp. 681-685.
9. Kozlowski, M. and Turowski, J. Stray losses and local overheating hazard in transformers, *CIGRE 1972*, Paper No. 12-10.
10. Girgis, R. S., teNyenhuis, E. G., and Beaster, B. Proposed standards for frequency conversion factors of transformer performance parameters, *IEEE Transactions on Power Delivery*, Vol. 18, No. 4, October 2003, pp. 1262-1267.
11. Hwang, M. S., Grady, W. M., and Sanders, H. W. Distribution transformer winding losses due to non-sinusoidal currents, *IEEE Transactions on Power Delivery*, Vol. PWRD-2, No. 1, January 1987, pp. 140-146.
12. Kulkarni, S. V. and Khaparde, S. A. Stray loss evaluation in power transformers - a review, *IEEE PES Winter Meeting 2000*, Singapore, January 2000, Paper No. 0-7803-5938-0/00.
13. Boyajian, A. Leakage reactance of irregular distributions of transformer windings by the method of double Fourier series, *AIEE Transactions - Power Apparatus and Systems,* Vol. 73, Pt. III-B, October 1954, pp. 1078-1086.
14. Brauer, J. R. Finite element analysis of electromagnetic induction in transformers, *IEEE PES Winter Power Meeting*, New York, January 1977, Paper A77-122-5.

15. Andersen, O. W. Finite element solution of skin effect and eddy current problems, *IEEE PES Summer Power Meeting*, Mexico City, July 1977, Paper A77-616-6.
16. Sato, T., Inui, Y., and Saito, S. Solution of magnetic field, eddy current and circulating current problems, taking magnetic saturation and effect of eddy current and circulating current paths into account, *IEEE PES Winter Power Meeting*, New York, January / February 1977, Paper A77-168-8.
17. Babare, A., Di Napoli, A., and Santini, E. A method for losses evaluation in large power transformer tanks, *Electromagnetic fields in electrical engineering*, Book edited by Savini, A. and Turowski, J., Plenum Press, New York, 1988, pp. 95-100.
18. Komulainen, R. and Nordman, H. Loss evaluation and the use of magnetic and electromagnetic shields in transformers, *CIGRE 1988*, Paper No. 12-03.
19. Pavlik, D., Johnson, D. C., and Girgis, R. S. Calculation and reduction of stray and eddy losses in core-form transformers using a highly accurate finite element modeling technique, *IEEE Transactions Power Delivery*, Vol. 8, No. 1, January 1993, pp. 239-245.
20. Sironi, G. G. and Van Hulse, J. R. Eddy current losses in solid iron pieces under three-dimensional magnetic field, *IEEE Transactions on Magnetics*, Vol. MAG-14, No. 5, September 1978, pp. 377-379.
21. El Nahas, I., Szabados, B., Findlay, R. D., Poloujadoff, M., Lee, S., Burke, P., and Perco, D. Three dimensional flux calculation on a 3-phase transformer, *IEEE Transactions on Power Delivery*, Vol. PWRD-1, No. 3, July 1986, pp. 156-159.
22. Junyou, Y., Renyuan, T., Yan, L., and Yongbin, C. Eddy current fields and overheating problems due to heavy current carrying conductors, *IEEE Transactions on Magnetics*, Vol. 30, No. 5, September 1994, pp. 3064-3067.
23. Renyuan, T., Junyou, Y., Feng, L., and Yongping, L. Solutions of three-dimensional multiply connected and open boundary problems by BEM in three-phase combination transformers, *IEEE Transactions on Magnetics*, Vol. 28, No. 2, March 1992, pp. 1340-1343.
24. Renyuan, T., Junyou, Y., Zhouxiong, W., Feng, L., Chunrong, L., and Zihong, X. Computation of eddy current losses by heavy current leads and windings in large transformers using IEM coupled with improved $\mathbf{R}-\psi$ method, *IEEE Transactions on Magnetics*, Vol. 26, No. 2, March 1990, pp. 493-496.
25. Higuchi, Y. and Koizumi, M. Integral equation method with surface impedance model for 3-D eddy current analysis in transformers, *IEEE Transactions on Magnetics*, Vol. 36, No. 4, July 2000, pp. 774-779.
26. Yongbin, C., Junyou, Y., Hainian, Y., and Renyuan, T. Study of eddy current losses and shielding measures in large power transformers, *IEEE*

Transactions on Magnetics, Vol. 30, No. 5, September 1994, pp. 3068-3071.

27. Krawczyk, A. and Turowski, J. Recent developments in eddy current analysis, *IEEE Transactions on Magnetics*, Vol. MAG-23, No. 5, September 1987, pp. 3032-3037.
28. Carpenter, C. J. Theory of flux penetration into laminated iron and associated losses, *Proceedings IEE*, Vol. 124, No. 7, July 1977, pp. 659-664.
29. Hemmings, R. F. and Wale, G. D. Heating in transformer cores due to radial leakage flux, Part-I: Experimental models and test results, *Proceedings IEE*, Vol. 124, No. 11, November 1977, pp. 1064-1072.
30. Carpenter, C. J., Sharples, K. O., and Djurovic, M. Heating in transformer cores due to radial leakage flux, Part-II: Computed results, *Proceedings IEE*, Vol. 124, No. 12, December 1977, pp. 1181-1186.
31. Kerenyi, D. Stray load losses in yoke-beams of transformers, *Electromagnetic fields in electrical engineering*, Book edited by Savini, A. and Turowski, J., Plenum Press, New York, 1988, pp. 113-118.
32. Turowski, J. *Technical Electrodynamics*, (in Polish), WNT, Warszawa, 1993.
33. Kerenyi, D. Approximate formula for the eddy current losses induced in rectangular metal blocks by magnetic fields, *GANZ Electric Review*, No. 16, 1987, pp. 13-22.
34. Koppikar, D. A., Kulkarni, S. V., Srinivas, P. N., Khaparde, S. A., and Jain, R. Evaluation of flitch plate losses in power transformers, *IEEE Transactions on Power Delivery*, Vol. 14, No. 3, July 1999, pp. 996-1001.
35. Kulkarni, S. V. *Stray losses in power transformer: evaluation and experimental verification*, Ph.D. Dissertation, Department of Electrical Engineering, IIT-Bombay, India, 2000.
36. Vogel, F. J. and Adolphson, E. J. A stray loss problem in transformer tanks, *AIEE Transactions*, August 1954, pp. 760-764.
37. Berezovski, A. A. The calculation of stray losses in transformer tanks, *Elektrichestvo*, No. 9, 1966, pp. 1-7.
38. Valkovic, Z. Calculation of the losses in three-phase transformer tanks, *Proceedings IEE.*, Vol. 127, Pt. C, No. 1, January 1980, pp. 20-25.
39. Szabados, B., El Nahas, I., El Sobki, M. S., Findlay, R. D., and Poloujadoff, M. A new approach to determine eddy current losses in the tank walls of a power transformer, *IEEE Transaction on Power Delivery*, Vol. PWRD-2, No. 3, July 1987, pp. 810-816.
40. Koppikar, D. A., Kulkarni, S. V., and Turowski, J. Industrial verification of 3D hybrid analytically-numerical calculation of stray field and losses in 3-phase power transformers, *International Symposium on Electromagnetic Fields, ISEF '95*, Thessaloniki, Greece, September 1995, pp. 345-348.
41. Koppikar, D. A., Kulkarni, S. V., and Turowski, J. Fast 3-D interactive computation of stray field and losses in asymmetric transformers,

Proceedings IEE - Generation, Transmission and Distribution, Vol. 147, No. 4, July 2000, pp. 197-201.

42. Zwolinski, G., Turowski, J., Kulkarni, S. V., and Koppikar, D. A. Method of fast computation of 3D stray field and losses in highly asymmetric transformers, *International Conference on Transformers, TRAFOTECH '98,* Mumbai, 23-24 January 1998, pp. I 51-58.
43. Dexin, X., Yunqiu, T., and Zihong, X. FEM analysis of 3D eddy current field in power transformer, *IEEE Transactions on Magnetics*, Vol. MAG-23, No. 5, September 1987, pp. 3786-3788.
44. Holland, S. A., O'Connell, G. P., and Haydock, L. Calculating stray losses in power transformers using surface impedance with finite elements, *IEEE Transactions on Magnetics*, Vol. 28, No. 2, March 1992, pp. 1355-1358.
45. Savini, A. and Turowski, J. Influence of structure geometry, screens and eddy currents on the critical distance of tank wall in power transformers, *Electromagnetic fields in electrical engineering*, Book edited by Savini, A. and Turowski, J., Plenum Press, New York, 1988, pp. 119-127.
46. Turowski, J. Overheating hazard in flanged bolt joints of transformers, *International Symposium on Electromagnetic Fields, ISEF'85,* Warsaw, Poland, 1985, pp. 271-274.
47. Saito, S., Inagaki, K., Sato, T., Inui, Y., Okuyama, K., and Otani, H. Eddy currents in structure surrounding large current bushings of a large capacity transformer, *IEEE Transactions on Power Apparatus and Systems*, Vol. PAS-100, No. 11, November 1981, pp. 4502-4509.
48. Furman, Y. I., Bereza, V. L., Ivankov, V. F., and Nizhnik, L. P. Losses in tanks of large power transformers caused by magnetic field, and methods of their reduction, *CIGRE 1988*, Paper No. 12-07.
49. Turowski, J. and Pelikant, A. Eddy current losses and hot spot evaluation in cover plates of power transformers, *Proceedings IEE – Electrical Power Applications*, Vol. 144, No. 6, November 1997, pp. 435-440.
50. Olivares, J. C., Escarela, R., Kulkarni, S. V., de Leon, F., Melgoza, E., and Hernandez, O. Improved insert geometry for reducing tank wall losses in pad-mounted transformers, Accepted for publication in *IEEE Transactions on Power Delivery*.
51. Poritsky, H. and Jerrard, R. P. Eddy current losses in a semi-infinite solid due to a nearby alternating current, *AIEE Transactions*, May 1954, pp. 97-106.
52. Koppikar, D. A., Kulkarni, S. V., Khaparde, S. A., and Jha, S. K. Evaluation of eddy losses due to high current leads in transformers, *Proceedings IEE - Science Measurement and Technology*, Vol. 144, No. 1, January 1997, pp. 34-38.
53. Krakowski, M. R., Kazmierski, M., and Kersz, I. Effect of current distribution in parallel bars on electromagnetic field at nearby steel wall, *Proceedings IEE*, Vol. 136, Pt. B, No. 4, July 1989, pp. 161-167.

54. Lowther, D. A. and Silvester, P. P. *Computer aided design in Magnetics*, Springer-Verlag, New York, 1985, pp. 213-215.
55. Olivares, J. C., Canedo, J., Moreno, P., Driesen, J., Escarela, R., and Palanivasagam, S. Experimental study to reduce the distribution transformer stray losses using electromagnetic shields, *Electric Power Systems Research*, 63 (2002), pp. 1-7.
56. Tagaki, T., Ishii, T., Okada, T., Kurita, K., Tamura, R., and Murata, H. Reliability improvement of 500 kV large capacity power transformer, *CIGRE 1978*, Paper No. 12-02.
57. Kerr, H. W. and Palmer, S. Developments in the design of large power transformers, *Proceedings IEE*, Vol. 111, No. 4, April 1964, pp. 823-832.
58. Beaumont R. Losses in transformers and reactors, *CIGRE 1988*, Paper No. 12-10.
59. Kozlowski, M. and Turowski, J. Stray losses and local overheating hazard in transformers, *CIGRE 1972*, Paper No. 12-10.
60. Kazmierski, M., Kozlowski, M., Lasocinski, J., Pinkiewicz, I., and Turowski, J. Hot spot identification and overheating hazard preventing when designing a large transformer, *CIGRE 1984*, Paper No. 12-12.
61. Bose, A. K., Kroon, C., and Hulsink, G. Some topics on designing transformers with low load losses, *CIGRE 1988*, Paper No. 12-05.
62. Inui, Y., Saito, S., Okuyama, K., and Hiraishi, K. Effect of tank and tank shields on magnetic fields and stray losses in transformer windings, *IEEE PES Summer Meeting*, Vancouver, Canada, July 1973, Paper No. C73-401-7.
63. Takahashi, N., Kitamura, T., Horii, M., and Takehara, J. Optimal design of tank shield model of transformer, *IEEE Transactions on Magnetics*, Vol. 36, No. 4, July 2000, pp. 1089-1093.
64. Bereza, V. L. Designing magnetic shields for transformer tanks, *Elektroteknika*, Vol. 52, No. 12, 1981, pp. 44-46.
65. Barton, M. L. Loss calculation in laminated steel utilizing anisotropic magnetic permeability, *IEEE Transactions on Power Apparatus and Systems*, Vol. PAS-99, No. 3, May/June 1980, pp. 1280-1287.
66. Silva, V. C., Meunier, G., and Foggia, A. A 3-D finite element computation of eddy current and losses in laminated iron cores allowing for electric and magnetic anisotropy, *IEEE Transactions on Magnetics*, Vol. 31, No. 3, May 1995, pp. 2139-2141.
67. Djurovic, M. and Carpenter, C. J. Three dimensional computation of transformer leakage fields and associated losses, *IEEE Transactions on Magnetics*, Vol. MAG-11, No. 5, September 1975, pp. 1535-1537.
68. Djurovic, M. and Monson, J. E. Three dimensional computation of the effect of the horizontal magnetic shunt on transformer leakage fields, *IEEE Transactions on Magnetics*, Vol. MAG-13, No. 5, September 1977, pp. 1137-1139.

69. Djurovic, M. and Monson, J. E. Stray losses in the step of a transformer yoke with a horizontal magnetic shunt, *IEEE Transactions on Power Apparatus and Systems*, Vol. PAS-101, No. 8, August 1982, pp. 2995-3000.
70. Rizzo, M., Savini, A., and Turowski, J. Influence of flux collectors on stray losses in transformers, *IEEE Transactions on Magnetics*, Vol. 36, No. 4, July 2000, pp. 1915-1918.
71. Harrison, T. H. and Richardson, B. Transformer loss reductions, *CIGRE 1988*, Paper No. 12-04.
72. Mullineux, N. and Reed, J. R. Eddy current shielding of transformer tanks, *Proceedings IEE*, Vol. 113, No. 5, May 1966, pp. 815-818.
73. Woolley, I. Eddy current losses in reactor flux shields, *Proceedings IEE*, Vol. 117, No. 11, November 1970, pp. 2142-2150.
74. Holman, J. P. *Heat transfer*, McGraw - Hill Company, Singapore, 1986.
75. Churchill, S. W. and Chu, H. H. S. Correlating equations for laminar and turbulent free convection from a vertical plate, *International Journal Heat Mass Transfer*, Vol. 18, 1975, pp. 1323-1329.
76. Niewierowicz, N. and Turowski, J. New thermometric method of measuring power losses in solid metal elements, *Proceedings IEE*, Vol. 119, No. 5, May 1972, pp. 629-636.
77. Alexander, G. W., Corbin, S L., and McNutt, W. J. Influence of design and operating practices on excitation of generator step-up transformers, *IEEE Transactions on Power Apparatus and Systems*, Vol. PAS-85, No. 8, August 1966, pp. 901-909.
78. Stuehm, D. L., Mork, B. A, and Mairs, D. D. Five-legged core transformer equivalent circuit, *IEEE Transactions on Power Delivery*, Vol. 4, No. 3, July 1989, pp. 1786-1793.
79. Mairs, D. D., Stuehm, D. L., and Mork, B. A. Overvoltages on five-legged core transformer on rural electric systems, *IEEE Transactions on Industry Applications*, Vol. 25, No. 2, March/April 1989, pp. 366-370.
80. Arthuri, C. M. Transient simulation and analysis of a three-phase five-limb step-up transformer, *IEEE Transactions on Power Delivery*, Vol. 6, No. 1, January 1991, pp. 196-203.
81. Chen, X. S. and Neudorfer, P. Digital model for transient studies of a three-phase five-legged transformer, *Proceedings IEE*, Pt. C, Vol. 139, No. 4, July 1992, pp. 351-358.
82. Koppikar, D. A., Kulkarni, S. V., Khaparde, S. A., and Arora, B. A modified approach to overfluxing analysis of transformer, *International Journal of Electrical Power and Energy Systems*, Elsevier Science Publication, Vol. 20, No. 4, 1998, pp. 235-239.
83. Koppikar, D. A., Kulkarni, S. V., and Khaparde, S. A. Overfluxing simulation of transformer by 3D-FEM analysis, *Fourth Conference on EHV Technology*, IISc Bangalore, India, 17-18 July 1998, pp. 69-71.

84. Specht, T. R., Rademacher, L. B., and Moore, H. R. Measurement of iron and copper losses in transformers, *AIEE Transactions,* August 1958, pp. 470-476.
85. Mehta, S. P. Measurement of transformer losses, *IEEE Winter Meeting*, Paper No. EHO209-7/84/0000-0043, 1983, pp. 43-52.
86. Lachman, M. F. and Shafir, Y. N. Influence of single-phase excitation and magnetizing reactance on transformer leakage reactance measurement, *IEEE Transactions on Power Delivery*, Vol. 12, No. 4, October 1997, pp. 1538-1545.
87. Bose, A. K., Kroon, C., and Wildeboer, J. The loading of the magnetic circuit, *CIGRE 1978*, Paper No. 12-09.
88. Mohseni, H. and Hamdad, F. Overvoltage caused by winding terminal arrangement of three-phase power transformer, *International Symposium on High Voltage Engineering*, Austria, 1995, pp. 6757:1-4.

6

Short Circuit Stresses and Strength

The continuous increase in demand of electrical power has resulted in the addition of more generating capacity and interconnections in power systems. Both these factors have contributed to an increase in short circuit capacity of networks, making the short circuit duty of transformers more severe. Failure of transformers due to short circuits is a major concern of transformer users. The success rate during actual short circuit tests is far from satisfactory. The test data from high power test laboratories around the world indicates that on an average practically one transformer out of four has failed during the short circuit test, and the failure rate is above 40% for transformers above 100 MVA rating [1]. There are continuous efforts by manufacturers and users to improve the short circuit withstand performance of transformers. A number of suggestions have been made in the literature for improving technical specifications, verification methods and manufacturing processes to enhance reliability of transformers under short circuits. The short circuit strength of a transformer enables it to survive through-fault currents due to external short circuits in a power system network; an inadequate strength may lead to a mechanical collapse of windings, deformation / damage to clamping structures, and may eventually lead to an electrical fault in the transformer itself. The internal faults initiated by the external short circuits are dangerous as they may involve blow-out of bushings, bursting of tank, fire hazard, etc. The short circuit design is one of the most important and challenging aspects of the transformer design; it has been the preferential subject in many CIGRE Conferences including the recent session (year 2000).

Revision has been done in IEC 60076-5 standard, second edition 2000-07, reducing the limit of change in impedance from 2% to 1% for category III (above 100 MVA rating) transformers. This change is in line with the results of many recent short circuit tests on power transformers greater than 100 MVA, in which an increase of short circuit inductance beyond 1% has caused significant deformation in windings. This revision has far reaching implications for transformer manufacturers. A much stricter control on the variations in materials and manufacturing processes will have to be exercised to avoid looseness and winding movements.

This chapter first introduces the basic theory of short circuits as applicable to transformers. The thermal capability of transformer windings under short circuit forces is also discussed. There are basically two types of forces in windings: axial and radial electromagnetic forces produced by radial and axial leakage fields respectively. Analytical and numerical methods for calculation of these forces are discussed. Various failure mechanisms due to these forces are then described. It is very important to understand the dynamic response of a winding to axial electromagnetic forces. Practical difficulties encountered in the dynamic analysis and recent thinking on the whole issue of demonstration of short circuit withstand capability are enumerated. Design parameters and manufacturing processes have pronounced effect on natural frequencies of a winding. Design aspects of winding and clamping structures are elucidated. Precautions to be taken during design and manufacturing of transformers for improving short circuit withstand capability are given.

6.1 Short Circuit Currents

There are different types of faults which result into high over currents, viz. single-line-to-ground fault, line-to-line fault with or without simultaneous ground fault and three-phase fault with or without simultaneous ground fault. When the ratio of zero-sequence impedance to positive-sequence impedance is less than one, a single-line-to-ground fault results in higher fault current than a three-phase fault. It is shown in [2] that for a particular case of YNd connected transformer with a delta connected inner winding, the single-line-to-ground fault is more severe. Except for such specific cases, usually the three-phase fault (which is a symmetrical fault) is the most severe one. Hence, it is usual practice to design a transformer to withstand a three-phase short circuit at its terminals, the other windings being assumed to be connected to infinite systems / sources (of constant voltage). The symmetrical short circuit current for a three-phase two-winding transformer is given by

$$I_f = \frac{V}{\sqrt{3}\,(Z_T + Z_S)} \quad \text{kA} \tag{6.1}$$

where V is rated line-to-line voltage in kV, Z_T is short circuit impedance of the transformer, and Z_S is short circuit impedance of the system given by

$$Z_S = \frac{V^2}{S_F} \text{ ohms} \quad \text{or} \quad Z_S = \frac{S}{S_F} \text{ per-unit} \tag{6.2}$$

where S_F is short circuit apparent power of the system in MVA and S is three-phase rating of the transformer in MVA. Usually, the system impedance is quite small as compared to the transformer impedance and can be neglected, giving an extra safety margin. In per-unit quantities using sequence notations we get

$$I_a = \frac{V_{pF}}{Z_1}\angle 0^\circ, \; I_b = \frac{V_{pF}}{Z_1}\angle 240^\circ \text{ and } I_c = \frac{V_{pF}}{Z_1}\angle 120^\circ \tag{6.3}$$

where Z_1 is positive-sequence impedance of the transformer (which is leakage impedance to positive-sequence currents calculated as per the procedure given in Section 3.1 of Chapter 3) and V_{pF} is pre-fault voltage. If the pre-fault voltages are assumed to be 1.0 per-unit (p.u.) then for a three-phase solid fault (with a zero value of fault impedance) we get

$$I_a = \frac{1}{Z_1}\angle 0^\circ, \; I_b = \frac{1}{Z_1}\angle 240^\circ, \; I_c = \frac{1}{Z_1}\angle 120^\circ \text{ and } V_a = V_b = V_c = 0 \tag{6.4}$$

The sequence components of currents and voltages are [3]

$$I_{a1} = \frac{1}{Z_1}, \; I_{a0} = I_{a2} = 0 \text{ and } V_{a1} = V_{a2} = V_{a0} = 0 \tag{6.5}$$

For a solid single-line-to-ground fault on phase a,

$$I_a = \frac{3}{Z_1 + Z_2 + Z_0}\angle 0^\circ, \quad I_b = I_c = 0, \quad V_a = 0 \tag{6.6}$$

$$I_{a0} = I_{a1} = I_{a2} = I_a/3 \tag{6.7}$$

where Z_2 and Z_0 are negative-sequence and zero-sequence impedances of the transformer respectively. For a transformer, which is a static device, the positive and negative-sequence impedances are equal ($Z_1 = Z_2$). The procedures for calculation of the positive and zero-sequence impedances are given in Chapter 3.

For a line-to-line fault (between phases b and c).

$$I_a = 0, \qquad I_b = -I_c = -\frac{j\sqrt{3}}{Z_1 + Z_2} \tag{6.8}$$

$$I_{a1} = -I_{a2} = \frac{1}{Z_1 + Z_2}\angle 0^\circ, \quad I_{a0} = 0 \tag{6.9}$$

and for a double-line-to-ground fault,

$$I_a = 0\,,\ I_b = -\frac{\left(\dfrac{3}{2} + j\dfrac{\sqrt{3}}{2}\right)Z_2 + j\sqrt{3}\,Z_0}{Z_1Z_2 + Z_1Z_0 + Z_2Z_0}\,,$$

$$I_c = -\frac{\left(\dfrac{3}{2} - j\dfrac{\sqrt{3}}{2}\right)Z_2 - j\sqrt{3}\,Z_0}{Z_1Z_2 + Z_1Z_0 + Z_2Z_0}\,,\ V_b = V_c = 0 \tag{6.10}$$

Since a three-phase short circuit is usually the most severe fault, it is sufficient if the withstand capability against three-phase short circuit forces is ensured. However, if there is an unloaded tertiary winding in a three-winding transformer, its design must be done by taking into account the short circuit forces during a single-line-to-ground fault on either LV or HV winding. Hence, most of the discussions hereafter are for the three-phase and single-line-to-ground fault conditions. Based on the equations written earlier for the sequence voltages and currents for these two types of faults, we can interconnect the positive-sequence, negative-sequence and zero-sequence networks as shown in figure 6.1. The solution of the resulting network yields the symmetrical components of currents and voltages in windings under fault conditions [4].

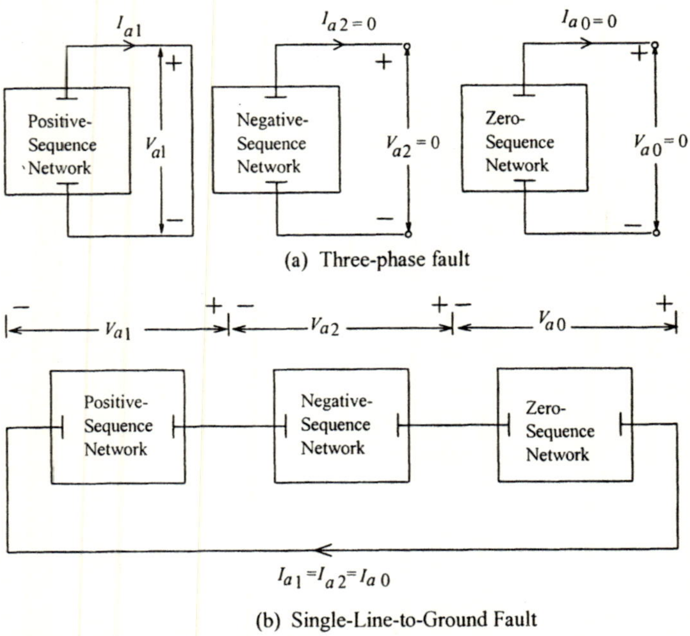

Figure 6.1 Sequence networks

The calculation of three-phase fault current is straight-forward, whereas the calculation of single-line-to-ground fault current requires the estimation of zero-sequence reactances and interconnection of the three sequence networks at the correct points. The calculation of fault current for two transformers under the single-line-to-ground fault condition is described now.

Consider a case of delta / star (HV winding in delta and LV winding in star with grounded neutral) distribution transformer with a single-line-to-ground fault on LV side. The equivalent network under the fault condition is shown in figure 6.2 (a), where the three sequence networks are connected at the points of fault (corresponding LV terminals). The impedances denoted with subscript S are the system impedances; for example Z_{1HS} is the positive-sequence system impedance on HV side. The impedances Z_{1HL}, Z_{2HL} and Z_{0HL} are the positive-sequence, negative-sequence and zero-sequence impedances respectively between HV and LV windings. The zero-sequence network shows open circuit on HV system side because the zero-sequence impedance is infinitely large as viewed / measured from a delta side as explained in Chapter 3 (Section 3.7). When there is no in-feed from LV side (no source on LV side), system impedances are effectively infinite and the network simplifies to that given in figure 6.2 (b). Further, if the system impedances on HV side are very small as compared to the inter-winding impedances, they can be neglected giving the sequence components and fault current as (fault assumed on a phase)

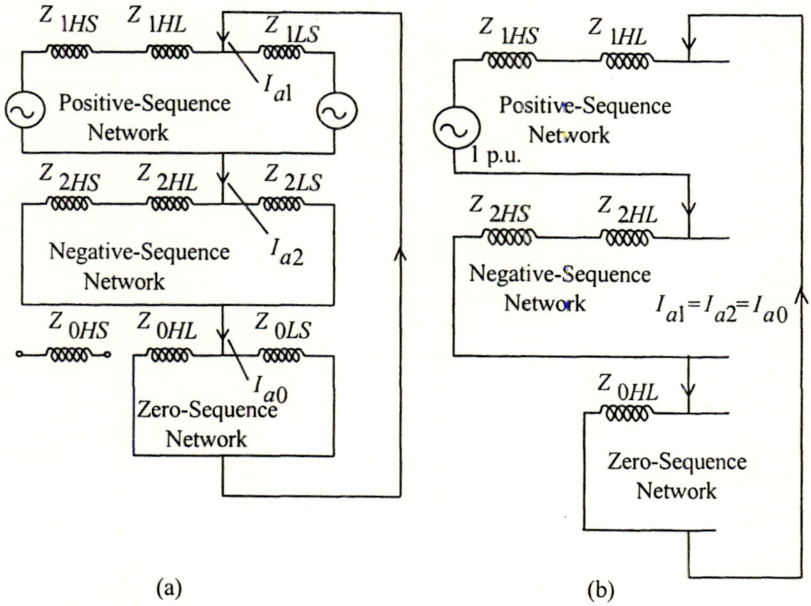

Figure 6.2 Single-line-to-ground fault on star side of delta/star transformer

$$I_{a0} = I_{a1} = I_{a2} = \frac{1}{Z_{1HL} + Z_{2HL} + Z_{0HL}} \tag{6.11}$$

$$I_a = I_{a0} + I_{a1} + I_{a2} = \frac{3}{Z_{1HL} + Z_{2HL} + Z_{0HL}} \tag{6.12}$$

Now, let us consider a three-winding transformer with an unloaded tertiary winding (HV and LV windings are star connected with their neutrals grounded, and tertiary winding is delta connected). The interconnection of sequence networks is shown in figure 6.3 (a). A single-line-to-ground fault is considered on phase a of LV winding. Since it is a three-winding transformer, the corresponding star equivalent circuits are inserted at appropriate places in the network. In the positive-sequence and negative-sequence networks, the tertiary is shown open-circuited because it is unloaded; only in the zero-sequence network the tertiary is in the circuit since the zero-sequence currents can flow in a closed delta. If the pre-fault currents are neglected, both the sources in positive-sequence network are equal to 1 per-unit voltage. The network gets simplified to that shown in figure 6.3 (b). The positive-sequence impedance is

$$Z_1 = (Z_{1HS} + Z_{1H} + Z_{1L}) // Z_{1LS} \tag{6.13}$$

where Z_{1HS} and Z_{1LS} are positive-sequence system impedances, and Z_{1H} and Z_{1L} are positive-sequence impedances of HV and LV windings respectively in the star equivalent circuit.

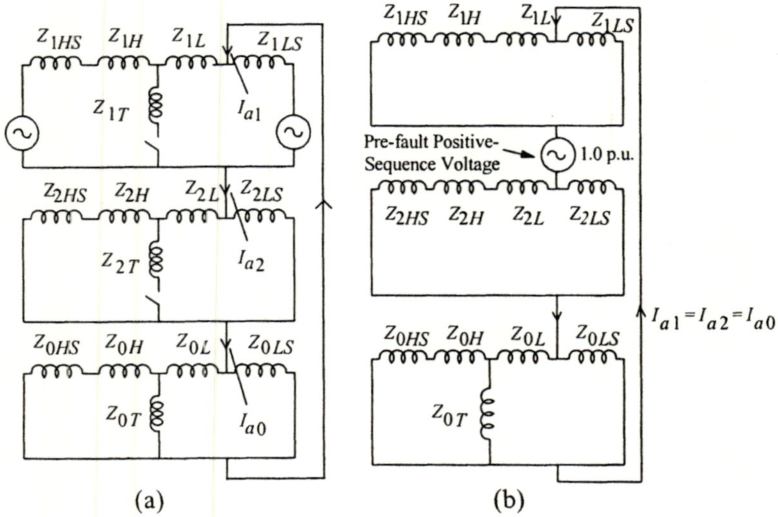

Figure 6.3 Single-line-to-ground fault in three-winding transformer

Similarly, the negative-sequence and zero-sequence impedances are given by

$$Z_2 = (Z_{2HS} + Z_{2H} + Z_{2L}) // Z_{2LS} \tag{6.14}$$

$$Z_0 = ([(Z_{0HS} + Z_{0H}) // Z_{0T}] + Z_{0L}) // Z_{0LS} \tag{6.15}$$

The impedances Z_1 and Z_2 are equal because the corresponding positive-sequence and negative-sequence impedances in their expressions are equal. The total fault current is then calculated as

$$I_f = 3/(Z_1 + Z_2 + Z_0) \tag{6.16}$$

The fault current in any of the windings is calculated by adding the corresponding sequence currents flowing in them in the three sequence networks. For example, the current in phase a of HV winding is sum of the currents flowing through the impedances Z_{1H}, Z_{2H} and Z_{0H} of the positive-sequence, negative-sequence and zero-sequence networks respectively. The tertiary winding current is only the zero-sequence current flowing through the impedance Z_{0T}.

An unloaded tertiary winding is used for the stabilizing purpose as discussed in Chapter 3. Since its terminals are not usually brought out, an external short circuit is not possible and it may not be necessary to design it for withstanding a short circuit at its own terminals. However, the above analysis of single-line-to-ground fault in a three-winding transformer has shown that the tertiary winding must be able to withstand the forces produced in it by asymmetrical fault on LV or HV winding. Consider a case of star / star connected transformer with a delta connected tertiary winding, in which a single-line-to-ground fault occurs on the LV side whose neutral is grounded. If there is no in-feed from the LV side (no source on the LV side), with reference to figure 6.3, the impedances Z_{1LS}, Z_{2LS} and Z_{0LS} will be infinite. There will be open circuit on the HV side in the zero-sequence network since HV neutral is not grounded in the case being considered. If these modifications are done in figure 6.3, it can be seen that the faulted LV winding carries all the three sequence currents, whereas the tertiary winding carries only the zero-sequence current. Since all the three sequence currents are equal for a single-line-to-ground fault condition (equation 6.7), the tertiary winding carries one-third of ampere-turns of the faulted LV winding. As explained in Chapter 3, an unloaded tertiary winding is used to stabilize the neutral voltage under asymmetrical loading conditions. The load on each phase of the tertiary winding is equal to one-third of a single-phase / unbalanced load applied on one of the main windings. Hence, the rating of the unloaded tertiary winding is commonly taken as one-third of the rating of the main windings. In single-line-to-ground fault conditions, the conductor of the tertiary winding chosen according to this rule should also help the tertiary winding in withstanding forces under a single-line-to-ground fault condition in most of the cases. This is particularly true for the case discussed previously in which the neutral terminal of one of the main windings is grounded (in this case the tertiary winding carries one-third of ampere-turns of the faulted winding). For the other connections of windings and neutral grounding conditions, the value of zero-sequence current flowing in the tertiary winding depends on the relative values of impedances of windings and system impedances in the zero-sequence network. For example, in the above case if the HV neutral is also grounded, the zero-sequence current has another path available, and the magnitude of zero-sequence current carried by LV, HV and tertiary windings depends on the relative impedances of the parallel paths (Z_{0T} in parallel with $(Z_{0HS} + Z_{0H})$ in figure 6.3). Hence, with the HV neutral also grounded, the forces on the tertiary winding are reduced.

As seen in Chapter 3, the stabilizing unloaded tertiary windings are provided to reduce the third harmonic component of flux and voltage by providing a path for third harmonic magnetizing currents and to stabilize the neutral by virtue of reduction in the zero-sequence impedance. For three-phase three-limb transformers of smaller rating with star / star connected windings having grounded neutrals, the tertiary stabilizing winding may not be provided. This is because the reluctance offered to the zero-sequence flux is high, which makes the zero-sequence impedance low and an appreciable unbalanced load can be taken by three-phase three-limb transformers with star / star connected windings. Also, as shown in Appendix A, for such transformers the omission of stabilizing winding does not reduce the fault current drastically, and it should get detected by the protection circuitry. The increase in zero-sequence impedance due to its omission is not significant; the only major difference is the increase in HV neutral current, which should be taken into account while designing the protection system. The removal of tertiary winding in three-phase three-limb transformers with both HV and LV neutrals grounded, eliminates the weakest link from the short circuit design considerations and reduces the ground fault current to some extent. This results in reduction of the short circuit stresses experienced by the transformers and associated equipment. Hence, as explained in Section 3.8, the provision of stabilizing winding in three-phase three-limb transformers should be critically reviewed if permitted by the considerations of harmonic characteristics and protection requirements.

The generator step-up transformers are generally subjected to short circuit stresses lower than the interconnecting autotransformers. The higher generator impedance in series with the transformer impedance reduces the fault current magnitude for faults on the HV side of the generator transformer. There is a low probability of faults on its LV side since the bus-bars of each phase are usually enclosed in a metal enclosure (bus-duct). But, since generator transformers are the most critical transformers in the whole network, it is desirable to have a higher safety factor for them. Also, the out-of-phase synchronization in generator transformers can result into currents comparable to three-phase short circuit currents. It causes saturation of the core due to which an additional magnetizing transient current gets superimposed on the fault current [5]. Considerable axial short circuit forces are generated under these conditions [6].

The nature of short circuit currents can be highly asymmetrical like inrush currents. A short circuit current has the maximum value when the short circuit is performed at zero voltage instant. The asymmetrical short circuit current has two components: a unidirectional component decreasing exponentially with time and an alternating steady-state symmetrical component at fundamental frequency. The rate of decay of the exponential component is decided by *X/R* ratio of the transformer. The IEC 60076-5 (second edition: 2000-07) for power transformers specifies an asymmetry factor corresponding to switching at the zero voltage instant (the worst condition of switching). For the condition $X/R > 14$, an asymmetrical factor of 1.8 is specified for transformers upto 100 MVA rating, whereas it is 1.9 for transformers above 100 MVA rating. Hence, the peak value of asymmetrical short circuit current can be taken as

$$I_{p(asym)} = \sqrt{2}\times 1.8\times I_{sym} = 2.55\times I_{sym} \quad \text{for ratings up to 100 MVA}$$

$$I_{p(asym)} = \sqrt{2}\times 1.9\times I_{sym} = 2.69\times I_{sym} \quad \text{for ratings} > 100\text{ MVA}$$

where I_{sym} is the r.m.s. value of the symmetrical three-phase short circuit current. The IEEE Standard C57.12.00-2000 also specifies the asymmetrical factors for various *X/R* ratios, the maximum being 2 for the *X/R* ratio of 1000.

6.2 Thermal Capability at Short Circuit

A large current flowing in transformer windings at the time of a short circuit results in temperature rise in them. Because of the fact that the duration of short circuit is usually very short, the temperature rise is not appreciable to cause any damage to the transformer. The IEC publication gives the following formulae for the highest average temperature attained by the winding after a short circuit,

$$\theta_1 = \theta_0 + \frac{2(\theta_0 + 235)}{\dfrac{106000}{J^2 t} - 1} \qquad \text{for copper} \qquad (6.17)$$

$$\theta_1 = \theta_0 + \frac{2(\theta_0 + 225)}{\dfrac{45700}{J^2 t} - 1} \qquad \text{for aluminum} \qquad (6.18)$$

where θ_0 is initial temperature in °C

J is current density in A/mm^2 during the short circuit based on the r.m.s. value of symmetrical short circuit current

t is duration of the short circuit in seconds

While arriving at these expressions, an assumption is made that the entire heat developed during the short circuit is retained in the winding itself raising its temperature. This assumption is justified because the thermal time constant of a winding in oil-immersed transformers is very high as compared to the duration of the short circuit, which allows us to neglect the heat flow from windings to the surrounding oil. The maximum allowed temperature for oil-immersed transformers with the insulation system temperature of 105°C (thermal class A) is 250°C for a copper conductor whereas the same is 200°C for an aluminum conductor. Let us calculate the temperature attained by a winding with the rated current density of 3.5 A/mm^2. If the transformer short circuit impedance is 10%, the current density under short circuit will be 35 A/mm^2 (corresponding to the symmetrical short circuit current). Assuming the initial winding temperature as 105°C (worst case condition), the highest temperature attained by the winding made of copper conductor at the end of the short circuit lasting for 2 seconds (worst case duration) is about 121°C, which is much below the limit of 250°C. Hence, the thermal withstand capability of a transformer under the short circuit conditions is usually not a serious design issue.

6.3 Short Circuit Forces

The basic equation for the calculation of electromagnetic forces is

$$\mathbf{F} = L\,\mathbf{I} \times \mathbf{B} \tag{6.19}$$

where **B** is leakage flux density vector, **I** is current vector and L is winding length. If the analysis of forces is done in two dimensions with the current density in the z direction, the leakage flux density at any point can be resolved into two components, viz. one in the radial direction (Bx) and other in the axial direction (By). Therefore, there is radial force in the x direction due to the axial leakage flux density and axial force in the y direction due to the radial leakage flux density, as shown in figure 6.4.

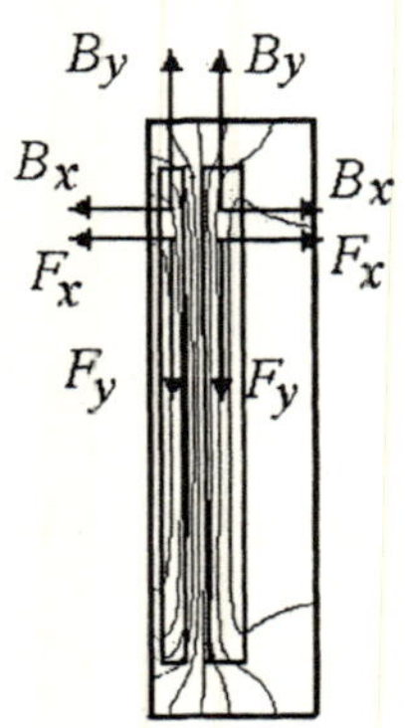

Figure 6.4 Radial and axial forces

The directions of forces are readily apparent from the Fleming's left hand rule also, which says that when the middle finger is in the direction of current and the second finger in the direction of field, the thumb points in the direction of force (all these three fingers being perpendicular to each other).

We have seen in Chapter 3 that the leakage field can be expressed in terms of the winding current. Hence, forces experienced by a winding are proportional to the square of the short circuit current, and are unidirectional and pulsating in nature. With the short circuit current having a steady state alternating component at fundamental frequency and an exponentially decaying component, the force has four components: two alternating components (one at fundamental frequency decreasing with time and other at double the fundamental frequency with a constant but smaller value) and two unidirectional components (one constant component and other decreasing with time). The typical waveforms of the short circuit current and force are shown in figure 6.5. Thus, with a fully offset current the fundamental frequency component of the force is dominant during the initial cycles as seen from the figure.

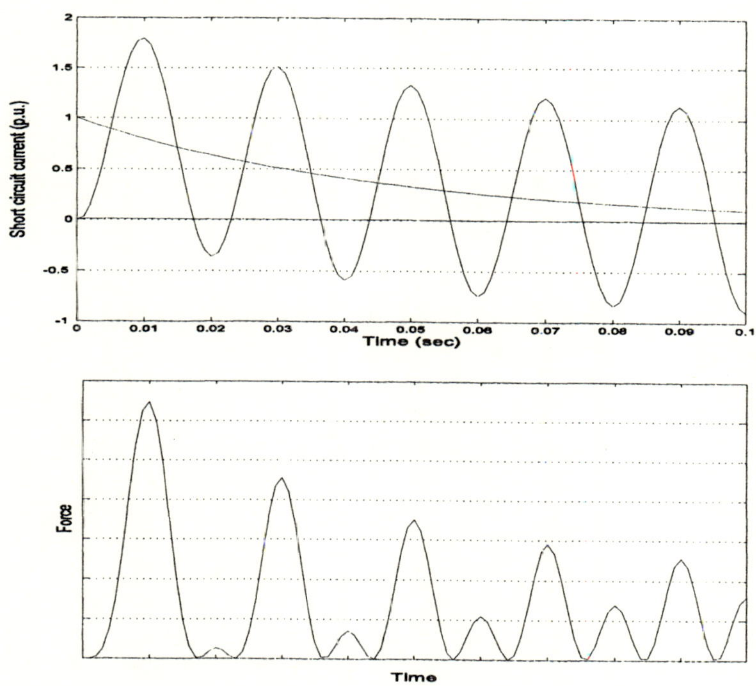

Figure 6.5 Typical waveforms of short circuit current and force

As described earlier, the short circuit forces are resolved into the radial and axial components simplifying the calculations. The approach of resolving them into the two components is valid since the radial and axial forces lead to the different kinds of stresses and modes of failures. There are number of methods reported in the literature for the calculation of forces in transformers. Once the leakage field is accurately calculated, the forces can be easily determined using equation 6.19. Over the years, the short circuit forces have been studied from a static consideration, that is to say that the forces are produced by a steady current. The methods for the calculation of static forces are well documented in 1979 by a CIGRE working group [7]. The static forces can be calculated by any one of the following established methods, viz. Roth's method, Rabin's method, the method of images and finite element method. Some of the analytical and numerical methods for the leakage field calculations are described in Chapter 3. The withstand is checked for the first peak of the short circuit current (with appropriate asymmetry factor as explained in Section 6.1).

A transformer is a highly asymmetrical 3-D electromagnetic device. Under a three-phase short circuit, there is heavy concentration of field in the core window and most of the failures of core-type transformers occur in the window region. In three-phase transformers, the leakage fields of adjacent limbs affect each other. The windings on the central limb are usually subjected to higher forces. There is a considerable variation of force along the winding circumference. Although, within the window the two-dimensional formulations are sufficiently accurate, the three-dimensional numerical methods may have to be used for accurate estimation of forces in the regions outside the core window [8].

6.3.1 Radial forces

The radial forces produced by the axial leakage field act outwards on the outer winding tending to stretch the winding conductor, producing a tensile stress (also called as hoop stress); whereas the inner winding experiences radial forces acting inwards tending to collapse or crush it, producing a compressive stress. The leakage field pattern of figure 6.4 indicates the fringing of the leakage field at the ends of the windings due to which the axial component of the field reduces resulting into smaller radial forces in these regions. For deriving a simple formula for the radial force in a winding, the fringing of the field is neglected; the approximation is justified because the maximum value of the radial force is important which occurs in the major middle portion of the winding.

Let us consider an outer winding, which is subjected to hoop stresses. The value of the leakage field increases from zero at the outside diameter to a maximum at the inside diameter (at the gap between the two windings). The peak value of flux density in the gap is

$$B_{gp} = \frac{\sqrt{2}\mu_0 NI}{H_w} \tag{6.20}$$

where NI is the r.m.s. value of winding ampere-turns and H_w is winding height in meters. The whole winding is in the average value of flux density of half the gap value. The total radial force acting on the winding having a mean diameter of D_m (in meters) can be calculated by equation 6.19 as

$$F_r = \left[\frac{1}{2}\frac{\sqrt{2}\mu_0 NI}{H_w}\right] \times \sqrt{2}\, NI \times \pi\, D_m = \frac{\mu_0 (NI)^2}{H_w} \times \pi\, D_m \quad \text{newton} \tag{6.21}$$

For the outer winding, the conductors close to gap (at the inside diameter) experience higher forces as compared to those near the outside diameter (force reduces linearly from a maximum value at the gap to zero at the outside diameter). The force can be considered to be transferred from conductors with high load (force) to those with low load if the conductors are wound tightly [9]. Hence, averaging of the force value over the radial depth of the winding as done in the above equation is justified since the winding conductors share the load almost uniformly. If the curvature is taken into account by the process of integration across the winding radial depth as done in Section 3.1.1 of Chapter 3, the mean diameter of the winding in the above equation should be replaced by its inside diameter plus two-thirds of the radial depth.

The average hoop stress for the outer winding is calculated as for a cylindrical boiler shell shown in figure 6.6. The transverse force F acting on two halves of the winding is equivalent to pressure on the diameter [10]; hence it will be given by equation 6.21 with πD_m replaced by D_m. If the cross-sectional area of turn is A_t (in m^2), the average hoop stress in the winding is

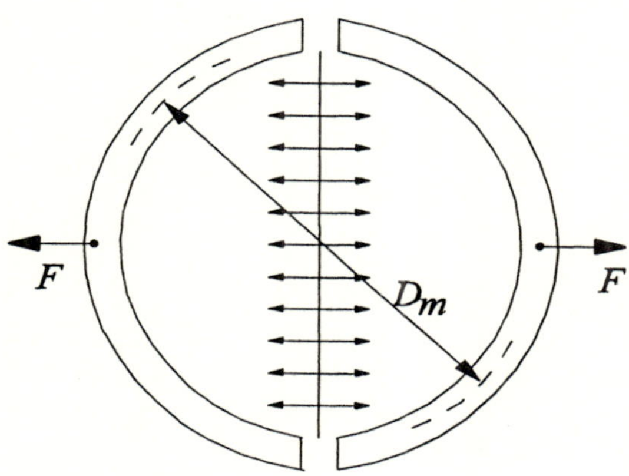

Figure 6.6 Hoop stress calculation

$$\sigma_{avg} = \frac{\mu_0 (NI)^2 D_m}{H_w (N \times 2 \times A_t)} = \frac{\mu_0 I^2}{2\rho\pi H_w} \times \frac{\rho\pi D_m N}{A_t} \tag{6.22}$$

Let I_r be the rated r.m.s. current and Z_{pu} be the per-unit impedance of a transformer. Under the short circuit condition, the r.m.s. value of current in the winding is equal to (I_r/Z_{pu}). To take into account the asymmetry, this current value is multiplied by the asymmetry factor k. If we denote copper loss per phase by P_R, the expression for σ_{avg} under the short circuit condition is

$$\sigma_{avg} = \frac{\mu_0}{2\rho\pi H_w} \times \frac{I_r^2 k^2}{Z_{pu}^2} \times \frac{\rho(\pi D_m N)}{A_t} = \frac{\mu_0}{2\left(\sqrt{2}\right)^2 \rho\pi H_w} \times \frac{P_R \left(k\sqrt{2}\right)^2}{Z_{pu}^2} \ \text{N/m}^2 \tag{6.23}$$

Substituting the values of μ_0 (=4 π × 10^{-7}) and ρ (resistivity of copper at 75°C = 0.0211 × 10^{-6}) we finally get

$$\sigma_{avg} = 4.74 \left(k\sqrt{2}\right)^2 \frac{P_R}{H_w Z_{pu}^2} \ \text{N/m}^2 \tag{6.24}$$

or

$$\sigma_{avg} = 0.48 \times 10^{-4} \times \left(k\sqrt{2}\right)^2 \frac{P_R}{H_w Z_{pu}^2} \ \text{kg/cm}^2 \ \ldots \text{Copper conductor} \tag{6.25}$$

where P_R is in watts and H_w in meters. It is to be noted that the term P_R is only the DC I^2R loss (without having any component of stray loss) of the winding per phase at 75°C. Hence, with very little and basic information of the design, the average value of hoop stress can be easily calculated. If an aluminum conductor is used, the numerical constant in the above equation will reduce according to the ratio of the resistivity of copper to aluminum giving,

$$\sigma_{avg} = 0.29 \times 10^{-4} \times \left(k\sqrt{2}\right)^2 \frac{P_R}{H_w Z_{pu}^2} \ \text{kg/cm}^2 \ \ldots \text{Aluminum conductor} \tag{6.26}$$

As mentioned earlier, the above value of average stress can be assumed to be applicable for an entire tightly wound disk winding without much error. This is because of the fact that although the stress is higher for the inner conductors of the outer winding, these conductors cannot elongate without stressing the outer conductors. This results in a near uniform hoop stress distribution over the entire winding. In layer / helical windings having two or more layers, the layers do not firmly support each other and there is no transfer of load between them. Hence, the hoop stress is highest for the innermost layer and it decreases towards the outer layers. For a double-layer winding, the average stress in the layer near the gap is 1.5 times higher than the average stress for the two layers considered together. Generalizing, if there are L layers, the average stress in k^{th} layer (from gap) is $[2-((2k-1)/L)]$ times the average stress of all the layers considered together. Thus, the design of outer multi-layer winding subjected to a hoop stress requires special considerations.

For an inner winding subjected to radial forces acting inwards, the average stress can be calculated by the same formulae as above for the outer winding. However, since the inner winding can either fail by collapsing or due to bending between the supports, the compressive stresses of the inner winding are not the simple equivalents of the hoop stresses of the outer winding. Thus, the inner winding design considerations are quite different, and these aspects along with the failure modes are discussed in Section 6.5.

6.3.2 Axial forces

For an uniform ampere-turn distribution in windings with equal heights (ideal conditions), the axial forces due to the radial leakage field at the winding ends are directed towards the winding center as shown in figure 6.4. Although, there is higher local force per unit length at the winding ends, the cumulative compressive force is maximum at the center of windings (see figure 6.7). Thus, both the inner and outer windings experience compressive forces with no end thrust on the clamping structures (under ideal conditions). For an asymmetry factor of 1.8, the total axial compressive force acting on the inner and outer windings taken together is given by the following expression [11]:

$$F_a = \frac{50.8 \times S}{Z_{pu} \times H_w \times f} \; kg \quad \cdots\cdots \textit{ ideal conditions} \tag{6.27}$$

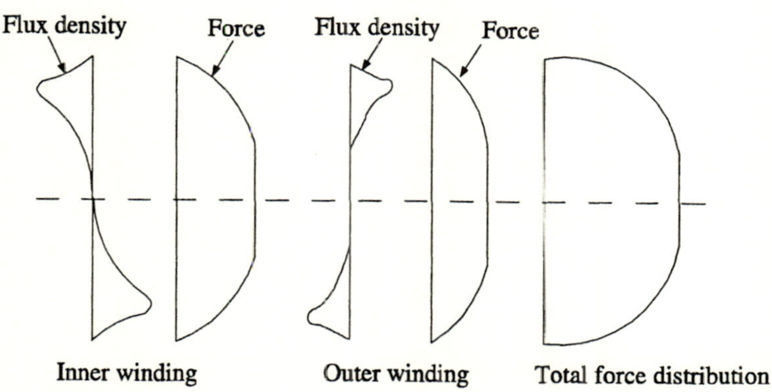

Figure 6.7 Axial force distribution

where S is rated power per limb in kVA, H_w is winding height in meters, Z_{pu} is per-unit impedance, and f is frequency in Hz. The inner winding being closer to the limb, by virtue of higher radial flux, experiences higher compressive force as compared to the outer winding. In the absence of detailed analysis, it can be assumed that 25 to 33% of force is taken by the outer winding, and the remaining 75 to 67% is taken by the inner winding.

Calculation of axial forces in the windings due to the radial field in non-ideal conditions is not straightforward. Assumptions, if made to simplify the calculations, can lead to erroneous results for non-uniform windings. The presence of tap breaks makes the calculations quite difficult. The methods discussed in Chapter 3 should be used to calculate the radial field and the resulting axial forces. The forces calculated at various points in the winding are added to find the maximum compressive force in the winding. Once the total axial force for each winding is calculated, the compressive stress in the supporting radial spacers (blocks) can be calculated by dividing the compressive force by the total area of the radial spacers. The stress should be less than a certain limit, which depends on the material of the spacer. If the pre-stress (discussed in Section 6.7) applied is more than the value of force, the pre-stress value should be considered while calculating the stress on the radial spacers.

The reasons for a higher value of radial field and consequent axial forces are: mismatch of ampere-turn distribution between LV and HV windings, tappings in the winding, unaccounted shrinkage of insulation during drying and impregnation processes, etc. When the windings are not placed symmetrically with respect to the center-line as shown in figure 6.8, the resulting axial forces are in such a direction that the asymmetry and the end thrusts on the clamping structures increase further. It is well known that even a small axial displacement of windings or misalignment of magnetic centers of windings can eventually cause enormous axial forces leading to failure of transformers [12, 13]. Hence, strict sizing / dimension control is required during processing and assembling of windings so that the windings get symmetrically placed.

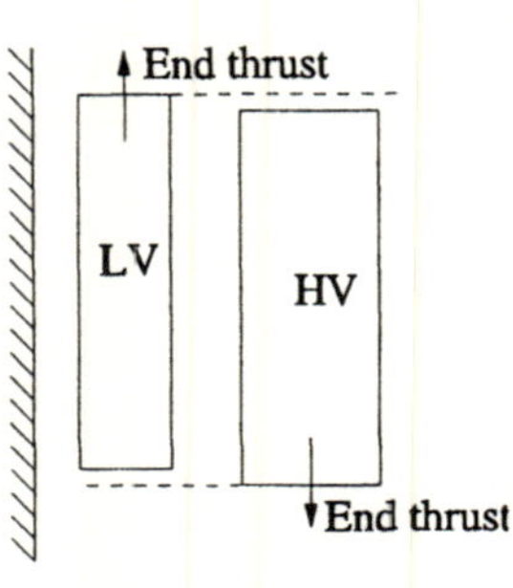

Figure 6.8 Axial asymmetry

6.4 Dynamic Behavior Under Short Circuits

The transformer windings along with the supporting clamping structure form a mechanical system having mass and elasticity. The applied electromagnetic forces are oscillatory in nature and they act on the elastic system comprising of winding conductors, insulation system and clamping structures. The forces are dynamically transmitted to various parts of the transformer and they can be quite different from the applied forces depending upon the relationship between excitation frequencies and natural frequencies of the system. Thus, the dynamic behavior of the system has to be analyzed to find out the stresses and displacements produced by the short circuit forces. The dynamic analysis, although quite complex, is certainly desirable which improves the understanding of the whole phenomenon and helps designers to enhance the reliability of the transformers under short circuit conditions. The dynamic behavior is associated with time-dependence of the instantaneous short circuit current and the corresponding force, and the displacement of the windings producing instantaneous modifications of these forces. The inertia of conductors, frictional forces and reactionary forces of the various resilient members of the system play an important role in deciding the dynamic response.

In the radial direction, the elasticity of copper is large and the mass is small, resulting into natural frequency much higher than 50 / 60 Hz and 100 / 120 Hz (the fundamental frequency and twice the fundamental frequency of the excitation force). Hence, there exists a very remote possibility of increase in displacements by resonance effects under the action of radial forces. Therefore, these forces may be considered as applied slowly and producing a maximum stress corresponding to the first peak of an asymmetrical fault current [10]. In other words, the energy stored by the displacement of windings subjected to radial forces is almost entirely elastic and the stresses in the windings correspond closely with the instantaneous values of the generated forces [14].

Contrary to the radial direction, the amount of insulation is quite significant along the axial direction, which is easily compressible. With the axial forces acting on the system consisting of the conductor and insulation, the natural frequencies may come quite close to the excitation frequencies of the short circuit forces. Such a resonant condition leads to large displacements and eventual failure of transformers. Hence, the dynamic analysis of mechanical system consisting of windings and clamping structures is essential and has been investigated in detail by many researchers.

The transformer windings, made up of large number of conductors separated by insulating materials, can be represented by an elastic column with distributed mass and spring parameters, restrained by end springs representing the insulation between the windings and yokes. Since there is heavy insulation at the winding ends, these springs are usually assumed as mass-less. When a force is applied to an elastic structure, the displacement and stress depend not only on the magnitude of force and its variation with time, but also on the natural frequencies of the structure.

The methods for calculating dynamic response are quite complex. They have to take into account the boundary conditions, viz. degree of pre-stress, stiffness of clamping structure and the proximity of tank / other windings. It should also take into account the effects of displacement of conductors. The method reported in [15] replaces a model of ordinary linear differential equations representing the system by an approximate equivalent model of linear difference equations with a constant time step-length. The non-linear insulation characteristics obtained from the experimental data are used to solve the difference equations by a digital computer. In [16, 17], the dynamic load and displacement at any point in the winding are calculated by using a generalized Fourier series of the normal modes (standing wave approach). The analysis presented can be applied to an arbitrary space distribution of electromagnetic forces with actual time variation of a fully asymmetric short circuit current taken into account. The dynamic forces are reported to have completely different magnitudes and waveshapes as compared to the applied electromagnetic forces.

A rigorous analytical solution is possible when linear insulation characteristics are assumed. The insulation of a transformer has non-linear insulation characteristics. The dynamic properties of pressboard are highly non-linear and considerably different from the static characteristics. The dynamic stiffness and damping characteristics can be experimentally determined [18, 19]. The use of static characteristics was reported to be acceptable [19], which leads to pessimistic results as compared to that obtained by using the dynamic characteristics. It was shown in [20] that the dynamic value of Young's modulus can be derived from the static characteristics. However, it is explained in [17] that this approximation may not be valid for oil-impregnated insulation. Oil provides hydrodynamic mass effect to the clamping parts subjected to short circuit forces, and it also significantly influences the insulation stiffness characteristics. These complexities and the non-linearity of the systems involved can be effectively taken into account by numerical methods. A dynamic analysis is reported in [21] which accounts for the difference in the electromagnetic forces inside and outside the core window. It is shown that a winding displacement inside the window is distinctly different and higher than that outside the window. A simplified model is proposed in [22] whereby the physical aggregation of conductors and supports is considered as a continuous elastic solid represented by a single partial differential equation.

Thus, a number of numerical methods are available for determining the dynamic response of a transformer under short circuit conditions. The methods have not been yet perfected due to the lack of precise knowledge of dynamic characteristics of various materials used in transformers. The dynamic calculations can certainly increase the theoretical knowledge of the whole phenomenon, but it is difficult to ascertain the validity of the results obtained. On the contrary, it is fairly easy to calculate the natural frequencies of windings and check the absence of resonance. Hence, a more practical approach can be to check the withstand for the worst possible peak value of an asymmetrical fault current (static calculation as explained in Section 6.3). In addition, the natural frequencies of windings should be calculated to check that they are far away from the power frequency or twice the power frequency. If the natural frequencies are close to either 50 or 100 Hz (60 or 120 Hz), these can be altered (to avoid resonance) by using a different pre-stress value or by changing the modes of vibration by a suitable sub-division of windings. Hence, the well-established static calculations along with the determination of natural frequencies could form a basis of short circuit strength calculations [23, 24] until the dynamic analysis is perfected and standardized.

In a typical core type power transformer, windings are commonly clamped between top and bottom clamping plates (rings) of insulating material. The construction of the winding is quite complicated consisting of many different materials like kraft paper, pre-compressed board, copper / aluminum conductor, densified wood, etc. The winding consists of many disks and insulation spacers. Thus, the winding is a combination of spacers, conductors and pre-compressed boards. Strictly speaking, the winding is having multiple degrees of freedom. The winding is considered as a distributed mass system in the analysis. The winding stiffness is almost entirely governed by the insulation only. The top and bottom end insulations are considered as mass-less linear springs. The winding can be represented by an elastic column restrained between the two end springs as shown in figure 6.9.

The equation of motion [16, 21] is given by

$$m\frac{\partial^2 y}{\partial t^2}+c\frac{\partial y}{\partial t}-k\frac{\partial^2 y}{\partial x^2}=F(y,t) \tag{6.28}$$

where $y(x, t)$ is the displacement (from a rest position) of any point at a vertical distance x, m is mass of winding per unit length, c is damping factor per unit length, k is stiffness per unit length, and F is applied electromagnetic force per unit length.

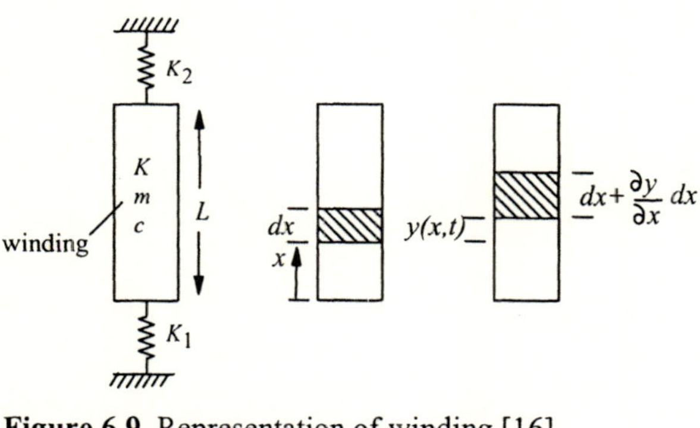

Figure 6.9 Representation of winding [16]

The expression for natural frequency, ω_n (in rad/sec) can be derived from equation 6.28 with the boundary conditions that the displacement and velocity at any position x are zero at $t = 0$, and the net force acting at positions $x = 0$ (winding bottom) and $x = L$ (winding top) is zero. The expression is

$$\omega_n = \lambda_n \sqrt{\frac{K}{M}} \tag{6.29}$$

where λ_n is eigen value corresponding to n^{th} natural frequency, $M\ (= mL\ )$ is total mass of winding, L is length (height) of winding, and K is winding stiffness ($= k / L$). The winding stiffness per unit length is given by

$$k = \frac{A\ E_{eq}}{L_{eq}} \tag{6.30}$$

where A is area of insulation, E_{eq} is equivalent Young's modulus of winding, and L_{eq} is equivalent length of winding. Thus, the natural frequency of a winding is a function of its mass, equivalent height, cross sectional area and modulus of elasticity. The conductor material (copper) is too stiff to get compressed appreciably by the axial force. Hence, all the winding compression is due to those fractions of its height occupied by the paper and press-board insulation. The equivalent Young's modulus can therefore be calculated from [20]

$$\frac{E_{eq}}{L_{eq}} = \frac{E_p E_b}{E_p L_b + E_b L_p} \tag{6.31}$$

where E_{eq} is modulus of elasticity of the combined paper and pressboard insulation system, E_p is modulus of elasticity of paper, and E_b is modulus of elasticity of pressboard. The terms L_p, L_b and L_{eq} represent thickness of paper, thickness of pressboard and total equivalent thickness of paper and pressboard respectively.

The eigen values (λ) are calculated [16] from the equation

$$\tan\lambda = \frac{\left(\frac{K_1}{K} + \frac{K_2}{K}\right)\lambda}{\lambda^2 - \frac{K_1 K_2}{K^2}} \tag{6.32}$$

where K_1 and K_2 are the stiffness values of bottom and top end insulation respectively. In equation 6.32, the only unknown is λ which can be found by an iterative method. Subsequently, the values of natural frequencies can be calculated from equation 6.29.

The natural frequencies can be more accurately calculated by numerical methods such as FEM analysis [24]. If any of the calculated natural frequencies is close to the exciting frequencies, they can be altered by making suitable changes in the winding configuration and / or pre-stress value.

6.5 Failure Modes Due to Radial Forces

The failure modes of windings are quite different for inward and outward radial forces. Winding conductors subjected to outward forces experience the tensile (hoop) stresses. The compressive stresses are developed in conductors of a winding subjected to the inward forces. In concentric windings, the strength of outer windings subjected to the outward forces depends on the tensile strength of the conductor; on the contrary the strength of inner windings subjected to the inward forces depends on the support structure provided. The radial collapse of the inner windings is common, whereas the outward bursting of the outer windings usually does not take place.

6.5.1 Winding subjected to tensile stresses

If a winding is tightly wound, the conductors in the radial direction in a disk winding or in any layer of a multi-layer winding can be assumed to have a uniform tensile stress. Since most of the space in the radial direction is occupied with copper (except for the small paper covering on the conductors), the ratio of stiffness to mass is high. As mentioned earlier, the natural frequency is much higher than the exciting frequencies, and hence chances of resonance are remote. Under a stretched condition, if the stress exceeds the yield strength of the conductor, a failure occurs. The conductor insulation may get damaged or there could be local bulging of the winding. The conductor may even break due to improper joints. The chances of failure of windings subjected to the tensile hoop stresses are unlikely if a conductor with a certain minimum 0.2% proof strength is used. The 0.2% proof stress can be defined as that stress value which produces a permanent strain of 0.2% (2 mm in 1000 mm) as shown in figure 6.10. One of the common ways to increase the strength is the use of work-hardened conductor; the hardness should not be very high since there could be difficulty in winding operation with such a hard conductor. A lower value of current density is also used to improve the withstand characteristics.

6.5.2 Windings subjected to compressive stresses

Conductors of inner windings, which are subjected to the radial compressive load, may fail due to bending between supports or buckling. The former case is applicable when the inner winding is firmly supported by the axially placed supporting spacers (strips), and the supporting structure as a whole has higher stiffness than conductors (e.g., if the spacers are supported by the core structure).

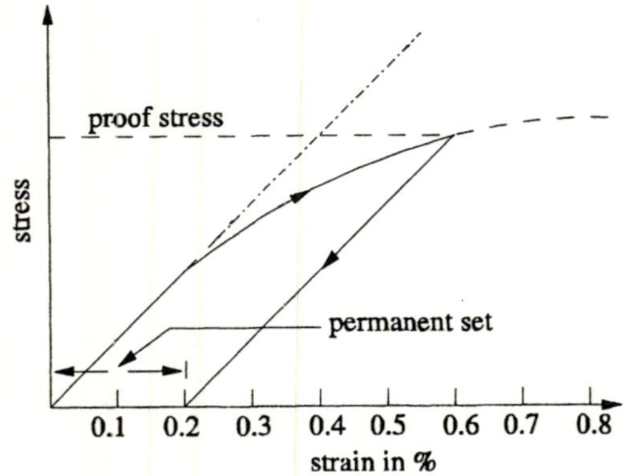

Figure 6.10 0.2% Proof stress

In that case, the conductors can bend between the supports all along the circumference as shown in figure 6.11 (a) if the stress exceeds the elastic limit of the conductor material. This form of buckling is termed as *forced buckling* [25, discussion of 26], which also occurs when the winding cylinder has a significant stiffness as compared to the winding conductors (i.e., when thick cylinders of a stiff material are used).

The latter case of buckling, termed as *free buckling*, is essentially an unsupported buckling mode, in which the span of the conductor buckle bears no relation to the span of axial supporting spacers as shown in figure 6.11 (b). This kind of failure occurs mostly with thin winding cylinders, where conductor has higher stiffness as compared to that of inner cylinders and / or the cylinders (and the axial spacers) are not firmly supported from inside. The conductors bulge inwards as well as outwards at one or more locations along the circumference. There are many factors which may lead to the buckling phenomenon, viz. winding looseness, inferior material characteristics, eccentricities in windings, lower stiffness of supporting structures as compared to the conductor, etc.

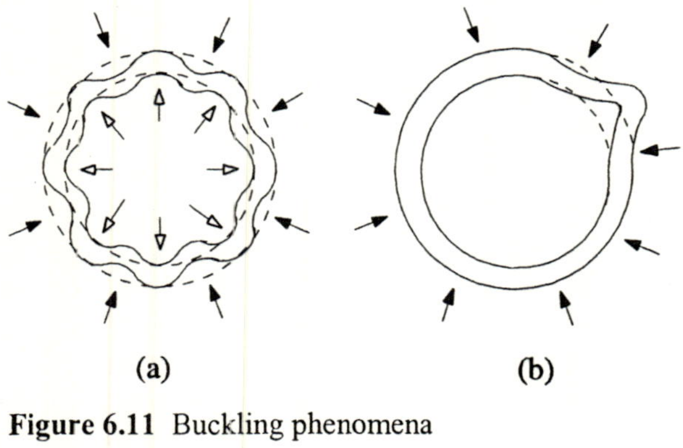

Figure 6.11 Buckling phenomena

The buckling can be viewed as a sequential chain of failures, initiated at the outermost conductor of the inner winding and moving towards the innermost conductor facing the core. The number of winding supports should be adequate for giving the necessary strength to the winding against the radial forces. When the supporting structures are in direct contact with the core, a winding can be taken as very rigidly supported. On the contrary, if there is no direct contact (fully or partly) with the core, the winding is only supported by the insulating cylinder made of mostly the pressboard material thereby reducing the effective stiffness of the support structure and increasing the chances of failure. The supports provided are effective only when the support structure as a whole is in firm contact with the core.

A winding conductor subjected to the inward radial forces is usually modeled as a circular loop under a uniformly distributed radial load. The critical load per unit length of the winding conductor is given by [27]

$$f_r = \frac{2E}{D_m^3}\,\frac{w\,t^3}{12}\,[N_s^2 - 4] \tag{6.33}$$

where E is modulus of elasticity of conductor material, N_s is total number of axially placed supports, w is width of conductor, t is thickness of conductor and D_m is mean diameter of winding. The compressive stress on the inner winding conductor is given as [10]

$$\sigma_{avg} = \frac{f_r\,D_m}{2A} \tag{6.34}$$

where A is area of conductor ($= w\,t$). Substituting the value of f_r from equation 6.33 we get

$$\sigma_{avg} = \frac{E}{12}\left(\frac{t\,N_s}{D_m}\right)^2\left[\frac{N_s^2 - 4}{N_s^2}\right] \tag{6.35}$$

For $N_s >> 1$, the expression for the minimum number of supports to be provided is

$$N_s = \frac{D_m}{t}\sqrt{\frac{12\,\sigma_{avg}}{E}} \tag{6.36}$$

The term σ_{avg} is the average value of the compressive stress (in an entire disk winding or in a layer of a multi-layer winding) calculated as per Section 6.3.1. It can be observed that the higher the conductor thickness, the lower the number of required supports will be. Adoption of higher slenderness ratio $(t\,N_s/D_m)$ allows higher critical radial compressive stresses [27]. For a winding with a low mean diameter, there is a limit up to which the conductor thickness can be increased (it is difficult to wind a thick conductor on a small winding diameter). Similarly, for a given winding diameter, there is a limit on the number of axially placed supports; the radially placed spacers (between disks) on these axially placed spacers reduce the surface area available for the cooling purpose. To avoid this problem, the intermediate axially placed spacers (between every two main spacers) are used, which do not have the radial spacers placed on them.

A more elaborate and accurate analysis of buckling behavior has been reported in the literature. The dynamic analysis of the buckling behavior of inner windings subjected to radial forces is reported in [13, 26]. The FEM analysis is used in [28] to evaluate the radial buckling strength of windings.

The design of windings for withstanding the tensile stresses is relatively easy as compared to the compressive stresses. This is because for the tensile stresses, the permissible stress depends on the yield strength of the conductor material. There are fewer ambiguities, once the calculated maximum stress (whose calculation is also usually straightforward) is kept below the yield stress. The design criteria for determining the withstand of the inner windings subjected to the compressive stresses are a bit complicated and may vary for different manufacturers. After the drying and oil-impregnation processes, the insulating components may shrink considerably. Hence, the lowering clearances and tolerances provided for the insulating components have to be properly decided based on the manufacturing practices and variations in dimensions of the insulating materials observed at various stages of manufacturing. If the inner winding is not assembled on the core-limb in tight-fit condition or if there is looseness, then the wedging of insulating components is necessary. If the total integrity of the support structure is ensured in this manner, the inner winding can be said to be supported from the inside and the number of supports calculated by equation 6.36 will be adequate to prevent the buckling. Some manufacturers [9] completely ignore the strength provided by the inner supporting structures and design the windings to be completely self-supporting. The current density used therefore has to be lower with the result that the material content and cost of the transformer increases. Nevertheless, the reliability of the transformer is enhanced and the extra material put can be easily justified for large transformers. It is reported (discussion of [26]) based on the model tests that the insertion of tight-fitting insulation spacers between the core and innermost insulating cylinder may not be an effective solution for increasing the strength because the clearances between the elements of the structure are larger than the disk displacements prior to the buckling. Thus, the concept of completely self-supporting design seems to be a better option.

6.6 Failure Modes Due to Axial Forces

There are various types of failures under the action of axial compressive forces. If a layer winding is not wound tightly, some conductors may just axially pass over the adjacent conductors, which may damage the conductor insulation leading eventually into a turn-to-turn fault. In another mode of failure, if a winding is set into vibration under the action of axial forces, the conductor insulation may get damaged due to a relative movement between the winding and axially placed insulation spacers.

High axial end thrusts could lead to deformations of the end clamping structures and windings. The end clamping structures play the most important role in resisting axial forces during short circuits. They have to maintain an effective pressure on the windings, applied usually on the clamping ring made of stiff insulating material (pre-compressed board or densified wood). The type of insulation material used for the clamping ring depends on the dielectric stress in the end insulation region of windings. The densified wood material is used for lower stresses and pre-compressed board, being a better grade dielectrically, is used for higher stresses and for complying stringent partial discharge requirements. When a clamping ring made of an insulating material is reinforced by the fiberglass material, an extra strength is provided. Some manufacturers use clamping rings made of steel material. The thickness of metallic clamping rings is smaller than that made from the insulating material. The metallic ring has to be properly grounded with a cut so that it does not form a short-circuited turn around the limb. The sharp edges of the metallic ring should be rounded off and covered with a suitable insulation.

In addition to above types of failures due to the axial forces, there are two principal types of failures, viz. bending between radial spacers and tilting.

6.6.1 Bending between radial spacers

Under the action of axial forces, the winding conductor can bend between the radially placed insulation spacers as shown in figure 6.12. The conductor bending can result into a damage of its insulation. The maximum stress in the conductor due to bending occurs at the corners of the radial spacers and is given by

$$\sigma_{\max} = \frac{F_{AL}\, S^2\, y}{12\, I_0} \quad \text{kg/cm}^2 \tag{6.37}$$

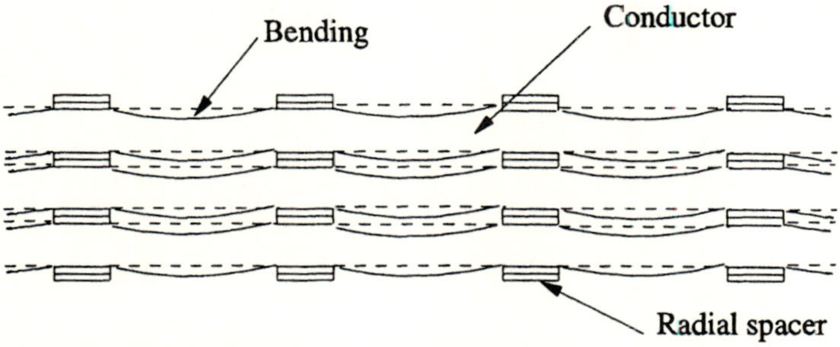

Figure 6.12 Bending between radial spacers

where,

F_{AL} is maximum axial bending load in kg/cm. It corresponds to the most highly stressed disk in a disk winding or turn in a helical winding (layer winding with radial spacers). The maximum axial load may usually lie in the region of non-uniform ampere-turn distribution (e.g., tap zone). The maximum axial load, calculated accurately by a method such as FEM, divided by the mean turn length (πD_m) gives the value of F_{AL}, where D_m is mean diameter of winding in cm.

S is span between two radial spacers in cm $\left(\frac{\pi D_m}{N_S} - spacer\ width\right)$;

N_S is number of radial spacers.

y is maximum distance from neutral axis for conductor in cm (i.e., half of conductor axial width: $w/2$).

I_0 is moment of inertia of disk or turn $\left(=\frac{n t w^3}{12}\right)$, n being number of conductors in radial direction, and t is conductor thickness in cm.

The maximum stress in the conductor calculated by the above formula should be less than the limiting value for the type of conductor used (about 1200 kg/cm^2 for soft copper).

6.6.2 Tilting under an axial load

The failure due to tilting under the action of axial compressive forces is one of the principal modes of failures in large power transformers. When these forces are more than a certain limit, a failure can occur due to tilting of conductors in a zigzag fashion as shown in figure 6.13 for a disk winding. In this mode of failure, there is turning of cross section of conductors around the perpendicular axis of symmetry. There are two kinds of forces that resist the tilting of the conductors. The first one is due to the conductor material, which resists being twisted. The second resisting force is the friction force (due to corners of conductors); during tilting the conductors at both ends must bite into the material of the radial spacer, producing a couple at the conductor ends which resists tilting. The two resisting forces are usually considered separately to arrive at the critical stress and load, causing the failure.

The critical load and critical stress for a disk winding, if the resistance offered by the conductor material alone to twisting is considered, are [10]:

$$F_{cr\_twist} = \frac{N \pi E t w^2}{6 R} \tag{6.38}$$

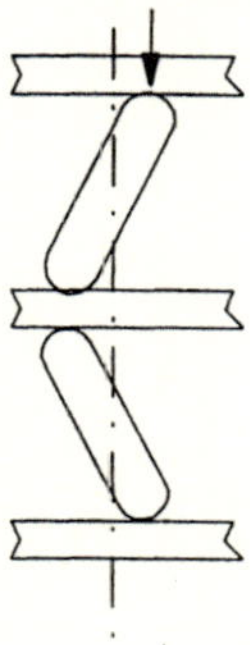

Figure 6.13 Conductor tilting in a disk winding

$$\sigma_{cr\_twist} = \frac{E}{12}\left(\frac{w}{R}\right)^2 \tag{6.39}$$

where R is mean radius of winding, w is conductor width (height in axial direction), t is conductor radial thickness, N is number of winding turns, and E is modulus of elasticity of conductor. For this case, rounded ends are assumed for the conductor so that the frictional resistance is absent. The tilting strength decreases inversely as the square of winding radius, suggesting that the large windings should be carefully designed.

If the conductor has sharp ends, the frictional force resists tilting. The critical load against the friction is [10]

$$F_{cr\_fric} = \frac{N_b\, N\, b\, c\, t^3}{6\, w} \tag{6.40}$$

where N_b is number of radial spacers (blocks), b is width of spacers and c is constant which depends on spacer material. Actually, due to the conductor corner radius, the contribution to tilting resistance (due to friction) reduces and this reduction should be considered suitably along with equation 6.40. The critical strength of a helical winding (a layer winding with radial spacers) is higher than a layer winding (which is without radial spacers) because of the additional strength offered by the spacers.

The total critical load F_{cr} is the addition of axial strengths against twisting and friction,

$$F_{cr} = \frac{N\,\pi\, E\, t\, w^2}{6R} + \frac{N_b\, N\, b\, c\, t^3}{6w} \tag{6.41}$$

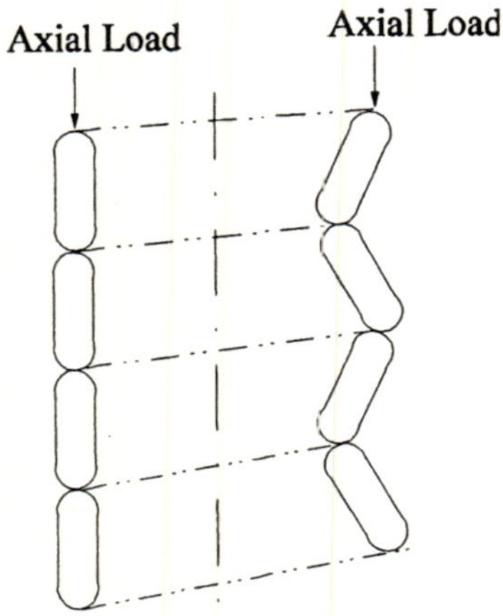

Figure 6.14 Half-tilting mode in a layer winding [29]

Analysis of a layer winding under a tilting load is given in [29]. It is reported that this winding can fail in the zigzag pattern on one side only as shown in figure 6.14 (termed as the half-tilting failure mode). It is also reported in the same paper that a disk winding can also fail in the half-tilting mode for which the winding strength is about 10% less than the corresponding full tilting strength. The mechanical resonance frequencies in the tilting mode of failure are usually much higher than the excitation frequencies, and therefore there is no chance of resonance in this mode. The conductor dimensions have a decisive role in the tilting strength. The twisting strength increases with the conductor width in the square proportion as per equation 6.39. Equation 6.41 indicates that the risk of tilting increases for thin conductors.

When a continuously transposed cable (CTC) conductor is used, although there are two axially placed rows of conductors in one common paper covering (see figure 4.9), it cannot be assumed that the effective tilting strength is higher. The tilting strength of the CTC conductor without epoxy bonding is analyzed for a layer winding in [30]. Two possible modes of failures are described. The first type of failure (termed as cable-wise tilting), in which two adjacent cables tilt against each other, is shown in figure 6.15 (a). If the inter-strand friction is higher, the winding is forced to tilt in pairs of strands in the CTC conductor. The critical stress for this mode of failure, with the consideration of conductor resistance alone, is

$$\sigma_{cr\_twist} = \frac{E}{3}\left(\frac{w_1}{R}\right)^2 \tag{6.42}$$

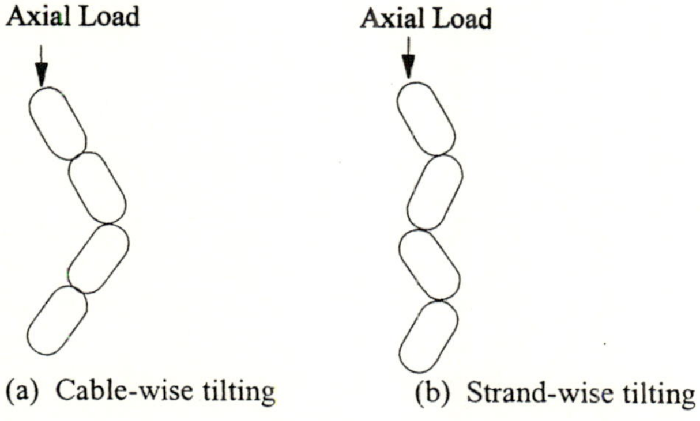

Figure 6.15 Tilting in layer winding with CTC conductor

where w_1 is width of individual strand in the CTC conductor. Comparison of equations 6.39 and 6.42 tells that for the same strand dimensions, the CTC conductor has four times greater tilting strength; the result is obvious because the effective width of its conductor is doubled increasing the strength by four times. This increase in strength is valid only when the two axially placed strands in the CTC conductor can be considered to act together under the tilting load.

In the second mode of failure (termed as strand-wise tilting), two axially placed strands in the CTC conductor tilt against each other as shown in figure 6.15 (b). The critical tilting load in this mode may be lower, reducing the effective overall tilting strength. This is because the lower of the cable-wise and strand-wise strengths triggers the axial instability. It is shown in [30] that while the critical stress in the cable-wise tilting is independent of number of strands in the cable (n), the critical stress in the strand-wise tilting is inversely proportional to n. As the number of strands in the CTC conductor increases, the critical load limit in the strand-wise tilting becomes lower than the cable-wise tilting. Hence, with the increase in number of strands in the CTC conductor, the mode of failure shifts from the cable-wise tilting to the strand-wise tilting.

The use of epoxy-bonded CTC conductor is quite common in which the epoxy coating effectively bonds the strands increasing the resistance against the strand-wise tilting. Each strand in the epoxy-bonded CTC conductor has, in addition to an enamel coating, a coat of thermosetting epoxy resin. The curing of this resin occurs at around 120°C during the processing of windings. After curing, the epoxy-bonded CTC conductor consisting of many strands can be considered as one conductor with an equivalent cross section for the mechanical strength consideration. Thus, the possibility of strand-wise tilting is eliminated, greatly increasing the strength of the CTC conductor against the tilting load. The epoxy-bonded CTC conductor not only reduces the winding eddy losses (as explained in Chapter 4) but it also significantly improves the short circuit withstand characteristics.

6.7 Effect of Pre-Stress

The clamping pressure applied on the windings after the completion of core-winding assembly is called as pre-stress. It has a significant impact on the response of windings during short circuits. It increases the stiffness of windings thereby increasing their mechanical natural frequencies. The relationship between the natural frequency and pre-stress is highly non-linear [31]. The pre-stress reduces oscillatory forces acting on the insulation. The winding displacements also decrease with the increase in the pre-stress value. It should be noted that the value of pre-stress should be judiciously chosen depending upon the characteristics of core-winding assembly [17, 32, 33]. The chosen value of pre-stress must get maintained during the entire life of a transformer. This means that the insulation stability should be fully realized during the processing of windings during manufacturing. If the natural frequency without pre-stress is higher than the excitation frequencies, a higher pre-stress value will significantly reduce the oscillatory forces. Contrary to this, if the natural frequency without pre-stress is lower than the excitation frequencies, a certain value of pre-stress will bring the natural frequency closer to the excitation frequencies leading to an increase in the oscillatory forces. The natural frequency is reported to vary as some function of square root of the ratio of pre-stress to maximum value of peak electromagnetic stress in the winding [17].

The natural frequency of a winding may change during the short circuit period due to changes in the insulation characteristics and ratio of pre-stress to total stress. Thus, the natural frequency measured from the free response may be different after the short circuit as compared to that before the short circuit. Also, during the short circuit the winding, which may be in resonance at some time experiencing a higher stress, may get detuned from the resonance due to change in insulation characteristics at some other instant.

6.8 Short Circuit Test

The short circuit test can be performed by one of the two techniques, viz. pre-set short circuit and post-set short circuit. In the pre-set short circuit test, a previously short-circuited transformer (i.e., with a short-circuited secondary winding) is energized from its primary side. If the secondary winding is the inner winding, the limb flux is quite low as explained in Section 5.12.2 (figure 5.38) resulting in an insignificant transient inrush current. Hence, the method will work quite well. If the primary is the inner winding, there is substantial flux density in the limb and hence the inrush current gets superimposed on the short circuit current. Since the inrush current flows through the primary winding only, it creates a significant ampere-turn unbalance between the primary and secondary windings resulting in high short circuit forces. Depending upon the instant of closing and the core residual flux, the magnitude of the inrush current varies (as explained in Chapter 2). In order to reduce inrush current and its effects during the test, the core can be deliberately pre-magnetized with the opposite polarity [34].

In the post-set short circuit test method, in which the transformer is in the energized condition, the secondary winding is short-circuited. Naturally, this method is preferred as there are no inrush currents and the related problems, and also due to the fact that it represents the actual fault conditions at site. However, the disadvantage of this method is that the short circuit capacity of test stations has to be much higher [9, 35] than the first method to maintain the rated voltage across the transformer terminals (by overcoming the voltage drop across the series impedance between the source and transformer) and establish the required value of short circuit current. If the source impedance is not negligible as compared to the transformer impedance, a higher voltage needs to be applied, subject to a limit of 1.15 p.u. (on no-load source voltage) as per IEC standard 60076-5 (second edition: 2000-07). Thus, the required short circuit capacity of the test stations increases. The capacity of the test stations should be at least 9 times the short circuit power of the transformer for a 15% over-excitation condition [35]. The test stations may not have such capability, and hence short circuit tests on large transformers are usually carried out by the pre-set method.

After the short circuit test, the performance is considered satisfactory if the following conditions are met as per the IEC standard:

- results of short circuit tests and measurements performed during tests do not indicate any condition of a fault
- routine tests including dielectric tests, repeated after the short circuit test, are successful
- out-of-tank inspection does not indicate any displacement / deformation of the active part and / or support structures
- there are no traces of internal electrical discharges
- change in impedance is not more than 2% for any phase after the test for transformers up to 100 MVA. The corresponding value is 1% for transformers above 100 MVA. The more stringent requirement for large transformers is in line with the experience that the variation in impedance more than 1% in large power transformers indicates a large deformation in one or more windings, whereas the change in impedance between 0.5% to 1% indicates a progressive movement of winding conductors [34].

To evaluate the test results, the most important and conclusive diagnostic tests seem to be the impedance measurements and visual inspection [34], although advanced techniques like FRA (frequency response analysis), vibration measurements, dynamic oil pressure measurements, etc. are also used.

6.9 Effect of Inrush Current

As explained in Chapter 2, when a transformer is switched on, the magnitude of inrush current depends on many factors, the predominant factors being the instant of switching and residual magnetism in the core. The inrush current can be as high as six to eight times the rated current, and transformer users are always apprehensive about the repeated switching of a transformer. The inrush transients are more frequent than the short circuits and they last for few seconds as compared to the short circuits which are usually cleared in tens of milliseconds. From the point of mechanical forces, only the energized winding is subjected to the mechanical stresses. The inrush currents are usually not taken very seriously from the mechanical design considerations. The inner windings have a lower value of air-core reactance. Therefore, a transformer switched from the inner winding experiences a higher inrush current. It is known that the layer / helical windings are quite vulnerable to short circuits. Hence, if a transformer is switched on a number of times in a day from its inner layer / helical winding, forces generated due to the inrush currents may weaken the winding over a period of time leading to a winding looseness and subsequent failure. When the transformer is switched from the outer HV winding having higher air-core reactance, the magnitude of the inrush current and corresponding forces are lower.

Recent insulation failures in larger transformers, which were frequently energized under no-load condition, have attracted attention of researchers. The impact of inrush currents on mechanical stresses of windings has been investigated in [36]. It is shown that the axial forces calculated with the maximum possible inrush current are of the same order of magnitude as that calculated with the short circuit currents, and hence the use of controlled switching strategies is recommended. In another paper [37], the force patterns under the short circuit and inrush conditions are compared and shown to be quite different from each other.

6.10 Split-Winding Transformers

Split-winding transformers have the advantage that splitting secondary windings into two parts obviates the need of having two double-winding transformers. There is a considerable saving in instrumentation (on the HV side) and space, since a transformer, with one HV winding and two LV windings, substitutes two double-winding transformers of half the power rating. The arrangement also results in a considerable reduction in the values of short circuit currents in the two separately supplied circuits decreasing the required rating of circuit breakers. The split-winding transformers are usually step-down transformers in which the secondary winding is split into two equally rated windings (half the primary rating). The two secondary windings are placed axially with respect to each other in order to have equal impedances to the primary winding. The primary winding is also split into two parts with center-line lead arrangement as shown in figure 6.16.

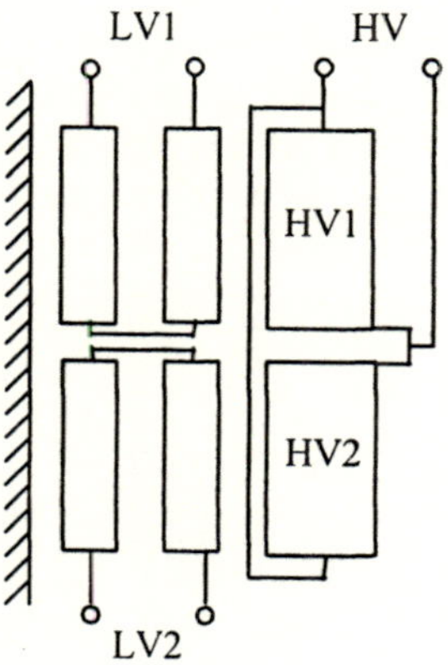

Figure 6.16 Split-winding transformer

The split-winding arrangement requires special short circuit design considerations. Consider the pre-set short circuit method, in which a voltage is applied to HV winding with one of the LV windings, say LV2, short-circuited. The flux density distribution is completely different in the lower and upper parts of the core limb. The flux density in the upper part is very high requiring high magnetizing inrush current for HV1 winding, whereas there is no demand for a significant magnetizing current by HV2 winding since the flux density in the bottom part of the limb is very low due to the short-circuited inner LV2 winding. This phenomenon results into heavy distortion of fields in the core and windings. There is a considerable amount of radial field due to an asymmetrical distribution of ampere-turns resulting into excessive axial forces and end thrusts. The phenomenon is analyzed in [35] by means of a non-linear magnetic model. It is recommended to do two set-up tests with the opposite polarity to magnetize the core in the reverse direction so that the inrush current is reduced to a harmless value during the pre-set method. If the short circuit capacity of the test station is high enough to allow the post-set method (which eliminates the inrush currents and related problems), there is still one more characteristic of the split-winding transformers which makes the short circuit test on them more severe than that on the conventional two winding transformers. With LV2 winding short-circuited, some current flows in HV1 winding not facing directly the short-circuited LV2 winding. The current in HV1 winding is small, in the range of 3 to 5% of that flowing in HV2 winding [38] due to much higher impedance between HV1 and LV2 windings, but it is sufficient to cause an ampere-turn unbalance along the height of windings. The ampere-turns of HV2 winding are smaller than that of LV2 winding, and corresponding to the ampere-turns of HV1 winding there are no balancing ampere-turns in LV1 winding (since it is open-circuited). Hence, there is a considerable distortion of the leakage field resulting into higher axial short circuit forces [39]. For the same value of short circuit current, the forces are higher for the case in which only one of the LV windings is short-circuited as compared to the case when both the LV windings are short-circuited. Thus, although the split-winding configuration helps in limiting the rating of circuit breakers, it poses problems for the short circuit withstand.

6.11 Short Circuit Withstand

The IEC 60076-5 (second edition: 2000-07) proposes that the ability to withstand dynamic effects of short circuits should be demonstrated either by the short circuit test or by calculations / design considerations. The latter method is used if there are constraints of cost, time, logistics and test limitations [23]. In this method, the manufacturers provide calculations, technological choices, design margins and adequacy of manufacturing processes. They have to demonstrate the validity of their dimensioning rules by reference to similar transformers having passed the test or by reference to the tests on representative models. A comparison of stress and strength values of the transformer with these other transformers / models should be done. The guidelines for the identification of similar transformers are given in the IEC standard. This method is a sort of comprehensive design review with the involvement of users and may cover review of calculations of short circuit currents and stresses for various types of faults, choice of particular type of material used, safety margins and quality control of manufacturing processes. The design review at different stages of design and manufacturing is a very important step in ensuring the reliability of transformers under short circuits. In order to improve the short circuit performance of transformers, the purchaser or his representative should get involved in reviewing and assessing the quality of design and manufacturing processes at few important stages during the execution of the whole contract. There is a lot of scope for cooperation between the transformer manufacturer and purchaser. Two major areas of cooperation are the improvement in technical specifications and design review. In many cases, the users may have confidence about the capability of manufacturers based on the design criteria limits and results of short circuit tests obtained on model coils as well as full size transformers over a period of time. A standardized calculation method for the demonstration of the ability to withstand the dynamic effects of a short circuit is under consideration in the IEC standard.

The short circuit test is the best way to ascertain the short circuit performance of a transformer. This is because the transformer is a highly labour intensive product and the short circuit performance may be greatly influenced by the quality of manufacturing processes. The transformer consists of different kinds of materials, whose responses to short circuit forces are quite different. Two transformers having an identical design may perform differently during the short circuit test if the quality of manufacturing processes is not consistent. Weaknesses in design and manufacturing processes get exposed when there is a failure during the test. The short circuit test may disclose some intricate failure modes based on which design and manufacturing processes can be improved. It should be remembered that in addition to a detailed theoretical analysis with the help of advanced computational techniques, it is very much essential to correlate the results of calculation with the experimental data of short circuit tests on transformers or their equivalent models. These tests give the most authentic confirmation of the assumptions or estimates of the material properties, transfer of forces, clearances, etc. used in the theoretical analysis.

There are a number of influencing factors which determine the short circuit stresses and withstand. These factors along with the general guidelines and precautions that can be taken at the specification, design and production stages of transformers for improving the short circuit strength are described below.

6.11.1 System configuration and transformer specification [1]

1. Limited extension of sub-transmission networks thereby reducing short circuit levels in the system.
2. High impedance grounding of the neutral of distribution and sub-transmission networks.
3. Specification of higher values of percentage impedances for critical transformers.
4. Use of transformers instead of autotransformers, if possible, even if it results in higher cost.
5. Neutral end tapping arrangement for transformers with on-load tap changer.
6. Specification of taps on more than one winding should be avoided.
7. Lower tapping range.
8. Removal of tertiary winding used for stabilizing purposes, from the specification for three-phase three-limb transformers upto a certain rating.
9. Specification and use of the split-winding configuration should be avoided, if possible.

The above recommendations have an impact on the power system protection and performance; hence they should be adopted after a thorough study.

6.11.2 Design

1. If the stabilizing tertiary winding is a must, its current density should be as low as possible (massive and stiff winding).
2. For withstanding radial forces, the conductor dimensions can be chosen such that the conductor can resist the compressive forces on its own, without relying on the supporting structures.
3. Whenever the CTC conductor is required to be used to achieve lower winding stray losses and for the ease of winding, it preferably should be epoxy-bonded. The epoxy-bonded CTC conductor greatly enhances the resistance of the winding against buckling and tilting because of the

bonding effect between strands due to the epoxy coating. The use of one CTC conductor, instead of a number of parallel strip conductors, has the advantage that the transpositions between parallel strands within the CTC conductor are made at regular intervals along the conductor length by automatic machines, and no manual transpositions are required to be done at the winding stage. Even for very large rating transformers, in which there are a number of CTC conductors in parallel (usually a maximum of three in the radial direction), the number of transpositions required are quite less ($n-1$, n being the number of parallel CTC conductors in the radial direction).

4. Windings can be made of high grade proof stress conductor material.
5. Use of lower current densities in windings for critical transformers.
6. Use of thicker insulating cylinders for supporting inner windings.
7. Adoption of higher slenderness ratio for the inner windings to increase the compressive strength against the radial forces.
8. Estimation of natural frequencies of windings and ensuring that there is no excited resonance.
9. Correct selection of winding arrangements to minimize short circuit forces.
10. It is usually preferable to have taps in a separate winding and not in the body of main winding from the short circuit strength consideration. Further, the turns should be so arranged in the separate tap winding that when one tap-step is cut out of the circuit, the turns get uniformly removed all along the height of the winding (e.g., interleaved tap winding in figure 7.15), minimizing ampere-turn unbalance between windings along the height.
11. If the taps need to be provided in the body of main winding, their placement at the winding end should be avoided. The short circuit forces are reduced when the taps are put in the center instead of one end. The forces are further reduced if they are put in two groups in the body and placed symmetrically around the center-line of the winding. The relative comparison of short circuit forces for various arrangements of tap positions is given in [10] using the residual ampere-turn diagram. The ampere-turns of the untapped winding should be reduced in the zone corresponding to the tapping zone of the tapped winding. The balancing of ampere-turns between the two windings in this zone should be done at the average tap position.
12. Understanding the service condition and installation environment from the point of short circuit duty.

6.11.3 Manufacturing processes

1. Proper alignment of axially placed spacers to give adequate support to the inner windings.
2. Accurate positioning of axial and radial support structures.
3. Winding should be wound tightly on the axial spacers placed on the cylinder; in which case a sufficiently thick cylinder can provide a

significant support to the winding for withstanding the forces that tend to buckle it.

4. Placement of tight-fitting wooden dowels on the core in close contact with the insulating cylinders, and radially in line with the axial supports.
5. Before use, special storage conditions are required for the epoxy-bonded CTC conductors as per the supplier's instructions. Also, it is a good practice to cut a sample of the CTC conductor used for the winding, process it along with the winding, and then check its mechanical properties through a suitable testing procedure [5].
6. Strictly controlled manufacturing processes for the windings.
7. When a double-layer winding is used to have an advantage of field cancellation due to go and return conductors (which reduces the stray losses), the forces under a short circuit are high both at the winding terminations and also in the leads. Hence, in the case of a double-layer winding, adequate precautions need to be taken for improving the short circuit withstand at the lead take-off points in the windings and also at the terminations. For a single-layer winding, one connection is made at the top and other at the bottom of the winding resulting in the manageable value of short circuit forces. But in this single-layer design, the route of high current leads from the bottom of winding should be carefully designed to minimize the stray losses as they run parallel to the tank and other structural components, and get terminated on the tank cover at the top.
8. Purchase of materials from qualified suppliers with clear material specifications and quality assurance procedures.
9. Use of high density pressboard for insulation components within windings, and between windings and yokes.
10. Judicious selection of pre-stress value and achieve required winding heights with no magnetic asymmetry between windings.
11. Use of clamping structures of adequate stiffness with appropriate fastening.
12. Adequate support and securing of leads at the winding ends.
13. Adequate fastening of connections to the tap-changer and bushings.
14. Use of fiberglass reinforced clamping rings, if required.
15. Use of winding cylinder made of fiberglass for inner windings.
16. Use of preshrunk and oil-impregnated spacers / special insulation components (like angle rings within the winding).
17. Vapour phase drying of windings before the final assembly in specific cases for a better dimensional control.
18. Burr-free edge rounding of spacers for eliminating the biting of the paper insulation on the winding conductor.
19. Ensuring tightness of the conductors in the radial direction.
20. For better sizing, windings are individually processed (heating and vacuum cycles) followed by an axial compression before the final assembly, which minimizes the possibility of any looseness in the windings. The designed winding height should be obtained at the final assembly stage before

applying the final clamping pressure. This is achieved by inserting / removing insulation spacers (blocks) so that the pre-determined pressure will get uniformly applied to all the assembled windings. Some manufacturers use the isostatic clamping process for each individual winding for better sizing, in which a constant pressure is applied on the winding uniformly and continuously throughout the drying process.

6.12 Calculation of Electrodynamic Force Between Parallel Conductors

A current carrying straight conductor of length l, placed in a uniform magnetic field of flux density B, experiences a mechanical force given by

$$F = B\,i\,l\sin\theta \tag{6.43}$$

where θ is the angle between the direction of the flux density vector and that of the current (i) in the conductor. The direction of the force is given by the Fleming's left hand rule. Figure 6.17 (a) shows two stranded conductors (P and Q) of equal length (l) carrying currents i_1 and i_2 respectively (in amperes) in the opposite directions and placed d distance apart (in meters). As per the Fleming's left hand rule, there is a repulsive force between the conductors. If the currents are in the same direction, then the force between the conductors is attractive in nature. The force per unit length of the conductors is given by

$$f = \frac{F}{l} = \mu_0 \frac{i_1 i_2}{2\pi\, d} = 2\times10^{-7}\,\frac{i_1 i_2}{d}\ \text{N/m} \tag{6.44}$$

This is the classical formula for an electrodynamic force between two current carrying wires. It is applicable to thin circular conductors only and is valid when the distance between conductors is considerably larger than their dimensions. The conductors are also assumed to be rectilinear and infinitely long.

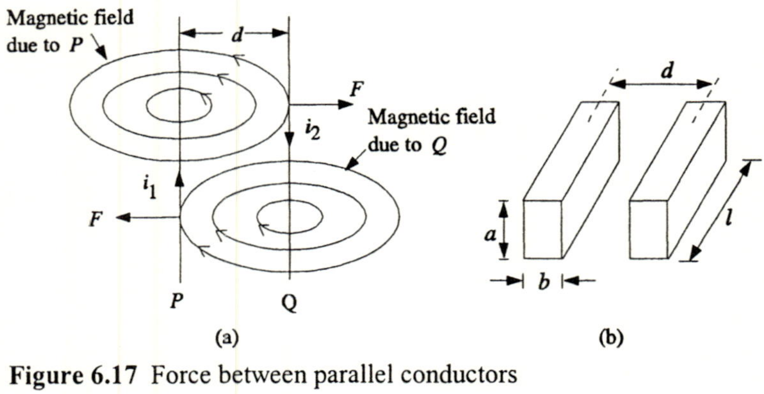

Figure 6.17 Force between parallel conductors

For rectangular conductors, an accurate force calculation is possible by considering the conductor cross section as a superimposition of interacting line current filaments. The force per unit length between parallel rectangular conductors shown in figure 6.17 (b), for which $l >> d$, is

$$f = 2\times10^{-7}\,\frac{i_1 i_2}{d}\,k_s \quad \text{N/m} \tag{6.45}$$

where k_s, a function of $\left(\dfrac{d-b}{a+b}\right)$, is shape factor that varies according to the dimensions and spacing between the conductors. The value of k_s for different values of a, b and d can be obtained from the curves given in [40, 41].

In transformers, rectangular conductors / flats are used between the windings (lead exits) and terminations (on tank). These conductors have to be supported at regular intervals to withstand the short circuit forces. Consider a rectangular conductor subjected to an electromagnetic force acting uniformly on it. Let the distance between two consecutive supports be S. The conductor structure, subjected to the short circuit loads, acts as a beam supported at the both ends as shown in figure 6.18. The maximum bending moment is

$$M = \frac{f\,S^2}{12} \tag{6.46}$$

where f is the load (force) per unit length (to be calculated by equation 6.45).

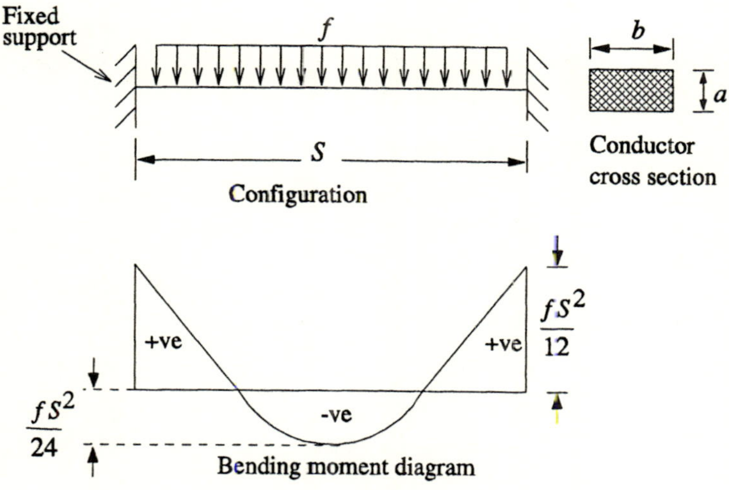

Figure 6.18 Calculation of maximum unsupported span

Now,

$$\frac{M}{I} = \frac{\sigma}{y} \tag{6.47}$$

where $I = \frac{b\,a^3}{12}$ is moment of inertia, $y = \frac{a}{2}$ is maximum distance from the neutral axis for the conductor, and σ is stress in the conductor. By putting the values of M, I and y in the equation 6.47 we get:

$$\sigma = \frac{f\,S^2}{2\,b\,a^2} \quad N/m^2 \tag{6.48}$$

If $\sigma_{\max}$ is the maximum allowable stress in the conductor in N/m², substituting the value of f from equation 6.45 for the same value of current (the peak value of an asymmetrical current) flowing in the two parallel conductors, we get the expression for the maximum allowable spacing between the consecutive supports of the conductor:

$$S = \frac{a}{i}\sqrt{1\times10^{7}\,\frac{\sigma_{max}\,b\,d}{k_s}} \quad \text{meters} \tag{6.49}$$

Similarly, the maximum allowable spacing between the supports of a rectangular conductor for any other support condition can be determined.

The analysis given above is applicable for a rectilinear conductor. For more complex arrangements (e.g., bends in a conductor or two conductors in different planes), the analytical formulation can be done [42] but it becomes more involved. Now, using numerical techniques like FEM, any complex arrangement of conductors can be easily analyzed for the calculations of forces.

6.13 Design of Clamping Structures

The clamping structures are provided in a transformer to prevent any movement of windings due to the forces produced at the time of a short circuit. Therefore, the clamping structures are designed such that they put the windings permanently under a desired pressure. The clamping structures consist of clamping ring, flitch plates and frames, as shown in figure 6.19.

Stresses in clamping ring:

For the clamping purpose, bolts are provided for maintaining a constant pressure on windings; the required pressure is applied on the clamping ring by tightening the bolts in small transformers. For large transformers, the pressure is usually applied by hydraulic jacks, and the bolts are then locked in positions or fiberglass / densified wood wedges are inserted. Under short circuit conditions, the forces produced in the windings try to bend the portion of the ring between two consecutive bolts. The stress on a circular clamping ring can be calculated by assuming that the structure is simply supported at the location of bolts. The maximum bending moment (*M*) and section modulus (*Z*) of the clamping ring are given by

$$M = \frac{F\,\pi\,D}{8\,n^2} \tag{6.50}$$

$$Z = \frac{1}{6}\,w\,t^2 \tag{6.51}$$

and the maximum bending stress is

$$\sigma_{max} = \frac{M}{Z} = \frac{6\pi\,F\,D}{8\,w\,t^2\,n^2} = \frac{3\pi\,F\,D}{4\,w\,t^2\,n^2} \tag{6.52}$$

where F is total axial force, D is mean diameter of the ring, w is width of ring (half of the difference between its outer diameter and inside diameter), t is thickness of ring, and n is number of clamping points.

If the stress on the clamping ring is calculated by assuming the fixed beam conditions, then the bending stress is given by

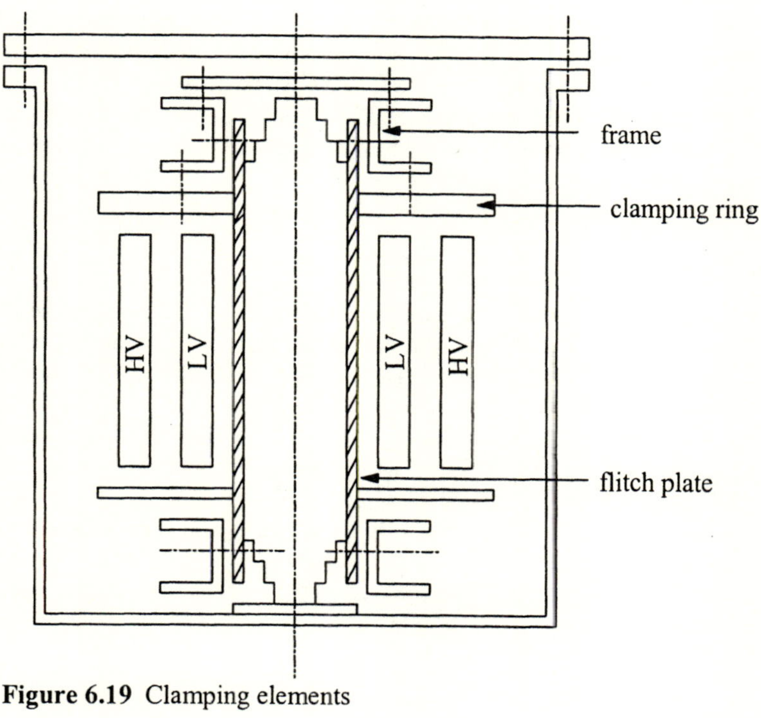

Figure 6.19 Clamping elements

$$\sigma_{max} = \frac{6\pi F D}{12\, w\, t^2\, n^2} = \frac{\pi F D}{2\, w\, t^2\, n^2} \tag{6.53}$$

The stress calculated by equation 6.52 is higher than that calculated by equation 6.53. For the design purpose, the stress can be calculated by assuming the boundary condition as the average of simply supported and fixed beam conditions, for which $M = \pi F D/(10 n^2)$ and

$$\sigma_{max} = \frac{6\pi F D}{10\, w\, t^2\, n^2} = \frac{3\pi F D}{5\, w\, t^2\, n^2} \tag{6.54}$$

the value of which should be within the limits based on material properties and the factor of safety.

Stresses in flitch plates and frames:

A flitch plate should be designed for withstanding the clamping force and core-winding weight (static loads). During a short circuit, the axial forces (end thrusts) developed in windings, act on the top and bottom frames; flitch plates help to keep the frames in position. The stresses produced in the flitch plates are the tensile stresses and shearing stresses. These stresses can be calculated by well-known formulae used in the structural analysis.

The frames are subjected to stresses while lifting core-winding assembly, during clamping of windings, or due to short circuit end thrusts. Usually, the short circuit stresses decide their dimensions. The stresses in the frames are determined from the calculated values of the short circuit forces acting on them and assuming the core bolt points and locking arrangements (pins, etc.) between flitch plates and frames as support locations.

References

1. Bergonzi, L., Bertagnolli, G., Cannavale, G., Caprio, G., Iliceto, F., Dilli, B., and Gulyesil, O. Power transmission reliability, technical and economic issues relating to the short circuit performance of power transformers, *CIGRE 2000*, Paper No. 12-207.
2. Bertula, T., Nordman, H., Saunamaki, Y., and Maaskola, J. Short circuit stresses in three-limb transformers with delta connected inner winding during single-phase faults, *CIGRE 1980*, Paper No. 12-03.
3. Anderson, P. M. *Analysis of faulted power systems*, Iowa State University Press, 1976.
4. Stevenson, W. D. *Elements of power system analysis*, Fourth Edition, McGraw-Hill Book Company, 1982.

5. Babare, A., Bossi, A., Calabro, S., Caprio, G., and Crepaz, S. Electromagnetic transients in large power step-up transformers: some design and testing problems, *CIGRE 1990*, Paper No. 12-207.
6. Arturi, C. M. Force calculations in transformer windings under unbalanced mmfs by a non-linear finite element code, *IEEE Transactions on Magnetics*, Vol. 28, No. 2, March 1992, pp. 1363-1366.
7. Sollergren, B. Calculation of short circuit forces in transformers, *Electra*, Report no. 67, 1979, pp. 29-75.
8. Salon, S., LaMattina, B., and Sivasubramaniam, K. Comparison of assumptions in computation of short circuit forces in transformers, *IEEE Transactions on Magnetics*, Vol. 36, No. 5, September 2000, pp. 3521-3523.
9. Bertagnolli, G. *Short circuit duty of power transformers – the ABB approach*, Golinelli Industrie Grafiche, Italy, 1996.
10. Waters, M. *The short circuit strength of power transformers*, Macdonald and Co. Ltd., London, 1966.
11. Waters, M. The measurement and calculation of axial electromagnetic forces in concentric transformer windings, *Proceedings IEE*, Vol. 101, Pt. II, February 1954, pp. 35-46.
12. Norris, E. T. Mechanical strength of power transformers in service, *Proceedings IEE*, Vol. 104, February 1957, pp. 289-306.
13. McNutt, W. J., Johnson, W. M., Nelson, R. A., and Ayers, R. E. Power transformer short circuit strength: requirements, design and demonstration, *IEEE Transactions on Power Apparatus and Systems*, Vol. PAS-89, No. 8, November/December 1970, pp. 1955-1969.
14. McNutt, W. J. Short circuit characteristics of transformers, *IEEE Summer Meeting*, Portland, July 1976, Paper No. CH1159-3/76, pp. 30-37.
15. Watts, G. B. A mathematical treatment of the dynamic behavior of a power transformer winding under axial short circuit forces, *Proceedings IEE*, Vol. 110, No. 3, March 1963, pp. 551-560.
16. Patel, M. R. Dynamic response of power transformers under axial short circuit forces, Part I: winding and clamp as individual components, *IEEE Transactions on Power Apparatus and Systems*, Vol. PAS-92, September/October 1973, pp. 1558-1566.
17. Patel, M. R. Dynamic response of power transformers under axial short circuit forces, Part II: windings and clamps as a combined system, *IEEE Transactions on Power Apparatus and Systems*, Vol. PAS-92, September/October 1973, pp. 1567-1576.
18. Swihart, D. O. and Wright, D. V. Dynamic stiffness and damping of transformer pressboard during axial short circuit vibration, *IEEE Transactions on Power Apparatus and Systems,* Vol. PAS-95, No. 2, March/April 1976, pp. 721-730.

19. Allan, D. J., Pratt, F. C., Sharpley, W. A., and Woollard, M. E. The short circuit performance of transformers – a contribution to the alternative to direct testing, *CIGRE 1980*, Paper No. 12-02.
20. Tournier, Y., Richard, M., Ciniero, A., Yakov, S., Madin, A. B., and Whitaker, J. D. A study of the dynamic behavior of transformer windings under short circuit conditions, *CIGRE 1964*, Report No. 134.
21. Hori, Y. and Okuyama, K. Axial vibration analysis of transformer windings under short circuit conditions, *IEEE Transactions on Power Apparatus and Systems,* Vol. PAS-99, No. 2, March/April 1980, pp. 443-451.
22. Quinney, D. A. Dynamic response of a power transformer winding under axial short circuit conditions, *Proceedings IEE*, Vol. 128, Pt. B, No. 2, March 1981, pp. 114-118.
23. Macor, P., Robert, G., Girardot, D., Riboud, J. C., Ngnegueu, T., Arthaud, J. P., and Chemin, E. The short circuit resistance of transformers: The feedback in France based on tests, service and calculation approaches, *CIGRE 2000*, Paper No. 12-102.
24. Lakhiani, V. K. and Kulkarni, S. V. Short circuit withstand of transformers - a perspective, *International Conference on Transformers*, *TRAFOTECH - 2002*, Mumbai, India, January 2002, pp. 34-38.
25. Steel, R. B., Johnson, W. M., Narbus, J. J., Patel, M. R., and Nelson, R. A. Dynamic measurements in power transformers under short circuit conditions, *CIGRE 1972*, Paper No. 12-01.
26. Thomson, H. A., Tillery, F., and Rosenberg, D. U. The dynamic response of low voltage, high current, disk type transformer windings to through fault loads, *IEEE Transactions on Power Apparatus and Systems*, Vol. PAS-98, No. 3, May/June 1979, pp. 1091-1098.
27. Saravolac, M. P., Vertigen, P. A., Sumner, C. A., and Siew, W. H. Design verification criteria for evaluating the short circuit withstand capability of transformer inner windings, *CIGRE 2000*, Paper No. 12-208.
28. Kojima, H., Miyata, H., Shida, S., and Okuyama, K. Buckling strength analysis of large power transformer windings subjected to electromagnetic force under short circuit, *IEEE Transactions on Power Apparatus and Systems,* Vol. PAS-99, No. 3, May/June 1980, pp. 1288-1297.
29. Patel, M. R. Dynamic stability of helical and barrel coils in transformers against axial short circuit forces, *Proceedings IEE*, Vol. 127, Pt. C, No. 5, September 1980, pp. 281-284.
30. Patel, M. R. Instability of the continuously transposed cable under axial short circuit forces in transformers, *IEEE Transactions on Power Delivery,* Vol. 17, No. 1, January 2002, pp. 149-154.
31. Madin, A. B. and Whitaker, J. D. The dynamic behavior of a transformer winding under axial short circuit forces: An experimental and theoretical investigation, *Proceedings IEE*, Vol. 110, No. 3, 1963, pp. 535-550.

32. Gee, F. W. and Whitaker, J. D. Factors affecting the choice of pre-stress applied to transformer windings, *IEEE Summer General Meeting*, Toronto, Canada, 1963, Paper No. 63-1012.
33. Dobsa, J. and Pirktl, E. Transformers and short circuits, *Brown Boveri Review*, Vol. 52, No. 11/12, November/December 1965, pp. 821-830.
34. Janssen, A. L. J. and te Paske, L. H. Short circuit testing experience with large power transformers, *CIGRE 2000*, Paper No. 12-105.
35. Leber, G. Investigation of inrush currents during a short circuit test on a 440 MVA, 400 kV GSU transformer, *CIGRE 2000*, Paper No. 12-104.
36. Steurer, M. and Frohlich, K. The impact of inrush currents on the mechanical stress of high voltage power transformer coils, *IEEE Transactions on Power Delivery*, Vol. 17, No. 1, January 2002, pp. 155-160.
37. Adly, A. A. Computation of inrush current forces on transformer windings, *IEEE Transactions on Magnetics,* Vol. 37, No. 4, July 2001, pp. 2855-2857.
38. Stuchl, P., Dolezel, I., Zajic, A., Hruza, J., and Weinberg, O. Performance of transformers with split winding under nonstandard operating conditions, *CIGRE 2000*, Paper No. 12-103.
39. Foldi, J., Berube, D., Riffon, P., Bertagnolli, G., and Maggi, R. Recent achievements in performing short circuit withstand tests on large power transformers in Canada, *CIGRE 2000*, Paper No. 12-201.
40. Dwight, H. B. *Electrical coils and conductors*, First Edition, McGraw-Hill Book Company Inc., New York, 1945.
41. Rodstein, L. *Electrical control equipment*, MIR Publishers, Moscow, Translated in English by G. Roberts, 1974.
42. Charles, E. D. Mechanical forces on current-carrying conductors, *Proceedings IEE*, Vol. 110, No. 9, September 1963, pp. 1671-1677.

7

Surge Phenomena in Transformers

For designing the insulation of a transformer suitable for all kinds of overvoltages, the voltage stresses within the windings need to be determined. For this purpose, voltage distributions within the transformer windings for the specific test voltages are calculated. For AC test voltages of power frequency, the voltage distribution is linear with respect to the number of turns and can be calculated exactly. For the calculation of the impulse voltage distribution in the windings, they are required to be simulated in terms of an equivalent circuit consisting of lumped R, L and C elements. There are a number of accurate methods described in the literature for computation of winding response to impulse voltages, some of which are discussed in this chapter. Electric stresses in the insulation within and outside the windings are obtained by analytical or numerical techniques which are described in the next chapter.

7.1 Initial Voltage Distribution

When a step voltage impinges on the transformer winding terminals, the initial distribution in the winding depends on the capacitances between turns, between windings, and those between windings and ground. The winding inductances have no effect on the initial voltage distribution since the magnetic field requires a finite time to build up (current in an inductance cannot be established instantaneously). Thus, the inductances practically do not carry any current, and the voltage distribution is predominantly decided by the capacitances in the network, and the problem can be considered as entirely electrostatic without any appreciable error. In other words, the presence of series capacitances between winding sections causes the transformer to respond to abrupt impulses as a network of capacitances for all frequencies above its lower natural frequencies of oscillations. When the applied voltage is maintained for a sufficient time (50 to 100 microseconds), appreciable currents begin to flow in the inductances eventually leading to the uniform voltage distribution. Since there is difference between the initial and final voltage distributions, as shown in figure 7.1, a transient phenomenon takes place during which the voltage distribution readjusts itself from the initial to final value. During this transient period, there is continual interchange of energy between electric and magnetic fields. On account of a low damping factor of the transformer windings, the transient is oscillatory. The voltage at any point in the winding oscillates about the final voltage value, reaching a maximum as shown by curve c. It is obvious that the strength of the transformer windings to lightning voltages can be significantly increased if the difference between the initial and final distributions can be minimized. This not only reduces the excessive stresses at the line end but also mitigates the oscillations thereby keeping voltage to ground at any point in the winding insignificantly higher than the final voltage distribution.

The differential equation governing the initial voltage distribution $u_0 = u(x,0)$, for the representation of a winding shown in figure 7.2 (and ignoring inductive effects), is [1]

$$\frac{d^2 u_0}{dx^2} - \frac{c_g}{c_s} u_0 = 0 \tag{7.1}$$

In figure 7.2, L_s, c_g and c_s denote self inductance per unit length, shunt capacitance per unit length to ground and series capacitance per unit length between adjacent turns respectively.

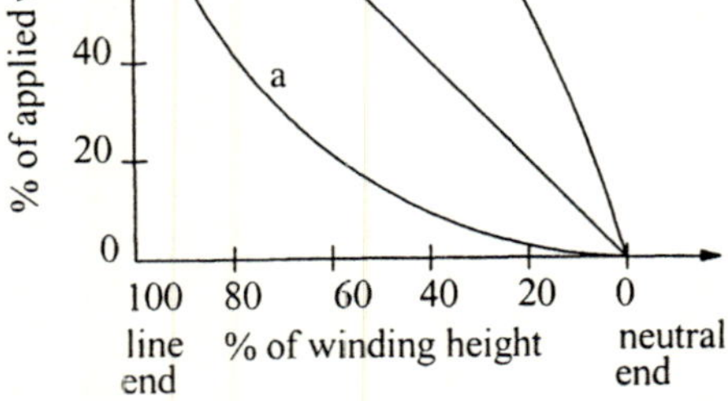

(a) initial distribution (b) final distribution (c) maximum voltage to ground

Figure 7.1 Impulse voltage distribution

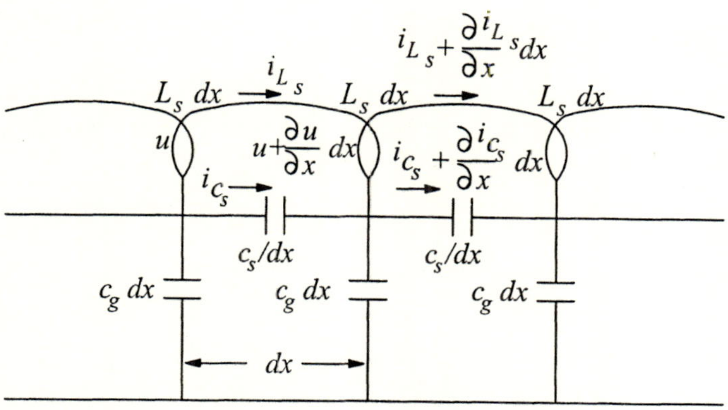

Figure 7.2 Representation of a transformer winding

Solution of the above equation is given by

$$u_0 = A_1 e^{kx} + A_2 e^{-kx} \tag{7.2}$$

where

$$k = \sqrt{\frac{c_g}{c_s}} \tag{7.3}$$

The constants of integration A_1 and A_2 can be obtained from the boundary conditions at the line and neutral ends of the winding. For the solidly grounded neutral, we have $u_0 = 0$ for $x = 0$. Putting these values in equation 7.2 we get

$$A_1 + A_2 = 0 \text{ or } A_1 = -A_2$$

Whereas at the line end, $x = L$ (L is the winding axial length) and $u_0 = U$ (amplitude of the step impulse voltage) giving

$$U = A_1 e^{kL} + A_2 e^{-kL} = A_1\left(e^{kL} - e^{-kL}\right)$$

$$\therefore A_1 = -A_2 = \frac{U}{e^{kL} - e^{-kL}} \tag{7.4}$$

Substituting the above expression in equation 7.2 we get

$$u_0 = U \frac{e^{kx} - e^{-kx}}{e^{kL} - e^{-kL}} = U \frac{\sinh kx}{\sinh kL} \tag{7.5}$$

The initial voltage gradient at the line end of the winding is given by

$$\left[\frac{du_0}{dx}\right]_{x=L} = \left[U\,\frac{k\cosh kx}{\sinh kL}\right]_{x=L} = U\,\frac{k\cosh kL}{\sinh kL} = k\,U\coth kL \tag{7.6}$$

The initial voltage gradient is maximum at the line end. Since $kL > 3$ in practice, $\coth kL \cong 1$, giving the initial gradient at the line end for a unit amplitude surge ($U = 1$) as

$$\left[\frac{du_0}{dx}\right]_{max} = k \tag{7.7}$$

The uniform gradient for the unit amplitude surge is $1/L$.

$$\therefore \left[\frac{du_0}{dx}\right]_{\max} = \frac{1}{L}\,k\,L = \frac{1}{L}\sqrt{\frac{c_g}{c_s}}\,L = \frac{1}{L}\sqrt{\frac{c_g L}{c_s/L}} = \frac{1}{L}\sqrt{\frac{C_G}{C_S}} = \frac{1}{L}\,\alpha \tag{7.8}$$

where C_G and C_S are the total ground capacitance and series capacitance of the transformer winding respectively. The ratio $\sqrt{C_G/C_S}$ has been denoted by the distribution constant α. Thus, the maximum initial gradient at the line end is α times the uniform gradient. The higher the value of ground capacitance, the higher are the values of α and voltage stress at the line end.

For the isolated neutral condition, the boundary conditions,

$$\frac{du_0}{dx} = 0 \text{ at } x = 0 \text{ and } u_0 = U \text{ at } x = L$$

give the following expression for the initial voltage distribution:

$$u_0 = U\,\frac{\cosh kx}{\cosh kL} \tag{7.9}$$

For the isolated neutral condition, the maximum initial gradient at the line end can be written as

$$\left[\frac{du_0}{dx}\right]_{x=L} = k\,U\,\tanh kL \tag{7.10}$$

For a unit amplitude surge and $(\alpha = kL) > 3$, $\tanh kL \cong 1$. Hence, the initial gradient becomes

$$\left[\frac{du_0}{dx}\right]_{max} = k = \frac{\alpha}{L} \tag{7.11}$$

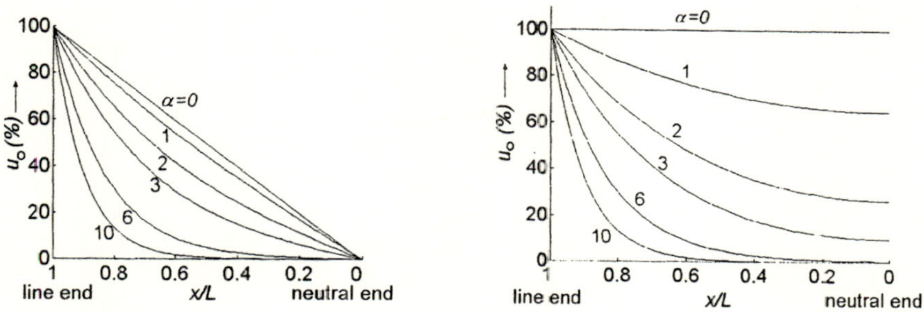

(a) Grounded neutral

(b) Isolated neutral

Figure 7.3 Initial voltage distribution

Hence, the value of maximum initial gradient at the line end is the same for the grounded and isolated neutral conditions for abrupt impulses or very steep wave fronts. The initial voltage distribution for various values of α is plotted in figure 7.3 for the grounded and isolated neutral conditions. The total series capacitance (C_S) and ground capacitance (C_G) of the transformer winding predominantly decide the initial stresses in it for steep fronted voltage surges. The total series capacitance consists of capacitance between turns and capacitance between disks / sections of the winding, whereas the total ground capacitance includes the capacitance between the winding and core / tank / other windings. Thus, the initial voltage distribution is characterized by the distribution constant,

$$\alpha = \sqrt{\frac{C_G}{C_S}} \tag{7.12}$$

This parameter indicates the degree of deviation of the initial voltage distribution from the final linear voltage distribution which is decided solely by winding inductances. The higher the value of α, the higher are the deviation and amplitudes of oscillations which occur between the initial and final voltage distributions. For a conventional continuous disk winding, the value of α may be in the range of 5 to 30. Any change in the transformer design, which decreases the distribution constant of the winding, results in a more uniform voltage distribution and reduces the voltage stresses between different parts of the winding. The initial voltage distribution of the winding can be made closer to the ideal linear distribution ($\alpha = 0$) by increasing its series capacitance and / or reducing its capacitance to ground. If the ground capacitance is reduced, more current flows through the series capacitances, tending to make the voltage across the various winding sections more uniform. The (ideal) uniform initial impulse voltage distribution will be achieved if no current flows through the (shunt) ground capacitances. Usually, it is very difficult and less cost-effective to reduce the ground capacitances. Insulation gaps between windings predominantly decide the ground capacitances. These capacitances depend on the radial gap and circumferential area between the windings. These geometrical quantities get usually fixed from optimum electrical design considerations. Hence, any attempt to decrease the distribution constant α by decreasing the ground capacitance is definitely limited. The more cost-effective way is to increase the winding series capacitance by using different types of windings as described in the subsequent sections.

7.2 Capacitance Calculations

In order to estimate the voltage distribution within a transformer winding subjected to impulse overvoltages, the knowledge of its effective series and ground capacitances is essential. The calculation of ground capacitance between a winding and ground or between two windings is straightforward. The capacitance between two concentric windings (or between the innermost winding and core) is given by

$$C_{gw} = \frac{\varepsilon_0 \pi D_m H}{(t_{oil}/\varepsilon_{oil}) + (t_{solid}/\varepsilon_{solid})} \tag{7.13}$$

where D_m is mean diameter of the gap between two windings, t_{oil} and t_{solid} are thicknesses of oil and solid insulations between two windings respectively, and H is height of windings (if the heights of two windings are unequal, average height is taken in the calculation).

Capacitance between a cylindrical conductor and ground plane is given by (appendix B, equation B30)

$$C = \frac{2\pi \varepsilon H}{\cosh^{-1}\left(\dfrac{s}{R}\right)} \tag{7.14}$$

where R and H are radius and length of the cylindrical conductor respectively and s is distance of center of the cylindrical conductor from the plane. Hence, the capacitance between a winding and tank can be given as

$$C_{gt} = \frac{2\pi \varepsilon_0 H}{\cosh^{-1}\left(\dfrac{s}{R}\right)} \left[\frac{t_{oil} + t_{solid}}{(t_{oil}/\varepsilon_{oil}) + (t_{solid}/\varepsilon_{solid})}\right] \tag{7.15}$$

In this case, R and H represent the radius and height of the winding respectively and s is the distance of the winding axis from the plane. The capacitance between the outermost windings of two phases is half the value given by above equation 7.15, with s equal to half the value of distance between the axes of the two windings (refer to equation B28).

7.3 Capacitance of Windings

7.3.1 Development of winding methods for better impulse response

In the initial days of transformer technology development for higher voltages, use of electrostatic shields was quite common (see figure 7.4). A non-resonating transformer with electrostatic shields was reported in [2, 3, 4]. It is a very effective shielding method in which the effect of the ground capacitance of individual section is neutralized by the corresponding capacitance to the shield. Thus, the currents in the shunt (ground) capacitances are supplied from the shields and none of them have to flow through the series capacitances of the winding. If the series capacitances along the windings are made equal, the uniform initial voltage distribution can be achieved. The electrostatic shield is at the line terminal potential and hence requires to be insulated from the winding and tank along its height. As the voltage ratings and corresponding dielectric test levels increased, transformer designers found it increasingly difficult and cumbersome to design the shields. The shields were found to be less cost-effective since extra space and material were required for insulating shields from other electrodes inside the transformer. Subsequent development of interleaved windings phased out completely the use of electrostatic shielding method. The principle of electrostatic shielding method is being made use of in the form of static end rings at the line end and static rings within the winding which improve the voltage distribution and reduce the stresses locally.

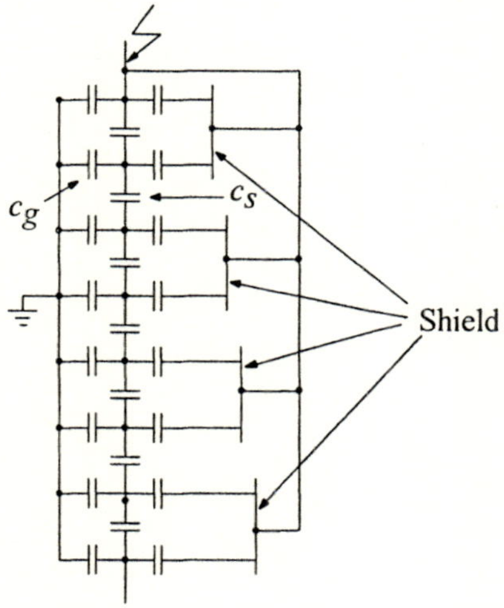

Figure 7.4 Electrostatic shields

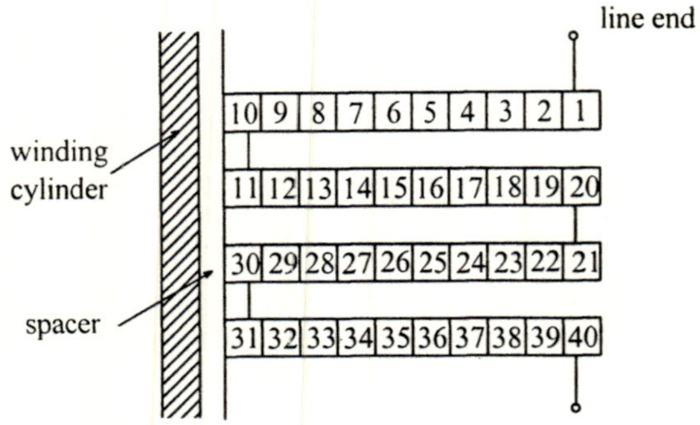

Figure 7.5 Continuous winding

In order to understand the effectiveness of an interleaved winding, let us first analyze a continuous (disk) winding shown in figure 7.5. The total series capacitance of the continuous winding is an equivalent of all the turn-to-turn and disk-to-disk capacitances. Although the capacitance between two adjacent turns is quite high, all the turn-to-turn capacitances are in series, which results in a much smaller capacitance for the entire winding. Similarly, all the disk-to-disk capacitances which are also in series, add up to a small value. With the increase in voltage class of the winding, the insulation between turns and between disks has to be increased which further worsens the total series capacitance.

The inherent disadvantage of low series capacitance of the continuous winding was overcome by electrostatic shielding as explained earlier till the advent of the interleaved winding. The original interleaved winding was introduced and patented by G. F. Stearn in 1950 [5]. A simple disposition of turns in some particular ways increases the series capacitance of the interleaved winding to such an extent that a near uniform initial voltage distribution can be obtained. A typical interleaved winding is shown in figure 7.6.

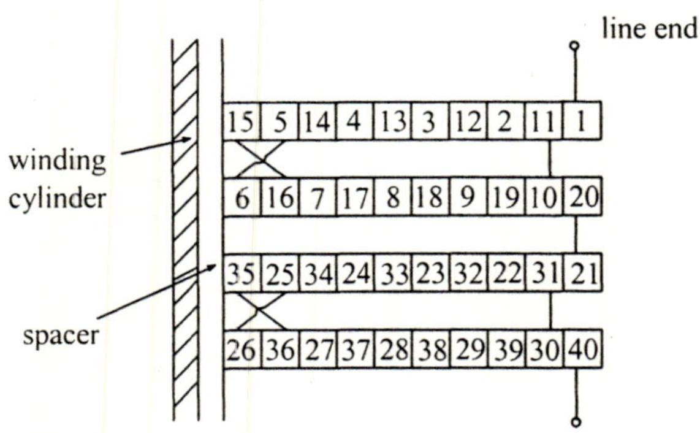

Figure 7.6 Interleaved winding

In an interleaved winding, two consecutive electrical turns are separated physically by a turn which is electrically much farther along the winding. It is wound as a conventional continuous disk winding but with two conductors. The radial position of the two conductors is interchanged (cross-over between conductors) at the inside diameter and appropriate conductors are joined at the outside diameter, thus forming a single circuit two-disk coil. The advantage is obvious since it does not require any additional space as in the case of complete electrostatic shielding or part electrostatic shielding (static ring). In interleaved windings, not only the series capacitance is increased significantly but the ground capacitance is also somewhat reduced because of the improvement in the winding space factor. This is because the insulation within the winding in the axial direction can be reduced (due to improvement in the voltage distribution), which reduces the winding height and hence the ground capacitance. Therefore, the distribution constant (α) is reduced significantly lowering stresses between various parts of the winding.

It can be seen from figure 7.6 that the normal working voltage between adjacent turns in an interleaved winding is equal to voltage per turn times the turns per disk. Hence, one may feel that a much higher amount of turn insulation may be required, thus questioning the effectiveness of the interleaved winding. However, due to a significant improvement in the voltage distribution, stresses between turns are reduced by a great extent so that % safety margins for the impulse stress and normal working stress can be made of the same order. Hence, the turn-to-turn insulation is used in more effective way [6]. Since the voltage distribution is more uniform, the number of special insulation components (e.g., disk angle rings) along the winding height reduces. When a winding has more than one conductor per turn, the conductors are also interleaved as shown in figure 7.7 (a winding with 6 turns per disk and two parallel conductors per turn) to get maximum benefit from the method of interleaving.

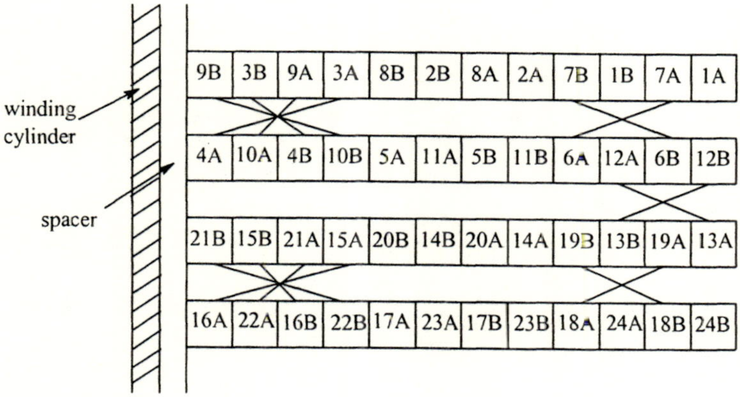

Figure 7.7 Interleaving with 2-parallel conductors per turn

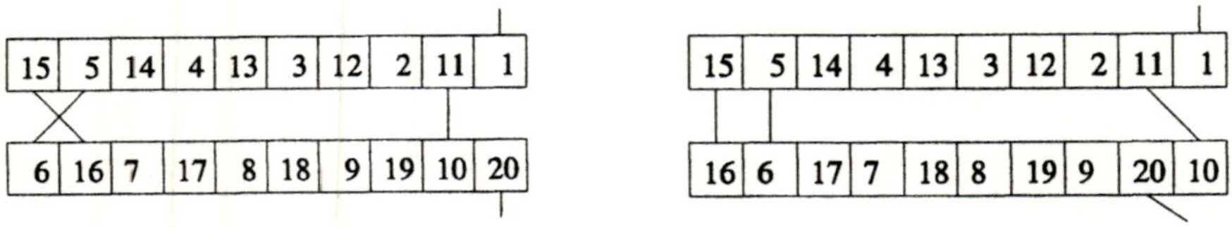

(a) Stern winding (b) Nuys winding

Figure 7.8 Two types of crossovers in interleaved winding

In [7], improved surge characteristics of interleaved windings are explained based on transmission line like representation of the disks with surge impedance, without recourse to the hypothesis of increased series capacitance.

There can be two types of interleaved windings as regards the crossover connections at the inside diameter as shown in figure 7.8. When steep impulse waves such as chopped waves or front-of-waves enter an interleaved winding, a high oscillatory voltage occurs locally between turns at the center of the radial build of the disk. This phenomenon is analyzed in [8, 9] for these two types of crossovers in the interleaved windings.

7.3.2 Turn-to-turn and disk-to-disk capacitances

For the calculation of series capacitances of different types of windings, the calculations of turn-to-turn and disk-to-disk capacitances are essential. The turn-to-turn capacitance is given by

$$C_T = \frac{\varepsilon_0 \varepsilon_p \times \pi\, D_m\, (w + t_p)}{t_p} \tag{7.16}$$

where D_m is average diameter of winding, w is bare width of conductor in axial direction, t_p is total paper insulation thickness (both sides), ε_0 is permittivity of the free space, and ε_p is relative permittivity of paper insulation. The term t_p is added to the conductor width to account for fringing effects.

Similarly, the total axial capacitance between two consecutive disks based on geometrical considerations only is given by

$$C_{DA} = \varepsilon_0 \left[\frac{k}{t_p/\varepsilon_p + t_s/\varepsilon_{oil}} + \frac{1-k}{t_p/\varepsilon_p + t_s/\varepsilon_s} \right] \times \pi\, D_m \left(R + t_s\right) \tag{7.17}$$

where R is winding radial depth, t_s and ε_s are thickness and relative permittivity of solid insulation (radial spacer between disks) respectively, and k is fraction of circumferential space occupied by oil. The term t_s is added to R to take into account fringing effects.

For continuous winding and its variations (with static end rings / static rings between disks), there are two approaches for calculating the series capacitance. In the first approach, the voltage is assumed to be evenly distributed within the disk winding, which makes the calculation quite easy. However, this is a major approximation for continuous disks having small effective inter-turn series capacitance. Hence, the second approach is more accurate in which the linear voltage distribution is not assumed within the disks for the capacitance calculation [10, 11, 12]. The corresponding representation of capacitances for this accurate method of calculation is shown in figure 7.9. The total series capacitance of the winding is given by [10, 13]

$$C_S = \frac{\left[\dfrac{2C_{DA}}{\alpha_d}\left(\tanh 2\alpha_d\right)\right]\left[\dfrac{\sqrt{2}\,C_{DA}}{\alpha_d}\left(\tanh\sqrt{2}\,\alpha_d\right)\right]}{\dfrac{4\,C_{DA}}{\alpha_d}\left(\tanh 2\alpha_d\right)+\left(N_{DW}-2\right)\dfrac{\sqrt{2}\,C_{DA}}{\alpha_d}\left(\tanh\sqrt{2}\,\alpha_d\right)} \tag{7.18}$$

where C_{DA} = disk-to-disk capacitance calculated based on geometrical considerations

α_d = distribution constant of disk = $\sqrt{\dfrac{C_{DA}}{C_T/(N_D-1)}}$

C_T = turn-to-turn capacitance

N_D = number of turns per disk

N_{DW} = number of disks in the winding

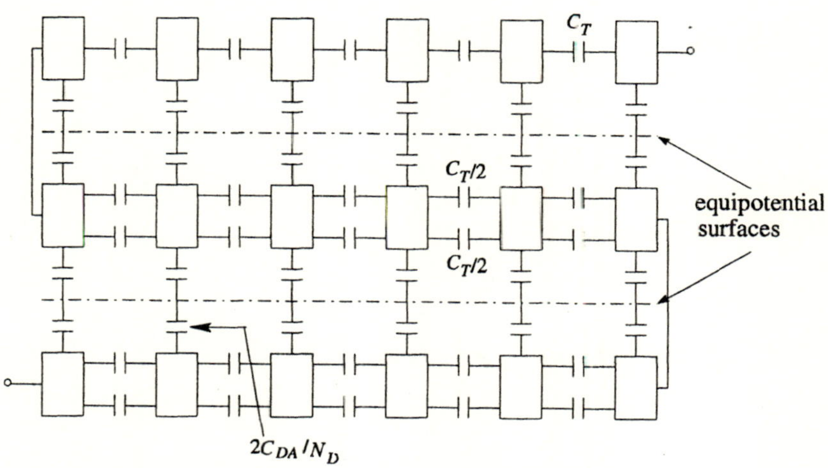

Figure 7.9 Representation of capacitances of a continuous winding

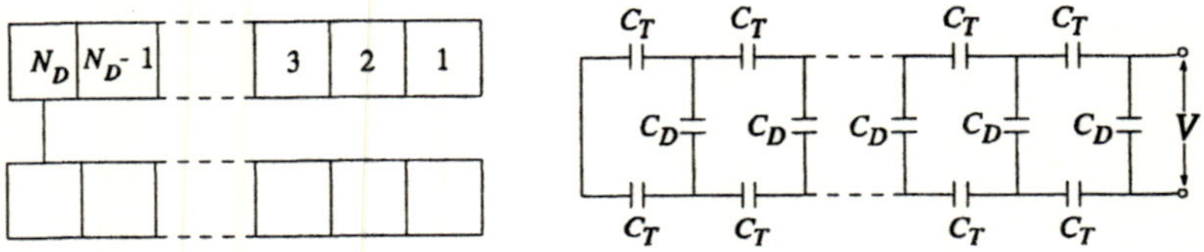

Figure 7.10 Disk-pair of a continuous winding

The first approach, in which linear voltage distribution is assumed for capacitance calculations, is definitely approximate for continuous windings. The total series capacitance of a disk is small and also the disk-to-disk capacitance (C_{DA}) is appreciable, making the distribution constant α_d for the disk larger. Hence, the voltage distribution within the disk and within the winding is non-linear. However, the approach is easier and the expressions obtained for the capacitances of various types of windings can be easily compared. The approach is used in the following sub-sections for the calculation of the series capacitance of various windings including continuous windings.

7.3.3 Continuous disk winding

Let us find the capacitance of a disk pair of a continuous winding shown in figure 7.10 with the assumption of linear voltage distribution. The term C_T denotes capacitance between touching turns and C_D denotes capacitance between a turn of one disk and the corresponding turn of the other disk. If N_D is number of turns in a disk, then number of inter-turn capacitances (C_T) in each disk is $(N_D - 1)$. Also, number of inter-section capacitances (C_D) between the two disks is $(N_D - 1)$. The series capacitance of the disk winding is the resultant of the inter-turn (turn-to-turn) and inter-disk (disk-to-disk) capacitances. The voltage per turn for the disk pair shown in figure 7.10 is $(V/2N_D)$. Using the principle that the sum of energies in the individual capacitances within the disk is equal to the entire energy of the disk coil, the following equation can be written:

$$\frac{1}{2}C_T\left(\frac{V}{2N_D}\right)^2 2(N_D - 1) = \frac{1}{2}C_{TR}\,V^2$$

where C_{TR} = resultant inter-turn capacitance.

$$\therefore C_{TR} = \frac{C_T}{2N_D^{\;2}}(N_D - 1) \tag{7.19}$$

Now, the voltages across the first, second and third inter-disk capacitances (C_D) from the inside diameter are

$$\left(\frac{2\,V}{2\,N_D}\right),\left(\frac{4\,V}{2\,N_D}\right) \text{ and } \left(\frac{6\,V}{2\,N_D}\right) \text{ respectively.}$$

Hence, the expression for C_D at the outside diameter is $2\left(N_D-1\right)\frac{V}{2\,N_D}$.

The total energy stored by all such capacitances is

$$En=\frac{1}{2}\,C_D\left[\left(\frac{2\,V}{2\,N_D}\right)^2+\left(\frac{4\,V}{2\,N_D}\right)^2+\left(\frac{6\,V}{2\,N_D}\right)^2+\ldots+\left(\frac{2\left(N_D-1\right)V}{2\,N_D}\right)^2\right] \tag{7.20}$$

Simplifying and using the identity: $1^2+2^2+3^2+\cdots+(n-1)^2=\frac{n(n-1)(2n-1)}{6}$

we get

$$En=\frac{\left(N_D-1\right)\left(2\,N_D-1\right)C_D\,V^2}{12\,N_D}=\frac{1}{2}\,C_{DR}\;V^2 \tag{7.21}$$

where C_{DR} is the resultant inter-disk capacitance.

$$\therefore C_{DR}=\frac{\left(N_D-1\right)\left(2N_D-1\right)C_D}{6\,N_D} \tag{7.22}$$

Instead of using the lumped parameter approach for the inter-disk capacitances, if they are represented by a distributed capacitance C_{DU} (capacitance per unit radial depth based on the geometrical considerations only), then the value of resultant inter-disk capacitance for a radial depth of R can be calculated as [14]

$$C_{DR}=\frac{C_{DU}\;R}{3}=\frac{C_{DA}}{3} \tag{7.23}$$

The previous two equations are equivalent, because if the number of turns per disk is much greater than 1 ($N_D >> 1$), equation 7.22 becomes

$$C_{DR}\cong N_D\;C_D/3=C_{DU}\;R/3$$

The resultant series capacitance of the disk pair is given as the addition of the resultant inter-turn capacitance and the resultant inter-disk capacitance,

$$C_{se}=\frac{C_T}{2{N_D}^2}\left(N_D-1\right)+\frac{C_D}{6N_D}\left(N_D-1\right)\left(2N_D-1\right) \tag{7.24}$$

or

$$C_{se} = \frac{C_T}{2N_D^{\,2}}(N_D - 1) + \frac{C_{DU}\ R}{3} \quad \cdots \textit{Continuous disk winding} \tag{7.25}$$

Now, if there are N_{DW} disks in the winding, the resultant inter-disk capacitance $(C_{DR})_W$ for the entire winding (with a voltage V_W across it) can be calculated as

$$\frac{1}{2}(C_{DR})_W\ V_W^2 = \frac{1}{2}(C_{DR})\left(\frac{V_W}{N_{DW}/2}\right)^2 (N_{DW} - 1) \tag{7.26}$$

$$\therefore (C_{DR})_W = \frac{4\,(N_{DW} - 1)}{N_{DW}^2}\ C_{DR} \tag{7.27}$$

Noting the fact that the expression for C_{TR} given by equation 7.19 is for two disks, the total series capacitance for the entire winding with N_{DW} disks can be given by using equations 7.19 and 7.27 as

$$C_S = \frac{C_T}{N_{DW}\ N_D^{\,2}}\,(N_D - 1) + \frac{4\,(N_{DW} - 1)}{N_{DW}^2}\ C_{DR} \tag{7.28}$$

The above expression gives the value of capacitance close to that given by equation 7.18 for the values of disk distribution constant α_d close to 1 (almost linear distribution within disk). For $N_{DW}, N_D >> 1$, the equation 7.28 becomes

$$C_S \cong \frac{1}{N_{DW}}\left[\frac{C_T}{N_D} + 4\ C_{DR}\right] \tag{7.29}$$

7.3.4 Continuous winding with SER and SR

As mentioned earlier, the concept of electrostatic shielding is used in a limited way by having a static end ring (SER) at line end or a static ring (SR) between disks as shown in figure 7.11.

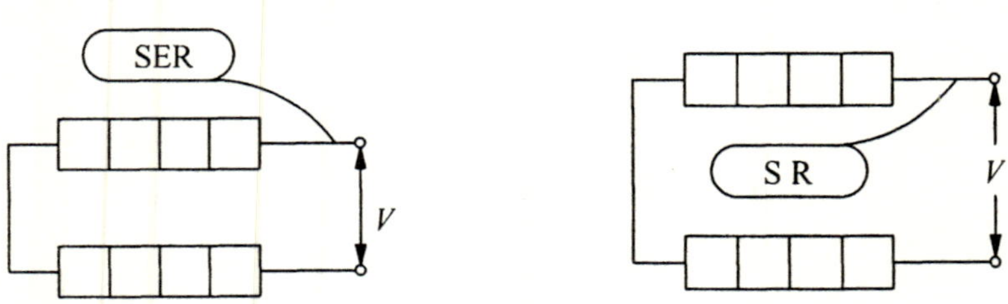

Figure 7.11 Static end ring (SER) and static ring (SR)

By providing a large equipotential surface with a good corner radius, SER reduces the stress concentration at the line end. It also improves the effective series capacitance at the line end as explained below. The closer the location of SER to the line end disk, the greater the increase in the series capacitance value is. This results in reduction of stresses appearing within the line end disk during the initial voltage distribution.

Let us calculate the increase in series capacitance of a disk pair with SER as per the method given in [14]. SER is usually connected to the first turn of the winding by means of a pig-tail; hence the potential of SER gets fixed to that of line terminal (V) as shown in figure 7.12. Let the winding radial depth be denoted by R.

The voltage at any point x of the upper section representing SER is

$$V_1(x) = V \tag{7.30}$$

and the voltage at any point x of the lower section representing the first disk is

$$V_2(x) = \frac{V}{2} + \frac{V}{2}\left[\frac{R-x}{R}\right] = V\frac{(2R-x)}{2R} \tag{7.31}$$

Let C_{SU} be the capacitance between SER and the first disk per unit depth of the winding (based on the geometrical considerations only). Therefore, the energy of the capacitance C_{SU} per unit depth at point x is

$$En_x = \frac{1}{2} C_{SU} \left[V_1(x) - V_2(x)\right]^2 \tag{7.32}$$

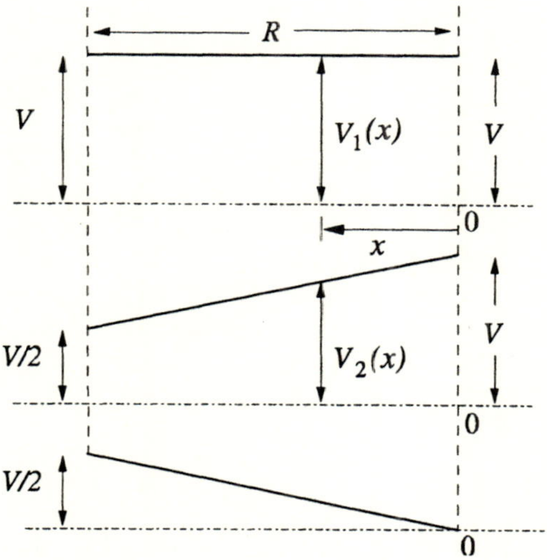

Figure 7.12 Calculation of capacitance between SER and line end disk

The total energy stored by the capacitance between the first disk and SER is

$$En = \int_0^R En_x \, dx = \int_0^R \frac{1}{2} C_{SU} [V_1(x) - V_2(x)]^2 \, dx \tag{7.33}$$

Substituting the values of $V_1(x)$ and $V_2(x)$ from equations 7.30 and 7.31, and simplifying we get

$$En = \int_0^R \frac{1}{2} C_{SU} \left[V \frac{x}{2R} \right]^2 dx = \frac{1}{2} \frac{1}{12} C_{SU} \; R \, V^2 \tag{7.34}$$

Thus, the resultant capacitance, C_{SER}, between SER and the first disk can be given by the equation

$$\frac{1}{2} C_{SER} \; V^2 = \frac{1}{2} \frac{1}{12} C_{SU} \; R \; V^2 \tag{7.35}$$

$$\therefore C_{SER} = \frac{C_{SU} \; R}{12} \tag{7.36}$$

Thus, the resultant capacitance between SER and the first disk is $(1/12)$ times the capacitance obtained purely from the geometrical considerations.

Using equations 7.25 and 7.36, the total series capacitance of the disk pair with SER is therefore given by

$$C_{se} = \frac{C_T}{2N_D^{\;2}} (N_D - 1) + \frac{C_{DU} \; R}{3} + \frac{C_{SU} \; R}{12} \quad \cdots \quad \textit{with SER} \tag{7.37}$$

Similarly, the expression for the series capacitance of a disk-pair with static ring (shown in figure 7.11) can be found as

$$C_{se} = \frac{C_T}{2N_D^{\;2}} (N_D - 1) + \frac{C_{SU} \; R}{12} + \frac{7 \, C_{SU} \; R}{12} \quad \cdots \quad \textit{with SR} \tag{7.38}$$

where the first, second and third terms on right hand side of the above equation represent the inter-turn capacitances, first disk to SR capacitance, and SR to second disk capacitance respectively. Here, it is assumed that the gap between the first disk and SR is equal to the gap between SR and the second disk.

7.3.5 Interleaved winding

As explained earlier, an interleaved winding results in a considerable increase of series capacitance. In this type of winding, geometrically adjacent turns are kept far away from each other electrically, so that the voltage between adjacent turns increases. By interleaving the turns in such a way, the initial voltage distribution can be made more uniform. The capacitance between the disks (inter-disk capacitance) has very little effect on the series capacitance of this type of winding since its value is relatively low. Therefore, it is sufficient to consider only the inter-turn capacitances for the calculation of series capacitance of the interleaved windings. It follows that for the interleaved windings, the second approach of capacitance calculation based on the assumption of linear voltage distribution is quite accurate as compared to the continuous windings.

For the interleaved winding shown in figure 7.6, the number of inter-turn capacitances per disk is $(N_D - 1)$. The total number of inter-turn capacitances in a disk-pair is $2(N_D - 1)$. As before, let V be the voltage applied across the terminals of the disk-pair. The voltage is assumed to be uniformly distributed over the disk-pair; the assumption is more appropriate for interleaved windings as explained earlier. For the interleaved winding shown in figure 7.6, the number of electrical turns between the first and second turn is 10, while that between the second and third turn is 9. This arrangement repeats alternately within the disks. Hence, the voltage across the N_D capacitances is $(V/2)$ and across the remaining $(N_D - 2)$ capacitances is $\left(\dfrac{N_D - 1}{2N_D} V\right)$. The energy stored in the disk-pair is given by

$$En = \frac{1}{2} C_T \left(\frac{V}{2}\right)^2 N_D + \frac{1}{2} C_T \left[\frac{(N_D - 1)}{2N_D} V\right]^2 (N_D - 2) = \frac{1}{2} C_{se} V^2$$

$$\therefore \; C_{se} = \frac{C_T}{4}\left[N_D + \left(\frac{N_D - 1}{N_D}\right)^2 (N_D - 2)\right] \tag{7.39}$$

For $N_D >> 1$, the expression simplifies to

$$C_{se} = \frac{C_T}{2}[N_D - 1] \quad \cdots \textit{interleaved winding} \tag{7.40}$$

The interleaving of turns can give a substantial increase in the series capacitance of a winding and hence interleaved windings are used widely in high voltage transformers. As the rating of power transformer increases, higher core diameters are used increasing the voltage per turn value. Hence, a high voltage winding of a large rating transformer has usually lower turns and correspondingly lower turns per disk as compared to a high voltage winding of the same voltage class in a lower rating transformer. Since the interleaved windings are more effective for higher turns per disk, they may not be attractive for use in high-voltage high-rating transformers. Added to this, as the rating increases, the current carried by the high voltage winding increases, necessitating the use of a large number of parallel conductors for controlling the winding eddy losses. The interleaved winding with a large parallel conductors is difficult from productivity point of view. Hence, an alternative method of increasing capacitance by use of shielded-conductor (wound-in-shields) is adopted for high voltage windings of large power transformers. This is because of the fact that the continuously transposed cable (CTC) conductor, which is ideally suited for such applications (as explained in Chapter 4), can be used with this shielded-conductor winding technology.

7.3.6 Shielded-conductor winding

A shielded-conductor winding gives a modest but sufficient increase in the series capacitance and is less labour intensive as compared to an interleaved winding. The number of shielded-conductors can be gradually reduced in the shielded disks from the line end, giving a possibility of achieving tapered capacitance profile to match the voltage stress profile along the height of the winding [15]. This type of winding has some disadvantages, viz. decrease in winding space factor, requirement of extra winding material (shields), possibility of disturbance in ampere-turn balance per unit height of LV and HV windings, and extra eddy loss in shields.

Let us calculate the total series capacitance of a shielded-conductor disk-pair shown in figure 7.13. For N_D turns per disk with an applied voltage of V across the disk-pair, the voltage per turn is $V/(2N_D)$. It is assumed that for shields also, the same value of voltage per turn is applicable. Out of N_D turns, the first k turns are shielded in each disk. The shield can be either floating or it can be connected to some turn. Here, the shield conductors are assumed to be floating. For the first disk the voltage of any turn is

$$V_c(i) = V - (i-1)\frac{V}{2\,N_D} \qquad i = 1, 2, \cdots, N_D \tag{7.41}$$

The voltage of ith shield turn is given by

$$V_s(i) = \frac{V}{2} - (i-1)\frac{V}{2\,N_D} \qquad i = 1, 2, \cdots, k \tag{7.42}$$

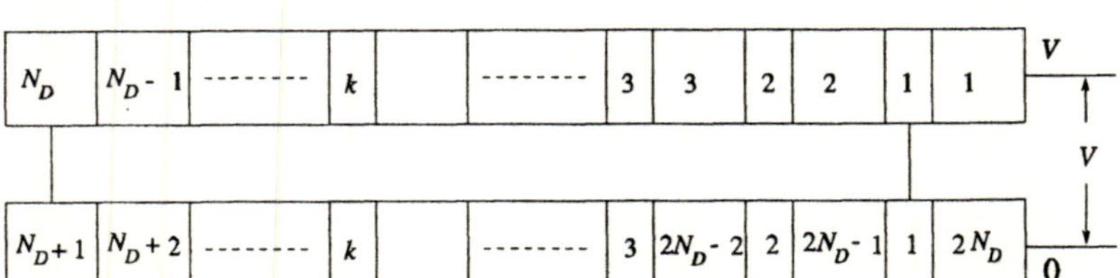

Figure 7.13 Shielded-conductor winding

If C_{sh} denotes the capacitance between a shield turn and adjacent disk turn, the energy between a shield turn i and touching adjacent disk turns is

$$En_{s1} = \frac{1}{2} C_{sh} \left[(V_c(i) - V_s(i))^2 + (V_c(i+1) - V_s(i))^2 \right] \tag{7.43}$$

Using the expressions from equations 7.41 and 7.42 we get

$$En_{s1} = \frac{1}{2} C_{sh} \left\{ \left(\frac{V}{2} \right)^2 + \left(\frac{V}{2} - \frac{V}{2N_D} \right)^2 \right\} \tag{7.44}$$

Similarly for the second disk, voltages of ith turn and i^{th} shield are given by

$$V_c(i) = V - (2N_D - i + 1) \frac{V}{2N_D} \qquad i = 1, 2, \cdots, N_D \tag{7.45}$$

$$V_s(i) = \frac{V}{2} + (i-1) \frac{V}{2N_D} \qquad i = 1, 2, \cdots, k \tag{7.46}$$

The energy between a shield turn i and touching adjacent disk turns for the second disk can be similarly calculated as

$$En_{s2} = \frac{1}{2} C_{sh} \left\{ \left(\frac{-V}{2} \right)^2 + \left(-\frac{V}{2} + \frac{V}{2N_D} \right)^2 \right\} \tag{7.47}$$

There are $2 \times (N_D - k - 1)$ turn-to-turn capacitances and the energy stored in each of these capacitances is

$$En_T = \frac{1}{2} C_T \left(\frac{V}{2N_D} \right)^2 \tag{7.48}$$

The expression for energy between the disks can be given by using equations 7.21 and 7.23 as

$$En_D = \frac{1}{2} \frac{C_{DU} R}{3} V^2 \tag{7.49}$$

For the type of shielded-conductor winding shown in figure 7.13, there is no contribution to the energy due to the capacitances between corresponding shield turns of the two disks, since they are at the same potential. For a precise calculation, the radial depth in the above equation should correspond to the radial depth of the winding excluding that of shield turns. The total energy stored in the disk-pair with shielded-conductors is

$$En = k\ En_{s1} + k\ En_{s2} + 2\left(N_D - k - 1\right) En_T + En_D \tag{7.50}$$

from which the effective capacitance of the disk-pair can be calculated. The similar procedure can be followed if, through an electrical connection, the shield is attached to some potential instead of the being in the floating condition. The calculation of capacitances of shielded-conductor winding has been verified in [15] by a circuit model and also by measurements on a prototype model.

7.3.7 Layer winding

For a simple layer (spiral) winding shown in figure 7.14, wherein an individual turn may have a number of parallel conductors depending upon the current rating, the series capacitance can be found as follows.

Let C_T be the inter-turn (turn-to-turn) capacitance and N_W be the total number of turns in the winding. As before, the voltage is assumed to be uniformly distributed within the winding. The energy stored in the winding is equal to the sum of the energies stored in the individual capacitances,

$$En = \frac{1}{2} C_T \left(\frac{V}{N_W} \right)^2 (N_W - 1) = \frac{1}{2} C_S V^2 \qquad (7.51)$$

$$\therefore C_S = \frac{C_T (N_W - 1)}{N_W^{\ 2}} \cong \frac{C_T}{N_W} \quad \cdots \quad \textit{layer winding} \qquad (7.52)$$

For a helical winding (layer winding with radial spacer insulation between turns), the above equation applies with C_T calculated by using equation 7.17 with the consideration of proportion of area occupied by spacers (solid insulation) and oil.

7.3.8 Interleaved tap winding

In high-voltage high-rating transformers, when a spiral winding is used as a tap winding, the tap sections are interleaved as shown in figure 7.15. The tap winding consists of 8 circuits (steps) giving a voltage difference between adjacent turns either corresponding to one-circuit difference or two-circuit difference. Thus, if there are 10 turns per circuit, the voltage difference between touching turns is either equal to 10 or 20 times the voltage per turn. This higher voltage difference necessitates the use of higher paper insulation reducing capacitance, but the reduction is more than compensated by the increased capacitive effect due to higher voltage between turns.

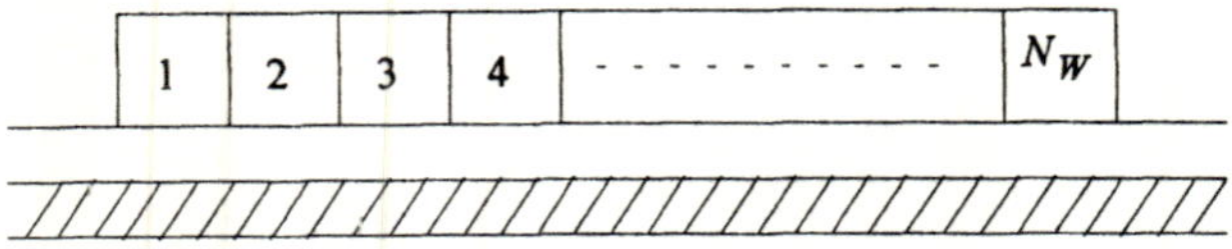

Figure 7.14 Layer winding

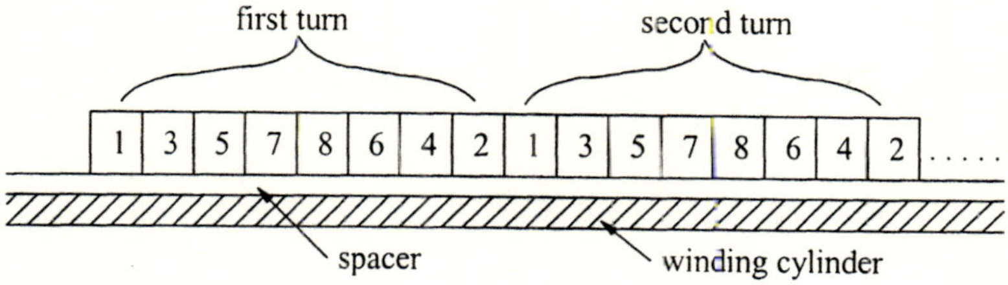

Figure 7.15 Interleaved tap winding

Let us calculate the value of series capacitance of an interleaved winding having 8 circuits with 10 turns per circuit, giving a total of 80 turns for the tap winding. Assuming again that the voltage is uniformly distributed within the tap winding with voltage per turn as $V/80$, the energy stored in the tap winding is

$$En = 10\left\{6\left[\frac{1}{2}C_T\left(2\times10\times\frac{V}{80}\right)^2\right]+\frac{1}{2}C_T\left(10\times\frac{V}{80}\right)^2\right\}+9\left\{\frac{1}{2}C_T\left(9\times\frac{V}{80}\right)^2\right\} \tag{7.53}$$

Simplifying and equating it to $(1/2)\,C_S\,V^2$, we get the effective series capacitance of the interleaved tap winding as

$$C_S \cong 4\;C_T \tag{7.54}$$

Comparing this value of series capacitance with that of layer winding of 80 turns as given by equation 7.52, it can be seen that the series capacitance has increased by about 320 times. The series capacitance for any other type of interleaved tap winding, with different turns per circuit and number of circuits, can be easily calculated by following the same procedure.

The method presented till now for the calculation of series capacitance of windings is based on the energy stored. There are a number of other methods reported in the literature. A rigorous analytical method is presented in [16] to calculate the equivalent series capacitance of windings. The method is also used to determine the natural frequencies and internal oscillations of windings.

The analytical methods have the disadvantage that the fringing effects and corresponding stray capacitances cannot be accurately taken into account. In this respect, numerical methods like Finite Element Method (FEM) can accurately give the value of capacitance which accounts stray effects also. In FEM analysis also, the capacitance is calculated from the stored energy (En) as

$$C = \frac{2\;En}{V^2} \tag{7.55}$$

The procedure is similar to that of the leakage inductance calculation by FEM analysis as described in Chapter 3.

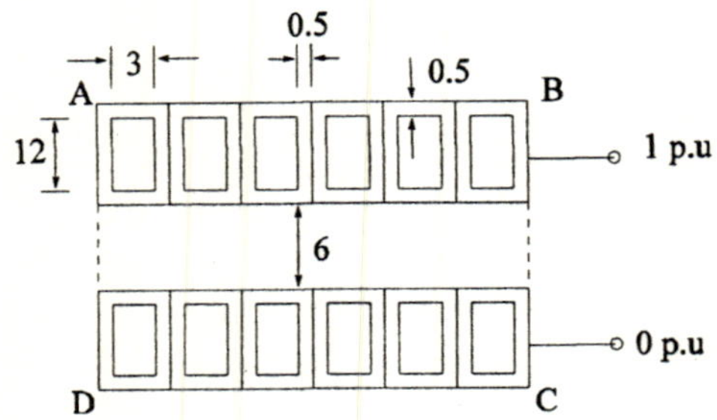

Figure 7.16 Capacitance calculation by FEM analysis

The series capacitance of a disk-pair of a continuous disk winding and interleaved winding has been calculated by FEM analysis for the geometry shown in figure 7.16 (dimensions are in mm). The gap between two disks is 6 mm. There are 6 turns per disk, and a uniform voltage distribution is assumed. The relative permittivities of oil and paper insulation are taken as 2.2 and 3.5 respectively. The geometry is enclosed in a rectangular boundary at a distance of 1 meter from the disks on all the sides, so that the boundary conditions do not affect the potential distribution in the disks. The energy is calculated for the rectangular area ABCD. The values of capacitance per unit length calculated by the analytical formulae (equations 7.25 and 7.40) and FEM analysis are given in table 7.1.

7.4 Inductance Calculation

Insulation design based on only initial voltage distribution (with inductances neglected) may be acceptable for transformers of smaller voltage rating. The difference between the initial and final (linear) distributions sets up oscillations in the winding. According to Weed's principle [17], a winding will be non-oscillating if the capacitive (initial) and inductive (final) distributions are alike, otherwise the difference will set up an oscillation under conditions favorable to it, and such an oscillation may result into much larger voltage gradients between different parts of the winding. Hence, the voltage distribution under impulse conditions should be calculated with the inclusion of inductances in the winding representation.

Table 7.1 Capacitance calculation by analytical method and FEM analysis

	Continuous winding	Interleaved winding
Analytical	57.4 pF/m	1006.8 pF/m
FEM	61.8 pF/m	1073.4 pF/m

The mutual inductance between two thin wire, coaxial coil loops (A and B) of radii R_A and R_B with a distance S between them is given in SI units as [15, 18, 19]

$$L_{AB} = \frac{2\mu_0}{k} N_A N_B \sqrt{R_A R_B} \left\{ \left[1 - \frac{k^2}{2} \right] K(k) - E(k) \right\} \tag{7.56}$$

where

$$k = \sqrt{\frac{4 R_A R_B}{(R_A + R_B)^2 + S^2}} \tag{7.57}$$

and N_A and N_B are the turns in sections A and B respectively, whereas $K(k)$ and $E(k)$ are the complete elliptic integrals of the first and second kinds respectively. The formula is applicable for thin circular filaments of negligible cross section. For circular coils of rectangular cross section, more accurate calculations can be done by using Lyle's method in combination with equation 7.56 [20, 21].

The self inductance of a single turn circular coil of square cross section with an average radius of a and square side length c is given in SI units as [15, 20]

$$L_{AA} = \mu_0 \, a \left[\frac{1}{2} \left(1 + \frac{1}{6} \left(\frac{c}{2a} \right)^2 \right) ln \left(\frac{8}{(c/2a)^2} \right) - 0.84834 + 0.2041 \left(\frac{c}{2a} \right)^2 \right] \tag{7.58}$$

The formula applies for relatively small cross section such that $(c/2a) < 0.2$. If the cross section is not square, it should be divided into a number of square cross sections, and then equations 7.56 and 7.58 can be used to compute the self inductance.

Accuracy of the calculated self and mutual inductances may significantly affect the results of computed impulse voltage distribution. The difference between the calculated and measured results is mainly due to effects of the field distortion and variation within the core at high frequencies [22]. For accurate results the field equations need to be solved which may not be practical. Hence, in practice correction factors are applied to the formulae for self and mutual inductances.

Some formulations reported in the literature use customary short circuit inductances (which are more easily and accurately calculated) in place of self and mutual inductances [23, 24]. Some others [25] use the network of inductances derived through the theory of magnetic networks, which avoids introduction of mutual inductances in the network of lumped parameters.

7.5 Standing Waves and Traveling Waves

The transient response of a winding subjected to impulse waves was initially obtained in the literature by two different methods: the standing wave and the traveling wave approach. The theory of electrical waves in transmission lines cannot be directly applied to transformers due to the fact that transformer, unlike transmission line, has series capacitances and mutual inductances between winding sections.

Consider a single layer winding having self inductance (L_s) per unit length, shunt capacitance (c_g) per unit length to ground and series capacitance (c_s) per unit length between adjacent turns (see figure 7.2). In this model, the mutual inductance between turns and the resistance of winding are neglected for the purpose of simplifying the calculations. The set of differential equations describing the transient process taking place in the winding can be given by applying Kirchhoff's laws as (notations as per figure 7.2)

$$\frac{\partial i_{L_s}}{\partial x} + \frac{\partial i_{c_s}}{\partial x} = -c_g \frac{\partial u}{\partial t} \tag{7.59}$$

$$i_{c_s} = -c_s \frac{\partial^2 u}{\partial x\, \partial t} \tag{7.60}$$

$$\frac{\partial u}{\partial x} = -L_s \frac{\partial i_{L_s}}{\partial t} \tag{7.61}$$

By eliminating the currents in the above 3 equations, these can be reduced to a single differential equation in terms of voltage as

$$\frac{\partial^2 u}{\partial x^2} - L_s\, c_g \frac{\partial^2 u}{\partial t^2} + L_s c_s \frac{\partial^4 u}{\partial t^2\, \partial x^2} = 0 \tag{7.62}$$

The solution of equation 7.62 gives the transient voltage distribution inside the winding. Let us assume the solution of this equation in the form [4, 26],

$$u = U e^{j\omega t} e^{j\psi x} \tag{7.63}$$

Since the solution contains exponential terms in both time and space, it includes both standing and traveling waves.

In the standing wave approach, the expression of assumed solution is put in equation 7.62 which after simplification becomes

$$\psi^2 - L_s c_g \omega^2 - L_s c_s \omega^2 \psi^2 = 0 \tag{7.64}$$

$$\therefore \psi = \sqrt{\frac{L_s\, c_g\, \omega^2}{1-\omega^2 L_s\, c_s}} \tag{7.65}$$

and

$$\omega = \frac{\psi}{\sqrt{L_s c_g \left(1+\frac{c_s}{c_g}\,\psi^2\right)}} \tag{7.66}$$

It can be seen from the equations 7.65 and 7.66 that both space frequency (ψ : number of standing wave cycles within the space interval of 2π) and angular frequency (ω) are related to each other. With $\psi \to \infty$, the critical angular frequency of the winding is obtained as

$$\omega_{cr} = \lim_{\psi\to\infty} \frac{\psi}{\sqrt{L_s c_g \left(1+\frac{c_s}{c_g}\,\psi^2\right)}} = \frac{1}{\sqrt{L_s\, c_s}} \tag{7.67}$$

This is the highest frequency in time with which the winding is capable of oscillating. It is equal to the natural frequency of a single turn with inductance L_s and capacitance c_s.

In the classical standing wave theory, the oscillations between the initial and final voltage distributions are resolved into a series of standing waves or harmonics both in space and time [1, 27, 28]. A study of the oscillations throughout the transition period from the initial to final distribution, allows the calculation of the surge voltage distribution within the winding. In this approach, the wave-shapes and frequencies of standing waves (eigen functions or natural modes) of the winding, not connected to any source, are determined for various terminal conditions. The natural frequencies of these free oscillations are computed and voltage distribution for each harmonic is obtained. The amplitudes of all these standing waves are then obtained for the applied wave shape, and the transient voltage distribution along the winding is finally obtained as the sum of all harmonics, a convergent infinite series. The contribution of each harmonic depends on its wave shape and natural frequency, and on its amplitude. The amplitude in turn depends on the difference between the initial and final distributions. Initially, mutual inductances were totally or partly neglected simplifying the analysis. Later on, the effect of mutual inductances was also included [29] for more accurate determination of standing waves.

In the traveling wave theory, the incident wave is represented as an infinite series of sinusoidal components, and the resulting differential equation is analyzed to determine the conditions under which these waves can enter the winding. The method was explained in [26, 30] by its application to windings with uniformly distributed insulation, and for incoming surges with vertical wave fronts and infinite tails. The traveling wave based analysis is superior and computationally less intensive than that based on the standing wave theory. The solution of the original differential equation 7.62 is assumed in the traveling wave theory as [26]

$$u = Ue^{j\omega(t - \frac{x}{v})} \tag{7.68}$$

which describes the traveling waves. These waves oscillate with time frequency ω and propagate at velocity v through the winding. The above solution and the previously assumed solution (equation 7.63) are equivalent for $\psi = (-\omega/v)$. By substituting equation 7.68 into equation 7.62 and differentiating we get

$$-\left(\frac{\omega}{v}\right)^2 + L_s c_g \omega^2 + L_s c_s \omega^2 \left(\frac{\omega}{v}\right)^2 = 0 \tag{7.69}$$

By solving the above equation, we get the equation for velocity of propagation,

$$v = \sqrt{\frac{1}{L_s c_g} - \frac{c_s}{c_g}\omega^2} \tag{7.70}$$

Equation 7.70 indicates that as the angular frequency (ω) increases, the velocity of traveling wave (v) decreases. For

$$\omega_{cr} = \frac{1}{\sqrt{c_s L_s}} \tag{7.71}$$

the velocity of propagation is zero, which means that at $\omega \geq \omega_{cr}$ the traveling waves cannot propagate inside the winding.

In case of standing waves, for $\omega > \omega_{cr}$, the ψ in equation 7.65 becomes imaginary and the solution according to equation 7.63 is transformed into

$$u = Ue^{j\omega t} e^{-\xi x} \tag{7.72}$$

where

$$\psi = j\xi = j\sqrt{\frac{L_s c_g \omega^2}{c_s L_s \omega^2 - 1}} \tag{7.73}$$

Thus, for supercritical frequencies ($\omega > \omega_{cr}$), no standing or traveling waves exist within the winding; there is an exponential attenuation of the voltage from the winding terminal towards the interior.

A transformer winding can propagate only those oscillations having a frequency below a certain critical value. The traveling wave generally gets flattened as it travels into the winding. Unlike in transmission lines, there is no simple relationship between the wavelength and frequency for a wave traveling through the transformer winding, and hence it cannot penetrate into the winding without distortion. Due to presence of mutual inductances and series capacitances between sections of the transformer winding, there is a continual change in the form of the complex wave as it penetrates inside the winding; the phenomenon is in marked contrast to that of transmission lines, where a complex wave of any shape propagates without distortion, except for the effect of resistances. A traveling wave does not change its shape when its velocity is independent of frequency, and all natural oscillations have the same decay coefficient [31]. While both these conditions are approximately satisfied for a transmission line (and the waveshape is maintained), these are not satisfied for a transformer resulting in a major distortion of traveling waves inside the winding. Higher frequency oscillations cannot penetrate deeply into the winding and establish a standing exponential distribution (exponential attenuation of voltage from the terminal towards the interior) similar to the initial distribution of the standing wave analysis [3]. In other words, the high frequency components form a standing potential distribution and the low frequency components form a traveling wave; the splitting of incoming surge into two parts is the characteristics of the traveling wave theory. In [32], the standing wave and traveling wave approaches are compared and correlated.

The traveling component moves along the winding conductor with a velocity governed by the fundamental equation,

$$v = \frac{1}{\sqrt{\mu\,\varepsilon}} = \frac{c}{\sqrt{\mu_r\,\varepsilon_r}} \text{ m/s} \tag{7.74}$$

where μ and ε are the permeability and permittivity of the medium respectively,

$$\mu = \mu_0\mu_r = 4\pi\times10^{-7}\mu_r\,, \qquad \varepsilon = \varepsilon_0\varepsilon_r = 8.85\times10^{-12}\varepsilon_r\,,$$

$$c = \text{velocity of light} = 3\times10^{8} \text{ m/s}$$

Hence, for a oil cooled transformer with $\varepsilon_r = 3.5$ (typical value of resultant dielectric constant of oil-paper-solid insulation system) and $\mu_r = 1$, the velocity of travel of a wave into the winding will be approximately equal to $160\ m/\mu s$.

7.6 Methods for Analysis of Impulse Distribution

Although the surge response of transformer windings was initially determined by two theories, the standing wave theory and the traveling wave theory, which helped in understanding and visualizing the surge phenomena, these methods had the disadvantage that they can be basically applied only to a uniform winding. Non-uniformities within the windings, presence of more than one winding per limb, windings of other phases, etc. are some of the complexities which cannot be handled by these two theories. Hence, it is impractical to do the analysis by purely analytical means for increasingly complicated present-day transformers. With the advent of computers, it became possible to solve the differential equations numerically and analyze practical transformer winding configurations. A lumped parameter network is particularly well suited for solution by computers. Using the Laplace transform, analysis of a ladder network having a finite number of uniform winding sections was done in [33]. Subsequently with the advent of digital computers, the network is solved by using both digital and analogue computers in [24]. The digital computer is used to calculate the coefficients of simultaneous integrodifferential equations which are then solved by an electronic analogue computer. A numerical analysis reported in [22] uses Runge-Kutta method to solve a second order differential equation. In [34] differential equations, formulated for the equivalent network using the state space approach, are solved by computer.

In [35], the transient response is calculated using the trapezoidal rule of integration through the companion network approach. It is easier to analyze a network for transient response calculations if there are no mutual couplings. Hence, the equivalent network with coupled elements is replaced by that with uncoupled elements by using the formulation given in [36]. The two circuits (coupled and uncoupled) are equivalent and have the same nodal admittance matrix. The advantage of using companion network is that it can be analyzed using well-known methods since it is purely resistive circuit.

The equivalent circuit representation is quite popularly used for finding response of a transformer winding to high voltage surges. During transients, the windings are coupled by electric and magnetic fields. The inductances (self and mutual) and capacitances are distributed along the windings. The transients can be described by partial differential equations, but their solution is very difficult. If the windings are subdivided into sections in which inductances and capacitances are lumped, the calculation becomes easier since the partial differential equations can be now replaced with close approximation by ordinary simultaneous differential equations. These ordinary differential equations can be solved by numerical analysis using computers for complex configurations of windings. The accuracy of results obtained from the circuit representation depends predominantly on the degree of sophistication used in the winding representation. In one of the most accurate representations, each turn of the winding is represented with corresponding turn-to-turn capacitances and inductances [37]. The knowledge of voltage distribution across the inter-turn insulation is important for transformers exposed to very fast transient overvoltages. Such a model, although very accurate, may be prohibitive from the point of time and memory of computers. Hence, from practical point of view, many simplifications are done in the detailed model. A sufficient accuracy can be obtained for the network model in which the windings are lumped into R, L and C circuit components. The windings are represented by as many elements as there are disks or groups of disks with the corresponding resistances, inductances (self and mutual) and capacitances (series and ground). Thus, the equivalent lumped parameter network is a series of π circuits with mutual magnetic couplings. In such a simplified model, although an individual turn voltage cannot be ascertained, disk voltages can be determined which can be used to decide the internal insulation between disks in the winding. Actually, the voltage stresses are the result of electric and magnetic fields which appear in the winding under surge conditions and are function of location and time. By representing the transformer winding as a network of elements, the field problem is effectively converted into a circuit problem.

An equivalent network for a multi-winding transformer has been reported in [19] in which the conventional ladder network used for a single winding consisting of lumped elements (self and mutual inductances, series and ground capacitances) is extended for multiple windings. The method takes into account the electrostatic and electromagnetic combinations of windings and therefore permits the analysis of not only the voltage response of the winding to which the impulse is applied but also the transferred voltage to other windings to which the impulse is not directly applied. It is very important to calculate the transferred voltages to other windings in the case of HV winding with the center-line lead arrangement. With both the ends of LV winding grounded, voltage at the center height (mid-height position) of LV winding could be oscillating and the net voltage difference between LV and HV windings at the center height could be more than the applied impulse magnitude. In such a case, the gap between LV and HV windings may get decided by the extra higher voltage stress. A transferred surge in two-winding transformers has four components, viz. the electrostatic component decided by the network of capacitances, the electromagnetic component due to mutual inductances between windings, free oscillations of the secondary winding, and forced oscillations of the secondary winding induced by the free oscillations of the primary winding. For the free oscillations of the secondary winding, the transferred electrostatic component represents the initial distribution and the electromagnetic component represents the final distribution. These four components are also present at the terminals of secondary and tertiary windings when the primary winding is subjected to the impulse condition in a three-winding transformer. In [38, 39], surge transfers in three-winding transformers have been analyzed in detail. It is shown that the third and fourth components are usually insignificant and their effect is not important. The electrostatic component can be reduced to a great extent if the secondary / tertiary winding terminals are connected to an equipment (like a cable of sufficient length) having a high capacitance to ground. Even the bushing and terminal bus capacitances have the effect of reducing the transferred electrostatic component. Simple formulae are given in [40] for calculation of transferred surge voltages in autotransformers.

In the initial works, damping of oscillations caused by core loss, copper loss and dielectric loss was generally not taken into account for simplifying the calculations. Subsequently, the effects of these losses were taken into account by shunt elements representing a conductance between each winding node to ground in the equivalent network [23]. In other words, the parameters such as winding resistances were not considered separately since their effect on damping was approximately taken into consideration by the shunt conductance elements. In case of distribution transformers, winding resistances significantly reduce the voltage peaks and hence they cannot be neglected. Due to availability of powerful computational facilities, complex models can be solved now. A detailed model of losses is incorporated in [37] for accurate calculations.

A method based on natural frequencies of windings is described in [41] for the calculation of impulse voltage distribution. A mathematical model representing capacitances and inductances of windings is analyzed to calculate eigenvalues and eigenvectors, based on which the temporal and spatial variations of the voltage in the windings are calculated.

Study of the effect of iron core on the impulse response of windings has been studied in [19]. The effect of core on the lightning impulse response of a single winding is significant for the case of ungrounded neutral as compared to the grounded one. It seems to indicate that there exists flux in the core when the neutral is not grounded. It has been also reported that with the inner non-impulsed winding short-circuited, the presence of core has negligible effect on the voltage response irrespective of the neutral grounding condition (of the impulsed winding). The main flux in the core will be cancelled by the flux produced by the inner short-circuited winding and only leakage flux contributes to the impulse response. When the winding connections / grounding conditions allow the flux to flow in the core, the iron losses in the core have a damping effect on the peaks of voltage response, and in this case voltages are lower in presence of core as compared to that with its absence. It has been reported in [42] that a considerable variation in core permeability gives a very moderate change in the voltage response. Hence, for finding impulse response of the winding, air core (self and mutual) inductances (which can be easily calculated as explained in Section 7.4) are generally used and suitable correction factors based on the experience / experimental measurements are applied.

In [43], a study of the behavior of transformer winding subjected to standard impulse voltage waves chopped at different instants (on the front as well as on the tail of the waves) is presented. It is well-known that a wave chopped at an unfavorable instant may result into higher voltage stresses at line end sections of the winding as compared to voltage stresses due to a full wave of the same steepness. The instant of switching and time to collapse of a chopped impulse predominantly decide the level of stresses.

7.7 Computation of Impulse Voltage Distribution Using State Variable Method

Accurate determination of impulse voltage distribution in a transformer winding is possible by using its equivalent circuit as explained in the previous section.

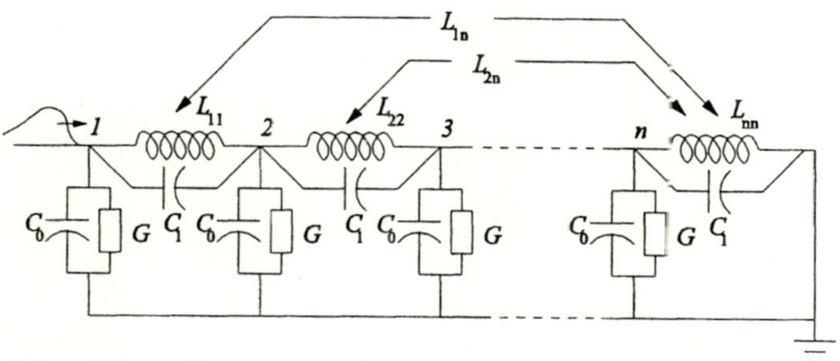

Figure 7.17 Equivalent circuit of a transformer winding

The evolution of the simplified equivalent network shown in figure 7.17 is explained in [44]. In this section, a method is elaborated for finding impulse distribution within a single winding, which can be easily extended for a case of multiple windings. The equivalent circuit of a transformer winding consists of a finite number of sections having elements C_0 , C_1, L_{ii}, L_{ij} and G which are the shunt capacitance, series capacitance, self inductance, mutual inductance and shunt conductance respectively.

Due to the advancements in the computational facilities, it is quite simple to calculate the impulse voltage distribution using the state space model of the lumped parameter network of the transformer winding.

7.7.1 Derivation of differential equations

The network equations for the circuit are formulated in the nodal form as [34]

$$\hat{C}\,\ddot{\hat{y}}(t) + \hat{G}\,\dot{\hat{y}}(t) + \hat{\Gamma}\,\hat{y}(t) = 0 \tag{7.75}$$

where $\hat{C}$ = nodal capacitance matrix with the inclusion of input node

$\hat{G}$ = nodal conductance matrix with the inclusion of input node

$\hat{\Gamma}$ = nodal matrix of inverse inductances with the inclusion of input node

$\hat{y}(t)$ = output vector of node voltages with the inclusion of input node

The relationship between the nodal matrices and branch matrices is defined by

$$\left.\begin{aligned} \hat{C} &= Q_c C_b Q_c^T \\ \hat{G} &= Q_G G_b Q_G^T \\ \hat{\Gamma} &= Q_L L_b^{-1} Q_L^T \end{aligned}\right\} \tag{7.76}$$

where Q_c, Q_G and Q_L are the incidence matrices for capacitive, conductive and inductive elements, and C_b, G_b and L_b are the branch matrices of capacitive, conductive and inductive elements of the network respectively.

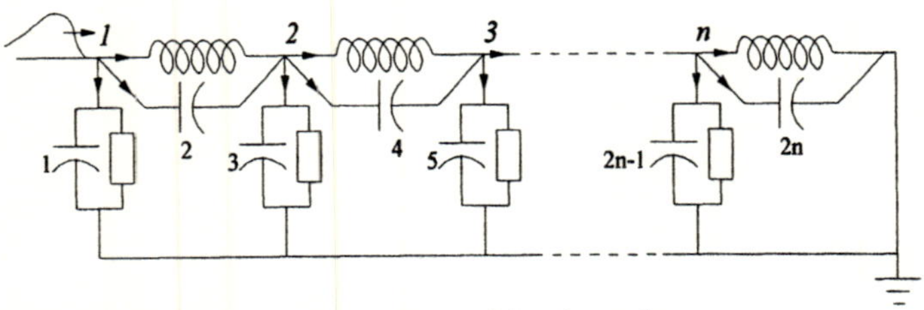

Figure 7.18 Numbering of capacitive branches

The number of equations is reduced by extracting the input node k because its voltage is known. Therefore, equation 7.75 can be rewritten as

$$C\,\ddot{y}(t) + G\,\dot{y}(t) + \Gamma\, y(t) = -\,C_k\,\ddot{x}(t) - G_k\,\dot{x}(t) - \Gamma_k\, x(t) \tag{7.77}$$

where, C = nodal capacitance matrix without the input node
G = nodal conductance matrix without the input node
Γ = nodal matrix of inverse inductances without the input node
$y(t)$ = output vector of node voltages without the input node
$x(t)$ = known voltage of the input node
$C_k,\ G_k,\ \Gamma_k = k^{th}$ column of $\hat{C},\ \hat{G},\ \hat{\Gamma}$ with the entry of kth row removed

7.7.2 Formation of $\hat{C}$ matrix

There are $2n$ capacitive branches in the network of figure 7.17. Therefore, the size of the branch capacitance matrix (C_b) is $2n \times 2n$. For the branches of the circuit numbered as shown in figure 7.18 it can be written as

$$C_b = \begin{bmatrix} C_0 & 0 & 0 & \cdots & 0 \\ 0 & C_1 & 0 & \cdots & 0 \\ 0 & 0 & C_0 & \cdots & 0 \\ \cdots & \cdots & \cdots & \cdots & \cdots \\ 0 & 0 & 0 & \cdots & C_1 \end{bmatrix} \tag{7.78}$$

The corresponding incidence matrix Q_c ($n \times 2n$) is

$$Q_c = \begin{bmatrix} 1 & 1 & 0 & 0 & 0 & 0 & \cdots & 0 & 0 & 0 \\ 0 & -1 & 1 & 1 & 0 & 0 & \cdots & 0 & 0 & 0 \\ 0 & 0 & 0 & -1 & 1 & 1 & \cdots & 0 & 0 & 0 \\ 0 & 0 & 0 & 0 & 0 & -1 & \cdots & 0 & 0 & 0 \\ \cdots & \cdots & \cdots & \cdots & \cdots & \cdots & \cdots & \cdots & \cdots & \cdots \\ 0 & 0 & 0 & 0 & 0 & 0 & \cdots & 1 & 1 & 0 \\ 0 & 0 & 0 & 0 & 0 & 0 & \cdots & -1 & 1 & 1 \end{bmatrix} \tag{7.79}$$

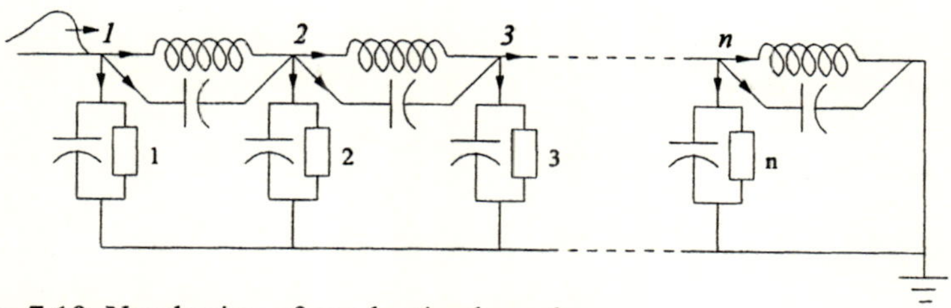

Figure 7.19 Numbering of conductive branches

We can get $\hat{C}$ from equation 7.76,

$$\hat{C} = Q_c C_b Q_c^T \tag{7.80}$$

7.7.3 Formation of $\hat{G}$ matrix

Similarly, the branch conductance matrix (G_b) and incidence conductance matrix Q_G of the order $(n \times n)$ for figure 7.19 can be given as

$$G_b = \begin{bmatrix} G & 0 & 0 & \dots & 0 \\ 0 & G & 0 & \dots & 0 \\ 0 & 0 & G & \dots & 0 \\ \dots & \dots & \dots & \dots & \dots \\ 0 & 0 & 0 & \dots & G \end{bmatrix} \tag{7.81}$$

$$Q_G = \begin{bmatrix} 1 & 0 & 0 & 0 & \dots & 0 & 0 \\ 0 & 1 & 0 & 0 & \dots & 0 & 0 \\ 0 & 0 & 1 & 0 & \dots & 0 & 0 \\ 0 & 0 & 0 & 1 & \dots & 0 & 0 \\ \dots & \dots & \dots & \dots & \dots & \dots & \dots \\ 0 & 0 & 0 & 0 & \dots & 1 & 0 \\ 0 & 0 & 0 & 0 & \dots & 0 & 1 \end{bmatrix} \tag{7.82}$$

and

$$\hat{G} = Q_G G_b Q_G^T \tag{7.83}$$

7.7.4 Formation of $\hat{\Gamma}$ matrix

Similarly, the branch inductance matrix (L_b) and incidence inductance matrix Q_L of the order $(n \times n)$ for figure 7.20 can be given as

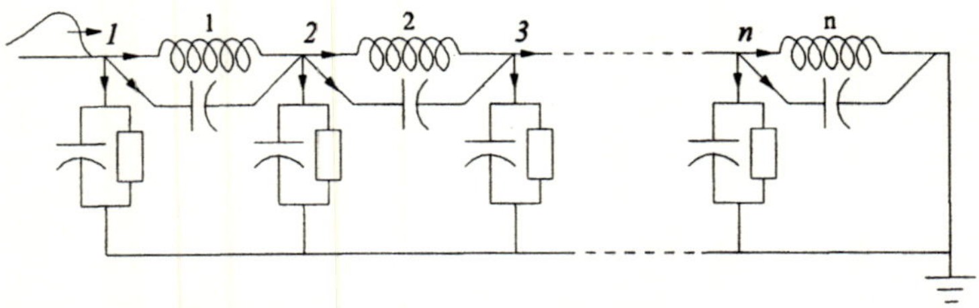

Figure 7.20 Numbering of inductive branches

$$L_b = \begin{bmatrix} L_{11} & L_{12} & L_{13} & \dots & L_{1n} \\ L_{21} & L_{22} & L_{23} & \dots & L_{2n} \\ L_{31} & L_{32} & L_{33} & \dots & L_{3n} \\ \dots & \dots & \dots & \dots & \dots \\ L_{n1} & L_{n2} & L_{n3} & \dots & L_{nn} \end{bmatrix} \tag{7.84}$$

$$\Gamma_b = L_b^{-1} = \begin{bmatrix} \Gamma_{11} & \Gamma_{12} & \Gamma_{13} & \dots & \Gamma_{1n} \\ \Gamma_{21} & \Gamma_{22} & \Gamma_{23} & \dots & \Gamma_{2n} \\ \Gamma_{31} & \Gamma_{32} & \Gamma_{33} & \dots & \Gamma_{3n} \\ \dots & \dots & \dots & \dots & .. \\ \Gamma_{n1} & \Gamma_{n2} & \Gamma_{n3} & \dots & \Gamma_{nn} \end{bmatrix} \tag{7.85}$$

$$Q_L = \begin{bmatrix} 1 & 0 & 0 & 0 & \dots & 0 & 0 \\ -1 & 1 & 0 & 0 & \dots & 0 & 0 \\ 0 & -1 & 1 & 0 & \dots & 0 & 0 \\ 0 & 0 & -1 & 1 & \dots & 0 & 0 \\ \dots & \dots & \dots & \dots & \dots & \dots & \dots \\ 0 & 0 & 0 & 0 & \dots & 1 & 0 \\ 0 & 0 & 0 & 0 & \dots & -1 & 1 \end{bmatrix} \tag{7.86}$$

and

$$\hat{\Gamma} = Q_L \Gamma_b Q_L^T \tag{7.87}$$

7.7.5 State space model

The state space model of the differential equation 7.77 (without input node) is

$$\dot{X}(t) = A\,X(t) + B\,v(t) \tag{7.88}$$

$$y(t) = F\,X(t) + D\,v(t) \tag{7.89}$$

where $X(t)$ = vector of state variables

A, F = matrices of constant coefficients
B, D = column matrices of constant coefficients
$v(t)$ = input vector of applied impulse voltage
$y(t)$ = output vector of node voltages

The equation 7.77 can be rewritten as

$$\ddot{y}(t)+C^{-1}G\,\dot{y}(t)+C^{-1}\Gamma\;y(t)=-C^{-1}C_k\;\ddot{x}(t)-C^{-1}G_k\,\dot{x}(t)-C^{-1}\Gamma_k\;x(t) \tag{7.90}$$

To get equation 7.90 in the desired form of state space equations, the state variables can be chosen as

$$X_1(t)=y(t)-\beta_0 x(t) \tag{7.91}$$

$$\begin{aligned} X_2(t)&=\dot{y}(t)-\beta_0\dot{x}(t)-\beta_1 x(t)\\ &=\dot{X}_1(t)-\beta_1 x(t) \end{aligned} \tag{7.92}$$

where $\beta_0=-C^{-1}C_k$

$\beta_1=-C^{-1}\left(G_k-G\,C^{-1}C_k\right)$

Now, rearranging equation 7.92 we get

$$\dot{X}_1(t)=X_2(t)+\beta_1\;x(t) \tag{7.93}$$

Replacing $y(t)$ and $\dot{y}(t)$ in equation 7.90 by their values from equations 7.91 and 7.92 respectively and simplifying we get

$$\dot{X}_2(t)=-\alpha_2\;X_1(t)-\alpha_1\;X_2(t)+\beta_2\;x(t) \tag{7.94}$$

where $\alpha_2=C^{-1}\Gamma$, $\alpha_1=C^{-1}G$,

$\beta_2=-C^{-1}\left(\Gamma_k-GC^{-1}\left(G_k-G\,C^{-1}C_k\right)-\Gamma C^{-1}C_k\right)$

Also, rearranging equation 7.91 we have

$$y(t)=X_1(t)+\beta_0 x(t) \tag{7.95}$$

Equations 7.93, 7.94 and 7.95 can be written in the matrix form as

$$\begin{bmatrix}\dot{X}_1(t)\\ \dot{X}_2(t)\end{bmatrix}=\begin{bmatrix}0 & I\\ -\alpha_2 & -\alpha_1\end{bmatrix}\begin{bmatrix}X_1(t)\\ X_2(t)\end{bmatrix}+\begin{bmatrix}\beta_1\\ \beta_2\end{bmatrix}x(t) \tag{7.96}$$

and $$[y(t)]=[I\quad 0]\begin{bmatrix}X_1(t)\\ X_2(t)\end{bmatrix}+[\beta_0]\,x(t) \tag{7.97}$$

Comparing equations 7.96 and 7.97 with equations 7.88 and 7.89 we get

$$X(t) = \begin{bmatrix} X_1(t) \\ X_2(t) \end{bmatrix}, \quad v(t) = x(t)$$

$$A = \begin{bmatrix} 0 & I \\ -\alpha_2 & -\alpha_1 \end{bmatrix}, \; B = \begin{bmatrix} \beta_1 \\ \beta_2 \end{bmatrix}, \; F = \begin{bmatrix} I & 0 \end{bmatrix} \text{ and } D = \begin{bmatrix} \beta_0 \end{bmatrix}$$

The above analysis has converted the original 2nd order circuit equation 7.77 to an equivalent 1st order state space system (equations 7.96 and 7.97). The solution of these state space equations can be written as [45]

$$X(t) = e^{At} X(0^-) + \int_{0^-}^{t} e^{A(t-\tau)} B\, x(\tau)\, d\tau \tag{7.98}$$

where $X(0^-)$ is the state vector at $t=0^-$ and is assumed to be zero. The above expression of $X(t)$ can be evaluated analytically for simple $x(\tau)$. Alternatively, a standard built-in function in MATLAB® for 1st order system can now be used to solve equations 7.96 and 7.97.

After getting the value of the state variables (X) of the circuit, the node voltages can be obtained from the equation 7.97. For the sample system given in [35], the impulse response is calculated by the above method. The winding consists of 12 sections; the details of lumped elements are given in table 7.2.

The input voltage is assumed to be the standard full wave defined by

$$x(t) = x_0\left(e^{-\beta t} - e^{-\delta t}\right) \tag{7.99}$$

For the standard (1/50) microsecond wave (which rises to its maximum value at 1 microsecond and decays to half the maximum value in 50 microseconds), when t is expressed in microseconds the values of the constants are

$$x_0 = 1.0167, \; \beta = 0.01423 \text{ and } \delta = 6.0691$$

The voltages calculated for various nodes are plotted in figure 7.21, which are in close agreement with that reported in [35].

Table 7.2 Inductance and capacitance parameters [35]

C_0	C_1
412.6 pF	120 pF

$L_{1\text{-}1}$	$L_{1\text{-}2}$	$L_{1\text{-}3}$	$L_{1\text{-}4}$	$L_{1\text{-}5}$	$L_{1\text{-}6}$
0.668 mH	0.526 mH	0.425 mH	0.344 mH	0.283 mH	0.232 mH
$L_{1\text{-}7}$	$L_{1\text{-}8}$	$L_{1\text{-}9}$	$L_{1\text{-}10}$	$L_{1\text{-}11}$	$L_{1\text{-}12}$
0.186 mH	0.139 mH	0.101 mH	0.060 mH	0.030 mH	0.000 mH

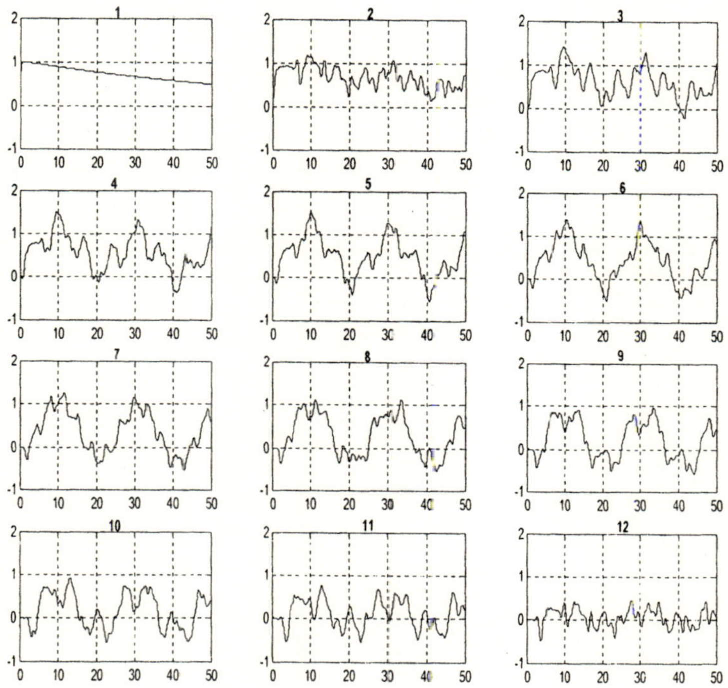

Figure 7.21 Voltage waveforms at all 12 nodes

The results obtained for another 10-section winding are also in close agreement with that given in [46] calculated for rectangular and chopped waves.

The voltages are also calculated using two software packages SPICE and SEQUEL. SPICE is a general-purpose circuit simulation program for nonlinear DC, nonlinear transient and linear AC analyses developed by University of Berkeley, California (http://bwrc.eecs.berkeley.edu/Classes/IcBook/SPICE/).

SEQUEL is a public-domain package (a Solver for circuit EQuations with User-defined ELements), developed at IIT Bombay (see the details at http://www.ee.iitb.ac.in/~microel/faculty/mbp/sequel1.html). It allows the user to incorporate new elements in the package by simply writing a "template" to describe the model equations. SEQUEL is based on the Sparse Tableau approach, in which all variables are treated in the same manner without separating them into "current-like" and "voltage-like" variables. This makes it particularly convenient to write new element templates. It solves a (generally nonlinear) system of equations of the form,

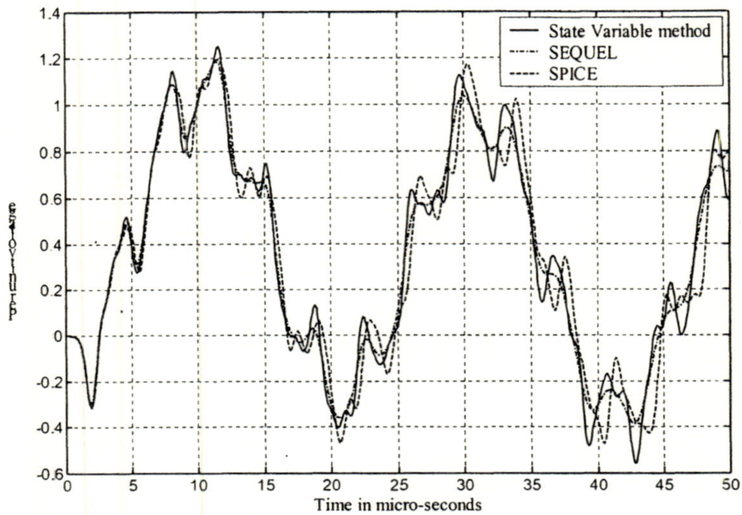

Figure 7.22 Comparison of voltage waveform at node 6 obtained using different methods

$$f_i = (x_1, x_2, x_3, \cdots, x_n) = g_i \tag{7.100}$$

where $g_i = 0$ or $g_i = dx_j/dt$. The method employed by SEQUEL to solve the nonlinear equations is the well-known Newton-Raphson (NR) iterative method. Four discretization schemes are offered for transient simulation: (i) Backward Euler, (ii) Trapezoidal, (iii) Gear's scheme (order 2) and (iv) TR-BDF2 scheme.

The results of all 3 methods (state variable method, SEQUEL and SPICE), shown in figure 7.22 for node 6 of the winding with 12 sections, closely agree with each other. The state variable method can be easily extended for multi-winding transformers. The matrices in equation 7.77 will get changed and the same procedure can be followed to get the impulse response.

7.8 Winding Design for Reducing Internal Overvoltages

7.8.1 Part winding resonance

Part winding resonance has been identified as the source of number of high voltage power transformer failures. If the frequency of an exciting oscillating voltage coincides with one of the fundamental natural frequencies of a winding or a part of winding, resonant overvoltages will occur. The failures of four single-phase autotransformers in 500 kV and 765 kV systems of American Electric Power between 1968 and 1971 led to the detailed investigations of winding resonance phenomenon [47, 48]. All these failures involved breakdown of off-circuit tap changers immediately after the occurrence of transmission system faults. The taps were at the neutral end of the common winding in all the four failed autotransformers. After the investigations, it was concluded that the traveling wave generated by the line fault contributed to the failures. Factory and field tests with nonstandard wave shapes and terminal conditions (simulating site conditions) revealed that transient voltages could be generated across the taps significantly in excess of those during the standard tests.

Because the tap changers are usually remote from the transformer line terminals, high frequency components of the incoming surges will not ordinarily find their way into the tap zone of the winding; instead the tap zone may experience transient overvoltages associated with the lower frequencies of winding resonance. It is also reported in [44, 48] that the part winding fundamental natural frequencies are proportional to the volt-ampere rating (per phase) raised to a power substantially less than one and inversely proportional to the voltage rating (phase value). Hence, as the voltage rating goes up in EHV transformers, natural frequencies may become substantially lower (few kHz), thus increasing the chances of part winding resonance.

Researchers have studied the generation of voltage transients in power systems leading to a winding resonance and the factors that might mitigate the phenomenon. The voltage transients are generated by switching operations of lines or other equipment nearby in the network. Use of closing resistors or point of wave switching during the switching operations may possibly mitigate the effects. A transformer should be designed such that it is as far as possible self-protecting against winding resonances. Some methods have been suggested to protect transformers from failures due to part winding resonance phenomenon involving tap windings [47, 49], viz. connection of external arrester to winding, use of shunt capacitors, and connection of nonlinear resistors in parallel with the tap winding.

Natural frequencies of core type transformers normally lie between 5 kHz to few hundred kHz, if one excludes the problem of very fast transients [50]. Values of natural frequencies do not vary much for transformers supplied by different manufacturers. There is always a chance that frequency of an external oscillating disturbance is close to any of the natural frequencies of the winding. The winding natural frequencies are determined by its parameters, and these cannot be changed beyond certain limits. In certain cases, where exact natural frequencies of the network can be determined (e.g., cable feeding transformer in a substation), it may be possible to change the winding type to avoid the possibility of a transformer natural frequency coming close to the external excitation frequency. However, every effort should be made to avoid network conditions which tend to produce oscillating voltages. If possible, parameters of the expected disturbances in the network should be made known to the transformer designer since unlimited requirements will make the transformer very costly. In this context, a closer cooperation between the users and manufacturers of transformers is desirable. The knowledge of precise transmission line propagation characteristics is essential for determining the amplitude of incoming surges. Also, the terminal conditions (loading, neutral grounding, etc.) of transformers have significant impact on the resonance phenomenon.

The resonance basically is the excitation of an oscillation in a winding by an external oscillating disturbance, both having frequencies very close to each other. Analysis of the failure of a generator step-up transformer due to internal resonance caused by an oscillating voltage excitation is reported in [51]. Oscillatory switching surges can be produced in a system by sudden voltage changes (switching operations, short line faults, etc.) at some distance from the transformer terminals. The natural frequency of the line is given by

$$f = \frac{v}{4L} \tag{7.101}$$

where v is the wave propagation velocity (300 $m/\mu s$ for overhead transmission lines and 100 $m/\mu s$ for cables) and L is length of line (distance of the transformer from the location where a switching operation or ground fault occurs). If the natural frequency of the line corresponds to the natural frequency of the winding, high internal overvoltages may develop. Hence, after knowing the dominant resonant frequency of the winding, the critical line length can be calculated from equation 7.101, at which the placement of circuit breaker should be avoided. It should be ensured as far as possible that a fault does not occur at this location.

The resonant overvoltages are basically determined by the winding design (arrangement and type of winding) and the damping (due to frequency dependent effective winding resistance). The calculation of the effective frequency dependent winding resistance, although quite laborious, is essential. Oscillations are significantly affected by the internal damping (winding and core losses) and external damping (line resistance); the amplitude of oscillations decreases with the increase in damping.

Possibility of resonant conditions in the windings can be known from the terminal and internal measurements. The resonances are of two types: terminal resonance and internal resonance. For a complicated non-uniform winding, the terminal response may not necessarily bear a direct relation to the internal response of a particular part of the winding. In other words, a part winding resonance may significantly influence transient oscillations of a major part of the winding but its effects may not be observed in the terminal impedance plot.

Although it was known long ago that high frequency resonances exist within the transformer windings, traditionally emphasis has been to check the response of transformers to pulse (unidirectional) test voltages since the sources of steady state high frequency excitation were not envisaged in the power system. Hence, the standards have been based on pulse shapes thought to be reasonably representing transient overvoltages in the system. Designers in earlier days worried only about the standard voltage shapes and not about the calculation of internal resonant frequencies which the nonstandard wave shapes may excite. The pattern of oscillations in transformer windings is not very orderly due to non-uniformities in insulation and turn distributions. It is obvious that in order to check the withstand capability against winding resonances, there is no point in increasing power frequency test voltage as it will not lead to a local high stress concentration in the winding. Also, it is observed in most of the cases that under standard test conditions, the front of standard switching impulse waveform may rise too slowly or the tail of standard lightning impulse waveform may fall too rapidly to excite the internal part winding resonance as compared to the possibility of such resonance in actual service with a lower magnitude of surge after the occurrence of a line fault. Hence, the withstand can be possibly checked by suitably modifying the impulse wave shape (front and tail). It has been reported in [51] that certain aperiodic overvoltage wave shapes such as a fast-front long-tail switching overvoltages can result into high internal voltage stresses in transformer windings.

7.8.2 Natural frequencies of windings

For eliminating the possibility of resonance in windings, an accurate determination of transformer's frequency response characteristics is essential. These characteristics can be determined by actual measurement which is the oldest and reliable technique. Its disadvantage is that the response cannot be predicted at the design stage and it is difficult to measure the internal winding response unless the winding insulation is pierced and damaged for the use of conductively coupled probes. Capacitively coupled non-destructive probes can be used after careful scrutiny of their accuracy and precision. In [52], authors have used an equivalent circuit of winding containing lumped inductances and capacitances for determining natural frequencies. It is shown that the mutual inductances between all sections of the winding must be taken into account to determine correctly the natural frequencies. Subsequently, the method of equivalent circuit has been used [53] for finding oscillations of coupled windings, in which the natural frequencies are determined for a primary winding with secondary winding short-circuited. The natural frequencies of three-phase transformer windings are calculated in [54] to take into account the effects of the capacitive and inductive couplings between the windings of different phases.

Another method is to use an electromagnetic model of the transformer [55], in which a scaled model is used to determine the natural frequencies and the voltage response; the obvious disadvantage being the high cost and time involved in building a new scale model for each transformer of interest. Hence, the most convenient and economical method is to determine the frequencies through simulations by computers. A numerical method has been presented in [56] for determining the terminal and internal impedance versus frequency characteristics for a general lumped parameter network, using which the resonant frequency characteristics and amplification factors are calculated. If a transformer winding is represented by a lumped parameter network, its response will be different than that of the actual one. In practice, it is sufficient to choose the number of sections in the winding representation somewhat larger than the number of required resonant frequencies [52, 56].

7.8.3 Graded capacitance winding

The development of interleaved windings is an important milestone in the history of power transformers. Although the interleaving improves dramatically the voltage distribution in the main winding, the tap (regulating) winding, if present, may get subjected to very high local voltages due to the part winding resonance. Thus, the improvement in surge voltage response obtained by interleaving may be offset by the overvoltages on account of the part winding resonance. When the main winding is fully interleaved, design of tap winding and tap changer becomes critical in high voltage transformers since the voltage across the tap winding may reach unacceptable levels. The surge performance of a power transformer having taps on HV winding has been analyzed in [57] for two cases, viz. interleaved winding and non-interleaved continuous disk winding. For the case of interleaved winding, where both HV main winding and its corresponding tap winding are of interleaved type, it is reported that the voltages (with respect to ground) of various points in the HV main winding are almost linearly distributed along the length of the winding indicating a marked improvement as compared to the non-interleaved type. For both designs, the voltage across the tap winding is shown to have a oscillatory behavior, but in the interleaved design there is no attenuation suggesting a part winding resonance in the tap winding. The peak value of voltage is practically limited by the winding resistance under the resonance conditions. It has been proved that because of high series capacitance due to interleaving, the resonant frequency of tap winding disks has reduced to a value of 22 kHz which is close to the excitation frequency for the standard 1.2/50 microseconds impulse wave (the impulse wave has a frequency of about 20 kHz when it reaches the tap winding). The study reported suggests that in order to eliminate the possibility of resonance conditions, resonant frequencies of different parts of the winding should be determined at the design stage, and also the winding response for a variety of input voltage waveforms (covering wide range of frequencies of practical importance) should be studied.

Thus, interleaving may not always be the right solution for high voltage windings and may lead to high voltages in some parts of the windings. Usually, it is thought that in order to improve the voltage distribution, it is always better to have the main winding as totally interleaved type. If the tap winding is of non-interleaved type, due to substantial increase of impedance (on account of less series capacitance), the voltage across the tap winding is observed to be higher. Hence, a designer may think to make the tap winding also as interleaved, in which case for specific winding parameters there is a possibility of a part winding resonance as explained earlier. One of the better options can be the use of winding with graded series capacitance. The series capacitance of the main winding can be gradually reduced in 2 or 3 steps (by the change in degree of interleaving) and its neutral end part (electrically adjacent to the tap winding) can be the continuous disk winding. In this case, the tap winding can also be of continuous disk nature. The part winding resonance phenomenon can be damped to a great extent by using the graded interleaving technique in some typical designs. The graded interleaving for a winding with two parallel conductors can be obtained by using the conductor interleaving (figure 7.7) at line end and the turn interleaving (figure 7.6) for subsequent disks. In the case of turns with only one conductor, interleaving schemes shown in figure 7.23 [58] and figure 7.24 [8] can be used. The first type of interleaving, in which four disks are required to complete the interleaving, results in a much higher capacitance as compared to that in figure 7.6 which is the two-disk interleaving method. Its capacitance can be easily calculated by the method given in Section 7.3.5. The second type is one-disk interleaving method, which results in less capacitance as compared to the two-disk interleaving method.

Although the series capacitance increases with the degree of interleaving, a marginal improvement in response may be achieved beyond a certain limit. Also the winding process becomes more difficult; hence interleaving methods involving more than four disks are rarely used in practice. The importance and usefulness of graded capacitance (higher capacitance at the line end which is reduced in steps towards the neutral end) in the case of interleaved windings and shielded-conductor windings have been verified in [59]. It may be more advantageous from the point of voltage distribution to have a graded series capacitance rather than a high series capacitance throughout the winding.

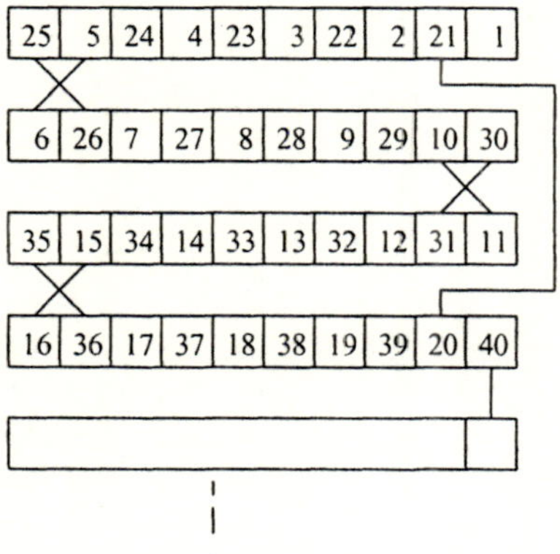

Figure 7.23 Four-disk interleaving

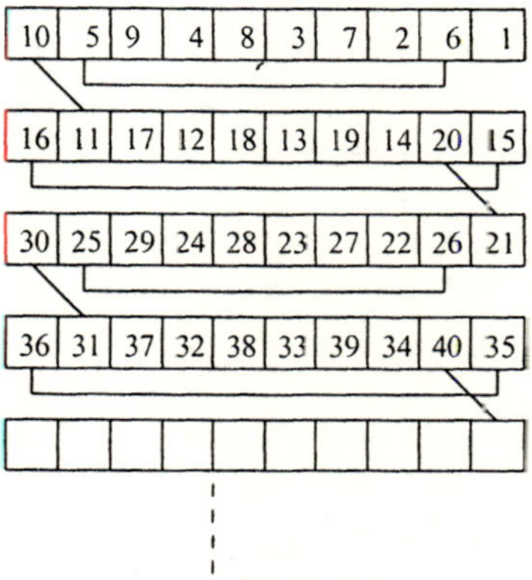

Figure 7.24 One-disk interleaving

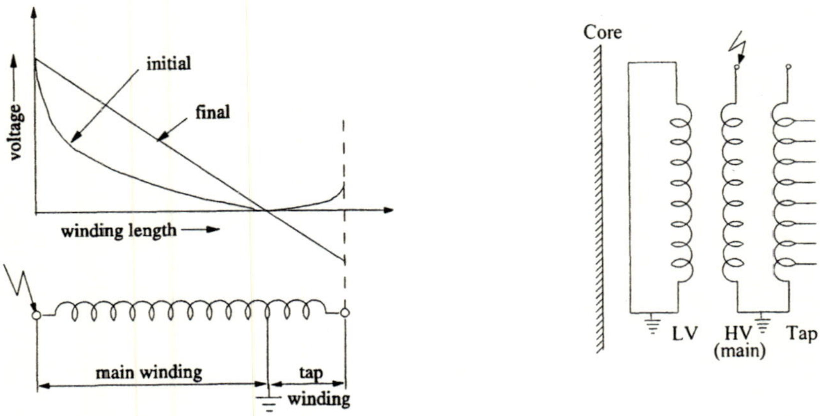

Figure 7.25 Tap winding with an open end

One useful arrangement could be two-disk interleaving at line end followed by one-disk interleaving followed by continuous disk type with the neutral end tap winding also as the continuous disk winding. The phenomenon of getting higher voltages due to sudden impedance (capacitance) change can be mitigated by means of a graded capacitance arrangement.

In the case of tap winding with a large tapping range, the problem of higher stress is more severe at its open end during the impulse test condition wherein the entire tap winding is out of circuit. The corresponding configuration and the typical voltage distribution are shown in figure 7.25. The large difference between the initial and final voltage distributions at the open end is responsible for large voltage oscillations. The types of HV main winding and tap winding can be judiciously selected to avoid a high voltage buildup in the tap winding as explained earlier. A reverse graded interleaving method is proposed in [60] to reduce the voltage buildup, in which the degree of interleaving is reduced from the line end to the tap end for the main winding, whereas for the tap winding it is increased from the main winding end to the open end. The increase of series capacitance at the open end reduces substantially the voltage stresses there.

7.8.4 Location of windings

The transient voltages appearing across the tap winding depend up on its design and position with respect to the main winding. One of the effective ways of reducing the high impulse voltages across the tap winding or between the tap winding and ground, is to have the tap winding located between the core and LV winding. The inner tap winding is usually of an interleaved type (as described in Section 7.3.8) having a high series capacitance reducing the impulse voltage across it.

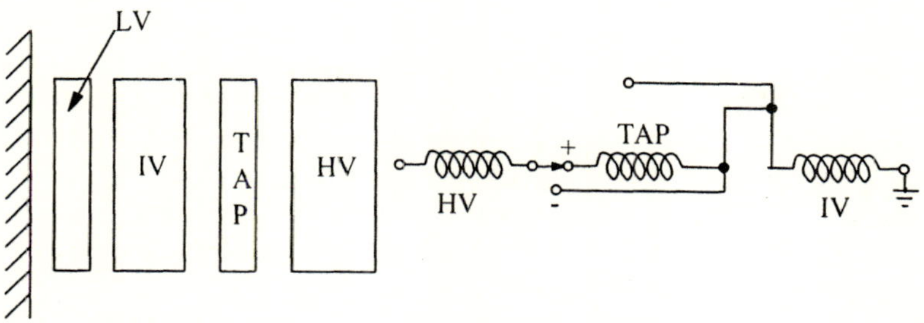

Figure 7.26 Effect of tap winding location in autotransformers

In autotransformers, if the tap winding is kept between HV (series) and IV (common) windings as shown in figure 7.26, the tap winding (with taps at the line end of IV winding) acts as a shield for IV winding and improves the voltage distribution significantly at the line end of IV winding (during the impulse test on IV winding). However, insulation of neutral end disks of IV winding may have to be strengthened (by special insulation components) since these face the tap winding which is at much higher potential.

References

1. Blume, L. F. and Boyajian, A. Abnormal voltages within transformers, *AIEE Transactions*, Vol. 38, February 1919, pp. 577-614.
2. Thomas, H. L. Insulation stresses in transformers with special reference to surges and electrostatic shielding, *Journal IEE*, Vol. 84, 1940, pp. 427-443.
3. Norris, E. T. The lightning strength of power transformers, *Journal IEE*, Vol. 95, Pt. II, 1948, pp. 389-406.
4. Heller, B. and Veverka, A. *Surge phenomena in electrical machines*, Iliffe Books Ltd., London, 1968.
5. Chadwik, A. T., Ferguson, J. M., Ryder, D. H., and Stearn, G. F. Design of power transformers to withstand surges due to lightning, with special reference to a new type of winding, *Proceedings IEE*, Pt. II, Vol. 97, 1950, pp. 737-750.
6. Grimmer, E. J. and Teague, W. L. Improved core form transformer winding, *AIEE Transactions*, Vol. 70, 1951, pp. 962-967.
7. Pedersen, A. On the response of interleaved transformer windings to surge voltages, *AIEE Transactions*, Vol. 82, June 1963, pp. 349-356.
8. Van Nuys, R. Interleaved high-voltage transformer windings, *IEEE Transactions on Power Apparatus and Systems*, Vol. PAS-97, No. 5, September/October 1978, pp. 1946-1954.
9. Teranishi, T., Ikeda, M., Honda, M., and Yanari, T. Local voltage oscillation in interleaved transformer windings, *IEEE Transactions on Power Apparatus and Systems*, Vol. PAS-100, No. 2, 1981, pp. 873-881.

10. Jayaram, B. N. The series capacitance of transformer windings, *Electrotechnics*, Indian Institute of Science, No. 28, 1961, pp. 69-87.
11. Jayaram, B. N. The equivalent series capacitance of a single disk-coil in a transformer winding, (in German), Elektrotechnische Zeitschrift – A, Vol. 94, 1973, pp. 547-548.
12. Kawaguchi, Y. Calculation of circuit constants for computing internal oscillating voltage in transformer windings, *Electrical Engineering in Japan*, Vol. 89, No. 3, 1969, pp. 44-53.
13. Jayaram, B. N. Determination of impulse distribution in transformers with a digital computer, (in German), Elektrotechnische Zeitschrift – A, Vol. 82, January 1961, pp. 1-9.
14. Karsai, K., Kerenyi, D., and Kiss, L. *Large power transformers*, Elsevier Publication, Amsterdam, 1987.
15. Del Vecchio, R. M., Poulin, B., and Ahuja, R. Calculation and measurement of winding disk capacitances with wound-in-shields, *IEEE Transactions on Power Delivery*, Vol. 13, No. 2, April 1998, pp. 503-509.
16. Chowdhuri, P. Calculation of series capacitance for transient analysis of windings, *IEEE Transactions on Power Delivery*, Vol. PWRD-2, No. 1, January 1987, pp. 133-139.
17. Weed, J. M. Prevention of transient voltage in windings, *AIEE Transactions*, February 1922, pp. 149-159.
18. Wilcox, D. J., Hurley, W. G., and Conlon, M. Calculation of self and mutual impedances between sections of transformer windings, *Proceedings IEE*, Vol. 136, Pt. C, No. 5, September 1989, pp. 308-314.
19. Miki, A., Hosoya, T., and Okuyama, K. A calculation method for impulse voltage distribution and transferred voltage in transformer windings, *IEEE Transactions on Power Apparatus and Systems*, Vol. PAS-97, No. 3, May/June 1978, pp. 930- 939.
20. Grover, F. W. *Inductance calculations: Working formulae and tables*, Van Nostrand Company, Inc., 1947.
21. Wirgau, K. A. Inductance calculation of an air-core disk winding, *IEEE Transactions on Power Apparatus and Systems*, Vol. PAS-95, No. 1, January/February 1976, pp. 394-400.
22. Okuyama, K. A numerical analysis of impulse voltage distribution in transformer windings, *Electrical Engineering in Japan*, Vol. 87, 1967, pp. 80-88.
23. Krondl, M. and Schleich, A. Predetermination of the transient voltages in transformers subject to impulse voltage, *Bulletin Oerlikon*, No. 342/343, December 1960, pp. 114-133.
24. McWhirter, J. H., Fahrnkopf, C. D., and Steele, J. H. Determination of impulse stresses within transformer windings by computers, *AIEE Transactions*, Vol. 75, Pt. III, February 1957, pp. 1267-1279.

25. Honorati, O. and Santini, E. New approach to the analysis of impulse voltage distribution in transformer windings, *Proceedings IEE*, Vol. 137, Pt. C, No. 4, July 1990, pp. 283-289.
26. Rudenberg, R. Performance of traveling waves in coils and windings, *AIEE Transactions*, Vol. 59, 1940, pp. 1031-1040.
27. Bewley, L. V. *Traveling waves on transmission lines*, John Wiley and Sons, Inc., New York, 1951.
28. Allibone, T. E., McKenzie, D. B., and Perry, F. R. The effects of impulse voltages on transformer windings, *Journal IEE*, Vol. 80, No. 482, February 1937, pp. 117-173.
29. Abetti, P. A. and Maginniss, F. J. Fundamental oscillations of coils and windings, *AIEE Transactions*, February 1954, pp. 1-10.
30. Rudenberg, R. Surge characteristics of two-winding transformers, *AIEE Transactions*, Vol. 60, 1941, pp. 1136-1144.
31. Glaninger, P. Modal analysis as a means of explaining the oscillatory behavior of transformers, *Brown Boveri Review*, 1-86, pp. 41-49.
32. Abetti, P. A. Correlation of forced and free oscillations of coils and windings, *AIEE Transactions*, December 1959, pp. 986-996.
33. Lewis, T. J. The transient behavior of ladder networks of the type representing transformer and machine windings, *Proceedings IEE*, Vol. 101, Pt. II, 1954, pp. 541-553.
34. Fergestad, P. I. and Henriksen, T. Transient oscillations in multi-winding transformers, *IEEE Transactions on Power Apparatus and Systems*, Vol PAS-93, 1974, pp. 500-509.
35. Kasturi, R. and Murty, G. R. K. Computation of impulse voltage stresses in transformer windings, *Proceedings IEE*, Vol. 126, No. 5, May 1979, pp. 397-400.
36. Carlin, H. J. and Giordano, A. B. *Network theory*, Prentice-Hall, Inc., 1964.
37. De Leon, F. and Semlyen, A. Complete transformer model for electromagnetic transients, *IEEE Transactions on Power Delivery*, Vol. 9, No. 1, January 1994, pp. 231-239.
38. Abetti, P. A. Electrostatic voltage distribution and transfer in three-winding transformers, *AIEE Transactions*, December 1954, pp. 1407-1416.
39. Abetti, P. A. and Davis, H. F. Surge transfer in three-winding transformers, *AIEE Transactions*, December 1954, pp. 1395-1407.
40. Koppikar, D. A. and Vijayan, K. Transferred surge voltage in transformers, *International Conference on Transformers, TRAFOTECH - 94*, Bangalore, January 1994, pp. I21-I24.
41. Gupta, S. C. and Singh, B. P. Determination of the impulse voltage distribution in windings of large power transformers, *Electric Power Systems Research*, Vol. 25, 1992, pp. 183-189.

42. Fergestad, P. I. and Henriksen, T. Inductances for the calculation of transient oscillation in transformers, *IEEE Transactions on Power Apparatus and Systems*, Vol. PAS-93, 1974, pp. 510-517.
43. Munshi, S., Roy, C. K., and Biswas, J. R. Computer studies of the performance of transformer windings against chopped impulse voltages, *Proceedings IEE,* Vol. 139, Pt. C, No. 3, May 1992, pp. 286-294.
44. McNutt, W. J., Blalock, T. J., and Hinton, R. A. Response of transformer windings to system transient voltages, *IEEE Transactions on Power Apparatus and Systems*, Vol. PAS-93, March/April 1974, pp. 457-466.
45. Phillips, C. L. and Harbor, R. D. *Feedback control systems*, Prentice-Hall, Inc., 1996.
46. Waldvogel, P. and Rouxel, R. A new method of calculating the electric stresses in a winding subjected to a surge voltage, *The Brown Boveri Review*, Vol. 43, No. 6, June 1956, pp. 206-213.
47. Margolis, H. B., Phelps, J. D. M., Carlomagno, A. A., and McElroy, A. J. Experience with part-winding resonance in EHV auto-transformers: diagnosis and corrective measures, *IEEE Transactions on Power Apparatus and Systems*, Vol. PAS-94, No. 4, July/August 1975, pp. 1294-1300.
48. McElroy, A. J. On the significance of recent EHV transformer failures involving winding resonance, *IEEE Transactions on Power Apparatus and Systems*, Vol. PAS-94, No. 4, July/August 1975, pp. 1301-1307.
49. Teranishi, T., Ebisawa, Y., Yanari, T., and Honda, M. An approach to suppressing resonance voltage in transformer tap windings, *IEEE Transactions on Power Apparatus and Systems,* Vol. PAS-102, No. 8, August 1983, pp. 2552-2558.
50. Preininger, G. Resonant overvoltages and their impact on transformer design, protection and operation, *International Summer School on Transformers*, *ISST'93*, Technical University of Lodz, Poland, 1993, Paper No. 11.
51. Musil, R. J., Preininger, G., Schopper, E., and Wenger, S. Voltage stresses produced by aperiodic and oscillating system overvoltages in transformer windings, *IEEE Transactions on Power Apparatus and Systems*, Vol. PAS-100, No. 1, January 1981, pp. 431-441.
52. Abetti, P. A. and Maginniss, F. J. Natural frequencies of coils and windings determined by equivalent circuit, *AIEE Transactions*, June 1953, pp. 495-503.
53. Abetti, P. A., Adams, G. E., and Maginniss, F. J. Oscillations of coupled windings, *AIEE Transactions*, April 1955, pp. 12-21.
54. Gururaj, B. I. Natural frequencies of 3-phase transformer windings, *AIEE Transactions,* June 1963, pp. 318-329.
55. Abetti, P. A. Transformer models for determination of transient voltages, *AIEE Transactions*, June 1953, pp. 468-480.

56. Degeneff, R. C. A general method for determining resonances in transformer windings, *IEEE Transactions on Power Apparatus and Systems,* Vol. PAS-96, No. 2, March/April 1977, pp. 423-430.
57. De, A. and Chatterjee, N. Part winding resonance: demerit of interleaved high-voltage transformer winding, *Proceedings IEE - Electric Power Applications*, Vol. 147, No. 3, May 2000, pp. 167-174.
58. Schleich, A. Behaviour of partially interleaved transformer windings subjected to impulse voltages, *Bulletin Oerlikon*, No. 389/390, pp. 41-52.
59. Okuyama, K. Effect of series capacitance on impulse voltage distribution in transformer windings, *Electrical Engineering in Japan*, Vol. 87, 1967, pp. 27-34.
60. De, A. and Chatterjee, N. Graded interleaving of EHV transformers for optimum surge performance, *International Symposium on High Voltage Engineering, ISH-2001*, Bangalore, Paper No. 6-30, pp. 916-919.

8

Insulation Design

Insulation design is one of the most important aspects of the transformer design. It is the heart of transformer design, particularly in high voltage transformers. Sound design practices, use of appropriate insulating materials, controlled manufacturing processes and good house-keeping ensure quality and reliability of transformers. Comprehensive verification of insulation design is essential for enhancing reliability as well as for material cost optimization.

With the steady increase in transmission system voltages, the voltage ratings of power transformers have also increased making insulation content a significant portion of the transformer cost. Also, insulation space influences the cost of active parts like core and copper, as well as the quantity of oil in the transformer, and hence has a great significance in the transformer design. Moreover, it is also environmentally important that we optimize the transformer insulation which is primarily made out of wood products. In addition, with the associated increase in MVA ratings, the weight and size of large transformers approach or exceed transport limits. These reasons together with the ever-increasing competition in the global market are responsible for continuous efforts to reduce insulation content in transformers. In other words, margin between withstand levels and operating stress levels is reducing. This requires greater efforts from researchers and designers for accurate calculation of stress levels at various critical electrode configurations inside the transformer under different test voltage levels and different test connections. Advanced computational tools (e.g., FEM) are being used for accurate calculation of stress levels. These stress levels are compared with withstand levels which are established based on experimental / published data.

For the best dielectric performance, reduction in maximum electric stress in insulation is usually not enough; the following factors affecting the withstand characteristics should be given due consideration, viz. waveform of applied voltage and corresponding response, volt-time characteristics of insulation, shape and surface condition of electrodes, partial discharge inception characteristics of insulation, types of insulating mediums, amount of stressed volume, etc. Minimization of non-uniform dielectric fields, avoiding creepage stress, improvement in oil processing and impregnation, elimination of voids, elimination of local high stresses due to winding connections / crossovers / transpositions, are some of the important steps in the insulation design of transformers. Strict control of manufacturing processes is also important. Manufacturing variations of insulating components should be monitored and controlled. Proper acceptance norms and criteria have to be established by the manufacturers for the insulation processing carried out before high voltage tests.

The transformer insulation system can be categorized into major insulation and minor insulation. The major insulation consists of insulation between windings, between windings and limb / yoke, and between high voltage leads and ground. The minor insulation consists of basically internal insulation within the windings, viz. inter-turn and inter-disk insulation. The chapter gives in details the methodology of design of the major and minor insulations in transformers. Various methods for field computations are described. The factors affecting the insulation strength are discussed. In transformers with oil-solid composite insulation system, two kinds of failures usually occur. The first kind involves a complete failure between two electrodes (which can be jump / bulk-oil breakdown, creepage breakdown along oil-solid interface or combination of both). The second one is a local oil failure (partial discharge), which may not immediately lead to failure between two electrodes. Sustained partial discharges lead to deterioration of the insulation system eventually leading to a failure. The chapter discusses these failures and countermeasures to avoid them. It also covers various kinds of test levels and method of conversion of these to an equivalent Design Insulation Level (DIL) which can be used to design major and minor insulation systems. Statistical methods for optimization and reliability enhancement are also introduced.

8.1 Calculation of Stresses for Simple Configurations

For uniform fields in a single dielectric material between bare electrodes, the electric stress (field strength) is given by the voltage difference between the electrodes divided by the distance between them,

$$E_u = \frac{V}{d} \tag{8.1}$$

The above equation is applicable to, for example, a parallel plate capacitor with one dielectric.

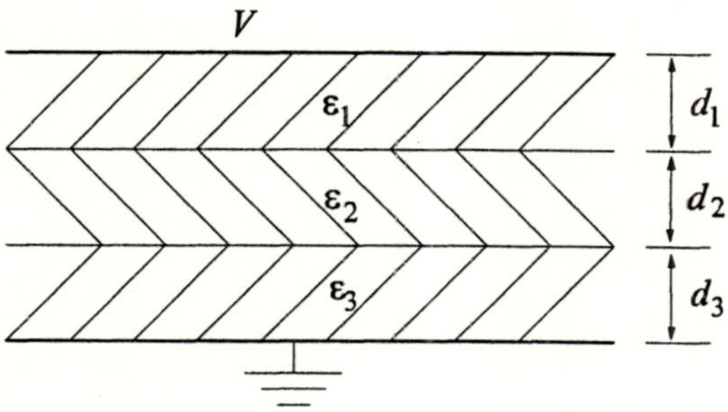

Figure 8.1 Multi-dielectric configuration

For non-uniform fields (e.g., cylindrical conductor – plane configuration), the stress (E_{nu}) is more at the conductor surface; the increase in stress value as compared to that under the uniform field condition is characterized by a non-uniformity factor (η),

$$\eta = \frac{E_{nu}}{E_u} \tag{8.2}$$

The non-uniformity factor is mainly a function of electrode configuration.

For a multi-dielectric case between two parallel plates shown in figure 8.1, the stress in any dielectric for a potential difference of V between the plates is

$$E_i = \frac{V}{\varepsilon_i\left(\frac{d_1}{\varepsilon_1}+\frac{d_2}{\varepsilon_2}+\frac{d_3}{\varepsilon_3}\right)} \qquad i = 1, 2, 3 \tag{8.3}$$

where ε_i is relative permittivity of ith dielectric. This expression for the configuration of parallel plates can be derived by using the fact that the stress is inversely proportional to permittivity. The stress value is constant within any dielectric.

For two concentric cylindrical electrodes of radii r_1 and r_2, with a single dielectric between them as shown in figure 8.2, the stress in the dielectric is not constant and varies with radius. The stress at any radius r ($r_1 < r < r_2$) is

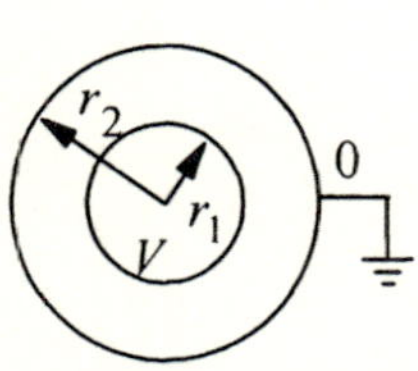

Figure 8.2 Concentric cylindrical electrodes

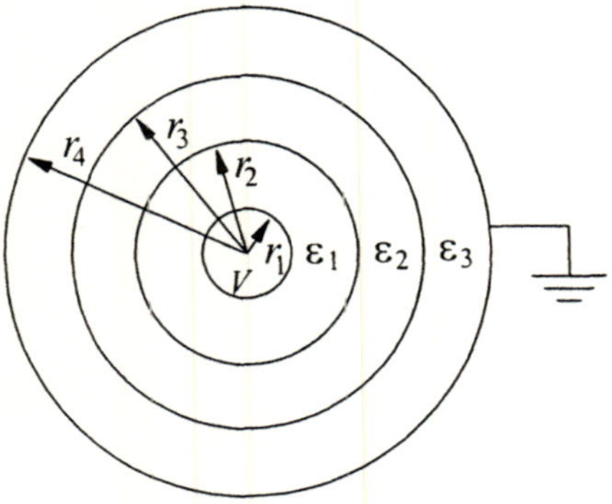

Figure 8.3 Concentric cylindrical electrodes with multiple dielectrics

$$E = \frac{V}{r \ln \frac{r_2}{r_1}} \tag{8.4}$$

and the maximum stress occurs at the inner electrode surface given by

$$E_{max} = \frac{V}{r_1 \ln \frac{r_2}{r_1}} \tag{8.5}$$

For the multi-dielectric case shown in figure 8.3, the stress at any radius r is

$$E_i = \frac{V}{r\varepsilon_i \left[\frac{1}{\varepsilon_1} \ln \frac{r_2}{r_1} + \frac{1}{\varepsilon_2} \ln \frac{r_3}{r_2} + \frac{1}{\varepsilon_3} \ln \frac{r_4}{r_3} \right]} \qquad i = 1, 2, 3 \tag{8.6}$$

For the cylindrical conductor - plane configuration (figure 8.4), formulae for stresses at the conductor surface and plane are derived in appendix B. The maximum stress on the conductor surface occurs at point P (along the shortest distance between the two electrodes) which is given by the expression (equation B14 of appendix B),

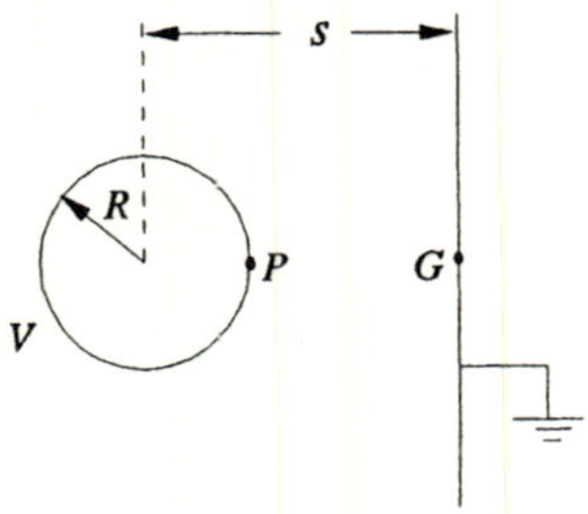

Figure 8.4 Cylindrical conductor - plane configuration

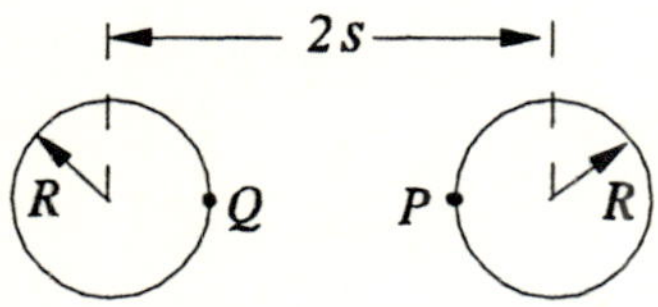

Figure 8.5 Stress between two bare cylindrical conductors

$$E_P = \frac{V}{(s-R)} \frac{\sqrt{(s/R)^2 - 1}}{\ln\left[\sqrt{(s/R)^2 - 1} + (s/R)\right]} \tag{8.7}$$

and the maximum stress at the plane occurs at point G, which is given by the expression (equation B19),

$$E_G = \frac{V}{(s-R)} \frac{2 \times \sqrt{((s/R)-1)/((s/R)+1)}}{\ln\left[\sqrt{(s/R)^2 - 1} + (s/R)\right]} \tag{8.8}$$

In the previous two equations, the factor multiplying the term $(V/(s-R))$ is non-uniformity factor. For calculation of stress at any other point along the shortest distance, equation B22 can be used.

For the configuration of two bare cylindrical conductors shown in figure 8.5 with a potential difference V between them, the maximum electric stress occurring at points P and Q is given as (from equations B12 and B13)

$$E = \frac{V}{2\,(s-R)} \frac{\sqrt{(s/R)^2 - 1}}{\ln\left[\sqrt{(s/R)^2 - 1} + (s/R)\right]} \tag{8.9}$$

The above equation is applicable when the electrostatic field between the two conductors is not influenced by any other boundary condition (the case of two isolated conductors).

Thus, for bare leads of equal radii, the configuration is equivalent to considering a potential difference of $(V/2)$ applied between one conductor and plane at a distance of s from the conductor center (cylindrical conductor - plane configuration of figure 8.4).

For the configuration of paper insulated cylindrical conductor (e.g., insulated high voltage lead in the transformer) and plane shown in figure 8.6, the maximum stress in oil at the surface of the covered conductor (at point A) is

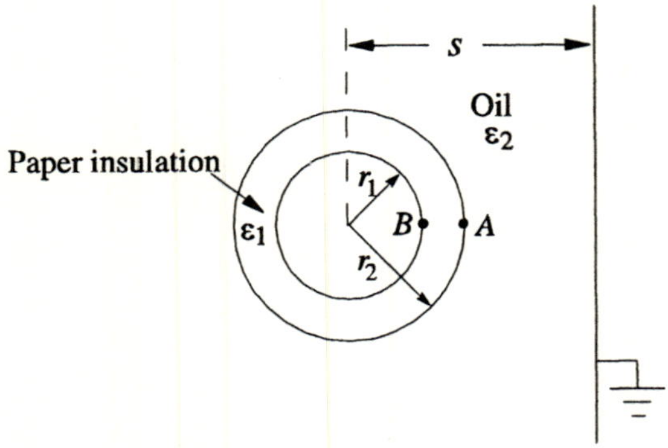

Figure 8.6 Stress between insulated lead and ground

$$E_A = \frac{V}{r_2\varepsilon_2\left\{\dfrac{\left(\dfrac{s}{r_2}-1\right)\ln\left[\left(\dfrac{s}{r_2}\right)+\sqrt{\left(\dfrac{s}{r_2}\right)^2-1}\right]}{\varepsilon_2\sqrt{\left(\dfrac{s}{r_2}\right)^2-1}}+\dfrac{\left(\dfrac{r_2}{r_1}-1\right)\ln\left[\left(\dfrac{r_2}{r_1}\right)+\sqrt{\left(\dfrac{r_2}{r_1}\right)^2-1}\right]}{\varepsilon_1\sqrt{\left(\dfrac{r_2}{r_1}\right)^2-1}}\right\}} \tag{8.10}$$

Similarly, the maximum stress in the paper insulation at the conductor surface (at point B) is given by the expression,

$$E_B = \frac{V}{r_1\varepsilon_1\left\{\dfrac{\left(\dfrac{s}{r_2}-1\right)\ln\left[\left(\dfrac{s}{r_2}\right)+\sqrt{\left(\dfrac{s}{r_2}\right)^2-1}\right]}{\varepsilon_2\sqrt{\left(\dfrac{s}{r_2}\right)^2-1}}+\dfrac{\left(\dfrac{r_2}{r_1}-1\right)\ln\left[\left(\dfrac{r_2}{r_1}\right)+\sqrt{\left(\dfrac{r_2}{r_1}\right)^2-1}\right]}{\varepsilon_1\sqrt{\left(\dfrac{r_2}{r_1}\right)^2-1}}\right\}} \tag{8.11}$$

At any other point in this geometry and for more complicated electrode configurations, analytical or numerical techniques should be used for accurate field computations as described in the next section.

8.2 Field Computations

8.2.1 Analytical methods

For estimating electric stress levels at various critical electrodes, it is necessary to find electrostatic field distribution. The field distribution can be found by a variety of methods. Classical methods such as method of images give quite accurate results whenever they can be applied. For complex configurations, which exist inside a transformer, these methods cannot be applied. Initially, transformer designers had to depend on analog methods in which conducting paper and electrolytic tank analogs were used [1]. Before the advent of computers and numerical methods, these methods were widely used for multi-electrode and multi-dielectric material systems of transformers with two-dimensional approximations of the problem. Stressed oil volume, required for estimation of strength, was also calculated by direct plotting of equigradient lines on a conducting paper analog by using suitable instrumentation [2]. Analog methods are inconvenient, inaccurate, expensive, and are limited in their application. They may not be relevant now due to the rapid development of computational techniques.

A conformal mapping technique such as the Schwarz-Christoffel transformation has also been widely used for relatively simple geometries within the transformers [3, 4]. In this method, the whole region of interest is mapped into a new plane in which the solution is constructed involving unknown constants in the transformation equation. The unknown constants are calculated by solving a set of nonlinear equations which describe the boundaries of the region in the original plane. Curved boundaries can also be handled in this method. Although the method is suitable for regions with a single dielectric material, for multi-dielectric problems an approximate solution can be obtained by converting them into a single dielectric region by using equivalent insulation distances. The method is best suited for a simply connected region containing few electrodes. For multiple connected regions with complicated electrode shapes and multiple dielectrics, this method is not suitable.

8.2.2 Numerical methods

In many cases, physical systems are so complex that analytical solutions are difficult or impossible, and hence numerical methods are commonly used for field computations. A numerical technique, Finite Difference Method (FDM), is used in [5, 6] for the field computations. It results into a set of linear equations which are solved by direct matrix methods or iterative methods. FDM gives accurate results and can handle curved boundaries accurately if large number of points (fine grid) is taken on the boundary. Its main disadvantage is that the solution (potential distribution) is available at discrete points only, and hence the method presents some difficulties where quantities like stressed areas / volumes are required to be calculated [7].

One of the most powerful and popular numerical techniques these days is FEM. It is in use for electrostatic field computations since the last three decades [8]. Usefulness of the method has already been demonstrated for magnetostatic and eddy current problems in the earlier chapters. At locations where the field is changing sharply, higher order polynomials can be used to approximate the potential distribution within the corresponding elements and / or fine mesh can be used. As the method yields a set of linear equations, solution can be obtained by direct matrix methods or iterative methods. The electric stress in any element is calculated by differentiating the approximated polynomial function. The stressed area between two equigradient lines can be derived by finding the elements in which the stresses are within the two limits of stress values.

Many adopt charge simulation method (CSM) for electric field computations because it can solve unbounded regions and has high accuracy [9]. In this method, physically distributed charges on the conductor surface are replaced by discrete fictitious line charges placed outside the space in which the field distribution is to be computed. The magnitude of these fictitious charges is calculated in order that their integrated effect satisfies the boundary conditions exactly, at some selected number of points on the boundary. The method requires proper selection and placement of a large number of charges for a good accuracy. For example, a distributed charge on the surface of high voltage electrode can be replaced by k line charges placed inside the electrode. For determining the magnitude of these charges, k points are chosen on the surface of the electrode and the condition to be satisfied is as follows. At each of these points on the electrode surface, the potential resulting from the superposition of these fictitious charges should be equal to the conductor electrode potential V_c,

$$\sum_{i=1}^{k} P_i \cdot Q_i = V_c \tag{8.12}$$

where P_i is potential coefficient and Q_i is discrete fictitious line charge. When the above equation is applied to all the selected points k, we get a system of k linear equations which are solved to get the magnitudes of k charges. The electric field value at any point in the domain of interest can be determined easily by the superposition method using these values of charges. Thus, although CSM has distinct advantages of its applicability to unbounded regions and reasonable computational efforts, it is not well suited for complex electrode configurations with a number of dielectric materials. On the contrary, FEM is most suitable for complex problems but for bounded regions. For electrodes with very small radius, because of the limitation on the smallest size of element that can be used and the approximation of curved path by small line segments, the accuracy of FEM may not be the best. Hence, advantages of CSM and FEM can be combined with elimination of their disadvantages in the combination method as reported in [10]. In this method, the entire problem space is divided into two parts; CSM is used mainly for the open space with infinite boundary and FEM is used for the finite enclosed space.

8.3 Factors Affecting Insulation Strength

The breakdown voltage of a dielectric material is a statistically distributed quantity which is a function of its physical / chemical properties and impurities present in it. Failures may not be always initiated by higher electrical stresses; interrelated thermal, chemical and mechanical factors may also have significant influence on the breakdown processes. As compared to metals, insulating materials exhibit an erratic behavior. With the ageing and / or deterioration of electrical and mechanical properties, it becomes even more difficult to predict their performance. In transformers, composite oil-solid insulation system is used. The erratic behavior of transformer oil is pronounced when used alone. There is a much larger scatter of breakdown voltage for oil as compared to a smaller scatter observed for air. The very large scatter of the oil gap breakdown voltage may be associated with the random path of streamers and variations in their progress in the oil [11]. Hence, larger oil ducts are always subdivided by solid insulation into smaller ducts due to which the transformer insulation system becomes more dependable and stable.

Compared with breakdown processes in gases, little is known about the processes which initiate and lead to breakdowns in the oil. General models using *micro-bubble* and *weak-link theory* have been attempted. It is reported in the literature that some micro-bubbles exist in the oil even in the absence of electric field, and the application of field creates additional bubbles. It is suggested that discharges are ignited in these micro-bubbles. Due to dielectrophoretic forces, particles / impurities are swept from surrounding oil regions to the points of highest stress in the oil gap [12]. These particles then tend to line up along the electric field lines to create a *weak-link* in the oil gap; this phenomenon is accentuated in the presence of moisture. Transformer designers use to a great extent semi-empirical data for the insulation design as there is still no coherent theory of oil breakdown.

8.3.1 Effect of moisture and impurities

Needless to say, moisture and other impurities have significant deteriorating effect on the dielectric strength of the transformer insulation. The moisture has deteriorating effect on both electrical and mechanical properties of the insulation. As the moisture content in oil increases, strength reduces drastically till the saturation point, after which there is no appreciable further deterioration of the strength. Hence, percentage saturation is the decisive factor influencing the dielectric strength of the transformer oil [13, 14]. The degrading effect of moisture content is also significantly affected by the amount of other impurities present in the oil [15]. The presence of solid impurities makes the deteriorating effect (of moisture on strength) more significant even at quite low moisture content in the oil. The solid insulation has more affinity (as compared to the oil) for moisture. It has been reported in [16] that at room temperature, the reduction in dielectric strength of the oil due to presence of cellulose particles gets amplified at higher moisture content.

An increase of pressure or temperature, increases the quantity of gas the oil can hold. If the oil temperature rises owing to increase of ambient temperature or load, the oil expands and the pressure increases. When the pressure falls, the oil has more gas content than it can hold. The excess gases eventually diffuse out of the oil after some time (few days or weeks) depending on the ratio of the oil surface exposed to gas and the total oil volume. If the pressure drops suddenly the gas bubbles may get formed in the oil, reducing the dielectric strength [15].

The dielectric strength of paper insulation is significantly decided by its mechanical properties. A brittle paper having lost mechanical strength has a low dielectric strength. Ageing of insulation affects its mechanical strength more significantly than the electrical strength [15]. The rate of ageing increases rapidly with the increase in temperature deteriorating mechanical properties.

A number of studies have been reported in the literature [13, 16-19] highlighting effects of various influencing factors, viz. temperature, pressure, impurities, moisture, electrode shape / surface, electrode metal, applied voltage and its duration, gap between electrodes, etc., on the oil breakdown strength.

8.3.2 Effect of time and frequency

Volt-time characteristics are specific curves representing the relationship between voltage and time to breakdown. These characteristics generally follow a law that some amount of energy is required to cause breakdown of a gap, and thus the breakdown voltage and time are interdependent [20]. The higher the voltage the lower the time is to cause the breakdown. A typical volt-time curve of air insulation is shown in figure 8.7.

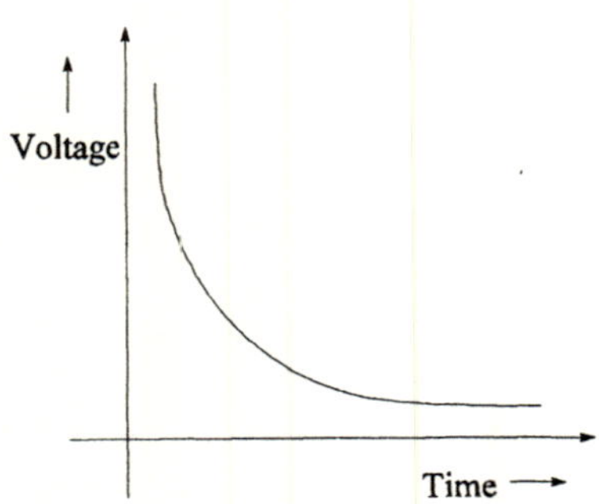

Figure 8.7 Typical volt-time curve

In short-time power frequency overvoltage tests, the breakdown strength of the solid insulation is influenced by temperature rise. When an alternating voltage is applied, initially the heat on account of dielectric losses is stored inside the insulation and temperature begins to rise. The heat is dissipated to surrounding ambient which is at a lower temperature than the insulation. The insulation temperature continues to rise until a state of equilibrium is reached, wherein the heat dissipated is equal to the heat generated. But with the increase in temperature, the resistance of the solid insulation decreases due to the negative temperature coefficient of resistance resulting in increase of current. Losses are further increased due to increase of current, and this leads to a run-away condition resulting in an eventual breakdown. Hence, with the increase in the time of application of voltage, the dielectric strength at power frequency reduces. For most of the insulating materials, the infinite time strength is approximately two-thirds of the one-minute strength, and the one-second strength is about 1.6 times the one-minute strength [21].

The dependence of the oil dielectric strength on the duration of voltage application is erratic as compared to the solid insulation. Although it is difficult to obtain definite volt-time characteristics, it can be said that the dielectric strength reduces rapidly after few seconds and remains more or less constant after few minutes for power frequency test voltages. It requires time for impurities to get lined up and bridge the gap; hence for very short time of application the strength is very high.

Like the duration of voltage application, the frequency of test voltage has a significant effect on the insulation strength; the increase in frequency results in reduction of strength, and thus increases rapidly the severity of the power-frequency overvoltage test. As the frequency increases, the dielectric loss and heating increase, reducing the strength of the solid insulation. The strength does not vary in the linear proportion to the reciprocal of frequency [22] but approximately with an exponent of 0.137, i.e., strength $\propto (1/f^{0.137})$. The effect of increase in frequency has much smaller effect on the oil strength as compared to the solid insulation. In general, it can be concluded that the increase of frequency has a harmful effect on the strength of transformer insulation during the power-frequency overvoltage tests. If the frequency is increased, the time of application of voltage should be reduced to produce the same amount of dielectric stress. Hence, the test standards specify the total number of cycles for the power frequency induced overvoltage test. When this test (in which the insulation is stressed to at least twice the voltage/turn) is conducted, frequency is also increased in order to avoid the core saturation. With the increase in frequency, time of application of voltage is reduced to maintain the same degree of severity. The IEC standard 60076-3, second edition: 2000, specifies that the test time should be 60 seconds for any frequency upto and including twice the rated frequency. For a test frequency higher than twice the rated frequency, the test time should be reduced and calculated in accordance with the formula,

$$test\ duration = 120 \times \frac{rated\ frequency}{test\ frequency}\ \text{seconds} \tag{8.13}$$

subject to a limit that it should not be less than 15 seconds. In other words, the number of cycles for the test is fixed; 6000 cycles for 50 Hz and 7200 cycles for 60 Hz transformer. Hence, for a 50 Hz transformer, if the test is conducted at 200 Hz, the test duration will be 30 seconds.

The effect of time and frequency of the voltage (during the power-frequency overvoltage test) on the creepage strength is not significant as compared to that on the puncture of the solid insulation [22]. The creepage breakdown in the absence of conducting material impurities is mostly decided by the oil failure, and the increase of frequency does not result in significant decrease of the oil strength due to smaller heat effect [23].

Regarding impulse volt-time curves, two types of curves are derived in the literature. For the first one, called full-wave volt-time curve, a full impulse wave is applied across the insulation and the breakdown may occur either on the wave-front or wave-tail portion or may not occur if the voltage is quite low. A volt-time curve is plotted using these breakdown points. For the second one, the wave-front slope is varied, and for each slope value the voltage is increased till the breakdown occurs. This front-of-wave volt-time curve is of similar nature but has somewhat different values as compared to the full-wave volt-time curve.

The impulse volt-time curve of the insulation used in a transformer is drastically different than that of the air insulation. Typical curves for oil and oil-impregnated pressboard are shown in figure 8.8. The curves indicate that the oil and solid insulations have flat characteristics after few microseconds. Composite oil-solid insulation generally has the volt-time characteristics close to that of the solid insulation alone. It is reported in [24] that the dependence of dielectric strength on the impulse duration in the range of 10^{-3} to 10^{-1} seconds is small for major insulation consisting of oil-barrier system (e.g., gap between windings).

The impulse ratio can be defined as the ratio of impulse full wave strength to one-minute r.m.s. AC (50 or 60 cycle) strength. Few of the earlier researchers have reported impulse and one-minute volt-time characteristics [25, 26, 27] and the corresponding impulse ratios. With the introduction of concept of partial discharge later on, volt-time curves for partial discharge inception voltage have also been reported. The partial discharge and breakdown volt-time curves, and corresponding impulse ratios are reported in [28] for three different cases, viz. turn-to-turn insulation, disk-to-disk insulation and oil duct insulation between barriers. It has been shown that the breakdown volt-time curve, as expected, is significantly above the partial discharge volt-time curve in the microsecond range. The two curves come very close to each other in the AC long-term region of several minutes as shown by the typical curves in figure 8.9. It means that for very short times, a partial discharge initiated is not sufficient to cause the breakdown, whereas in the longer duration of several minutes its magnitude is sufficient to cause the breakdown of the insulation.

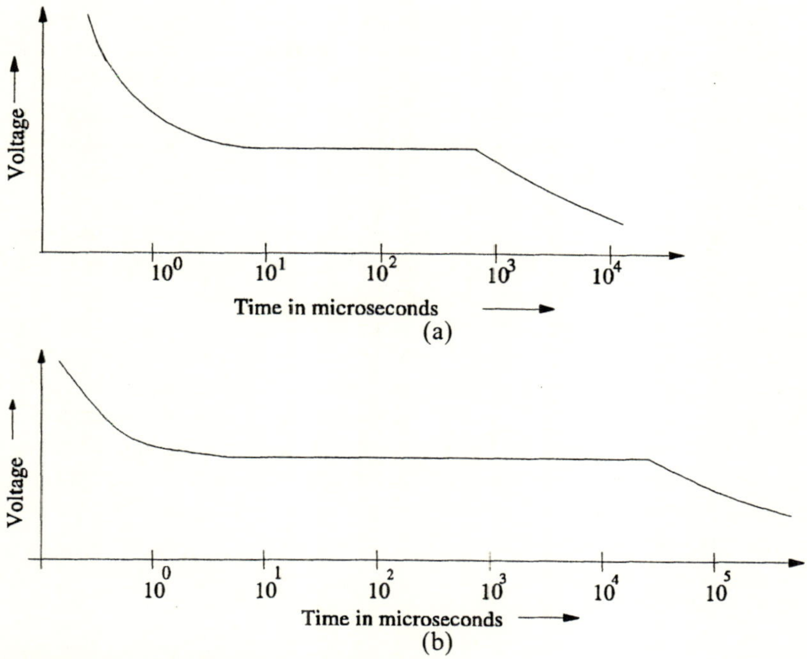

Figure 8.8 (a) Volt-time curve of oil-gap
(b) Volt-time curve of oil-impregnated pressboard

In transformer works, different dielectric tests, viz. lightning impulse, switching impulse and short-time / long-time power frequency tests, are carried out separately independent of each other. In actual power system operation, the transformer may be subjected to superimposed AC and lightning impulse voltages. The dielectric strength under such superimposed AC and lightning stress levels is reported in [29]. The breakdown voltage under the superimposed stress condition can be significantly lower than the corresponding value for the lightning impulse alone.

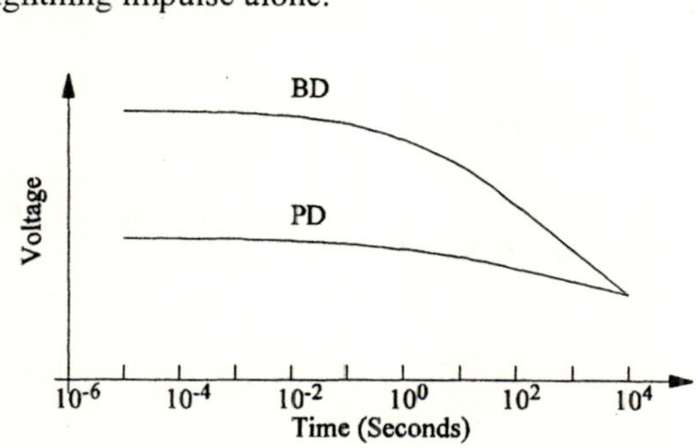

Figure 8.9 Breakdown (BD) and Partial Discharge (PD) volt-time curves [28]

Gas Insulated Substations (GIS) are being widely used all over the world. Disconnecting switch operations in GIS generate steep front transient overvoltages characterized by a rise time of few nanoseconds (5 to 20 ns), a short duration of several microseconds and amplitude as high as 2.5 per-unit. The 50% breakdown probability voltage of oil-paper insulation is reported [30] lower for steep fronted GIS transients than for lightning impulses. The oil-paper insulated equipment like transformers or their bushings, subjected to GIS transients, may fail at voltages below the lightning impulse level. Hence, the insulation of transformers for the GIS application must be designed with due consideration to these steep fronted transients.

Sometimes transformers are subjected to high frequency oscillatory overvoltages. In such cases, the damping of oscillation (defined as the ratio of two consecutive amplitudes of the same polarity) has significant impact on the dielectric strength of insulation; the strength increases with the increase in damping. For example, the strength for an undamped oscillating voltage of frequency of 0.9 MHz is below the dielectric strength for the lightning impulse, whereas with a damping ratio of 0.9 the strength is in the same range of that for the lightning impulse voltage [31].

8.3.3 Effect of temperature

As temperature increases, the dielectric strength of most of the solid insulations reduces. Due to increase in dielectric loss (and power factor), insulation temperature goes up further. The insulation ohmic resistance reduces with the increase of temperature, which results in flow of more current in the insulation. It may finally lead to the current run-away condition and eventual breakdown. The deterioration of the solid insulation strength with increase of temperature is opposite to the effect usually observed for the transformer oil.

The oil dielectric strength usually increases with temperature in the operating range. A marked improvement in the strength with the temperature increase is observed for the oil containing high moisture content. The temperature effects are dynamic in the sense that a considerable amount of time is required for establishing equilibrium between moisture in the oil and that in the solid insulation made of cellulose material. During different thermal loading conditions, there is a continuous interchange of moisture affecting the strength to some extent. For a reasonable temperature rise, the amount of moisture in the oil reduces, and thus this helps to keep the transformer insulation system in a healthy condition. It is known that an increase of temperature usually increases the mobility of carriers and conductivity. Hence, the breakdown voltage of oil should decrease with the increase in temperature; experiments conducted by many researchers show the opposite trend. The variation of carriers and mobility, therefore, may not possibly be used to explain the results of experiments. There is some amount of gas bubbles present in the oil and their solubility increases with temperature; this explains the increase of strength with the temperature [32]. Hence, it is generally not preferable to keep a transformer idle for a long time. Even a spare transformer should be kept in no-load condition for a reasonable amount of time periodically. The strength increases with temperature from -5°C to about 80 to 100°C; above which it reduces [21, 33]. Below -5°C, the strength increases rapidly as moisture particles in suspension get frozen.

8.3.4 Effect of thickness

It is known that dielectric strength of insulation does not generally increase in direct proportion to its thickness in a non-uniform field. The strength of the insulating materials can be expressed by the simple exponential formula [34],

$$E \propto (thickness)^n \tag{8.14}$$

The value of numerical constant n varies from 0.5 to 1.0 depending up on the processing / treatment of material and degree of non-uniformity of the field. For an untreated insulation, n is lower as compared to a treated insulation. For better shaped electrodes giving uniform field, its value is higher than that for a non-uniform field. For an electrode with a small radius, there is more crowding of equipotential lines at the electrode surface resulting in higher stress and lower value of breakdown voltage. The above equation indicates that with the increase in voltage rating, the insulation content of the transformer increases more rapidly.

8.3.5 Stressed volume effects

It is well-known that the breakdown strength of the transformer oil decreases statistically with the increase in its stressed volume [35]. This is also in line with the general fact that insulation strength reduces with the increasing area of insulation under test [36]. The oil dielectric strength reduces with the increase in stressed volume for both power frequency and impulse voltages [13]. Although, designers take either stressed area or stressed volume for strength considerations, it is generally agreed that the size of the structure is instrumental in the breakdown process. The strength calculations based on stressed oil volume are more commonly used in the industry. A breakdown in a gap is usually initiated at the weakest spot under a high stress condition. If in some other gap, a greater volume of oil is subjected to the same level of stress, it is quite probable that a still weaker spot will be present, resulting in breakdown at a lower voltage. Amount of impurities and electrode protrusions become quite important considerations for the area / volume effects. Usually the relationship between the breakdown strength and stressed volume is obtained experimentally from a number of breakdown studies for various types of electrode configurations. The breakdown strength for a plain oil gap is expressed in terms of the stressed oil volume (for power frequency test voltages) as [12]

$$E = 17.9\,(SOV)^{-0.137}\ \text{kVrms/mm} \tag{8.15}$$

where stressed oil volume (SOV) is in cm^3. In [37], the 50% breakdown probability stress for one-minute power frequency voltage is calculated as

$$E = 11.5\,(SOV)^{(-1/9.5)} + 2.5\ \text{kVrms/mm} \tag{8.16}$$

where SOV is in cm^3. It can be verified that the strengths given by the formulae 8.15 and 8.16 give the values of E of the same order for practical values of SOV (for the cases such as high voltage lead to ground configuration). The design should be such that the calculated stress value should be lower than E by some margin which is decided based on experience. It is assumed in above equations that electrodes are covered with some minimum thickness of paper insulation.

Thus, if a failure is predominantly decided by particles / impurities, a larger oil volume will provide more number of particles which are drawn into high stress zones, which may subsequently lead to a breakdown. However, a point is reached after which a further increase in volume will have insignificant effect on the strength since there is little influence of field on remote zones unless the field divergence is quite low. Hence, the volume in the above equations is taken as corresponding to the region in which the calculated electric stress values are between the maximum value and 90% of the maximum value.

Equations 8.15 and 8.16 tell us that with the increase in stressed volume, the dielectric strength of the insulation system reduces. If the electrode radius is increased, the stress values reduce; but at the same time the stressed oil volume (between maximum value and 90% of maximum value) increases reducing withstand. Hence, the optimum electrode contour can be determined by studying the relative variation of stress and strength due to the changes made in the electrode contour [38].

8.3.6 Creepage phenomenon

The solid insulation is used inside a transformer at a number of places, viz. between turns, between layers, between disks, between winding and ground, and between windings. The designer is confronted with mainly two types of electrical failures, viz. puncture and creepage. The puncture strength of the solid insulation is significantly higher than the creepage strength. Its creepage strength is maximum when it is along equipotential lines, i.e., when electric field (**E**) is normal to the insulation surface (equipotential lines are at right angles to the electric field). Due to complications of winding construction and connections, it is difficult to keep the field normal to the insulation surface everywhere. Also, it is not always possible to bend the solid insulation components to any desired radius or contour. In any case, considering the fact that the electric field is actually a three-dimensional one, difficulties of having shaped insulation components normal to the field are obvious. These places, where there is a field component parallel to the insulation surface, the strength is significantly reduced. When there is a failure along the surface of a insulation, it is termed as a creepage failure phenomenon. If the placement of the solid insulation results in stresses along its surface, much of the purpose of using it, viz. subdivision of oil ducts for higher overall strength, will get lost.

The creepage flashover characteristics of oil-pressboard interfaces have been analyzed and reported in a number of papers [39, 40]. A permittivity mismatch of two insulation materials usually assists flashover phenomenon at the interface [41]. The electric field in the oil immediately adjacent to the pressboard gets distorted due to the permittivity mismatch. Significant improvements can be gained by matching permittivities of the oil and pressboard insulations [42]. Thus, a pressboard having permittivity close to that of the oil not only reduces the oil stress but also results in higher value of flashover voltage along the oil-pressboard interface [43]. Hence, with a low permittivity pressboard, there is significant scope for optimization of insulation content since the electric field distribution will become more uniform in the oil-paper-pressboard insulation system of transformers. A low permittivity pressboard, manufactured by blending polymethylpentene fiber with cellulose fiber, has been used in a 765 kV, 500 MVA transformer [43].

8.3.7 Cumulative stress calculations

Using calculations based on SOV, the strength of an oil gap can be calculated as discussed in section 8.3.5 (SOV based calculations are more relevant for larger oil gaps). Another approach, in which cumulative stress calculations are done, is commonly used for the design of the insulation system of transformers. The approach is used for both creepage withstand assessment and oil gap design.

For estimating creepage withstand characteristics, the cumulative stress distribution is determined along the oil-solid interface. For two electrode case, finding the cumulative stress distribution is easy. The maximum stress is usually at one of the electrodes, and it reduces as we go towards the other electrode. Hence, the cumulative stress at any point is the difference between the electrode voltage (which is maximum) and voltage at that point divided by its distance from the electrode. For complex electrode configurations, with more than two electrodes, the maximum stress may not be at one of the extremities of the path (interface) under consideration, in which case the cumulative stress calculation starts from this maximum stress point. The creepage stress distribution along a solid-oil interface can be calculated by using the procedure described in [44]. Let us consider a solid-oil interface shown in figure 8.10.

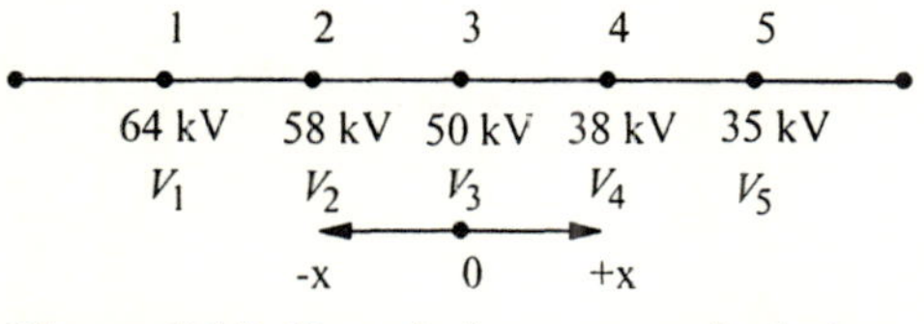

Figure 8.10 Cumulative stress calculation

Let us assume that the potential values, calculated by some method at points 1, 2, 3, 4 and 5, spaced at 2 mm distance, are as given in the figure. It is also assumed that the stress is maximum at point 3 (having value 7 kVrms/mm). The creepage stress is calculated for a unit length of 2 mm in either direction from point 3, and the path is extended in the direction of higher stress. Next, the cumulative stress is calculated for 4 mm length in either direction as shown in table 8.1. The calculation procedure is continued till the entire path is traced. The calculated cumulative stress values are plotted in figure 8.11. The withstand for each of these creepage distances (for power frequency overvoltages) can be calculated by [45]

$$E_{creep} = 15\, d_2^{-0.37} \text{ kVrms/mm} \tag{8.17}$$

where d_2 is the creepage path length in mm. The equation is valid for a degassed oil and a good quality solid insulation with clean surface. The above equation for creepage strength is in line with the curve arrived in [46, 12] by considering creepage field strengths for four different configurations.

It should be remembered that the calculated cumulative stress levels should be lower than the value given by equation 8.17 by a margin depending on quality of components and manufacturing processes, i.e., the margin between the creepage withstand and creepage stress for any length should be more than a certain value fixed by the transformer designer based on experience / established practices.

Usually, the creepage strength is considered about 30% lower than the bulk oil (jump) strength [47, 48]. The reference withstand curve for bulk oil (for power frequency overvoltages) is described by the following equation [12],

$$E_{oil} = 18\, d_1^{-0.38} \text{ kVrms/mm} \tag{8.18}$$

where d_1 is the oil gap distance in mm between covered electrodes, and the oil is considered without gases.

Table 8.1 Cumulative stress calculation

Path length (mm)	Cumulative stress calculation		
	+ x direction	− x direction	Path extended to point
2	$\frac{V_3 - V_4}{2} = 6.0$ kV/mm	$\frac{V_2 - V_3}{2} = 4.0$ kV/mm	4
4	$\frac{V_3 - V_5}{4} = 3.75$ kV/mm	$\frac{V_2 - V_4}{4} = 5.0$ kV/mm	2
6	$\frac{V_2 - V_5}{6} = 3.83$ kV/mm	$\frac{V_1 - V_4}{6} = 4.33$ kV/mm	1

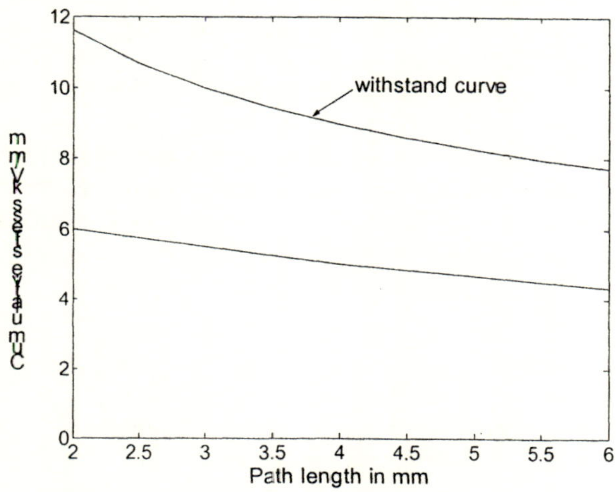

Figure 8.11 Plot of cumulative stress

It is always advisable to cover the electrodes with some insulation as it improves the strength by about 15 to 20% [12]. The creep withstand calculated by equation 8.17 is about 17 to 13% lower than the bulk oil withstand given by equation 8.18 (for a length of 1 to 100 mm), but as a conservative value it can be taken 30% lower as mentioned earlier for any distance. The creep withstand has been taken about 21% lower than the bulk oil strength in [49].

Thus, we have seen that there are two distinct approaches for determining withstand for bulk oil breakdown phenomenon; one is based on distance (equation 8.18 above) and other is based on stressed oil volume (equation 8.15 or 8.16). These two approaches are not contradicting each other, and the consistency of strength given as a function of gap length with the well established theory of stressed oil volume has been elaborated in [12].

8.3.8 Effect of oil velocity

Oil velocity has noticeable effect on its breakdown characteristics. It is reported in [37] that the power frequency breakdown voltage of moving oil at a velocity of about 25 cm/s is equal to that of stationary oil. It is higher than stationary oil by 10 to 15% at about 5 cm/s and lower than stationary oil when the velocity exceeds 100 cm/s. The explanation for this particular behavior is as follows. If oil movement is more dominant than the force by which impurities are swept and drawn in high stress zones as observed in the case of stationary oil, the breakdown voltage tends to be higher with the increase in velocity. When the velocity increases beyond a certain value, reduction of the breakdown voltage may be linked to the stressed oil volume effect, wherein the probability of large number of impurities passing through a high stress zone between the electrodes increases. This phenomenon is in line with the *weak-link* theory; the chances that *weak-link* of the oil (particle / impurity) may initiate a discharge are higher due to the fact that more volume of oil passes through the stressed zone of an insulation arrangement [14]. Contrary to this theory, a higher velocity shortens the time for which impurities will remain in high stress zones and the breakdown voltage should increase as per volt-time characteristics. Hence, the breakdown voltage at higher velocities will depend on which of the two effects, volt-time or oil volume effect, is the deciding factor. For impulse conditions, the breakdown voltage does not seem to be affected by the oil velocity.

8.3.9 Processing of insulation

Removal of moisture and impurities from the insulation is one of the most important processes of transformer manufacture. With the increase in the size of transformers, the time taken for processing of their insulation also increases. The time taken by a conventional hot air - vacuum process is considerably higher for large transformers with high voltage ratings, and it may be unacceptable to the transformer manufacturer. The conventional drying method may take more than 7 days for a 220 kV class transformer. The method consists of heating core-windings assembly with air as a medium and applying vacuum for extracting the moisture. The moisture content of the insulation can be reduced by raising its temperature and / or by reducing water vapour partial pressure, i.e. by vacuum. The application of vacuum speeds up the moisture extraction process; heating alone will take more time to remove a given amount of moisture from the solid insulation. Depending on the rating and voltage class of the transformer, several cycles of alternate heating and vacuum are required till the transformer insulation is dried. A moisture content of less than 0.5% is usually taken as the acceptance criterion for ending the process.

Requirement of a faster and more efficient process along with the need for better insulation performance resulted in the development and use of kerosene Vapour Phase Drying (VPD) method [50]. It is a fast and efficient method in which kerosene vapour at a high temperature is used as the heating medium instead of hot air (used in the conventional method). A special grade of kerosene, heated to about 130°C in an evaporator and converted into vapour form, is injected into the autoclave or transformer tank housing the core-windings assembly. As the process of heating is done under vacuum, the moisture extraction starts taking place during the heating period itself. When the insulation reaches a certain desired temperature, a fine vacuum is applied to remove the remaining moisture. In the VPD process, the windings and insulation are almost uniformly heated, whereas in the conventional process the inner insulation may not get heated to the desired temperature. The total process time in the VPD method is less than half the conventional method. In the conventional method, as heating is done through the medium of air with the presence of oxygen, the temperature is limited to about 110°C. On the contrary, in the VPD process, the drying is done in an oxygen free atmosphere (virtually in vacuum), and hence there are no harmful effects or loss of insulation life even at 130°C. Thus, the uniform heating at higher temperature under vacuum results in faster removal of moisture. The VPD process has the other advantage of vapour condensing on colder winding and insulation parts, thus washing out the impurities improving the dielectric properties. This aspect of VPD is extremely useful when one wants to clean the windings of transformers which come back to factory for repair or when windings are inspected to find a fault in a transformer failed in a test. With the uniform heating of all insulation parts, the shrinkage is also uniform and no further shrinkage is likely during the transformer operation at site, ensuring the mechanical strength of the windings.

Presence of small cellulose particles in oil has pronounced effect on its dielectric strength. Due to hygroscopic nature of cellulose, moisture in the oil plays a decisive role as it gets absorbed into the cellulose. The cellulose particles have deteriorating effect on the strength even when the ppm content of moisture in the oil is well within limits; the water content (saturation) in cellulose insulation is the deciding factor. Hence, it is absolutely essential to minimize content of suspended cellulose particles and fibers in the oil. These particles usually originate from the surface of paper / pressboard insulation. The major sources of these particles are the edges of radial spacers (pressboard insulation between disks) and axial spacers (pressboard strips in major insulation between windings) if these components do not have a machined finish. Punching operations required for making these components involve shearing of pressboard. Burrs, micro-delaminations and subsequent formation of cellulose fibers will be unavoidable if punching tools are not of good quality or their maintenance is inadequate. Hence, it is always recommended to have an additional operation of milling so that the edges of these pressboard components are smooth and free from fibers.

Once properly processed and impregnated with dry oil, it is important to prevent moisture from getting access to the oil impregnated solid / paper insulation through the oil. Simple and widely followed method for small and medium power transformers is to provide a breather with a dehydrating material (like silica gel) and an oil seal. The oil seal provides isolation to the dehydrating material from the atmosphere. Thus, the transformer breathes through the dehydrating material, and hence the moisture from the atmosphere cannot get into the transformer oil. Advanced breather systems (e.g., drycol breather) are also popular in some countries.

Transformer oil deteriorates to some extent when exposed to the atmosphere. The air acts as an oxidizing agent forming sludge in the oil. Hence, many of the large transformers, nowadays, are equipped with an air bag fitted inside the conservator, so that the transformer oil does not come in direct contact with the outside atmosphere. The changes in oil volume on account of temperature variations are absorbed by this flexible bag thus maintaining a constant pressure. One side of this bag is in contact with the atmosphere through a dehydrating agent so that if the bag ruptures, the oil does not get exposed to the atmosphere since the dehydrating breather steps into the conventional mode of operation. Action can be initiated to replace the ruptured bag in the mean time.

Once the transformer (insulation) is processed, it is filled with dry and degassed oil (having dielectric characteristics within the acceptable limits) under vacuum. Immediately after the oil filling, some air bubbles are formed in the insulation which may lead to partial discharges. Hence, a certain hold / settling time needs to be provided before the commencement of tests. During the settling period, heavier particles / impurities settle down at the bottom and the air bubbles move up or get absorbed. After the hold time, air bubbles should be released by bleeding (through air release plugs) at the top so that the transformer is free of (trapped) air. With this, dielectrically critical areas get cleared. The hold time increases with the increase of voltage rating of transformers. For 132 kV class transformers it should be minimum 24 to 36 hours, for 220 kV class it should be about 48 hours, and for 400-500 kV class transformers it is desirable to have the hold time of about 72 hours [51]. A lower hold time may be adopted by the transformer manufacturers based on their experience.

8.4 Test Methods and Design Insulation Level (DIL)

In service, the transformer insulation is subjected continuously to operating voltages and occasionally to overvoltages. The former decides the working voltage stress on the insulation. The overvoltages can be broadly divided into lightning overvoltages (aperiodic surges with duration of one to tens of microseconds), switching overvoltages (oscillatory surges with duration up to thousands of microseconds), and temporary overvoltages (lasting for few minutes) at or close to the power frequency. The standards on transformers have defined voltage test levels for various voltage classes of transformers. There are basically four different types of tests, viz. lightning impulse test, switching impulse test, short duration power frequency test and long duration power frequency test with partial discharge measurement. While the first three tests check the insulation's overvoltage withstand capability, the long duration test is mainly intended for checking the insulation behavior under the working voltage stress. The design of insulation between various electrodes is decided by one or more of these test levels. The inter-disk spacing is decided by impulse stresses, whereas the end insulation (between windings and yoke) may get predominantly decided by the power frequency test voltages. The conductor insulation is decided by either the impulse stress or working voltage stress.

When oil and solid are stressed together, two kinds of failures can usually occur. In one type, there could be a complete catastrophic failure between two electrodes (can be jump breakdown, creepage breakdown or combination of both). This kind of failure is quite uncommon if adequate clearances are provided and maintained. The second one is a local oil failure (partial discharge). At one time, the partial discharge (PD) measurement test was not conducted due to non-availability of proper instrumentation. Also, many partial discharges used to result into undetected damages (since the solid insulation acted as a barrier to propagation of discharges). With the development of instrumentation and awareness of slow but damaging effect of discharges, the partial discharge test has become one of the most important tests for high voltage transformers. The international standards on transformers have defined the voltage levels and corresponding PD limits. Some manufacturers set their internal norms much lower than that specified in the standards in order to ensure long-term reliability of transformers. In order to get very low PD levels, all the sharp electrodes inside the transformer have to be rounded or electrostatically shielded. Processing and impregnation procedures have to be excellent for minimizing impurities and air voids. Also, the manufacturers should have proper instrumentation for the low PD measurement. Ambient partial discharges in the test set-up area and surrounding environment have to be minimized. The partial discharges lead to a progressive damage of insulation resulting into eventual failure after a time period which depends on their severity and location inside the transformer. Hence, it is advisable to eliminate all the sources and causes of partial discharges, so that the progressive deterioration of insulation becomes a remote possibility.

Aim of a transformer engineer should be to design and manufacture a *partial discharge free* transformer. It should be noted that the impregnation of insulation may not be perfect and also it may not be possible to control the dielectric field uniformly at all locations. Hence, in practice some amounts of partial discharges are always there; endeavor should be to minimize them. The electric stress values below the PD inception levels do not result into ageing of the mechanical strength of the pressboard. In such a case, the life of pressboard insulation is decided by the mechanical strength as affected by thermal conditions [43]. The PD is usually measured either in pico-coulombs or micro-volts. Detection and measurement of PD at its origin is very difficult. The two major approaches of PD detection, viz. acoustic and electrical signal based, are discussed in chapter 12.

***Design Insulation Level*:** After calculating the voltages at critical places inside a transformer (within the windings, at the ends of windings, on the surface of leads, etc.) under all the test conditions, it is quite meaningful to convert these voltages to one equivalent voltage, which is usually the (one-minute) short duration power frequency voltage. This equivalent voltage level is called as Design Insulation Level (DIL) which is expressed in kVrms. Thus, at any point inside the transformer, there is only one DIL, which is the maximum of equivalent one-minute power frequency voltage levels during the four different tests. This approach is widely followed by transformer designers simplifying the insulation design process. The conversion factors used by different manufacturers for converting the calculated voltages (under various tests) to one-minute r.m.s. value may not be identical but they fall in a narrow range.

Table 8.2 Factors for conversion to one-minute (r.m.s.) power frequency level

Test voltage	Multiplication factor
Lightning Impulse Level (BIL)	~ (1/2.30)=0.44
Switching Impulse Level (SIL)	~ (1/1.80)=0.55
Long duration (one hour) power frequency voltage	~ (1/0.80)=1.25

The factors reported in [48] are given in table 8.2. The conversion factor for one hour power frequency test voltage is in agreement with the following equation given in [37],

$$V_t = (1/t)^{0.055}\ V_1 \tag{8.19}$$

where V_1 and V_t are the 50% breakdown probability voltages for 1 minute and t minutes respectively. The equation gives V_{60}/V_1 as 0.8, which is in line with that given in table 8.2.

The multiplication factor for BIL in table 8.2 is in line with the impulse ratio (the ratio of lightning impulse level to power frequency level) of about 2.3 as evident from table 8.3 which gives commonly used BIL and power frequency test levels for various voltage classes. An impulse ratio in the range of 2.3 to 2.5 is being used in the industry [11, 12] while calculating the equivalent DIL corresponding to a BIL value. This value of impulse ratio represents an average correction factor for the impulse voltage duration in the range of 10 to 100 microseconds. The overvoltages stressing the inter-disk or inter-turn insulation during the impulse test may be shorter than 10 microseconds in which case the correction factor seems conservative. The impulse ratio of 2.68 has been suggested in [24] for 1.5/40 microsecond impulse waveform while designing inter-winding and inter-disk insulation, giving the corresponding multiplication factor of 0.37 for BIL in table 8.2 (in place of 0.44). A lower multiplication factor may open avenues for insulation optimization since in that case the equivalent DIL value is lower. On the other hand for impulse overvoltages of more than hundreds of microseconds duration (close to switching transients), it is risky to use the impulse ratio between 2.3 to 2.5 (it should be lower). Thus, the correction factor should be judiciously selected based on detailed analysis and experimental data. For AC voltages and non-oscillatory impulses having duration other than that of lightning, switching and long duration power frequency voltages, an interpolation may be done with caution for finding DIL [47].

Table 8.3 Impulse ratio

Voltage class	BIL (kVp)	Power frequency test level (kVrms)	Impulse ratio
132 kV	550	230	2.39
220 kV	950	395	2.40
400 kV	1300	570	2.28

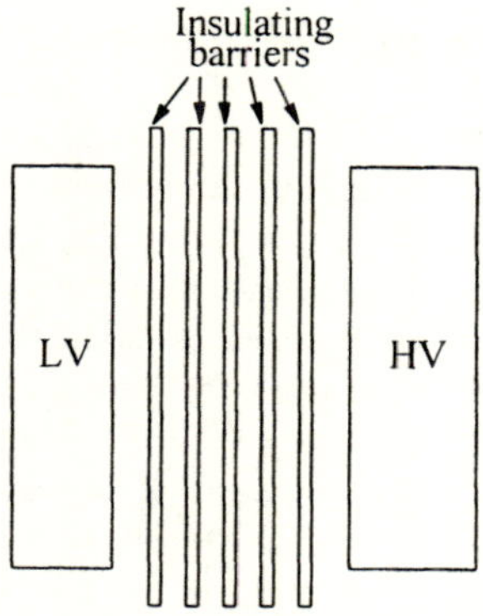

Figure 8.12 Subdivided duct between LV and HV windings

Having estimated the Design Insulation Level at critical electrodes inside the transformer, one can proceed with the design of insulation system.

8.5 Insulation Between Two Windings

The gap between low voltage (LV) and high voltage (HV) windings is subdivided into many oil ducts by means of solid insulating barriers (figure 8.12). The insulation system of oil cooled power transformers consists of combination of oil and solid insulations (paper and pre-compressed board). The oil and solid insulations are cheaper than most other insulating materials, and as a combination they give much higher dielectric strength than individually.

In an oil-pressboard insulation system, since the maximum admissible stress in a good quality pressboard for one-minute power frequency overvoltage is about 40 kVrms/mm [47], the limiting design stress values are primarily due to stress in oil. In a good quality system with degassed oil and near uniform field distribution (e.g., between barriers in the HV-LV gap), typical PD inception field strength (stress) value for a 8 mm oil duct is 10 kVrms/mm for insulated winding conductors [46, 52]. Since the stress is high at winding conductor corners, the PD inception field strength of 8 mm oil duct next to windings for degassed oil is lower, i.e., about 8.4 kVrms/mm. The allowed (design) stress values can be lower than these PD inception strength values depending on the required safety margin and quality of manufacturing processes. Thus, the dielectric strength of the solid insulation is much higher than that of the oil. Since the electric stress (E) is inversely proportional to the permittivity, the stress in the oil is significantly higher than that in the solid insulation. The oil permittivity ($\cong 2.2$) is about half that of the solid insulation, hence the oil stress is twice the solid insulation stress in uniform fields. Since the oil has to bear higher stress, it is recommended to make barriers as thin as possible and give more space for the oil ducts. In the oil-paper-pressboard insulation system, the strength of the insulation arrangement is predominantly decided by the strength of the oil ducts. Hence, the design of transformer insulation system essentially means the design of oil ducts and oil-solid interfaces. For a given total gap between two windings, there is no advantage gained in increasing solid insulation thickness just because its strength is higher. On the contrary, higher solid insulation thickness results in more stress in oil ducts. The barriers should be as thin as possible, permitted from mechanical strength considerations. Barriers with a thickness of 1.5 to 2 mm are generally used for getting good mechanical stability. The lower the oil duct width, the higher the withstand stress (kV/mm) is. The number of barriers in the inter-winding gap should be adequate so that larger oil gaps are avoided (width of oil duct more than 12 mm is usually not recommended in the inter-winding gap). Insulating barriers have an additional function of acting as barriers against the propagation of a discharge streamer in the oil between electrodes. The barriers break the oil path into smaller ducts and prevent lining up of impurities in the oil.

The size of first duct next to windings should be properly chosen. Although a small width is desirable as it gives higher kV/mm withstand value, thermal considerations may not allow the use of duct width below 6 mm. The first duct is usually of 6 to 10 mm size. Withstand for the ducts (first duct and other ducts in HV-LV gap) can be found out by referring published curves [46, 52] for four types of gaps, viz. degassed oil between barriers, gas saturated oil between barriers, degassed first oil duct and gas saturated first oil duct. The maximum stress in the first duct has to be accurately determined (say, by FEM analysis). The breakdown gradients obtained for a typical configuration are given in [24]. The values of breakdown gradients for yoke-end-line-lead arrangement (line terminal at yoke end of winding) are lower than that for center-line-lead arrangement (line terminal taken at the midpoint of the winding stack with two parallel branches). This is because the field distribution in the first duct at the line terminal is quite non-uniform for the former as compared to the latter arrangement. Actual design values should be smaller than the breakdown values depending on the margin to be kept. The HV-LV (*HILO*) gap value is usually based on the average stress (which is of the order of 5.5 to 6 kVrms/mm) given by [53]

$$E_{av} = \frac{DIL}{(HILO\ gap - 0.5 \times solid\ insulation)} \tag{8.20}$$

A manufacturing tolerance is added to the *HILO* gap calculated by the above equation. For winding to end limb clearance (e.g., in single-phase three-limb construction) or innermost winding to core clearance, a lower value of E_{av} is used due to the presence of sharp corners of core. In some critical cases, an electrostatic shield is used around the core / end limb facing the windings. The phase-to-phase insulation design is very similar but the stresses due to switching impulse and power frequency test levels need to be considered; most of the times the voltages between phases are 150% of phase to ground voltage levels.

8.6 Internal Insulation

Internal / minor insulation consists of all the insulation components within the winding, viz. conductor paper covering, insulation between layers in the radial direction, insulation between turns or disks in the axial direction, and special insulating components that are placed close to the insulated conductors.

In addition to withstanding short and long duration overvoltages during tests at manufacturer's works and operation at site, the conductor insulation has to withstand a continuous working voltage stress. This is particularly important criterion for deciding the conductor insulation in the case of interleaved main and tap windings. In such windings, physically adjacent turns are electrically many turns apart, resulting in significantly higher working stress. The conductor paper covering in such cases gets predominantly decided by this working stress. For example, in a disk winding the total stressed paper area can be calculated as

$$A_s = M_t \times k \times (N_t - 1) \times w \times N_d \times N_p \tag{8.21}$$

where M_t = mean turn length of winding,

N_t = number of turns per disk, w = width of conductor,

k = multiplying factor which depends on degree of interleaving (e.g., it has a value of 2 for conductor interleaving shown in figure 7.7),

N_d = number of disks per phase, N_p = number of phases.

The higher the stressed paper area, the lower the allowed working voltage stress is. Figure 8.13 gives the allowed voltage per mm of paper insulation thickness between turns as a function of the stressed paper insulation area. The voltage per mm allowed is usually in the range of 2400 to 4000 V/mm. The voltage per turn of the transformer, corresponding to the continuous overvoltage condition, is divided by the paper insulation thickness between two turns. The value obtained should be less than that given by figure 8.13 (the y axis values are not marked since they depend on individual manufacturer's established design practices).

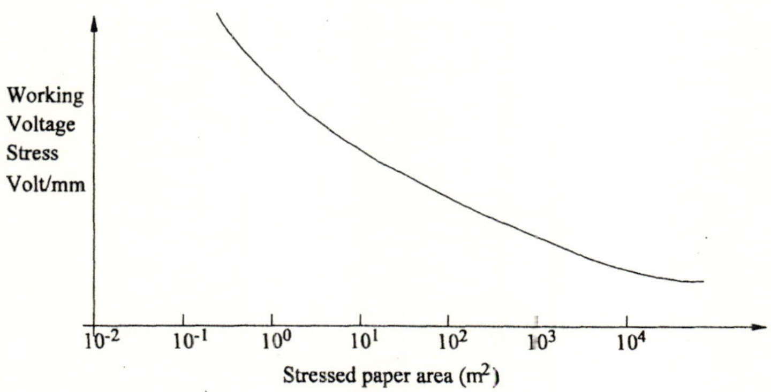

Figure 8.13 Working voltage stress

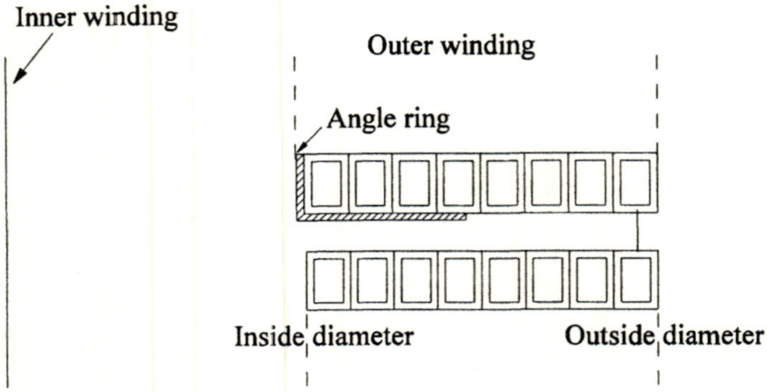

Figure 8.14 Angle ring placed at inside diameter of outer winding

Considering primarily the lightning impulse stresses, special insulating components are used closely hugging the paper insulated winding conductors. These are usually placed at line end disks or turns, where the resultant stress at the disk corner is quite high. This solution is beneficial as compared to increasing the conductor paper insulation which would reduce the series capacitance, increase the winding temperature rise and reduce the winding space factor. These special insulating components inside the winding are called as disk angle rings (figure 8.14) or disk angle caps (figure 8.15). Since they are closely fitted to the winding conductor, the oil stress is reduced (since the oil duct shifts away from the conductor). Thus, they are doing the same function as achieved by increasing the conductor radius. If these are made from layers of paper insulation, they become flexible and can easily take the shape of the conductor eliminating oil voids between them and the paper covered conductors. On the contrary, if they are made from molded solid insulation, small oil voids may exist which then will have higher stress. Being of very small length, the withstand strength of the voids is also high. In any case, it is always preferable to have angle rings / caps closely fitting the conductors to eliminate voids.

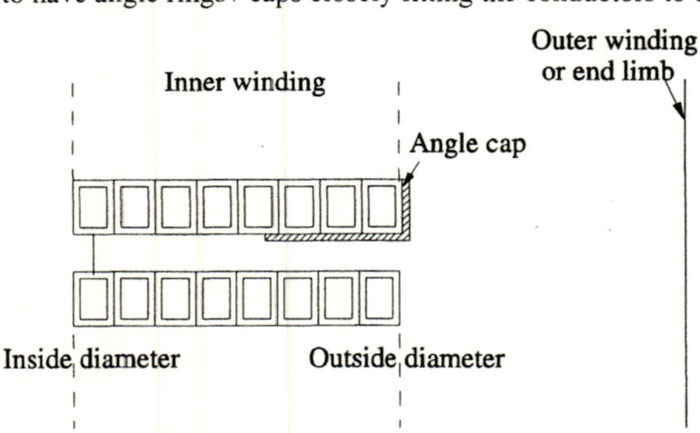

Figure 8.15 Angle cap placed at outside diameter of inner winding

The angle caps are used to protect the winding corners from adjacent winding or end limb as shown in figure 8.15. Angle rings / caps made from molded material are usually available in sector shaped form, which have to be overlapped by doing proper chamfering. The overlapping distance is usually kept as 30 to 40 times the thickness of the angle ring / cap.

Although it is essential to achieve the uniform initial impulse voltage distribution, what is more important from the insulation design point of view is getting uniform dielectric stresses in relation to strength values. Generally, it is very difficult to get the initial impulse voltages between disks / turns equal throughout the winding; the insulation should be designed such that the margin between the strength and stress is in the same range of values along the winding height. This is an ideal design; but it may not be possible since for cooling purposes adequate inter-disk or inter-turn gaps are kept at the neutral end even though the stresses are low. It is not advisable to increase turn or disk insulation, where stresses are high, without detailed calculations. There could be failure even if extra insulation is provided because the increase in strength may not be commensurate with the increase in stress due to a reduced series capacitance.

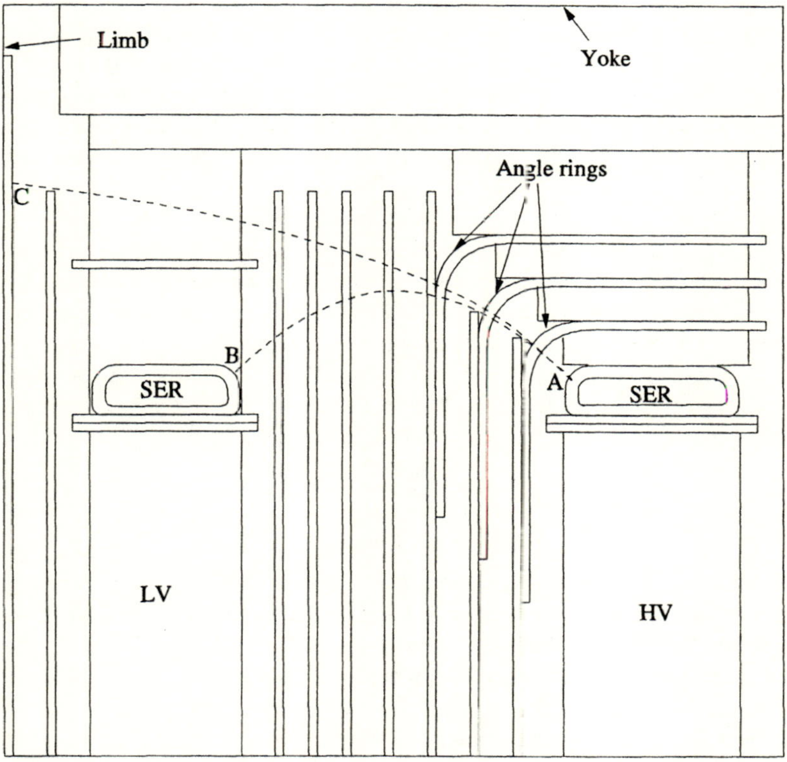

Figure 8.16 Typical end insulation arrangement of a power transformer

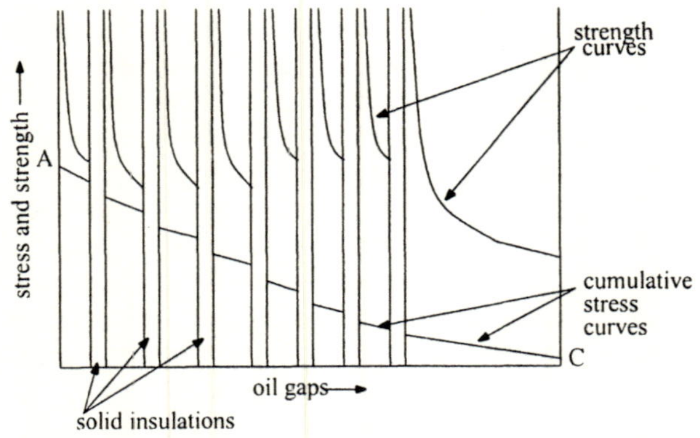

Figure 8.17 Analysis of oil ducts in end insulation region

8.7 Design of End Insulation

In the composite oil-solid insulation system, the strength of the insulation arrangement is predominantly determined by the strength of oil gaps. The solid insulation is used to subdivide long oil gaps (having lower dielectric strengths) into smaller ones. While assessing the end insulation design, the strengths of oil gaps and solid-oil interfaces are evaluated. The FEM analysis is commonly used for estimation of electrostatic field distribution. A typical end insulation arrangement is shown in figure 8.16. The electric stress distributions along various contours between windings and between yoke and windings are examined [54]. Two such contours are shown in the figure with dotted lines.

For each oil duct within the selected contour (e.g., A-C), the cumulative field distribution is determined by integrating the field along the contour starting from the maximum value as described in section 8.3.7. If in an oil gap, the maximum stress is at one of the extremes, the calculations are greatly simplified since the cumulative stress at any point with a distance of d from the point of maximum stress on the contour is calculated as the voltage difference between the two points divided by d. In this case the cumulative stress is

$$E_{(d)} = \frac{1}{d} \int_{0}^{d} E_{(x)} \, dx \tag{8.22}$$

The cumulative oil stress values for each oil gap along the contour A-C are plotted and compared with a reference strength / withstand curve (described by formula such as equation 8.18) as shown in figure 8.17.

The end insulation arrangement would be optimum if the minimum margin between the withstand and stress curves is approximately same for all the ducts. The task of insulation designers is to adjust the barriers / angle rings in such a way that the minimum margin in each resulting oil duct is greater than some fixed safety margin value and that these minimum margins are in the same range of values. The minimum safety margin is decided by insulation designers depending on the quality of oil-solid insulation and technology of drying and oil impregnation processes. Usually a 25 to 35% safety margin in critically stressed parts of the transformer insulation is considered as sufficient to pass successfully the long duration (one hour) induced overvoltage test without partial discharges and also the impulse test [55].

The placement of angle rings / caps is an important aspect of end insulation design. They divide the larger oil ducts into smaller ones increasing the withstand of the total gap and resulting oil ducts. The angle rings / caps should be placed along the equipotential lines; otherwise higher creepage stress along their surface may reduce the dielectric strength. As we go away from the winding corner, the radius of angle ring / cap should be more so that it lies along an equipotential line and the creepage stress is minimized. The angle rings / caps with fixed and small radius of 2 to 3 mm are cheaper than the contoured ones with much higher radius. But the creepage stress on the angle rings / caps with small radius can be high, reducing the overall strength of the insulation arrangement. In case one wants to optimize end insulation clearance, contoured angle rings / caps need to be used which reduce not only the creepage stress but also the oil duct size in high stress zones thereby improving the withstand. The effectiveness of a contoured angle ring can be clearly seen from figure 8.18. The creepage stress along portion AB of the contoured angle ring is quite less than the corresponding portion CD of the non-contoured angle ring. The first angle ring / cap should be close to the winding for having smaller first oil duct width.

Apart from effective subdivision of oil ducts by barriers / angle rings (caps) / solid washers, it is necessary to reduce the electric stress at the winding corners. The stress is reduced by placing at winding ends a static end ring (SER) with a certain minimum corner radius. In lower voltage LV windings, if SER is dispensed with, turn or disk angle ring / cap is usually placed on the last turn / disk, which reduces the stress in the oil duct immediately next to it. The inner and outer diameters of SER have to be flush with the corresponding diameters of the winding, and SER should be as close to winding end disk / turn as possible all along the circumference.

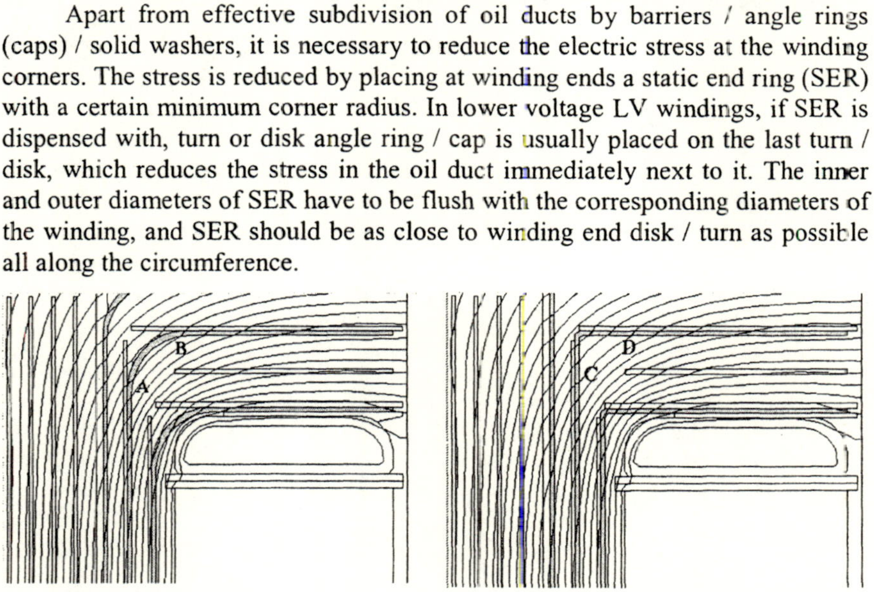

(a) Contoured angle ring (b) Non-contoured angle ring

Figure 8.18 Effectiveness of contoured angle ring

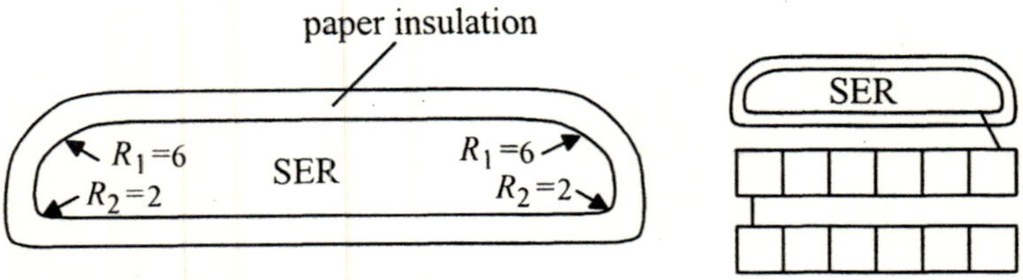

Figure 8.19 Static end ring

A typical sketch of SER is shown in figure 8.19. The inner radius close to winding (R_2) is usually standardized to 2 mm, whereas outer radius (R_1) is minimum 6 mm and it may be more as the voltage rating and radial depth of the winding increases. The paper insulation thickness is of the order of 3 mm. Sometimes a static ring (SR) may be placed between two disks of a winding, particularly between second and third disks and between fourth and fifth disks from the line end of a continuous disk winding since the lightning impulse voltage between these disks can be quite high.

Sometimes, LV winding height is kept more (by few mm) than HV winding height. The stress at LV corner facing HV winding is lower in this case as compared to the case wherein LV winding height is equal or shorter than HV winding height.

8.8 High Voltage Lead Clearances

In high voltage transformers, the clearances between leads and ground, between leads, etc. have to be judiciously selected. Consider a high voltage insulated lead near a ground plane as shown in figure 8.20. The maximum electric stress in the paper insulation and oil can be calculated by the well-known formulae given in section 8.1. The complete stress distribution can be accurately calculated by FEM analysis.

The oil duct strength can be calculated on the basis of the stressed oil volume (SOV) theory. The stressed oil volume between the maximum stress and 90% of the maximum stress (for cylindrical lead - plane configuration) can be calculated by the following approximate formula [56],

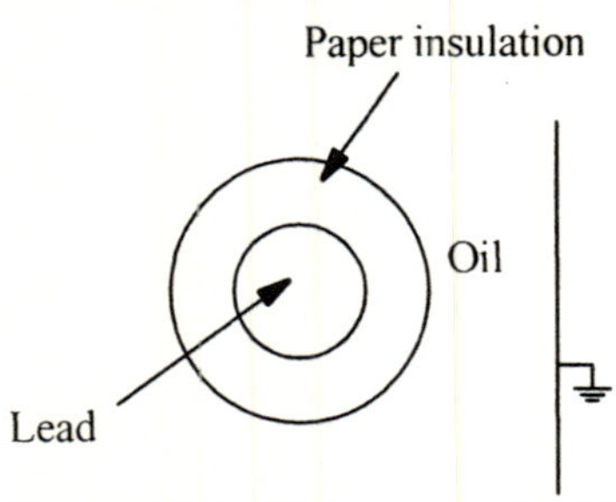

Figure 8.20 High voltage lead insulation

$$SOV = \frac{\pi r^2}{18} l \tag{8.23}$$

where r is bare lead radius and l is effective lead length. A more accurate formula for a bare lead is given in [57]. The accurate SOV for a covered lead and / or for a complex configuration of electrodes can be calculated by FEM analysis. The 50% breakdown probability stress for one-minute power frequency voltage can be calculated from the SOV by using equation 8.16. The safe withstand value is lower than this breakdown stress value and can be decided by the transformer manufacturer depending on the acceptable value of % breakdown probability and the quality of manufacturing processes (which decide the level of impurities) [37, 58]. The margin kept is also a function of manufacturing practices and working tolerances required. For example, lead to tank clearance may actually be less than that specified in the drawing. For lower voltages the mechanical tolerances, lowering clearances and magnetic clearances may dominate the dielectric design considerations.

An increase in breakdown voltage for the lead to ground configuration can be obtained by one or more of the following means, viz. increase in clearance, increase of bare lead diameter, increase in thickness of paper insulation and placement of barriers to subdivide oil duct. The first option is obviously not cost effective as it increases the material cost and size of the transformer. Out of the remaining three options, increase in paper covering has significantly more impact in improving the withstand [59]. With the increase in paper insulation, more voltage is dropped across it, reducing the stress in oil. Although a higher paper insulation results in significant reduction in the oil stress at the covered lead surface, thermal considerations and manufacturing time required may also be the deciding factors while standardizing bare diameter, paper insulation thickness and clearances for various Design Insulation Levels. When the electrode is not sharp, barriers are put (along equipotential lines) to enhance the strength.

It should be noted that the increase of either bare lead diameter or paper thickness increases SOV, but reduction in strength (due to increased SOV) is not significant as compared to reduction in stress value (because of the logarithmic relationship between the strength and SOV).

When there is a row of tap leads facing either winding or tank, the stress at the edge of the last lead in the row is less compared to the case of an isolated lead. The extent of reduction has to be assessed for the worst configuration (giving minimum reduction) and then accordingly the electrical clearances can be standardized using the isolated lead to ground clearances. The clearance of the row of tap leads to ground is usually lower as compared to that of the isolated lead to ground case.

The above analysis has assumed smooth ground plane. When the ground electrode is not smooth and is having sharp corners, it becomes difficult to standardize the clearances. It is preferable to have at least 2 mm radius at the corners. With such a minimum provision of the corner radius, in the absence of detailed analysis, 1 mm of oil gap per kVrms of Design Insulation Level seems to be a safe design criterion (i.e., the average stress of 1 kVrms/mm). A value of about 1.4 kVrms/mm or more can be used based on experience.

In some cases due to the nearness of high voltage electrodes to the tank surface, the values of electric field at the tank surface may exceed the permitted values. In this case, the tank should be covered with a proper insulation [47].

The above analysis holds good for jump clearances in oil between lead and ground. There is one more design consideration: when the leads are supported by solid insulation from the grounded part as shown in figure 8.21 (equipotential lines obtained from FEM analysis are displayed). In this case, the cumulative creepage stress on the pressboard and the corresponding withstand are calculated using the procedure given in section 8.3.7. For non-homogeneous field distributions (involving sharp electrodes), the rule of 1 mm of creepage path per kVrms may be followed in the absence of detailed analysis.

Sometimes insulating barriers are placed close to high voltage leads in high stress zones. The combined effects of oil velocity and particles sweeping towards high stress zones have definite role to play in the power frequency breakdown. The barriers make the oil flow restricted in the high stress zones so that the probability of occurrence of breakdown phenomenon (in which floating particles are drawn into high stress points helping breakdown processes) is reduced to a great extent. Hence, the barriers can be quite effective when used to subdivide large oil gaps in the high stress zone. The barriers are effective when they are placed close to the electrode in the high stress zone along the equipotential lines. If there is appreciable creepage stress along their surface (due to improper placement), the overall strength may not improve.

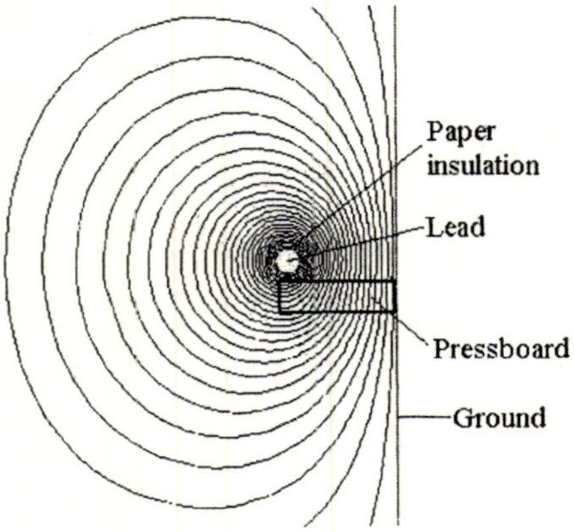

Figure 8.21 Optimum creepage distance calculation by FEM analysis

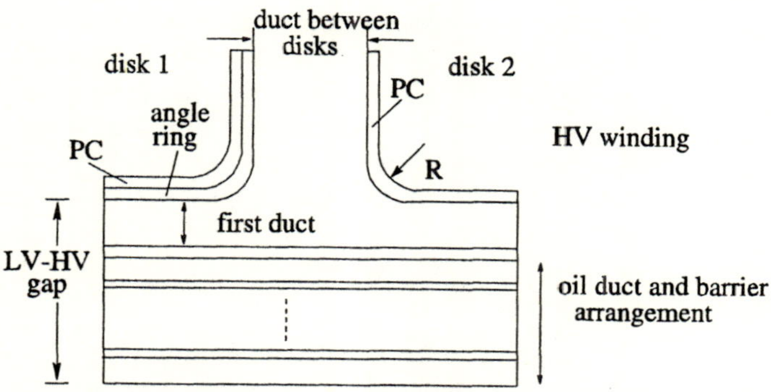

Figure 8.22 Insulation arrangement between two disks

Very non-uniform field patterns exist in the vicinity of high voltage leads between winding terminations and lower bushing connections. Conventionally, the leads can be covered by a thick paper insulation. This may not help much, as the stress in the bigger oil duct after thick insulation is still exposed to a high non-uniform field. Also, too much insulation on the lead limits its thermal capability. The scatter of breakdown voltage for such a wide oil duct can be much higher resulting in lower dielectric strength. The modern lead exit systems [48] consist of pressboard barrier systems dividing the oil gap into smaller gaps. The lead is covered by a thin layer of molded pressboard.

8.9 Statistical Analysis for Optimization and Quality Enhancement

8.9.1 Parameter design

The optimization of insulation requires identification of components having significant influence on the insulation stress and strength. Quantification of effects of different factors affecting the stress and strength helps in optimization. *Orthogonal array design of experiments* is one of those techniques which allows the effect of several parameters to be determined efficiently. The results of these experiments can be used to judge the relative magnitude of the factor effects. The analysis of relative effects of the different factors can be obtained by decomposition of variance which is commonly called as *Analysis of Variance* (ANOVA).

An approach for optimization of major insulation of a 400 kV class of transformer has been described in [60]. Figure 8.22 shows a sectional view of transformer windings, in which the insulation arrangement of two disks of HV winding is depicted. The factors affecting the maximum oil stress and strength are, viz. conductor radius (R), conductor paper covering (PC), disk angle ring thickness (DAR), HV-LV gap, duct between disks, first oil duct width and voltage difference between electrodes. The quantification of effect of these factors has been reported in [60], for a particular arrangement of 400 kV transformer winding, which is based on 36 FEM simulations for specific combinations of above seven factors. Such an analysis helps designer to do the parameter design, i.e., choosing the optimum combination of these factors. A quadratic surface, generated by using multiple regression analysis to express the oil stress in terms of the factors, can be used further for the tolerance design as explained below.

8.9.2 Tolerance design

Insulation optimization cannot be done in isolation without due consideration to variations usually observed during the manufacturing processes. The tolerance design methodology can be used to identify critical insulation components whose manufacturing variation needs to be controlled. For this, the deviations in dimensions of all (control) factors are first found out by choosing the random samples from the manufacturing site. The standard deviation (SD) can be calculated from the data. For the tolerance design, three levels of factors can be taken [61] around the mean (m) as $m-\sqrt{3/2}\ SD$, m, and $m+\sqrt{3/2}\ SD$. Either FEM simulations can be used to obtain the oil stress values for the specific combinations of the factors or the quadratic surface generated earlier (in the parameter design stage) can be used expediting the analysis. The subsequent ANOVA analysis reveals the relative effects of manufacturing variations of the factors on the stress values. This can help the transformer manufacturer to identify manufacturing processes which need to be controlled on priority to help reliability enhancement / optimization efforts. The extra investment needed to improve a particular process can also be justified by such an analysis.

References

1. McDonald, D. The electrolytic analogue in the design of high voltage power transformers, *Proceedings IEE*, Vol. 100, Pt. II, 1953, pp. 145-166.
2. Birke, P. V., Lackey, J., and Palmer, S. Determination of highly stressed volumes in oil dielectrics, *IEEE Transactions on Electrical Insulation*, Vol. EI-7, No. 3, September 1972, pp. 139-144.
3. Cockcroft, J. D. The effect of curved boundaries on the distribution of electrical stress around conductors, *Journal IEE*, Vol. 66, 1928, pp. 385-409.
4. Fogaras, L. and Lampe, W. Calculation of electrical field strength around transformer winding corners, *IEEE Transaction on Power Apparatus and Systems*, Vol. PAS-87, No. 2, February 1968, pp. 399-405.
5. Galloway, R. H., Ryan, H. M., and Scott, M. F. Calculation of electric fields by digital computer, *Proceedings IEE*, Vol. 114, No. 6, June 1967, pp. 824-829.

6. Wexler, A. Computation of electromagnetic fields, *IEEE Transactions on Microwave Theory and Techniques*, Vol. MTT-17, No. 8, August 1969, pp. 416-439.
7. Rele, A., Palmer, S., and Birke, P. V. Dielectric stress calculations for EHV transformer winding, *IEEE PES Summer Meeting*, Vancouver, Canada, 1973, Paper No. C-73-396-9.
8. Andersen, O. W. Laplacian electrostatic field calculations by finite elements with automatic grid generation, *IEEE Transactions on Power Apparatus and Systems*, Vol. PAS-92, September/October 1973, pp. 1485-1492.
9. Singer, H., Steinbigler, H., and Weiss, P. A charge simulation method for the calculation of high voltage fields, *IEEE Transactions on Power Apparatus and Systems*, Vol. PAS-93, 1974, pp. 1660-1668.
10. Okubo, H., Ikeda, M., Honda, M., and Yanari, T. Electric field analysis by combination method, *IEEE Transactions on Power Apparatus and Systems*, Vol. PAS-101, October 1982, pp. 4039-4048.
11. Tschudi, D. J., Krause, C., Kirch, H. J., Franchek, M. A., and Malewski, R. Strength of transformer paper-oil insulation expressed by the Weidmann oil curves, *CIGRE 1994*, WG 33.03.
12. Nelson, J. K. An assessment of the physical basis for the application of design criteria for dielectric structures, *IEEE Transactions on Electrical Insulation*, Vol. 24, No. 5, October 1989, pp. 835-847.
13. Nelson, J. K., Salvage, B., and Sharpley, W. A. Electric strength of transformer oil for large electrode areas, *Proceedings IEE*, Vol. 118, No. 2, February 1971, pp. 388-393.
14. Krause, C. Dielectric strength of transformer oil: Impact of oil flow, moisture and cellulose particles, *WICOR Insulation Conference*, Rapperswil, Switzerland, September 1996.
15. Norris, E. T. High voltage power transformer insulation, *Proceedings IEE*, Vol. 110, No. 2, February 1963, pp. 428-440.
16. Miners, K. Particles and moisture effect on dielectric strength of transformer oil using VDE electrodes, *IEEE Transactions on Power Apparatus and Systems*, Vol. PAS-101, No. 3, March 1982, pp. 751-756.
17. Zein El-Dine, M. E. and Tropper, H. The electric strength of transformer oil, *Proceedings IEE*, Vol. 103, Pt. C, 1956, pp. 35-45.
18. Maier, G. The electric strength of insulating oil with power-frequency voltage, switching surges and standard impulses, *The Brown Boveri Review*, Vol. 54, No. 7, July 1967, pp. 368-376.
19. Kamata, Y. and Kako, Y. Flashover characteristics of extremely long gaps in transformer oil under non-uniform field conditions, *IEEE Transactions on Electrical Insulation*, Vol. EI-15, No. 1, February 1980, pp. 18-26.
20. Peek, F. W. The effect of transient voltages on dielectrics: Law of impulse spark-over and time lag, *AIEE Transactions*, October 1930, pp. 1456-1469.

21. Blume, L. F., Boyajian, A., Camilli, G., Lennox, T. C., Minneci, S., and Montsinger, V. M. *Transformer engineering*, John Wiley and Sons, New York, and Chapman and Hall, London, 1951.
22. Montsinger, V. M. Effects of time and frequency on insulation test of transformers, *AIEE Transactions*, Vol. 43, February 1924, pp. 337-347.
23. Vogel, F. J. Insulation tests of transformers as influenced by time and frequency, *AIEE Transactions*, Vol. 43, February 1924, pp. 348-355.
24. Lokhanin, A. K., Morozova, T. I., Voevodin, I. D., Beletsky, Z. M., Kuchinsky, G. S., and Kaplan, D. A. Problems of coordination of dielectric strength of extra high-voltage power transformer major insulation, *CIGRE 1970*, Paper No. 12-06.
25. Vogel, F. J. Factors influencing the insulation coordination of transformers, *AIEE Transactions*, June 1933, pp. 411-416.
26. Montsinger, V. M. Coordination of power transformers for steep-front impulse waves, *AIEE Transactions – Electrical Engineering*, Vol. 57, April 1938, pp. 183-195.
27. Bellaschi, P. L. and Teague, W. L. Dielectric strength of transformer insulation, *AIEE Transactions – Electrical Engineering*, January 1937, pp. 164-171.
28. Ikeda, M., Yanari, T., and Okubo, H. PD and BD probability distribution and equi-probabilistic V-t characteristics of oil-filled transformer insulation, *IEEE Transaction on Power Apparatus and Systems*, Vol. PAS-101, No. 8, August 1982, pp. 2728-2735.
29. Kamata, Y., Endoh, K., Furukawa, S., Endoh, F., Nonomura, K., Iwata, Y., Horiuchi, S., and Takasu, N. Dielectric strength of oil-immersed transformer insulation with superimposed AC and lightning impulse voltage, *IEEE Transaction on Electrical Insulation*, Vol. 25, No. 4, August 1990, pp. 683-687.
30. Vandermaar, A. J., Wang., M., Neilson, J. B., and Srivastava, K. D. The electrical breakdown characteristics of oil-paper insulation under steep front impulse voltages, *IEEE Transaction on Power Delivery*, Vol. 9, No. 4, October 1994, pp. 1926-1933.
31. Breitfelder, D., Buckow, E., Knorr, W., and Peschke, W. Dielectric strength of transformer oil under impulse and high frequency voltage stress, *International Symposium on High Voltage Engineering, ISH'87*, Braunschweig, Germany, August 1987.
32. Rasaruk. M. and Petcharaks, K. Thermal effect on AC breakdown characteristic of transformer oil under inhomogeneous fields, *IEEE Power Engineering Society Winter Meeting, 2000*, Vol. 3, 23-27 January 2000, pp. 2176 -2180.
33. Clark, F. M. *Insulating materials for design and engineering practice,* John Wiley and Sons, Inc., New York, 1962.

34. Clark, F. M. and Montsinger, V. M. Dielectric strength – thickness relation in fibrous insulation, *General Electric Review*, Vol. 28, No. 5, May 1925, pp. 286-290.
35. Wilson, W. R. A fundamental factor controlling the unit dielectric strength of oil, *AIEE Transactions*, February 1953, pp. 68-74.
36. Hill, L. R. and Schmidt, P. L. Insulation breakdown as a function of area, *AIEE Transactions*, Vol. 67, 1948, pp. 442-446.
37. Ikeda, M., Teranishi, T., Honda, M., and Yanari, T. Breakdown characteristics of moving transformer oil, *IEEE Transaction on Power Apparatus and Systems*, Vol. PAS-100, No. 2, February 1981, pp. 921-928.
38. Kato, K., Han, X., and Okubo, H. Insulation optimization by electrode contour modification based on breakdown area/volume effects, *IEEE Transactions on Dielectrics and Electrical Insulation*, Vol. 8, No. 2, April 2001, pp. 162-167.
39. Anderson, J. G. and Liao, T. W. The propagation mechanism of impulse creepage discharges over oil-immersed surfaces, *AIEE Transactions*, May 1955, pp. 218-226.
40. Okubo, H., Okamura, K., Ikeda, M., and Yanabu, S. Creepage flashover characteristics of oil/pressboard interfaces and their scale effects, *IEEE Transactions on Power Delivery*, Vol. PWRD-2, No. 1, January 1987, pp. 126-132.
41. Anker, M. U. Effect of test geometry, permittivity matching and metal particles on the flashover voltage of oil/solid interfaces, *IEEE Transactions on Power Apparatus and Systems*, Vol. PAS-102, No. 12, December 1983, pp. 3796-3802.
42. Taylor, R. J. Effect of permittivity matching on the flashover of solid/liquid interfaces, *Proceedings IEE*, Vol. 124, No. 10, October 1977, pp. 899-904.
43. Kamata, Y., Ohe, E., Endoh, K., Furukawa, S., Tsukioka, H., Maejima, M., Fujita, H., Nozaki, M., Ishizuka, F., and Hyohdoh, K. Development of low-permittivity pressboard and its evaluation for insulation of oil-immersed EHV power transformers, *IEEE Transactions on Electrical Insulation*, Vol. 26, No. 4, August 1991, pp. 819-825.
44. Nelson, J. K. Some steps toward the automation of the design of composite dielectric structures, *IEEE Transactions on Dielectrics and Electrical Insulation*, Vol. 1, No. 4, August 1994, pp. 663-671.
45. Derler, F., Kirch, H. J., Krause, C., and Schneider, E. Development of a design method for insulating structures exposed to electric stress in long oil gaps and along oil / transformerboard surfaces, *International Symposium on High Voltage Engineering, ISH'91*, Dresden, Germany, August 1991.
46. Moser, H. P. *Transformerboard*, Scientia Electrica, 1979.
47. Tschudi, D. J. AC insulation design: paper-oil insulation systems, *WICOR Insulation Conference*, Rapperswil, Switzerland, September 1996.

48. Dahinden, V., Schultz, K., and Kuchler, A. Function of solid insulation in transformers, *TRANSFORM 98*, April 1998, Germany, pp. 41-54.
49. Aubin, J. and Malewski, R. Safety margin of internal insulation of EHV power transformers, *International Symposium on High Voltage Engineering, ISH'95*, Graz, Austria, August 1995.
50. Grieder, H. Drying transformers with kerosene vapour, *Brown Boveri Review,* 2/3-73, 1973, pp. 117-122.
51. Kawamura, T. Proposals on standardization on the site installation to secure the reliability of transformer insulation, *CIGRE 1976*, Paper 12-01.
52. Weidmann, H. Catalogue, pp. A41-5, A41-6.
53. Lakhiani, V. K. and Kulkarni, S. V. Insulation design of EHV transformers – a review, *International Insulation Conference, INSUCON 2002*, Berlin, Germany, June 2002, pp. 283-287.
54. Tschudi, D., Dahinden, V., and Derler, F. Dielectric strength assessment of power transformer insulation, *International Symposium on High Voltage Engineering, ISH'95*, Graz, Austria, August 1995.
55. Vailles, C., Malewski, R., Dia-Do, X., and Aubin, J. Measurement of dielectric stress in EHV transformer insulation, *CIGRE 1994*, WG 33.03.
56. Kawaguchi, Y., Murata, H., and Ikeda, M. Breakdown of transformer oil, *IEEE Transactions on Power Apparatus and Systems*, Vol. PAS-91, January/February 1972, pp. 9-19.
57. Palmer, S. and Sharpley, W. A. Electric strength of transformer insulation, *Proceedings IEE*, Vol. 116, No. 12, December 1969, pp. 2029-2037.
58. Ikeda, M. and Menju, S. Breakdown probability distribution on equi-probabilistic V-T characteristics of transformer oil, *IEEE Transaction on Power Apparatus and Systems*, Vol. PAS-98, No. 4, July-August 1979, pp. 1430-1437.
59. Rengarajan, S. and Satyanarayana, S. Computation and analysis of HV lead clearance to tank wall in large power transformers using stressed oil volume concept, *International Symposium on High Voltage Engineering, ISH'95*, Graz, Austria, August 1995, pp. 1042:1-4.
60. Koppikar, D. A., Kulkarni, S. V., and Dubey, A. K. Optimization of EHV Transformer Insulation by Statistical Analysis, *International Symposium on High Voltage Engineering, ISH'97,* Montreal, August 1997, Vol. 6, pp. 289-292.
61. Phadke, M. S. *Quality engineering using robust design*, Prentice-Hall International, Inc, 1989.

9

Cooling Systems

The magnetic circuit and windings are the principal sources of losses and resulting temperature rise in various parts of a transformer. Core loss, copper loss in windings (I^2R loss), stray loss in windings and stray loss due to leakage / high current field are mainly responsible for heat generation within the transformer. Sometimes loose electrical connections inside the transformer, leading to a high contact resistance, cause higher temperatures. Excessive temperatures due to heating of curb bolts, which are in the path of stray field, can damage gaskets (refer to Chapter 5). The heat generated due to all these losses must be dissipated without allowing the core, winding and structural parts to reach a temperature which will cause deterioration of insulation. If the insulation is subjected to temperatures higher than the allowed value for a long time, it looses insulating properties; in other words the insulation gets aged, severely affecting the transformer life. There are two principle characteristics of insulation: dielectric strength and mechanical strength. The dielectric strength of insulation aged in oil remains high up to a certain temperature after which it drops rapidly. At this point the insulation material becomes brittle and looses its mechanical strength. Thus, it is primarily the mechanical strength which gets affected by the higher temperatures and aging, which in turn affects the dielectric strength. Hence, the dielectric strength alone cannot always be depended upon for judging the effect of temperature on the insulation [1].

Accurate estimation of temperatures on all surfaces is very critical in the design of transformers to decide the operating flux density in core and current densities in windings / connections. It helps in checking the adequacy of cooling arrangements provided for the core and windings. It also helps in ensuring reliable operation of the transformer since the insulation life can be estimated under overload conditions and corrective actions can be taken in advance.

The values of maximum oil and winding temperatures depend on the ambient temperature, transformer design, loading conditions and cooling provided. The limits for ambient temperature and the corresponding limits for oil temperature rise and winding temperature rise are specified in the international standards. As the ambient temperature varies from one country to another, the limits could be different for different countries. For example in IEC 60076-2 (second edition: 1993), a maximum ambient temperature of 40°C is specified with a limit on top oil temperature rise of 60°C. In a country where the maximum ambient temperature is 50°C, the top oil temperature rise limit may be correspondingly reduced to 50°C. If the installation site is more than 1000 m above the sea level, the allowable temperature rise for transformers is reduced as per the guidelines given in the standards because of the fact that air density reduces with the increase in altitude lowering the effectiveness of cooling. Altitude basically affects the convective heat transfer (because of lower buoyancy effect) and not the radiation. A corresponding reverse correction is applied when the altitude of factory location is above 1000 m and the altitude of installation site is below 1000 m.

In oil cooled transformers, the oil provides a medium for both cooling and insulation. Heat from core, windings and structural components is dissipated by means of the oil circulation. The heat is finally transmitted either to atmospheric air or water. In the subsequent sections, modes of heat transfer and their application in different cooling configurations in a transformer are discussed.

9.1 Modes of Heat Transfer

The heat transfer mechanism in a transformer takes place by three modes, viz. conduction, convection and radiation. In the oil cooled transformers, convection plays the most important role and conduction the least important. Rigorous mathematical treatment for expressing these modes of heat transfer is quite difficult and hence designers mostly rely on empirical formulae.

9.1.1 Conduction

Almost all the types of transformers are either oil or gas filled, and heat flows from the core and windings into the cooling medium. From the core, heat can flow directly, but from the winding it flows through the insulation provided on the winding conductor. In large transformers, at least one side of insulated conductors is exposed to the cooling medium, and the heat flows through a small thickness of the conductor insulation. But in small transformers the heat may have to flow through several layers of copper and insulation before reaching the cooling medium.

The temperature drop across the insulation due to the conduction heat transfer mechanism can be calculated by the basic thermal law:

$$\Delta\theta = Q \times R_T \tag{9.1}$$

where Q is heat flow (power loss) in W and R_T is thermal resistance in °C/W. The thermal resistance is given by

$$R_T = \frac{t_i}{k \times A} \tag{9.2}$$

where t_i is the insulation thickness in m, A is cross-sectional area in m^2, and k is thermal conductivity in W/(m °C). If q denotes heat flux per unit transfer area, the temperature drop across the insulation can be rewritten as

$$\Delta\theta = \frac{q \times t_i}{k} \tag{9.3}$$

It should be noted that the thermal conductivity of oil-impregnated paper insulation is temperature dependent and its proper value should be taken in the calculations [2].

9.1.2 Radiation

Any body, at a raised temperature compared to its surroundings, radiates heat energy in the form of waves. The heat dissipation from a transformer tank occurs by means of both radiation and natural convection. The cooling of radiators also occurs by radiation, but it is far less as compared to that by convection. Because of closeness of radiator fins, the entire radiator surface does not participate in the heat transfer mechanism by radiation. Thus, the effective area for radiation can be taken as the outside envelope surface of the radiator. Therefore, for the case of tank with radiators connected to it, actual radiating surface area is that area on which a tightly stretched string would lie. The emissivity of the radiating surface affects the radiation. The heat transfer in watts by radiation is expressed by the Stephan-Boltzmann law:

$$P_R = \eta\, E\, A_R \left(T_s^4 - T_a^4\right) \tag{9.4}$$

where $\eta = 5.67 \times 10^{-8}$ W/(m^2 °K^4) is the Stephan-Boltzmann constant, E is surface emissivity factor, A_R is surface area for radiation in m^2, T_s is average temperature of radiating surface in °K, and T_a is ambient air temperature in °K.

Surface emissivity is a property, which depends on several factors like surface finish, type of paint applied on the surface, etc. When the emissivity factor is less than unity, the effective radiating surface is correspondingly less (as indicated by the above equation). For tank and radiators painted with grey colour having emissivity of 0.95, the effective radiating area is usually assumed to be that of outside envelope without introducing much error.

9.1.3 Convection

The oil, being a liquid, has one important mechanical property that its volume changes with temperature and pressure [3]. The change of volume with temperature provides the essential convective or thermosiphon cooling. The change of volume with pressure affects the amount of transferred vibrations from the core to tank.

The heat dissipation from the core and windings occurs mainly due to convection. When a heated surface is immersed in a fluid, heat flows from the surface to the cooling medium. Due to increase in the fluid temperature, its density (or specific gravity) reduces. The fluid (oil) in oil-cooled transformers, rises upwards and transfers its heat to outside ambient through tank and radiators. The rising oil is replaced by the colder oil from the bottom, and thus the continuous oil circulation occurs. The convective heat transfer is expressed by the relationship:

$$Q = h\,A\left(T_{surface} - T_{fluid}\right) \tag{9.5}$$

where Q is heat flow in W, h is heat transfer coefficient in W/(m^2 °C), A is surface area in m^2, and temperatures $T_{surface}$ and T_{fluid} are in °C. Since h depends on both geometry as well as fluid properties, its estimation is very difficult. However, a lot of empirical correlations are available, which can be used in majority of design calculations. In one such correlation, the heat dissipated per unit surface area is expressed as equal to a constant multiplied by temperature rise raised to an empirical coefficient.

The heat dissipation from the transformer tank to ambient air occurs similarly but the warmed air after cooling does not come back and its place is occupied by new quantity of fresh air. In the case of tank, heat dissipation by convection and radiation mechanisms are comparable since the surface area available for the convective cooling is same as that for the radiation cooling. The heat dissipated by the tank through the convection and radiation is also usually calculated by empirical relations in which the resultant effect of both the mechanisms is taken into account.

9.2 Cooling Arrangements

9.2.1 ONAN / OA cooling

In small rating transformers, the tank surface area may be able to dissipate heat directly to the atmosphere; while the bigger rating transformers usually require much larger dissipating surface in the form of radiators / tubes mounted directly on the tank or mounted on a separate structure. If the number of radiators is small, they are preferably mounted directly on the tank so that it results in smaller overall dimensions.

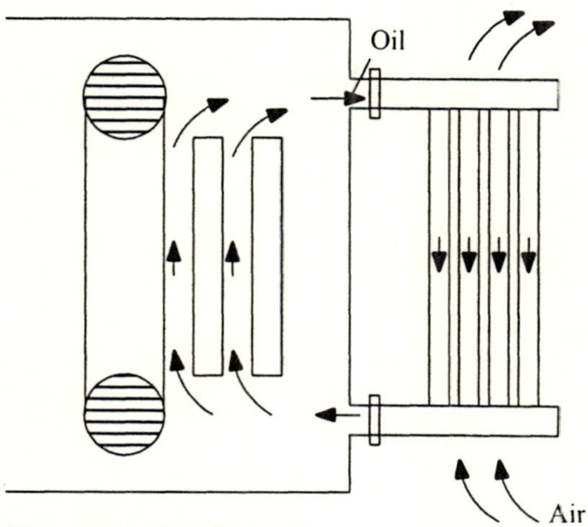

Figure 9.1 ONAN cooling

When number of radiators is large, they are mounted on a separate structure and the arrangement is called as radiator bank. The radiators are mounted on headers, which are supported from the ground. In this case, strict dimensional control of pipes and other fittings is required in order to avoid oil leakages.

Oil is kept in circulation by the gravitational buoyancy in the closed-loop cooling system as shown in figure 9.1. The heat developed in active parts is passed on to the surrounding oil through the surface transfer (convection) mechanism. The oil temperature increases and its specific gravity drops, due to which it flows upwards and then into the coolers. The oil heat gets dissipated along the colder surfaces of the coolers which increases its specific gravity, and it flows downwards and enters the transformer tank from the inlet at the bottom level. Since the heat dissipation from the oil to atmospheric air is by natural means (the circulation mechanism for oil is the natural thermosiphon flow in the cooling equipment and windings), the cooling is termed as ONAN (Oil Natural and Air Natural) or OA type of cooling.

In the arrangement consisting of radiator banks, higher thermal head can be achieved by adjusting the height of support structures. The thermal head can be defined as the difference between the centers of gravity of fluids in the tank and radiator bank. Although it is difficult to get higher thermal head for the case of tank mounted radiators, reasonable amount of thermal head is achieved by the arrangement shown in figure 9.2. When the radiators are mounted at higher height, the buoyancy effect on the cooling-loop increases resulting in increase of the rate of oil flow and heat dissipation in the cooling equipment. However, it is to be noted that the increase in flow rate results in increased frictional pressure loss, thereby offsetting the thermal head gained by the height difference.

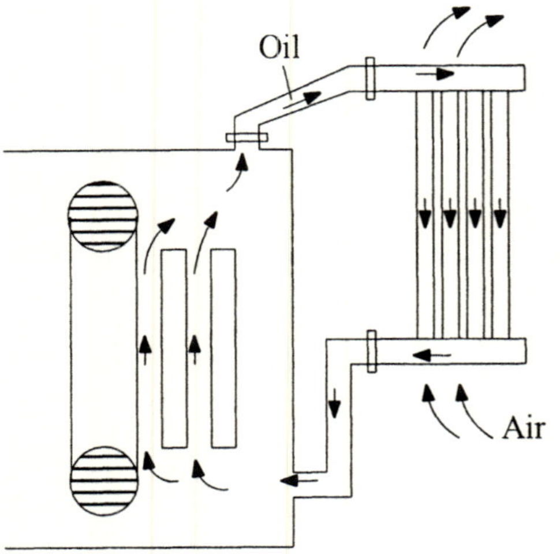

Figure 9.2 Arrangement for higher thermal head

9.2.2 ONAF / FA cooling

As the transformer rating increases, the total loss to be dissipated also increases. One way of increasing the heat transfer is to increase the heat transfer coefficient between the radiator outside surface and air (equation 9.5). In this equation, for a radiator $T_{surface}$ corresponds to its outside wall surface temperature. However, the temperature drop across the radiator plate is very small, hence $T_{surface}$ can be considered as the oil temperature itself. If fans are used to blow air on to the cooling surfaces of the radiators, the heat transfer coefficient is significantly increased. For a given set of ambient air temperature and oil temperature, a compact arrangement is possible since less number of radiators is required to cool the oil. This type of cooling is termed as ONAF (Oil Natural and Air Forced) or FA type of cooling.

If there is a particular case in which either ONAN or mixed ONAN/ONAF cooling can be specified; the ONAN cooling has the following advantages (although it may take more space):

- it is more reliable as no cooler controls are involved and it requires less maintenance.
- the cost increase due to extra radiators is, to a large extent, compensated by the reduction in cost due to the absence of fans and control system.
- it is particularly useful when low noise transformers are required. Absence of fans makes it easier to achieve the required low noise level.
- there is no cooler loss.
- winding losses also reduce (although marginally) because of lower winding temperature rise at fractions of rated load as compared to the mixed cooling.

- most of the time, when load on the transformer is less than its full rating, temperature rise inside the transformer is low and its life increases (gain of life).

Thus, in cases where the ONAN rating is 75% or more (it is closer to the ONAF rating), ONAN cooling can be specified instead of mixed ONAN/ONAF cooling based on cost-benefit analysis.

There are two typical configurations for mounting fans in ONAF cooling. One method is to mount the fans below the radiators, which blow air from bottom to top. Larger capacity fans can be used since it is easy to design the support structures for them. In this system the fans can be either supported directly from the radiators or they can be ground mounted. Care should be taken that the fans mounted on radiators do not produce appreciable vibrations. Usually, sufficient surface of radiators is covered in the air-flow cone created by the fan; the remaining surface is taken to be naturally cooled. In the second method, fans are mounted on the side of radiators. These fans are relatively smaller in size compared to the first arrangement since the number of fans is usually more for this configuration. Both the configurations have their own advantages and disadvantages, particular selection depends on the specific design requirement.

9.2.3 OFAF / FOA cooling

As discussed previously, the flow rate inside the windings under ONAN and ONAF cooling arrangements is governed by the natural balance between the viscous resistance and the thermosiphon pressure head. Normally this flow rate is relatively low. Because of this, the heat carrying (or dissipating) capacity of the oil is low. The heat carrying capacity can be defined as

$$Q = \dot{m}\, C_p \left(T_{out} - T_{in}\right) \tag{9.6}$$

where Q is heat flow in W, $\dot{m}$ is mass flow rate in kg/s, C_p is specific heat in J/(kg °C), and temperatures T_{out} and T_{in} are in °C. For the given transformer oil inlet (T_{in}) and top oil (T_{out}) temperatures, the only way to increase the heat dissipation capability is to increase $\dot{m}$. This necessitates the use of an external pump to circulate the oil in high rating transformers. Also, in order to get a higher heat transfer rate, fans have to be always operating at the radiator sections. This type of cooling is called as OFAF (Oil Forced and Air Forced) or FOA cooling. There are basically two types of pump designs: axial flow in-line type and radial flow type for circulating oil against low and high frictional head losses respectively. The axial flow type is used with mixed cooling (ONAN / ONAF / OFAF) since it offers less resistance when switched-off. The radial flow type pumps, which offer very high resistance to oil flow under the switched-off condition, are used with oil-to-air heat exchangers (unit cooler arrangement) or oil-to-water heat exchangers in which no natural cooling is provided. The head required to be developed for these two types of compact heat exchangers is quite high and the radial flow pump can cater to this requirement quite well.

In OFAF cooling arrangement, when fans are mounted on the sides of radiators, they should be uniformly distributed over the radiator height, whereas for ONAF cooling more fans should be mounted at the top of radiator height. This is because in OFAF condition, the temperature difference between top and bottom portions of radiators is small as compared to that under ONAF condition.

When the oil is forced into the transformer (figure 9.3), its flow is governed by the least resistance path as well as the buoyancy. Hence, part of the oil may not enter either windings or core, and may form a parallel path outside these two. Thus, the top oil temperature may reduce because of the mixture of hot oil coming from the windings and the cool oil coming from the pump. This in turn reduces the effectiveness of radiators. The heat dissipation rate can be improved if the oil is forced (by use of pumps) and directed in the windings through the predetermined paths as shown in figure 9.4. This type of cooling is termed as ODAF (Oil Directed and Air Forced) type of cooling. ODAF type of cooling is used in most of the large rating power transformers. One disadvantage of ODAF cooling is the increased pressure loss because of the ducting system used for directing the oil flow. For each winding, the oil flow rate is required to be determined accurately. In the absence of proper oil flow rates, an unreasonable temperature rise will result. Additionally, any blockage or failure of the ducting system leads to higher temperature rise.

Generally, the higher the pump capacity (and the greater the oil velocity) the higher the rate of heat dissipation is. Hence, during the early development, there was a general trend for using higher capacity pumps permitting higher loss density (use of higher current density in windings and / or higher flux density in core), leading to lower material cost and size of transformers. The trend continued till a number of large transformers failed due to the phenomenon called static electrification (explained in Section 9.6). Hence, the oil pump capacity should be judiciously selected.

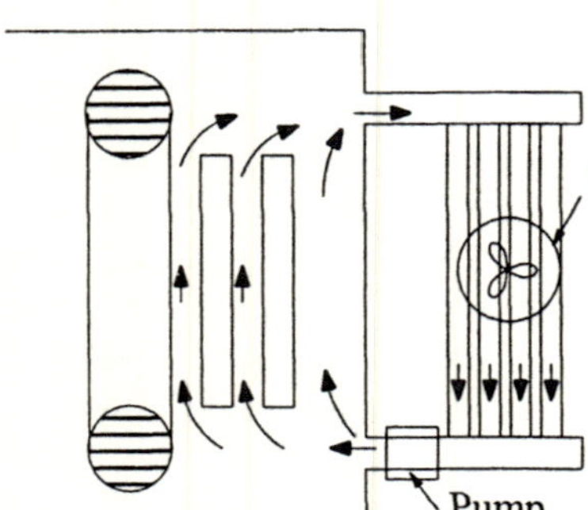

Figure 9.3 OFAF cooling

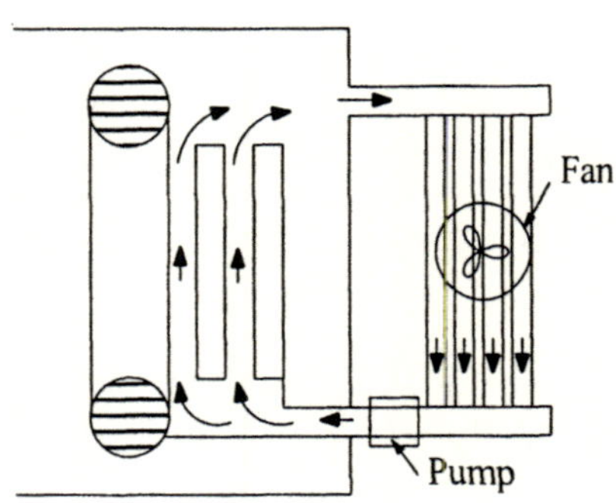

Figure 9.4 ODAF cooling

9.2.4 Unit coolers

As mentioned earlier, sometimes OFAF cooling is provided through the use of compact heat exchangers when there is space constraint at site. In this small box type structure, an adequate surface area is provided by means of finned tubes. Usually, about 20% standby cooling capacity is provided. Disadvantage of these coolers is that there is only one rating available (with running of fans and pumps). If the system of fans and pumps fails (e.g., failure of auxiliary supply), ONAN rating is not available. Hence, the continuity of auxiliary supply to fans and pumps is required to be ensured.

9.2.5 OFWF cooling

For most of the transformers installed in hydropower stations, where there is abundance of water, oil-to-water heat exchangers are used. As the surface heat transfer coefficient of water is more than air, such type of cooling results in smaller radiators. This type of cooling is termed as water forced (WF) cooling. Depending on the type of oil circulation, the transformer cooling system is termed as OFWF or ODWF type of cooling. During operation, it is very important to ensure that the oil pressure is always more than the water pressure so that the possibility of water leaking into the oil is eliminated. A dedicated differential pressure gauge and the corresponding protection circuit are used to trip the transformer if a specific value of pressure difference between the oil and water is not maintained during the operation.

9.3 Dissipation of Core Heat

As the transformer core size increases, it becomes more important to decide the positions of cooling ducts in it. These cooling ducts (shown in figure 9.5) reduce both the surface temperature rise of the core relative to that of oil and the temperature rise of the interior of the core relative to that at the surface.

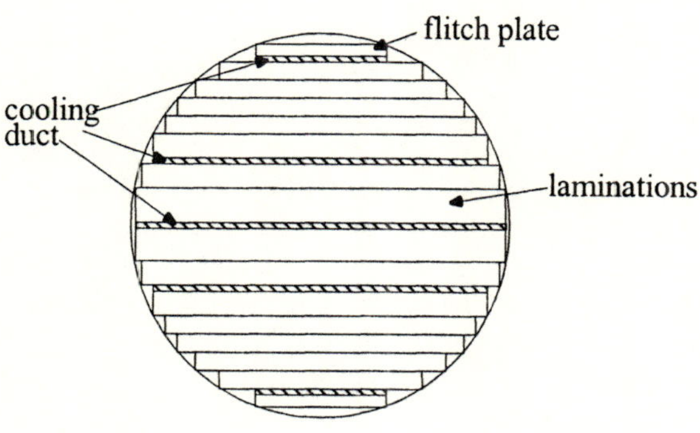

Figure 9.5 Core cooling ducts

It is necessary to maximize core area (net iron area) to get an optimum design. The cooling ducts reduce the core area, and hence their number should be as minimum as necessary. This requires accurate determination of temperature profile of the core and effective placement of the cooling ducts. The complicated geometry of the boundary surface between the core and oil, and the anisotropy of the thermal conductivity of the laminated core are some of the complexities involved in the computations. A general formulation of the approximated two-dimensional problem of temperature distribution in rectangular cores subjected to linear boundary conditions (thermal resistance being independent of heat flow and oil temperature) is given in [4]. The method described in [5] solves the two-dimensional problem by transforming Poisson's equation of heat conduction into Laplace's equation. The method can be applied to any arbitrary shape due to use of a functional approximation. The paper also reports the use of electrical analog method which uses the analogy between electrical potential difference and temperature difference, between electrical current and heat flow, and between electrical conductivity and thermal conductivity. The calculation of temperature distribution in the transformer core is a complex three-dimensional problem with non-uniform heat generation. Furthermore, the thermal properties of core are anisotropic in the sense that the thermal conductivity along the plane of laminations is quite different from that across them. The problem can be solved by using three-dimensional finite element thermal formulation with the anisotropic thermal material properties taken into account.

The surface of core is normally in contact with the insulation (between core and frame). Hence, the limit on the core surface temperature is the same as that for the windings. For the interior portions of the core which are in contact with only the oil (film), the limit is 140°C. In most cases, the temperature difference between the core interior (e.g., mid-location between two cooling ducts) and surface is about 15 to 20°C.

9.4 Dissipation of Winding Heat

Radial spacers (pressboard insulations between disks / turns) cover about 30 to 40% of the winding surface, making the covered area ineffective for the convective cooling. The arrangement is shown in figure 9.6. Thus, although higher spacer width may be required from the short circuit withstand considerations, it is counterproductive for cooling. Hence, while calculating the gradients, only the uncovered winding surface area is taken into account.

Heat from the covered winding area is transferred to the uncovered area by thermal conduction process increasing thermal load on the uncovered surfaces. Contrary to the width of radial spacer, cooling is improved with the increase in its thickness. Hence, radial spacers may not be required from insulation considerations in low voltage windings, but they are essential for providing the cooling ducts.

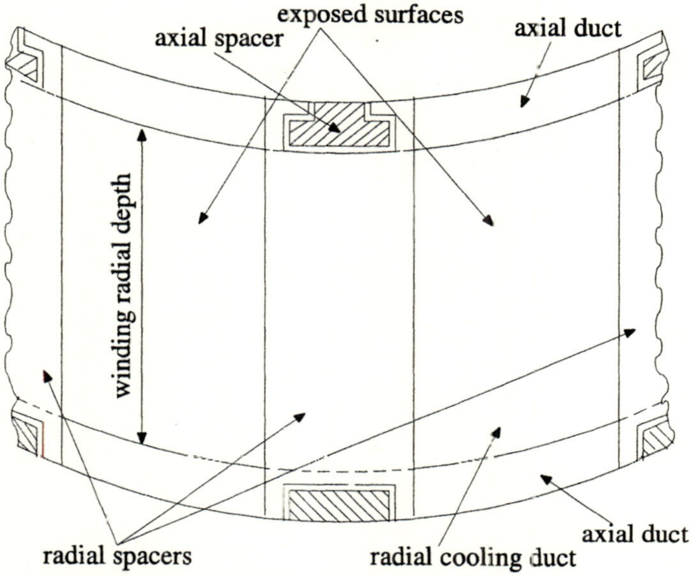

Figure 9.6 Effect of radial spacers on cooling

The required spacer thickness bears a specific relationship with the radial depth of winding. For a given radial depth, a certain minimum thickness of radial spacers is required for effective cooling (otherwise resistance to oil flow in the duct between two disks / turns is higher and the oil largely flows through the axial ducts at inside and outside diameters resulting in higher temperature rise at the middle portion of the radial depth).

When the winding radial depth is quite high, the usual practice of providing two axial ducts at the inside and outside diameters (along with the radial ducts) may not be enough. Hence, some manufacturers provide an additional axial cooling duct in the middle of the radial depth as shown in figure 9.7, so that the thickness of the radial spacers can be lower. With this arrangement, the axial space factor of the winding improves (due to reduction of insulation along the winding height), but the radial space factor worsens. Hence, the design and dimensioning of the axial and radial spacers have to be judiciously done, which may also depend on the manufacturing practices.

Axial ducts play an important role in dissipation of heat from the windings. The higher the axial duct width, the better the oil flow conditions are; this is more valid for windings without radial ducts. In large transformers with radial cooling ducts between disks / turns, thicknesses of the axial ducts (at the inside and outside diameters of the winding) and radial ducts decide the oil velocity within the winding and the rate of heat dissipation.

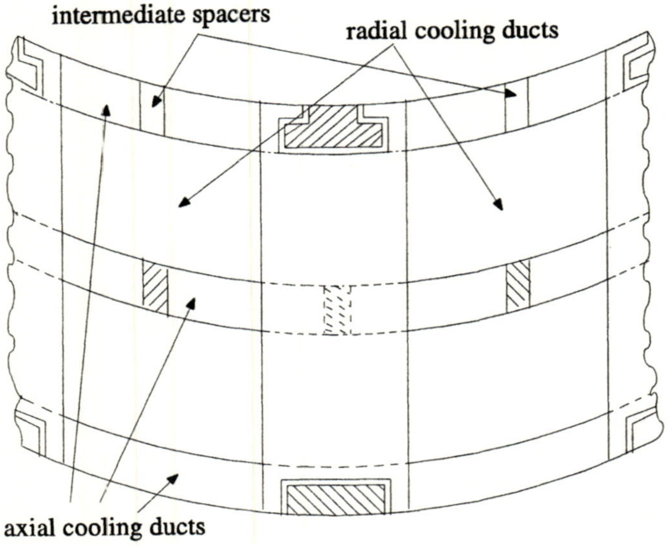

Figure 9.7 Axial cooling ducts

The axial duct at the winding outside diameter is usually of the order of 10 to 12 mm, whereas the duct at the winding inside diameter is kept equal to or below 8 mm from the dielectric strength consideration (as explained in Chapter 8). Thus, the thermal and dielectric requirements are in conflict while designing the thickness of the inner axial duct. The size of axial duct less than 6 mm is not preferred from the thermal consideration. Narrow ducts or manufacturing deficiencies result in higher flow resistance, thereby leading to an unacceptable temperature rise within the winding.

In power transformers, a guided oil flow is commonly used to cool the windings effectively. The oil guiding is achieved by use of washers as shown in figure 9.8. The washers have to be accurately cut to the required dimensions so that there is proper sealing at the desired locations eliminating oil leakages. If oil guiding strips are used, they have to be securely held to the innermost conductor (sealing at the inside diameter) or the outermost conductor (sealing at the outside diameter) along the circumference. The location and number of these oil guiding components have to be properly selected. Oil flow through multiple passages has been studied in [6], in which the oil flow rates are shown to be different in upper ducts as compared to lower ducts. Similar trends are reported in [7] for SF6 flow rate in gas cooled transformers. The effects of number of radial ducts between two consecutive oil guiding washers, width of radial duct and width of axial duct on the value of hot spot temperature have been reported in [8] based on a number of FEM simulations.

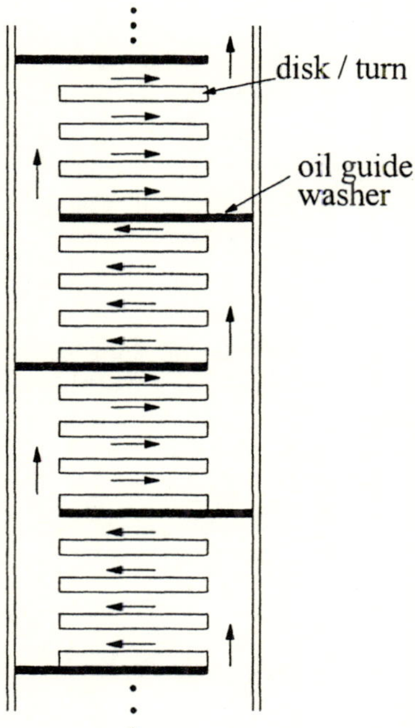

Figure 9.8 Directed oil flow in a winding

When winding gradients are calculated by simple formulae, average loss per disk (in disk winding) or per turn (in layer winding) is considered, which is calculated by adding the average eddy loss in the disk or turn to its I^2R loss. The effects of winding curvature can be neglected. The average temperature rise of the winding above the average oil temperature is then calculated as the sum of the temperature drop across the conductor paper insulation and the surface temperature drop. It is to be noted that the heat flux per unit surface area used in these calculations should be determined by considering only the exposed horizontal surface area of the winding (not covered by the radial spacers).

Although the temperature rise calculations in transformers have been heavily relying on empirical factors, there have been continuous efforts for determining the temperatures through accurate formulations. Equations governing the distributions of interlayer temperature and duct-oil velocity under thermal and hydrodynamic conditions are derived in [9] for layer-type transformer winding. A resistive-network analog computer has been used in the iterative solution of equations. Numerical solution using finite difference method is presented in [10] for disk-type transformer winding. A general network method is described in [11], in which interconnecting flow paths or ducts are represented by a network diagram, each element of which corresponds to a single path with nodes placed at the junctions. Solution is obtained by a numerical procedure which predicts the temperature distribution in ducts. The temperature distribution in the winding does not vary linearly with its height as usually assumed. Variation in temperature with winding height is close to the linear distribution in forced oil cooling, whereas for naturally oil cooled transformers (ONAN and ONAF) it can be quite nonlinear [12].

9.5 Aging and Life Expectancy

As explained earlier, the insulation properties and hence the life of a transformer is basically decided by the mechanical strength of insulation under the normal aging process. Brittleness in the paper insulation is strongly conducive to dielectric failure by various mechanisms. The period of time, until deterioration of an insulating material becomes critical, is called the service life or life expectancy. The service life of a transformer can be calculated by using the Montsinger formula applicable to the 80 to 140°C range of temperatures,

$$\text{Life} = D e^{-p\theta} \text{ years} \tag{9.7}$$

where D is a constant expressed in years, p is a constant expressed in $°\mathrm{C}^{-1}$, and θ is temperature in °C. It is generally recognized that the rate of aging as measured by tensile strength of class A insulation doubles for each 5 to 10°C increase in temperature [1]. It was reported in year 1930 [13] that the aging of tensile strength gets doubled for approximately each 8°C increase in the temperature. Subsequent tests reported [14] on paper insulation confirmed this aging rate. During discussions in CIGRE Transformer Working Group in 1961, 6°C was considered to be a more correct value [15] and is now used widely by transformer designers and users.

Hence, for the temperature range of 80 to 140°C, the life expectancy is taken to be halved for the increase in temperature of 6°C. This means that if a service life of N years applies for a temperature θ °C, the temperature of $(\theta + 6)$ °C will reduce the life by $(N/2)$ years. With these figures, the constant p in the Montsinger formula can now be determined. Rewriting equation 9.7 as

$$N = D e^{-p\theta} \text{ years} \tag{9.8}$$

$$\frac{N}{2} = D e^{-p(\theta+6)} \text{ years} \tag{9.9}$$

and taking the ratio of the above two equations we get

$$2 = e^{p6}$$

$$\therefore \; p = 0.1155 °C^{-1} \tag{9.10}$$

If the life expectancy at some temperature θ_b is chosen to be normal, then that applicable to an arbitrary temperature θ can be related to this normal life expectancy. The ratio of these two life expectancies is called as aging factor or relative reduction of life, and is given using the Montsinger relation as

$$K_{ag} = \frac{De^{-0.1155\,\theta_b}}{De^{-0.1155\theta}} = e^{0.1155(\theta-\theta_b)} \tag{9.11}$$

The maximum limits on current and temperature are given in standards, which are applicable to loading beyond the nameplate rating. For example, as per the IEC standard 60354: 1991, the limits are given separately for three categories of transformers: distribution, medium power and large power transformers. For each of these, the limits are specified for three types of loadings, viz. normal cyclic loading, long-time emergency cyclic loading and short-time emergency loading. The limits are lower for large power transformers. Under normal cyclic loading conditions a current of 150% of the rated value and hot spot temperature of 140°C for metallic parts in contact with insulating material are allowed for distribution and medium power transformers, whereas for large power transformers the corresponding limits are 130% and 120°C. For all transformers the top oil temperature limit of 105°C is specified under normal cyclic loading conditions, whereas the limit of 115°C is specified for long-time emergency cyclic loading and short-time emergency loading.

Pure oil, free from impurities and sealed from the atmosphere, can withstand temperature up to 140°C [3] which is the flash point of oil. Generally, the temperature of structural components and other metallic parts should be limited up to 135 to 140°C, provided this temperature occurs over a small surface area of few cm^2 which is in contact with a bulk quantity of oil [2].

It is generally accepted that a continuous operation with the hot spot temperature of 98°C results into the normal use of life or rate of degradation. This is the temperature at which the insulation of a transformer deteriorates at the normal rate. Thus, if we choose 98°C as the temperature corresponding to the normal life expectancy (IEC standard 60354: 1991), then the aging factor applicable to an arbitrary temperature θ can be given as

$$K_{ag} = e^{0.1155(\theta-98)} \tag{9.12}$$

If the ambient temperature (θ_a) is appreciably varying during the period under consideration, the weighted ambient temperature should be taken for the hot spot temperature calculation. The weighted ambient temperature θ_{awt} is that temperature which lasts for the same time t and causes the same loss of expectancy as the ambient temperature $\theta_a(t)$ which is varying in magnitude with time. For equal reduction of life expectancy the following equation holds

$$t\,e^{0.1155\,\theta_{awt}} = \int_0^t e^{0.1155\,\theta_a(t)}\,dt \tag{9.13}$$

The weighted ambient temperature can be fixed once for all in the given climatic conditions. This temperature, when calculated by equation 9.13, is such that the extra loss of life during the summer season is compensated by the gain in life in the winter season.

The temperature rise limits set by the standards are in line with the hot spot limit for the normal aging of a transformer. The hot spot limit of 98°C is applicable to the average winding temperature rise of 55°C above ambient temperature (for transformers without thermally upgraded insulation). The corresponding limit is 95°C as per the IEEE standard C57.91: 1995. As per this standard the limit is 110°C for a thermally upgraded insulation (for transformers with 65°C average winding rise), i.e., the maximum (hottest spot) winding temperature rise of 80°C is allowed above the average ambient temperature of 30°C for the normal aging.

The steady-state hot spot temperature is calculated as the sum of ambient external cooling medium temperature, temperature rise of top oil above the cooling medium and hot spot winding gradient. Let us assume that the hot spot gradient is 1.1 times the average gradient (which is the difference between the average winding temperature rise and average oil temperature rise). For yearly weighted average ambient temperature of 20°C, and top oil rise and average winding rise limits of 60°C and 65°C respectively (as per IEC standard 60076-2: 1993), the hot spot temperature is

$$\text{hot spot temperature} = 20 + 60 + (65 - 0.8 \times 60) \times 1.1 \cong 98\,^{\circ}\mathrm{C}$$

where the average oil temperature rise is taken as 80% of the top oil temperature rise, which is usually true for the natural oil circulation. For forced oil condition, the hot spot temperature is lower since the average oil temperature rise is closer to the top oil temperature rise. Thus, it can be seen that if the temperature rise of oil and windings are within the limits set by the standards, the hot spot temperature will not be exceeded and a certain transformer life (of few decades), with the normal aging process, can be expected. If the hot spot gradient is 1.3 times the average gradient, the top oil rise and average winding rise should be lower so that the hot spot temperature limit is not exceeded. Thus, based on the calculated value hot spot gradient, designer may have to limit the top oil rise and average winding rise values.

Example 9.1

The loading of a transformer is such that the hot spot temperature does not exceed 92°C for 18 hours in a day. For remaining 6 hours, during peak load periods, the transformer can be overloaded. What will be the temperature corresponding to the allowed overload condition?

Solution:

Here, the objective is to have a normal use of life over the whole period of the day. Transformer gains life during the lower load period of 18 hours and this gain of life can be utilized to allow an overload. The temperature θ during the overload period of 6 hours is given by

$$24 = 18\, e^{0.1155\,(92-98)} + 6\, e^{0.1155\,(\theta-98)} = 18 \times 0.5 + 6\, e^{0.1155\,(\theta-98)}$$

$$\therefore e^{0.1155\,(\theta-98)} = \frac{24 - 18 \times 0.5}{6} = 2.5$$

$$\therefore \theta = \frac{\ln(2.5)}{0.1155} + 98 = 106°\mathrm{C}$$

If the thermal time constant of the transformer is considered, the duration of the overload can be higher than the one calculated above. The usual values of thermal time constants for transformers lie in the range of 1 hour to 8 hours; smaller values are for the forced cooled and larger values are for the naturally cooled transformers. The thermal time constants of windings considered alone are much smaller (in the range of few minutes to tens of minutes).

Example 9.2

Estimate the relative extent of insulation aging of the transformer winding subjected to a temperature of 136°C for 3 hours during a day.

Solution:

The aging factor calculated by equation 9.11 is

$$K_{ag} = e^{0.1155(136-98)} = 80.6$$

Hence, the operation of the transformer at 136°C for 3 hours is equivalent to $3 \times 80.6 = 241.8$ hours (approximately 10 days) of operation with the normal aging process.

Here again, the thermal time constant is ignored which results into a very conservative calculation. For accurate calculations the IEC standard 60354: 1991 or IEEE standard C57.91: 1995 can be used.

It is well-known that the paper aging is highly dependent on temperature and the presence of water and oxygen. The thermally upgraded paper is less affected by the presence of water and oxygen as compared to the normal paper, and hence its use is technically beneficial [16]. The thermally upgraded paper can be used for increasing the overloading capability and life of a transformer. Many transformer manufacturers are using it but the use is still not widespread. Furfural analysis of commonly available thermally upgraded kraft papers has been reported in [17]; it is suggested to have a detailed knowledge of the thermally upgraded paper and its behavior during the aging process before its use is made in transformers.

Most of the earlier work is based on the use of tensile strength as a measure to assess the remaining useful life; reduction of tensile strength to 50% of its original value is used as the criterion. The more precise DP (degree of polymerization) method, which is now widely getting acceptance, can give the same information more conveniently.

Transformers subjected to a non-sinusoidal duty deserve special attention. The hot spot temperature and loss of life should be accurately estimated [18]. A suitable derating factor has to be used to compensate the effects of harmonics.

9.6 Direct Hot Spot Measurement

The rate of deterioration of the winding insulation increases as the conductor temperature increases. Hence, it is necessary to know the hottest conductor temperature in order to ensure a reasonable and expected life for the insulation and the transformer. The oil temperature is higher at the top. Also, there is usually higher loss density in the winding at the top because of eddy loss due to radial leakage field. In addition, an extra insulation may have been provided to line end turns at the top. All these reasons lead to having the hot spot at disks / turns in the top portion of the winding. The hot spot is usually assumed to be present at the second disk / turn from the top. The hot spot winding gradient can be about 1.1 to 1.3 times the average winding gradient over the average oil rise. The winding temperature is traditionally measured by a winding temperature indicator (WTI) which uses the principle of thermal imaging. Thermometer type sensor placed in a tank pocket, which measures the hot oil temperature at the top of tank, is surrounded by a heater coil carrying a current in proportion of load current. After proper adjustment of settings, the device reads the hot spot temperature of the winding. The settings are specified by designers based on the calculated value of the hot spot gradient. Since the actual value of hot spot gradient is a function of many design and manufacturing parameters, viz. winding eddy loss density, effectiveness of cooling ducts provided, etc., the accuracy of winding hot spot temperature measurement by WTI may not always be good. Hence, the direct hot spot measurement technique is being increasingly specified and used by transformer users. In this method a sensor, made of photo-luminescent material and attached to the end of optical fiber, is in thermal contact with the winding. The sensor is usually placed between insulated conductor and radial spacer. The fiber-optic cable is brought out of the tank up to the instrument through a hole made in the tank with a proper oil-sealing arrangement. A pulse of light from the light emitting diode (LED) in the instrument is sent to the sensor through the fiber-optic cable, which stimulates the sensor material to fluoresce. Depending on the decay time of the returning fluorescent signal, which is a function of conductor temperature, the instrument calculates and displays the corresponding hot spot temperature by lookup table approach. By using the fiber-optic sensor, accurate measurement of hot spot temperature can be done but the sensor insertion method is critical. The requirements of the measurement system are: the sensor should be sufficiently small, signal transmission system should not degrade dielectric strength of the transformer, and the components used should withstand thermal, mechanical, and chemical rigors of the transformer environment [19].

It is important to accurately calculate the loss density and the corresponding gradients at various critical locations in windings and thereafter predict the locations of hot spots. The hot spot temperatures measured at these locations by direct hot spot measurement technique should be reasonably close to the calculated values. The accurate calculation of hot spot temperature is complicated by the fact that the resistivity of winding conductor changes as the temperature along the winding height changes. The I^2R loss of the winding is directly proportional to resistivity, whereas the winding eddy loss is inversely proportional to resistivity. Hence, generally as the oil temperature increases from bottom to top, the I^2R loss increases while the eddy loss tends to reduce. Since the I^2R loss is the dominant component, losses are higher at the top. Stray leakage field and corresponding winding eddy losses are different along the winding radial depth, as explained in Chapter 4, and this variation should be taken into account during the accurate calculation of hot spot temperatures.

9.7 Static Electrification Phenomenon

Generation of static charges, caused by oil streaming on a solid insulation, is responsible for streaming or static electrification phenomenon in the transformers. This phenomenon occurs due to the friction between the oil and solid dielectric components of the transformer. Depending on the type of oil and its velocity, high levels of localized electrostatic charges (due to charge separation) can be generated leading to very high voltage inside the transformer. This overvoltage, depending on where it occurs inside the transformer, could trigger a sequence of electrical discharges and arcing. Failures in some of the large high voltage transformers and autotransformers have been attributed to the occurrence of electrostatic charges. When the voltage and power ratings of the transformer increase, tendency is to use higher rate of oil flow for the cooling purpose and to improve the insulation resistance. From the standpoint of static electrification, these improvements result in increase of charging tendency. The accumulation of charges leads to production of a strong DC field, which may stress the insulation to an unacceptable level. If high voltage transformers are manufactured with reduced dimensions and kg per MVA value, chances of electrostatic charging are higher. The reduction in the weight to power ratio usually results in greater oil velocity and more labyrinths aiding the static electrification phenomenon.

When oil flows through insulation ducts, charge separation occurs at the interface of the solid insulation and oil. The charge separation also occurs in the other regions of the flow system such as radiator pipes and pumps. It has been observed that the paper / pressboard insulation structure acquires a negative charge and the oil carries a positive charge. The lower part of the insulation arrangement (bottom end insulation) may accumulate a high negative charge leading to the development of excessive DC voltage [20, 21]. As the oil flows up through the windings, it becomes more and more positively charged, and the upper tank may act as a reservoir for the positive charge. There is charge relaxation in every part of the flow system which mitigates the effect. The static charge distribution in the system is determined by the balance of the charge separation and charge relaxation processes [22]. The radiator pipes are the efficient charge dissipating devices.

There are a number of factors that influence the static electrification phenomenon:

1) *Moisture content*: The moisture has a significant effect on the charging tendency; drying out causes the charge density to increase, while addition of moisture reduces it. Since transformers are operated with low moisture levels in oil (below 10 ppm at the time of new oil filling) the highest charging tendency may be experienced [23].
2) *Temperature*: There is rise in charging tendency with the temperature since the dryness of oil increases. Hence, it may be advisable to reduce the flow rate during the warming-up process.
3) *Flow rate*: The charging tendency increases with greater flow rates. The increase varies somewhere between the second and fourth power of the oil flow velocity. The average flow rate involved in one of the failures was 20 cm/s over an average typical cross section [24]; the flow rate in windings was 45 to 60 cm/s, and in the pumps and piping in the heat exchangers it was more than 4.5 m/s. The consideration of static electrification decides the upper limit of oil flow rate in forced oil cooled transformers, and thus impacts the cooling system design.
4) *Turbulence*: The charge motion or generation depends on turbulence in oil.
5) *Surface condition*: The charge generation / separation process is enhanced with increase in roughness of solid insulation.
6) *Pumps*: Pumps can be substantial sources of charge generation [25].
7) *Orifices*: Orifice effects have been demonstrated to generate charges.
8) *Fields*: The AC and DC fields have definite impact on static electrification, which is investigated in [26].

Some of the methods reported for reducing charging tendency are clay filtration of oil, addition of charge suppressors to oil, etc. [23]. Charge reduction by addition of charge suppressers is not a viable solution as it increases the electrical conductivity of oil. The oil flow can be reduced in the susceptible temperature range by operating the cooling system using automatic control.

In addition to incorporating pumps that operate at lower flow velocities, the effect of changing the location of pumps may also be a consideration if the pumps prove to be the prime sources of charge generation. Some transformers now have the pumps mounted at the top of radiators to allow more distance for charge relaxation in the oil prior to entering the bottom of the transformer.

References

1. Blume, L. F., Boyajian, A., Camilli, G., Lennox, T. C., Minneci, S., and Montsinger, V. M. *Transformer engineering*, John Wiley and Sons, New York, and Chapman and Hall, London, 1951.
2. Karsai, K., Kerenyi, D., and Kiss, L. *Large power transformers*, Elsevier Publication, Amsterdam, 1987.
3. Norris, E. T. High voltage power transformer insulation, *Proceedings IEE*, Vol. 110, No. 2, February 1963, pp. 428-440.
4. Higgins, T. J. Formulas for calculating temperature distribution in transformer cores and other electrical apparatus of rectangular cross section, *AIEE Transactions - Electrical Engineering*, Vol. 64, April 1945, pp. 190-194.
5. Rele, A. and Palmer, S. Cooling of large transformer cores, *IEEE Transactions on Power Apparatus and Systems*, Vol. PAS-91, No. 4, July/August 1972, pp. 1527-1535.
6. Nakamura, Y., Jia, W., and Yasuhara, M. Incompressible flow through multiple passages, *Numerical Heat Transfer*, Pt. A, Vol. 16, 1989, pp. 451-465.
7. Nakadate, M., Toda, K., Sato, K., Biswas, D., Nakagawa, C., and Yanari, T. Gas cooling performance in disk winding of large-capacity gas-insulated transformer, *IEEE Transactions on Power Delivery*, Vol. 11, No. 2, April 1996, pp. 903-908.
8. Kamath, R. V. and Bhat, G. Numerical simulation of oil flow through cooling ducts of large transformer winding, *International Conference on Transformers, TRAFOTECH-1998*, Mumbai, India, 1998, pp. I1-I5.
9. Allen, P. H. G. and Allan, D. J. Layer-type transformer-winding cooling factors derived from analogue solution of the governing equations, *Proceedings IEE*, Vol. 110, No. 3, March 1963, pp. 523-534.
10. Preiningerova, S. V. and Pivrnec, M. Temperature distribution in the coil of a transformer winding, *Proceedings IEE*, Vol. 124, No. 3, March 1977, pp. 218-222.
11. Oliver, A. J. Estimation of transformer winding temperatures and coolant flows using a general network method, *Proceedings IEE*, Vol. 127, Pt. C, No. 6, November 1980, pp. 395-405.
12. Pierce, L. W. An investigation of the thermal performance of an oil filled transformer winding, *IEEE Transactions on Power Delivery*, Vol. 7, No. 3, July 1992, pp. 1347-1358.

13. Montsinger, V. M. Loading transformers by temperature, *AIEE Transactions*, Vol. 49, 1930, pp. 776-792.
14. Clark, F. M. Factors affecting the mechanical deterioration of cellulose insulation, *AIEE Transactions - Electrical Engineering*, Vol. 61, October 1942, pp. 742-749.
15. Hochart, B. *Power transformer handbook*, Butterworths and Co. Publishers Ltd., London, 1987.
16. Shroff, D. H. A review of paper aging in power transformers, *Proceedings IEE*, Vol. 132, Pt. C, No. 6, November 1985, pp. 312-319.
17. Morais, R. M., Mannheimer, W. A., Carballeira, M., and Noualhaguet, J. C. Furfural analysis for assessing degradation of thermally upgraded papers in transformer insulation, *IEEE Transactions on Dielectrics and Electrical Insulation*, Vol. 6, No. 2, April 1999, pp. 159-163.
18. Emanuel, A. E. and Wang, X. Estimation of loss of life of power transformers supplying nonlinear loads, *IEEE Transactions on Power Apparatus and Systems*, Vol. PAS-104, No. 3, March 1985, pp. 628-636.
19. McNutt, W. J., McIver, J. C., Leibinger, G. E., Fallon, D. J., and Wickersheim, K. A. Direct measurement of transformer winding hot spot temperature, *IEEE Transactions on Power Apparatus and Systems*, Vol. PAS-103, No. 6, June 1984, pp. 1155-1162.
20. Higaki, M., Kako, Y., Moriyama, M., Hirano, M., Hiraishi, K., and Kurita, K. Static electrification and partial discharges caused by oil flow in forced oil cooled core type transformers, *IEEE Transactions on Power Apparatus and Systems*, Vol. PAS-98, No. 4, July/August 1979, pp. 1259-1267.
21. Shimizu, S., Murata, H., and Honda, M. Electrostatics in power transformers, *IEEE Transactions on Power Apparatus and Systems*, Vol. PAS-98, No. 4, July/August 1979, pp. 1244-1250.
22. Tamura, R., Miura, Y., Watanabe, T., Ishii, T., Yamada, N., and Nitta, T. Static electrification by forced oil flow in large power transformers, *IEEE Transactions on Power Apparatus and Systems*, Vol. PAS-99, No. 1, January/February 1980, pp. 335-343.
23. Oommen, T. V. Static electrification properties of transformer oil, *IEEE Transactions on Electrical Insulation*, Vol. 23, No. 1, 1988, pp. 123-128.
24. Crofts, D. W. The static electrification phenomena in power transformers, *IEEE Transactions on Electrical Insulation*, Vol. 23, No. 1, 1988, pp. 137-146.
25. Howells, E., Zahn, M., and Lindgren, S. R. Static electrification effects in transformer oil circulating pumps, *IEEE Transactions on Power Delivery*, Vol. 5, No. 2, April 1990, pp. 1000-1006.
26. Miyao, H., Higaki, M., and Kamata, Y. Influence of AC and DC fields on streaming electrification of transformer oil, *IEEE Transactions on Electrical Insulation*, Vol. 23, No. 1, 1988, pp. 129-135.

10

Structural Design

10.1 Importance of Structural Design

The tank of a transformer is a closed structure which is made by steel plates. It behaves like a plate structure. Stiffeners are usually provided on all the sides and also on the top cover of the tank to reduce stresses and deflections in plates under various types of loads. The transformer tanks are designed for a pressure higher than the operating one, as specified by the standards. The tank design and fabrication are complicated due to limitations imposed by transportation (weight and size), requirement that the oil quantity should be optimum, etc. Apart from pressure and vacuum loads, the transformer structure has to withstand other loads such as lifting, jacking, haulage, etc. Depending on the location of transformer installation, the strength of the transformer structure against a seismic load may also need to be ascertained.

Design of the transformer tank becomes complicated due to number of accessories and fittings connected or mounted on it. These include: conservator and radiator mounting arrangements, cooler pipes, turrets which house bushings, support arrangement for control box housing controls for fans and pumps, support structures for tap changer drive mechanism, valves for sampling / draining / filtration, cable trays or conduits for auxiliary wiring, inspection covers for getting access to important parts inside the transformer such as bushings (for making connections) and tap changer, cable box, bus duct termination, etc. Certain simplifying assumptions are done while analyzing the strength of the tank with all these fittings under various loading conditions.

The stress analysis of a transformer structure can be done by mainly two methods, viz. analytical methods and numerical methods. The analytical methods are used for determining the stiffening requirements to limit stresses and deflections for simple tank constructions. The tank shapes are usually complex and the application of analytical methods is difficult. For example, if the tank is not rectangular and if there are many pockets (extruding structures) or openings, numerical methods such as FEM are used to determine the stresses and deflections under various loading conditions.

10.2 Different Types of Loads and Tests

10.2.1 Loads

The transformer tank should be capable of withstanding the following loads:

Lifting and jacking: The tank is designed to facilitate handling of the transformer. For this purpose, lifting lugs and jacking pads (as shown in figure 10.1 (a)) are provided on the tank. Lifting lugs, provided towards the top of the tank, are used to lift the structure by a crane. Jacking pads provided towards the base of the tank, are used for handling the transformer in the absence of crane, especially at the site. Generally, four jacking pads / lifting lugs are used.

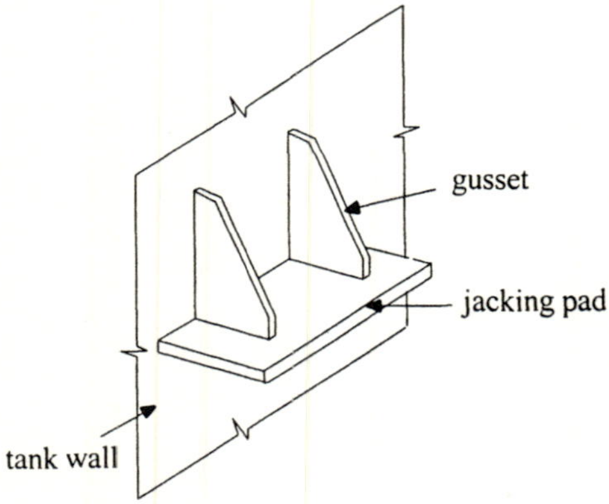

(a) Jacking pad

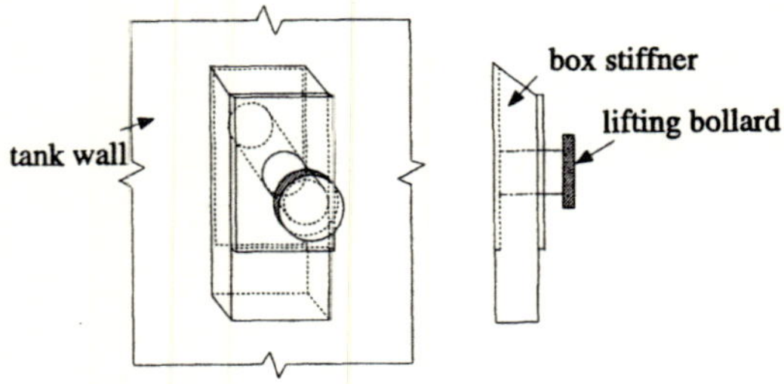

(b) Lifting bollard

Figure 10.1 Jacking pad and lifting bollard

Lifting lugs are used for distribution transformers, where the loads are less. Lifting bollards are used for medium and large power transformers as shown in figure 10.1 (b). Ride-over (transport) lugs are provided for the purpose of supporting the transformer on a floorless wagon during transport.

Haulage load: For local movements of the transformer at the place of installation, rollers and haulage lugs are provided. The haulage lugs are provided on the lower portion of the tank, whereas the rollers are provided under the base plate. Usually, four rollers are provided but for large transformers six or eight rollers may be provided. In place of rollers, a solid under-base is sometimes provided to facilitate skidding over rails or pipes.

Seismic and wind load: The transformer has to be designed for a specified seismic acceleration and wind load. Seismic and wind loads are very important design considerations for bushings, supporting structures of conservator and radiators, etc. It is very difficult, if not impossible, to conduct the seismic test on a transformer. Seismic tests on bushings are usually specified and can be done. Special care has to be taken for bushings because they have high cantilever load

Transient pressure rise: When an internal fault takes place in an oil filled transformer, a large volume of decomposed gases may get generated due to arcing. Under these conditions, the tank structure has to withstand a rapid rise of pressure if the pressure relief device does not act in such a short time. If the tank is not designed with adequate factor of safety, it may rupture leading to fire hazard and serious environmental impact due to outflow of oil. The tank should be designed in such a way that it should be in an elastic limit under the pressure rise conditions. The tank should not be too rigid or too flexible, otherwise it may burst. Special devices such as sudden pressure relays are used which can act quickly under such transient pressure rise conditions.

10.2.2 Tests

The following tests are conducted to check the strength of the transformer structure:

Leak test: This test is meant to check whether the welded joints of the tank structure are leak-proof or not. The test is conducted by pressurizing the tank using air pressure. A soap solution is sprayed over all the welded joints under specified pressure conditions. Any leak due to weld defects (crack, pin hole, etc.) leads to bubble formation.

Vacuum test: The leak test (done with pressurized air) is followed by the vacuum test. A specified vacuum is applied to the tank for at least an hour. The permanent deflections measured after removal of the vacuum should be within the limits (which depend on the size of tank) specified by the users / standards. The tank is then cleared for shot-blasting and painting processes. This test is important because oil filling is done under specified vacuum conditions (either at works or site). In addition, the drying and impregnation may be done in the tank itself (e.g., in vapour phase drying process). The vacuum may be partial or full depending on the voltage class and size of the transformer, and the user specifications.

Pressure test: This test is usually done after all the dielectric tests are completed in the manufacturer's works. The accessories like bushings are removed and a pressure of 5 psi higher than the maximum operating pressure is generally applied to check the pressure withstand capability of the tank. All the welded joints are checked manually; if any oil leakage is noticed, the oil is drained and the defective welding is rectified. The gasket leaks, if any, are also rectified.

Dye-penetration test: This test is conducted for load bearing members to detect weld defects. In this test, the surface to be tested is cleaned thoroughly and a dye (usually of pink colour) is applied to the weld surface. The dye is left there for some time, typically 30 minutes, and then it is wiped clean. During this period, if there are any weld defects in the surface being tested, the dye due to capillary action penetrates through. After this, another solution known as developer is sprayed on the surface. This developer brings out the dye that has penetrated inside and leaves the pink marks on the locations where the weld defects are present. This test is useful to detect weld integrity of load bearing members like jacking pads and lifting lugs / bollards.

10.3 Classification of Transformer Tanks

Depending upon the position of joint between upper and lower parts of the tank, we have two types of tank construction, viz. conventional tank and bell tank.

Conventional tank: This type of construction has a top cover as shown in figure 10.2 (a). Since the joint is usually above the top yoke level, it facilitates a proper placement of magnetic shunts on the tank wall for an effective stray loss control. The disadvantage is that the core and windings are not visible at site when the cover is removed. Hence, for inspection of core-winding assembly, a crane with higher capacity is required to remove the core-winding assembly from the tank.

Bell tank: In this type of tank construction, shown in figure 10.2 (b), the joint between the two parts is at the bottom yoke level to facilitate the inspection of core-winding assembly at site after the bell is removed. Thus, it consists of a shallow bottom tank and a bell shaped top tank. The bell tank construction may not be convenient for a proper placement of magnetic shunts if the joint is at such a height from the bottom that it comes in the path of leakage field. This may lead to a bolt overheating problem (discussed in Chapter 5).

For the above two types of tank, either plain or shaped tank can be used.

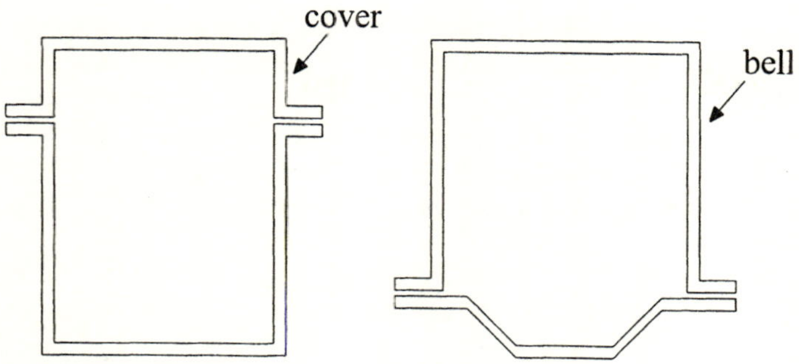

(a) Conventional tank (b) Bell tank

Figure 10.2 Types of tank

Plain tank: The plain tank of rectangular shape is quite simple in construction. It is easy from design and manufacturing points of view since it facilitates standardization. The design of stiffeners is also quite simple. It usually leads to higher oil and steel quantity in high voltage transformers. If special detachable (bolted) bushing pockets are used for center-line lead HV winding arrangement, some saving in oil quantity can be achieved. This is usually done in large high voltage transformers.

Shaped tank: In order to save oil quantity, tank is shaped so that its volume reduces. The tank shaping is mainly influenced by electrical clearances (between the high voltage leads and grounded tank), transport considerations, tap changer mounting arrangements, etc. The lower portion of the tank may be truncated in order to facilitate the loading of a large transformer on some specific type of wagon (in case of rail transport) and / or to reduce the oil quantity. The tank walls may be curved / stepped to reduce the tank size and volume. The shaped tank has the advantage that the curved portions of its walls give a stiffening effect. But the design of the shaped tank is more complex leading to higher engineering and manufacturing time. Also, it may not be conducive for putting magnetic shunts or eddy current shields on it for an effective stray loss control.

The joint between the two parts of the tank can be either bolted or welded type, which gives the following two types of construction.

Bolted construction: The joint between top and bottom tanks can be of bolted type. The bolted construction, though preferred for easy serviceability, has the disadvantage of developing leaks if gaskets deteriorate over a period of time. The oil leakage problem can occur if there is unevenness in the plates which are bolted or if the gaskets are over-compressed. The bolted joint may lead to overheating hazard in large transformers.

Welded construction: This type of construction eliminates the possible leakage points since the two parts of the tank are welded together. It can thus ensure leak-proof joints throughout the life of the transformer. But if a problem or fault develops inside the transformer, de-welding operation has to be done and there is a limit on the number of times the de-welding and subsequent welding operations that can be done. The C-shaped clamps are used during the welding operation and a thin gasket is provided between the two curbs so that the welding spatters do not enter inside the tank. Some arrangement is provided inside the tank at the top for arresting the buckling of cover under the lifting loads.

Depending on whether the tank is totally sealed from the outside atmosphere or is in contact with the atmosphere, the following two types of construction exist.

Breathing tank construction: Ambient temperature and load variations result in change of oil volume. The conservator fitted on the tank top allows these volume changes. The conservator is partially filled with oil and the space in the conservator communicates with the atmosphere through a breather containing a moisture absorbing material. In order to eliminate the contact of oil with the atmosphere (to avoid moisture absorption by it), constant oil pressure system is used in which a flexible bag (membrane) fitted inside the conservator communicates with the outside air. The air bag contracts or expands depending on changes in the oil volume. This construction is commonly used for large power transformers.

Sealed tank construction: In this type of arrangement, free space (filled usually with nitrogen gas) is provided in the tank for oil expansion based on the maximum expected oil temperature. The contact of oil with the outside atmosphere gets totally eliminated. The tank is designed to withstand the pressure variations due to changes in the oil volume. The construction has the disadvantage that with a sudden fall in temperature, gases may get released from the oil seriously affecting the dielectric strength of the insulation system. Higher clearances have to be provided between electrodes separated by the combined oil and gas spaces (as compared to the conventional clearances for the oil immersed electrodes).

There are some special types of tank construction based on the application and features as given below.

Corrugated tank: This construction is used in small distribution transformers to obviate the need of providing radiators separately. The corrugations are made by folding a steel sheet continuously on a special purpose machine. These corrugations are then welded to a steel frame to form a tank wall. The corrugations provide an adequate cooling surface and also play the roll of stiffeners. In small distribution transformers, the use of corrugated tanks is common; it can reduce the manufacturing (fabrication) time substantially.

Cover-mounted construction: In this type of construction, core and windings are attached to the tank cover. Lifting lugs / bollards are provided on frames. The construction facilitates connections from the windings to cover mounted accessories like in-tank type OLTC and small bushings. Access to the lifting lugs / bollards is provided through the inspection openings on the cover. The complete core-winding assembly with the top cover can be lifted by means of lifting lugs / bollards and lowered into the tank. The whole arrangement can be made compact and simpler. For servicing purpose, the un-tanking of the core-winding assembly is possible without removing the bushing connections.

Perforated tank: This type of tank is used in dry-type transformers, where the tank is used just as an enclosure to house the active parts. The perforations allow the flow of air cooling the inside active parts. The construction generally consists of detachable panels which cannot take any lifting load. The absence of oil and the presence of perforations usually lead to higher noise level in dry transformers as compared to oil cooled transformers, and special measures need to be taken to reduce the noise level.

10.4 Tank Design

The mechanical design is taken up after the electrical design of a transformer is finalized. The mechanical design requires following inputs: core dimensions (diameter, center-to-center distance, etc.), winding details, design insulation levels at various electrodes (as described in Chapter 8), details of tap changer, details of accessories (bushings, radiators, fans, pumps, protection devices, etc.), weight and size limitations during transport and at site, etc. The designer has to keep in mind the requirements of tank shielding arrangements. The tank dimensions and profile are decided in a layout drawing drawn to scale considering electrical clearances, magnetic clearances, transport size limits and manufacturability. The design of stiffeners is a very important aspect of tank design. An effective stiffening arrangement can reduce the tank plate thickness. The stiffeners are designed in such a way that the tank weight is minimum, and at the same time it should be able to withstand the specified loads. The stiffeners used are of following types (shown in figure 10.3):

Flat stiffeners: These are used in small rating transformers. These stiffeners, which have low section modulus, are suitable for small tanks. They are more compact as compared to the other types of stiffeners.

T stiffeners: These stiffeners offer higher section modulus as compared to the flat stiffeners but lower than the box stiffeners (for the same cross-sectional area). They occupy more space than the flat stiffeners but less than the box stiffeners. They are useful in the cover area where less space is available due accessories like bushings, turrets, etc. because of which the box stiffeners cannot be used. These are also useful for stiffening a dome shaped cover / irregular cover where stiffening is difficult with the other types of stiffeners.

Box stiffeners: For large power transformers, the flat and T type stiffeners are not suitable because their number increases. The box stiffeners give much higher value of section modulus, and hence they are used in large power transformers. Aesthetically they look better than the other types of stiffeners. The box stiffeners can also be used for other purposes. A lifting bollard can be embedded into a box stiffener for the lifting purpose. A jacking arrangement can be achieved if a plate is provided (with gussets) at the bottom of a box stiffener. It can also be used to provide an extra gas space in sealed transformers.

Usually, the stiffening is done vertically. Sometimes horizontal stiffeners are also provided. The stiffeners are designed to distribute the lifting load properly (more uniformly). The location of stiffeners on the tank may be affected by space restrictions. The stiffener dimensions and location depend not only on the strength considerations but also on the various fittings and accessories which have to be mounted on the tank.

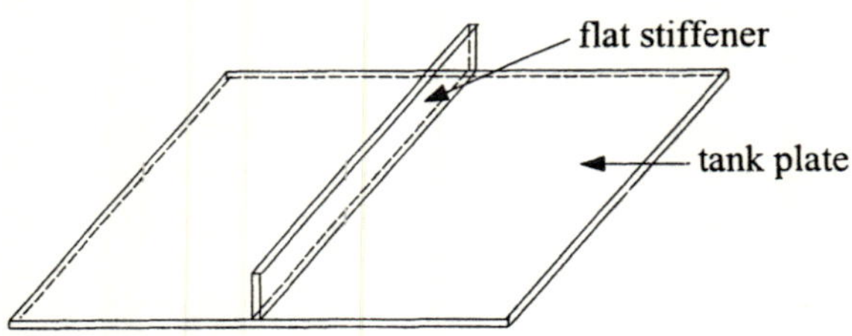

(a) Flat stiffener

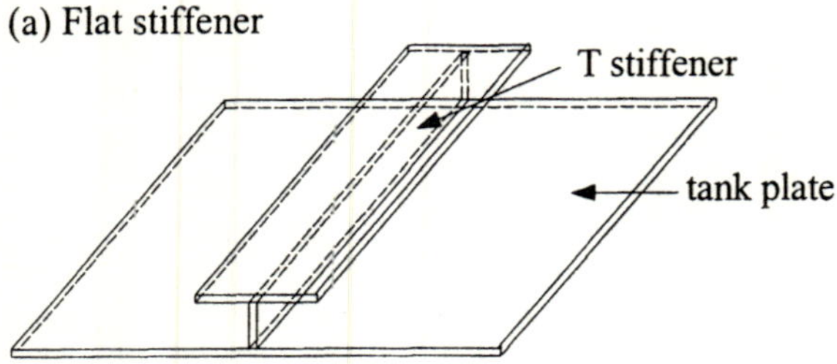

(b) T stiffener

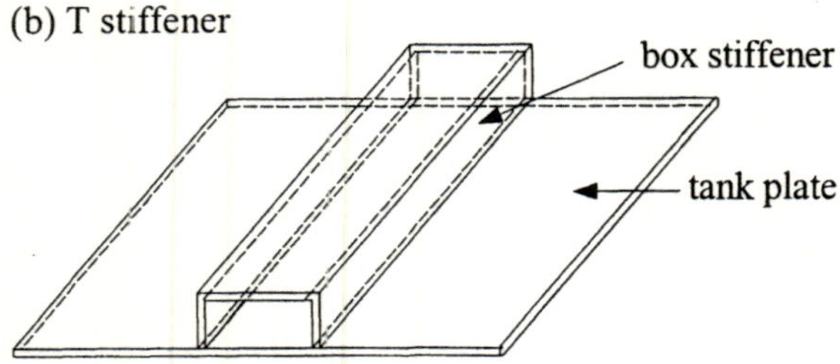

(c) Box stiffener

Figure 10.3 Types of stiffeners

The stiffeners can be designed as simply supported or fixed support structures. In the simply supported case, the stiffeners are terminated at some distance from the top or bottom edge of the tank plates, which may result in higher deflection. If the stiffener ends are anchored to the top curb and bottom plate (in a conventional tank) then it is termed as the fixed support stiffener, and this arrangement gives lower deflection. The T stiffeners and flat stiffeners can be terminated on the curb whereas box stiffeners can not be terminated because of the space requirement for bolting operations. For practical reasons one has to leave some space between the termination of a box stiffener and curb. In such cases, the box stiffener can be tied to the curb by means of a gusset.

Since many accessories are mounted on the top cover, an adequate space may not be available for its stiffening. In such cases, higher cover plate thickness needs to be used with the application of flat or T stiffeners wherever possible.

The base plate of a tank is usually much thicker than its vertical plates. It is designed to carry a total load corresponding to the sum of entire core-winding assembly weight, oil head and test pressure. The base plate can be stiffened by cross channels to reduce its thickness. The box stiffeners may also be used sometimes for stiffening of the base plate.

A number of local small stiffeners are provided under extended projections / pockets and shaped tank parts.

10.5 Methods of Analysis

The design of transformer tank structure comprises mainly the analysis of the combined behavior of plates and stiffeners.

10.5.1 Analytical method

In an analytical method, which can be applied to plain rectangular tanks, each side (plate) of tank is divided into number of plate panels. One side of a rectangular tank with three vertical stiffeners is shown in figure 10.4. The center line of a stiffener is taken as the panel boundary. Hence, for the purpose of analysis there are four panels. These panels are subjected to loads such as pressure, vacuum, etc. as described earlier.

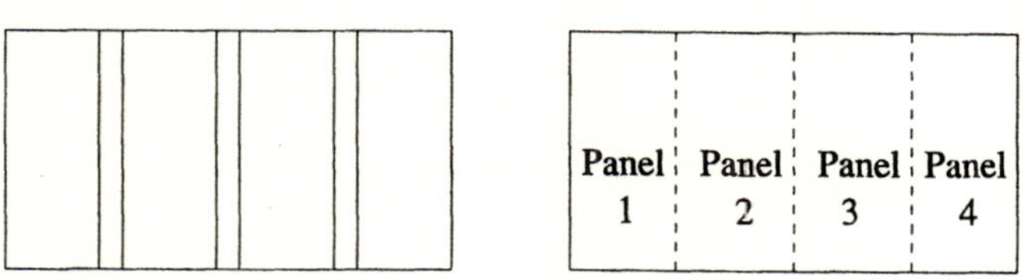

Figure 10.4 One side of a rectangular tank

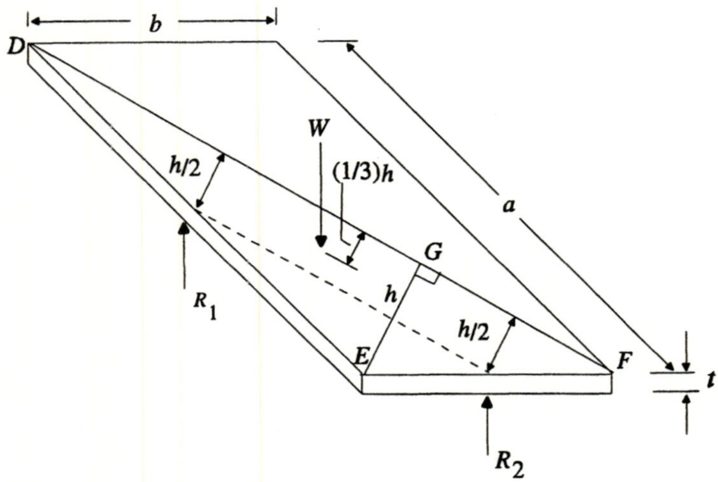

Figure 10.5 Rectangular simply supported plate under uniform load

The stress analysis of each panel can be done by using theory of plates. The stress calculation for simply supported and fixed type of rectangular plates is an integral part of the transformer tank design. Let us first analyze a simply supported plate.

Consider a rectangular plate of dimensions $a \times b$ and thickness t as shown in figure 10.5. Let the load per unit area be w; hence the total load on the plate is wba. The load on the plate area on one side of the diagonal is (1/2) wba, which is denoted by W. This load acts on the centroid of the triangular area DEF. The centroid is at a distance of (1/3) h from DF.

Experiments on the simply supported rectangular plate show that the plate has a tendency to curl up at the corners, and the resultant pressure on each edge acts at its mid-point. The diagonal DF is the most critical section when one side of the plate is not very much longer than the other side [1]. The moment arm for the two reactions R_1 and R_2 is same. From the conditions of symmetry and equilibrium, their sum is equal to (1/2) wba.

The bending moment about DF is

$$M = (R_1 + R_2)\left(\frac{1}{2}h\right) - W\left(\frac{1}{3}h\right) \tag{10.1}$$

Substituting the expressions for reactions and load we get

$$M = \frac{1}{2}\, wba\left(\frac{1}{2}h - \frac{1}{3}h\right) = \frac{1}{12}\, wbah \tag{10.2}$$

The length of *DF* is $\sqrt{a^2+b^2}$. Therefore, the average bending moment per unit length of the diagonal is

$$M_1 = \frac{1}{12}\,\frac{wbah}{\sqrt{a^2+b^2}} \tag{10.3}$$

From the similarity of triangles *FGE* and *FED* we have

$$\frac{EG}{EF} = \frac{ED}{DF}$$

$$\therefore h = \frac{ab}{\sqrt{a^2+b^2}} \tag{10.4}$$

Substituting the expression of *h* in equation 10.3 we get

$$M_1 = \frac{1}{12}\left[\frac{a^2}{(a^2+b^2)}\right] wb^2 \tag{10.5}$$

The section modulus (z) of the plate per unit length along the diagonal is equal to $(1/6)\,t^2$. Accordingly, the bending stress at the surface of the plate across the diagonal *DF* in the simply supported case is

$$\sigma_{ss} = \frac{1}{2}\left[\frac{w\,a^2\,b^2}{(a^2+b^2)\,t^2}\right] \quad \text{................ Simply supported case} \tag{10.6}$$

The analysis for a plate with fixed edges is quite involved. The deflection for a fixed plate is symmetrical and maximum at its centre. The ratio of two adjacent sides play an important role in deciding the deflection, and the bending moment at various locations is calculated by using analytical methods [2]. The maximum bending moment divided by the section modulus gives the maximum bending stress in the plate. For the panels of a tank side as shown in figure 10.4, three sides can be approximately considered as fixed and the fourth (top) side can be considered as simply supported. The analysis for this case is given in [2].

In order to simplify the calculations, it can be assumed that the behavior of the tank plate is in between the simply supported and fixed edge conditions. The stresses calculated under these two conditions are multiplied by empirical factors to calculate the resultant stress (σ_r),

$$\sigma_r = k_1\,\sigma_{ss} + k_2\,\sigma_{fe} \tag{10.7}$$

where σ_{ss} and σ_{fe} are stresses for simple supported and fixed edge conditions respectively. The constants k_1 and k_2 are empirical factors such that $k_1 + k_2 = 1.0$.

The numerical methods give accurate stress and deflection values without having to do simplifying assumptions as done in the analytical methods.

10.5.2 Numerical method

The analytical solution is quite accurate and is a mathematical expression that gives the value of a desired unknown quantity at any location in a body. The analytical methods can be used for simplified situations. For problems involving complex geometries, material properties and boundary conditions, the designer depends on numerical methods which give sufficiently accurate solutions.

The finite element method (FEM) is a very effective numerical analysis tool for the simulation of structural components under various loading conditions. The FEM analysis can be used for material optimization, reliability enhancement, failure analysis and corrective action, verification of new designs, etc. Before the advent of FEM, many approximations had to be made in the analytical methods for complex tank geometries, and it was almost impossible to predict the exact performance of the structures under the given loading conditions. It was not possible for the designer to know the margin or factor of safety for new designs. Due to this ignorance factor, the designer had to put extra material thereby increasing the transformer cost. Using FEM analysis, it is possible to detect high stress zones and take suitable corrective / preventive actions. The FEM analysis can be used to investigate problems like vibrations, buckling, non-linear behaviour, etc.

The steps of FEM analysis have been given in Section 3.4 (while discussing reactance calculation). A given problem domain is divided into a number of elements, that are straight lines for 1-D domains, triangular or quadrilateral elements for 2-D domains, and tetrahedral or cubical elements for 3-D domains. The transformer tanks can be considered as 2-D shells and can be discretized into 2-D shell elements. The required solution function is approximated over an element by interpolation between the values at its nodes. The interpolating functions may be linear or higher order polynomials. Let u_i and v_i be the x and y components of displacements at the nodes of a triangular element respectively (i = 1, 2, 3). The vector $U = (u_1\ v_1\ u_2\ v_2\ u_3\ v_3)^T$ then represents nodal displacements. The displacements over the entire element area may be given as [3]

$$\begin{bmatrix} u(x,y) \\ v(x,y) \end{bmatrix} = \begin{bmatrix} N_1 & 0 & N_2 & 0 & N_3 & 0 \\ 0 & N_1 & 0 & N_2 & 0 & N_3 \end{bmatrix} U \tag{10.8}$$

where $N_i(x,y)$ are the interpolating functions. Using standard symbols [3], the strain vector $\varepsilon = (\varepsilon_x \quad \varepsilon_y \quad \gamma_{xy})^T$ is related to the displacements as

$$\begin{bmatrix} \varepsilon_x \\ \varepsilon_y \\ \gamma_{xy} \end{bmatrix} = \begin{bmatrix} \dfrac{\partial u}{\partial x} \\ \dfrac{\partial v}{\partial y} \\ \dfrac{\partial u}{\partial y} + \dfrac{\partial v}{\partial x} \end{bmatrix} = \begin{bmatrix} \dfrac{\partial N_1}{\partial x} & 0 & \dfrac{\partial N_2}{\partial x} & 0 & \dfrac{\partial N_3}{\partial x} & 0 \\ 0 & \dfrac{\partial N_1}{\partial y} & 0 & \dfrac{\partial N_2}{\partial y} & 0 & \dfrac{\partial N_3}{\partial y} \\ \dfrac{\partial N_1}{\partial y} & \dfrac{\partial N_1}{\partial x} & \dfrac{\partial N_2}{\partial y} & \dfrac{\partial N_2}{\partial x} & \dfrac{\partial N_3}{\partial y} & \dfrac{\partial N_3}{\partial x} \end{bmatrix} U$$

or $\varepsilon = [B]U$ (10.9)

The potential energy of the element (e) is dependent on the displacements over the element area. As per equation 10.8, it becomes a function of the nodal displacement vector U. In the absence of internal (body) forces, initial strains and initial stresses, it is given by

$$E = \frac{1}{2}\int_V (\varepsilon^T D\, \varepsilon)\, dV - F^T U = \frac{1}{2}\int_V (U^T B^T D B U)\, dV - F^T U \qquad (10.10)$$

where D is an elasticity matrix containing the appropriate material properties, F is the nodal load vector, and V is the element volume. Minimizing the above expression gives the governing equation which determines the solution, i.e.,

$$\partial E/\partial U = [K]^{(e)} U - F = 0 \qquad (10.11)$$

where $[K]^{(e)} = \int_V B^T D B\, dV$. This forms the element equation.

All such element equations are combined by first replacing the element-wise nodal quantities by the corresponding global nodal quantities, and adding all the element equations. This forms the global system of linear equations,

$$K\,\overline{U} = \overline{F} \qquad (10.12)$$

where $\overline{U}$ and $\overline{F}$ are the global displacement and load vectors, and K is the global stiffness matrix. The above linear system of equations is a large and sparse system of equations, solved normally by the iterative methods especially suited for solving such systems. The solutions can be refined for obtaining more accurate solutions by either using a finer mesh with smaller elements or using higher order interpolating functions over the elements or both.

10.6 Overpressure Phenomenon in Transformers

The problem of explosions of oil insulated equipment as a result of low-impedance internal faults has been a major concern. When an internal fault takes place inside an oil filled transformer, arcing produces a large amount of decomposed gas increasing the tank pressure rapidly. The pressure relief device may not be able to keep up with the gas generation rate and the tank can rupture [4]. The severity of an internal fault depends mainly on the arc energy and the tank expansion coefficient. The higher the arc energy and the lower the tank expansion coefficient, the higher the severity is. The arc voltage is not related very much to the arc current and is mainly a function of arc length, electrode shape, pressure, etc. [5, 6]. Since the oil is relatively incompressible and since the bottom plate and side plates of the tank act together as a rigid structure, the tank cover is usually subjected to the overpressures [7]. In order to reduce such consequences, it is necessary to determine the resulting overpressures for different faults and geometrical parameters of the transformer.

Use of flange reinforcing measures such as C-shaped clamps and joint reinforcement beams [5] is made to increase the strength of tank structures against excessive overpressures. In a 3-phase split type transformer the major part of the tank is divided into three parts (one part per phase), with a common ducting for connections, to take care of transport limits. The pressure rise in such a configuration may reach an excessive level due to a small expansion coefficient and the effect of kinetic energy of the oil. Use of a diaphragm type conservator as a pressure reducing space is suggested as a countermeasure.

The phenomenon has been studied both analytically and experimentally. Different formulations are proposed for predicting the overpressures during a low impedance fault. In [8], the results of analysis and experimental work are combined to get a semi-empirical equation for the peak pressure in the air space of a pole-type distribution transformer. It is reported that the arc length and $(i^2t)_{arc}$ are the most significant variables, former being generally beyond control. However, this semi-empirical equation is probably only valid for geometrically similar transformers and hence may not be generalized. A comprehensive explanation of the different failure modes of distribution transformers is given in [9]. The overpressure phenomenon is studied using high speed photography and it is shown experimentally that the arc depth under oil plays an important role in the oil motion, compression of the air space and the resulting overpressure. It is also reported that the maximum pressure exerted against the transformer tank cover depends on two principal parameters: one is the arc energy expended per unit volume of the air space, and another is the efficiency of the process by which the arc energy is converted into the kinetic energy.

The equations for the static pressure within the faulted oil filled distribution transformer as a result of arcing and gas generation are given in [4]. The equations are derived under known conditions such as tank dimensions, air space, specific fuses and specific pressure relief devices, and unknown quantities such as fault current, arc length, arc location and gas temperatures. The finite difference approach is proposed in [10] to study the phenomenon of arcing in oil insulated equipment. The solution of proposed method is compared with that of an analytical formulation for infinite cylinder filled with oil. The application of the finite difference method for analysis of low-impedance faults in a cylindrical pole-type distribution transformer is given in [11].

10.7 Seismic Analysis

Earth quake is a dynamic phenomenon which occurs due to release of energy below the ground because of instability of the earth's internal structure. The source of earthquake is a sudden displacement of ground on both sides of a fault which results from a rupture of a crystal rock. The size of earthquake is measured by the amount of strain energy released at the source. The earthquake produces random ground motions which are characterized by simultaneous but statistically independent horizontal and vertical components. A moderate earthquake may persist for 15 sec to 30 sec and a severe one for 60 sec to 120 sec. The vibration of ground motion may be magnified many fold in the equipment. The magnification depends on the characteristic frequency of vibration of the system consisting of soil, foundation and equipment.

Transformers are important elements of power supply systems. It is very essential that utmost care is taken while designing their tank and accessories for seismic withstand. If they are not adequately designed, it could result into anchorage failure, bushing failure, conservator bracket deformation, oil leakage and other miscellaneous damages. Certain accessories and protection devices (e.g., buchholz relay) may malfunction during an earthquake giving a false indication of fault in the transformer. The main principle for improving strength under earthquake conditions is that the natural frequency of the transformer and its parts should be above 30 Hz ensuring a lower acceleration factor.

Design for seismic conditions is based on the seismic zone where the equipment would be installed. The transformer user should provide information to the transformer manufacturer about the seismic activity in terms of maximum accelerations, response spectra or time histories The seismic zone of a place defines the intensity of an earthquake which is likely to hit that place. As per IEEE C57.114-1990 (IEEE seismic guide for power transformers and reactors), typical values of maximum ground acceleration range from 0.1 g (zone 1) to 0.5 g (zone 4), where g is the acceleration due to gravity. If the transformer is not ground mounted, the acceleration at the mounting location has to be considered.

Although seismic withstand can be most accurately checked by a laboratory test on an equipment, it is very difficult to conduct the test on a product like transformer. Hence, the following three calculation methods are commonly used for checking the seismic withstand of transformers.

***Seismic coefficient method*:** This is an approximate method in which normal static stress calculations are done with certain seismic accelerations applied to the center of gravity of structures. Seismic coefficients are applied separately to various vulnerable components such as bushings, conservators, radiators, etc., which are mostly the overhanging or extended portions of the transformer structure. This method does not take into consideration the natural frequencies of the structure or its components.

Response spectrum method: As per IEEE C57.114-1990, when the natural frequencies of a transformer are lower than about 30 Hz, the static method should not be used and one has to take into account the natural frequencies of the structure. The response spectrum method determines the dynamic response which depends on the natural frequencies of the structure. The transformer needs to be analyzed as a spring-mass model using response spectrum curve with an appropriate damping factor. The response of the structure to an earthquake due to each mode of vibration is calculated, and the total response is determined by combining the individual modal responses (square root of sum of squares technique). A numerical method like FEM needs to be used for this purpose. The FEM analysis gives stresses, accelerations and displacement plots which help in identifying weak structures that need to be strengthened.

Time history method: This method is computationally very intensive and requires actual earthquake data. This method can be used for analysis of structures which underwent an earthquake whose time history is known.

Since it is not possible to test the seismic withstand of transformers by an actual test, experimental investigations have been done to evaluate their natural frequencies and mode shapes of vibrations. The results of multi-point random excitation test and forced vibration test are compared with that of FEM analysis in [12]. A significant global deformation mode is reported at a frequency of about 3.5 Hz. The results of experimental tests on a buchholz relay are compared with that of the numerical analysis. In another reference [13], the efficacy of amplification factor and response factor of a bushing given in IEC standard (IEC 61463, Ed. 1.1, Bushings - Seismic qualification) is examined.

The design precautions suggested by IEEE C57.114-1990 are: placing of transformer and interconnected accessories / equipment on a strong and common foundation to reduce a differential movement during an earthquake, firm anchoring of the transformer by welding its base to the structural steel members embedded in or firmly fixed to the concrete foundation, etc.

10.8 Transformer Noise: Characteristics and Reduction

With the growing consciousness on the ill effects of noise pollution, many users are specifying lower noise levels for transformers. While the trend of ever increasing transformer ratings implies a corresponding rise in noise level, noise-reducing measures have to be adopted to make the transformer quieter. By using modern design methods and materials, noise emissions from the transformer can be economically lowered to the acceptable levels. In order to reduce noise level, it is very important to know and understand the sources of noise. The noise pressure generated by vibration of core and windings is transmitted to tank surfaces though the oil medium. Since the oil is relatively incompressible, the noise is transmitted without appreciable damping. The tank responds to these noise waves depending on its natural frequencies and mode shapes of vibrations.

The principal source of transformer noise, the magnetic core, has been elaborated in Chapter 2. In this chapter the other two sources of noise, viz. load-controlled noise and equipment noise are discussed.

10.8.1 Load-controlled noise

This noise is emitted by a loaded transformer in addition to its no-load noise. It is caused by electromagnetic forces between the windings resulting from the leakage fields and is proportional to the square of the load current. These forces cause the winding vibrations and acoustic radiations having frequency of 100 or 120 Hz (twice the power frequency). It is mainly the axial vibrations of the winding which contribute to the noise (the radial vibrations can be significant only for winding diameters greater than 6 meters [14]). The contribution of load-controlled noise to the overall noise level of the transformer becomes significant when the operating flux density in the core is lower than 1.4 T. At such a low value of flux density, noise from the core is considerably reduced. The other sources of load-controlled noise are the vibrations of tank walls and the magnetic shunts placed on them. If the magnetic shunts are rigidly anchored to the tank wall, the noise due to their vibrations is usually low.

The vibration amplitudes produced by a given axial compressive force (corresponding to a load current flowing in a winding) depend on the winding properties, viz. mass, modulus of elasticity and damping. Pressboard and other insulating materials play an important role in deciding the winding response. The winding noise can be kept as low as possible by using a pressboard material with a high damping coefficient and applying a proper value of pre-stress to the winding. The winding natural frequencies should be quite away from the frequencies of the exciting compressive forces (twice the power frequency and its multiples), since a resonance will amplify the vibrations and noise.

The loaded transformer represents a typical magneto-mechanical system immersed in a fluid (oil). For developing the numerical method for accurate calculation of the load-controlled noise, the electromagnetic field, mechanical displacement field, acoustic pressure field and their couplings have to be considered as one system. Due to the complexity of this multi-field problem, a combination of the finite element and boundary element methods is used in [15] for the prediction of the load-controlled noise of power transformers.

10.8.2 Noise due to cooling equipment

Fan noise is a result of vortex flows in the vicinity of its blades. The noise is a function of air delivery, blade size and speed. While the noise due to core produces frequencies in the range of 100 to 600 Hz, the frequencies associated with the noise due to cooling equipment (fans and pumps) are usually below and above this frequency range in the sound spectrum. In general, high flow speed of cooling medium of fans and pumps should be avoided.

Since the fan noise is a function of its speed and circumferential velocity, a low speed fan has a smaller noise level. As the speed is lowered, air delivery also reduces necessitating an increase in number of fans. Many times, the noise level specified is so low that it may not be possible to get such a low noise fan. Therefore, ONAN (OA) cooling should be specified / used in place of mixed ONAN / ONAF (FA) cooling for small and medium rating power transformers, even if it results in increase of number of radiators.

A radiator noise is caused by the tank vibration transmitted through cooler pipes connecting the tank and radiator (structure borne vibration). Pipe-work and supporting structures should be designed such that there is no resonance.

10.8.3 Noise level reduction

It is pointed out in Chapter 2 that a reduction in noise level is not significant if the operating peak flux density in the core is reduced. A lower value of the operating flux density also results in higher material cost and size of the transformer. Hence, other cost-effective noise reduction methods are commonly used which are now described.

There are different ways by which the noise can be reduced. Methods like stiffening the bracing or supporting parts and adding cushions between parts of a transformer have long been known and used [16] for reducing vibrations and noise. Barrier walls and total sound-proof enclosures have also been commonly used [17]. An easy but expensive way would be to put the transformer in a closed room whose walls and floor are massive. The noise reduces as it tries to pass through a massive wall. The noise can also be reduced by building a free-standing enclosure of concrete and steel plates around the transformer. However, this method has some disadvantages (e.g., a large area is needed for the transformer installation). Use of sound insulation panels is another way of getting reduced noise levels without any additional space requirement. The closely fitting sound insulation panels described in [18] are mounted between reinforcing channels (stiffeners). The assembly consists of a resilient steel sheet, a steel plate and weights. The steel sheet connects the steel plate to the stiffeners. The weights are placed at the boundaries of the plate and sheet to avoid the transmission of structure borne vibrations from the stiffeners to the steel plate. The noise level reduction of 14 dB is reported by the use of these insulation panels. Development of a vibration controlled sound insulation panel, capable of reducing the noise generated from a transformer by 12 to 13 dB, is reported in [19]. The panel consists of highly damped plates which are mounted on the side walls of the transformer tank with isolation rubber pieces.

A substantial reduction of noise (of the order of 15 dB) can also be obtained by using a double tank design. The transformer is contained in the inner tank which is supported inside the outer tank. Both the tanks are suitably insulated from each other to reduce the structure borne sound. Glass wool is placed in the space between the two tanks for the effective noise reduction.

Active noise control is one more technique for the noise level reduction, in which an anti-phase noise is generated and superimposed on the noise emitted by the transformer. It requires very sophisticated instrumentation and computational facilities. The active control scheme implemented with Digital Signal Processing (DSP) is reported in [20]. It is reported that a noise level reduction of 5 to 15 dB can be achieved depending upon the effectiveness of implementation of the technique.

In dry type distribution transformers (resin impregnated or cast resin), due to the absence of oil and presence of openings / perforations on the tank (for effective air circulation and cooling), the noise level can be higher. Hence, the core limbs can be of bolted construction in addition to the bolted yokes to give more rigidity to the core structure and reduce the noise emanating from it.

The noise reduction techniques can be summarized as below.

1. Reduction in core flux density: This gives noise reduction of 3 to 5 dB for a reduction in flux density by 10% (or approximately 2 dB per flux density reduction of 0.1 T). The method has adverse effects on the cost and size of transformers.
2. Hi-B grade and scribed core materials give 2 to 3 dB reduction as compared to non Hi-B grades.
3. Avoidance of core resonance by calculation of core resonant frequencies: The core natural frequencies should not coincide with the excitation frequencies as discussed in Chapter 2.
4. Increased core damping: By application of suitable viscoelastic or adhesive coating to the core laminations, the noise level can be reduced.
5. It should be ensured that any links or attachments to the core are flexible so that they do not transmit the vibrations.
6. Use of step-lap joint: It gives reduction by about 4 to 5 dB as compared to the mitred construction for the commonly used flux densities (1.6 to 1.7 T).
7. The corner protrusions of the built core should be cut since they may contribute to noise due to vibrations (also they are not useful as they do not generally carry any flux).
8. The clamping pressure on the core should be adequately distributed so that no appreciable length of the core is left unclamped. If limbs / yokes are clamped with resin-glass or fiber-glass tapes, the pitch (distance between two tapes) should be small so that an adequate uniform pressure is applied.
9. To reduce the structure borne vibrations, the core-winding assembly should be isolated from the tank base by use of oil compatible anti-vibration pads between them. Use of anti-vibration pads is also made between frames and tank. Such isolations can give a noise level reduction of 2 to 4 dB.
10. Use of sound insulation panels between tank stiffeners can give 5 to 15 dB reduction.

11. An increased tank wall mass, by use of sand in hollow braces on the wall, can give appreciable noise level reduction.
12. Use of double tank design: Inner and outer tanks are suitably insulated from each other to eliminate structure borne vibrations. Also, suitable sound absorbent wool is placed between the two tanks. The noise reduction is about 15 dB.
13. Complete concrete or brick wall enclosures: The noise reduction is about 20 to 30 dB, but the method is quite expensive.
14. Use of active phase cancellation technique: The sound emitted by a transformer is overlaid by externally applied anti-phase sound. A noise level reduction of 5 to 15 dB may be possible.
15. If the transformer noise level required is too low to get a fan with a lower noise level, ONAN (OA) cooling may be specified / used in place of mixed ONAN / ONAF (FA) cooling for small and medium rating transformers.

Some precautions which need to be taken at the site for noise level control are:

16. The reflecting surfaces should not coincide with half the wavelength of frequencies of noise emitted by the transformer [21] (to avoid standing waves and reverberations / echoes).
17. Fire walls are sometimes placed adjacent to the transformer. It may not be possible to place them at a location so that no undesirable reflections occur. In such cases a sound-absorbent material, suitable for outdoor use, may have to be applied on the walls.
18. Dry type distribution transformers are mostly located in a room inside a building. With the walls of the room having a low sound absorption coefficient, the sound emitted by the transformer reflects back and forth between the walls. This may lead to a considerable increase of noise level. These aspects should be duly considered by the users (while designing the room) and manufacturers (while designing the transformer).
19. If simple barrier walls are used for obstructing the noise, they are not effective at the edges. The walls have to be extended at right angles on one or both ends with an application of sound-absorbent material for better results.
20. The transformer should not be mounted on a foundation on which adjacent walls are also mounted because the vibrations from the transformer may get transmitted through the foundation to the walls. The vibration of these walls will increase the overall noise level.
21. A solid connection between a vibrating transformer and any solid structure in the vicinity should be avoided (flexible connections can be used as far as possible).
22. The tank base can be isolated from the supporting ground / foundation by a suitable vibration-damper to reduce structure borne vibrations.

10.8.4 Noise level measurement

A noise level is commonly measured in decibels (dB) by comparing the pressure generated by a noise source with some standard level. The noise level is measured on 'A' weighted scale which closely follows the sensitivity of human ear. There are basically two methods of noise measurement: sound pressure measurement and sound intensity measurement. The details of test methods and acceptable test environment conditions are given in IEC standard 60076-10 (Determination of sound levels, First Edition, 2001). Sound pressure level is a scalar quantity and requires simple instrumentation.

Sound intensity is a vector quantity and the method measures directional sound. It is therefore less affected by a background noise. Hence, the sound intensity method can give more accurate measurements in the presence of background noise. However, sound intensity measurements require higher skill and more sophisticated instrumentation. Information about the location and characteristics of noise sources can be obtained by studying the frequency spectrum.

Apart from design challenges, the measurement of low noise poses a difficult problem. The minimum level of noise which can be measured is limited by the ambient noise conditions in the test area. Special enclosures may have to be used to shield the instruments (test set-up) and transformer from the high ambient noise.

References

1. Seely, F. B. and Smith, J. O. *Advanced mechanics of materials,* John Wiley and Sons, New York, London, 1952.
2. Timoshenko, S. and Woinowsky-Krieger, S. *Theory of plates and shells,* McGraw-Hill Inc., Singapore, 1959.
3. Zienkiewicz, O. C. *Finite element method*, McGraw Hill, London, 1977.
4. Goodman, E. A. and Zupon, L. Static pressures developed in distribution transformers due to internal arcing under oil, *IEEE Transactions on Power Apparatus and Systems*, Vol. PAS-95, No. 5, September/October 1976, pp. 1689-1698.
5. Kawamura, T., Ueda, M., Ando, K., Maeda, T., Abiru, Y., Watanabe, M., and Moritsu, K., Prevention of tank rupture due to internal fault of oil-filled transformer, *CIGRE 1988*, Paper No. 12-02.
6. Tagaki, T., Ishii, T., Okada, T., Kurita, K., Tamura, R., and Murata, H. Reliability improvement of 500 kV large capacity power transformer, *CIGRE 1978*, Paper No. 12-02.
7. Hamel, A., Dastous, J. B., and Foata, M. Estimating overpressures in pole-type distribution transformers - Part I: Tank withstand evaluation, *IEEE Transactions on Power Delivery*, Vol. 18, No. 1, January 2003, pp. 113-119.

8. Mahieu, W. R. Prevention of high-fault rupture of pole-type distribution transformers, *IEEE Transactions on Power Apparatus and Systems*, Vol. PAS-94, No. 5, September/October 1975, pp. 1698-1707.
9. Barkan, P., Damsky, B. L., Ettlinger, L. F., and Kotski, E. J. Overpressure phenomena in distribution transformers with low impedance faults: experiment and theory, *IEEE Transactions on Power Apparatus and Systems*, Vol. PAS-95, No. 1, January/February 1976, pp. 37-47.
10. Foata, M., Iordanescu, M., and Hardy, C. Computational methods for the analysis of explosions in oil-insulated equipment, *IEEE Transactions on Power Systems*, Vol. 3, No. 1, February 1988, pp. 286-293.
11. Dastous, J. B., Foata, M., and Hamel, A. Estimating overpressures in pole-type distribution transformers - Part II: Prediction tools, *IEEE Transactions on Power Delivery*, Vol. 18, No. 1, January 2003, pp. 120-127.
12. Bellorini, S., Bettinali, F., Salvetti, M., Gatti, F., Zafferani, G., and Monzani, O. Mechnical Seismic behavior of power transformers, *CIGRE 1998,* Paper No. 12-212.
13. Bellorini, S., Bettinali, F., Salvetti, M., and Zafferani, G. Seismic qualification of transformer high voltage bushings, *IEEE Transactions on Power Delivery*, Vol. 13, No. 4, October 1998, pp. 1208-1213.
14. Reiplinger, E. Study of noise emitted by power transformers based on today's viewpoint, *CIGRE 1988,* Paper No. 12-08.
15. Rausch, M., Kaltenbacher, M., Landes, H., Lerch, R., Anger, J., Gerth, J., and Boss, P. Combination of finite and boundary element methods in investigation and prediction of load-controlled noise of power transformers, *Journal of Sound and Vibration*, 250 (2), 2002, pp. 323-338.
16. George, R. B. Power transformer noise: Its characteristics and reduction, *AIEE Transactions,* March 1931, pp. 347-353.
17. Murray, C. S. Transformer audio noise problems on an electric power system, *AIEE Transactions,* Vol. 68, 1949, pp. 740-752.
18. Kanoi, M., Hori, Y., Maejima, M., and Obta, T. Transformer noise reduction with new sound insulation panel, *IEEE Transactions on Power Apparatus and Systems*, Vol. PAS-102, No. 9, September 1983, pp. 2817-2825.
19. Ebisawa, Y., Hirai, K., Suda, K, and Ikeda, M. Development of new type sound insulation panel for transformers, *International Conference on Transformers, Transformer-97*, Kolobrzeg, May 1997, pp. 37-42.
20. Teoh, C., Soh, K., Zhou, R., Tien, D., and Chan, V. Active noise control of transformer noise, *International Conference on Energy Management and Power Delivery*, EMPD '98. Vol. 2, March 1998, pp. 747 -753.
21. http://www.federalpacific.com/: Understanding transformer noise.

11

Special Transformers

11.1 Rectifier Transformers

Duties of rectifier transformers serving special industrial loads are more stringent than conventional transformers. Electrical energy in the form of direct current is required in electrolytic processes used in aluminum smelters and chemical plants (production of chlorine, soda, etc.). Various methods used for converting AC into DC in earlier days included use of motor-generator set, rotary converters and mercury arc rectifiers. With the rapid development in power electronic converters and switching devices, transformers with modern static converters (rectifiers) are being widely used for current ratings as high as hundreds of kilo-amperes. Design and manufacture of transformers with the rectifier duty poses certain challenges. Complex winding arrangements, high currents and associated stray field effects, additional losses and heating effects due to harmonics, necessity of maintaining constant direct current, etc. are some of the special characteristics of rectifier transformers.

11.1.1 Bridge connection

One of the most popular rectifier circuits is three-phase six-pulse bridge circuit as shown in figure 11.1. It gives a 6-pulse rectifier operation with the r.m.s. value of the secondary current for ideal commutation (zero overlap angle) as

$$I = \sqrt{\frac{1}{2\pi}\int_0^{2\pi} i^2 \, d\theta} = \sqrt{\frac{1}{2\pi}\left[I_d^2\,\frac{2\pi}{3} + I_d^2\,\frac{2\pi}{3}\right]} = I_d\sqrt{\frac{2}{3}} \tag{11.1}$$

where I_d is the direct current. For a transformer with unity turns ratio, the r.m.s. value of the primary current is also given by the above expression.

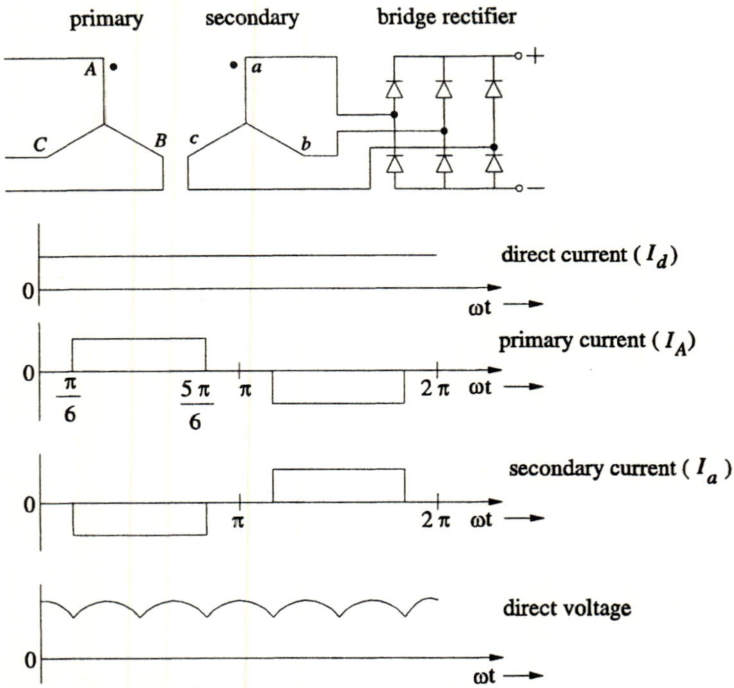

Figure 11.1 Bridge connection

The average value of direct voltage is

$$V_d = \frac{1}{\pi/3} \int_{-\pi/6}^{+\pi/6} \sqrt{2}\; E \cos\theta \, d\theta = \frac{3\sqrt{2}}{\pi}\; E \tag{11.2}$$

where E is line-to-line r.m.s. voltage.

The secondary winding does not carry any direct current (the average value over one cycle is zero). The ratings of both primary and secondary windings are equal, which can be obtained by using equations 11.1 and 11.2 as

$$P = \sqrt{3}\, E\, I = 1.047\; V_d\, I_d = 1.047\; P_d \tag{11.3}$$

Thus, in the bridge connection the capacity of a transformer is well utilized because the required rating of ($1.047\; P_d$) is the minimum value for a 6-pulse operation. The bridge connection is simple and quite widely used.

11.1.2 Interphase transformer connection

When the current rating increases, two or more rectifier systems may need to be paralleled. The paralleling is done with the help of an interphase transformer which absorbs at any instant the difference between the direct voltages of the individual systems so that there are no circulating currents. Two 3-pulse rectifier systems (operating with a phase displacement of 60°) paralleled through an interphase transformer are shown in figure 11.2.

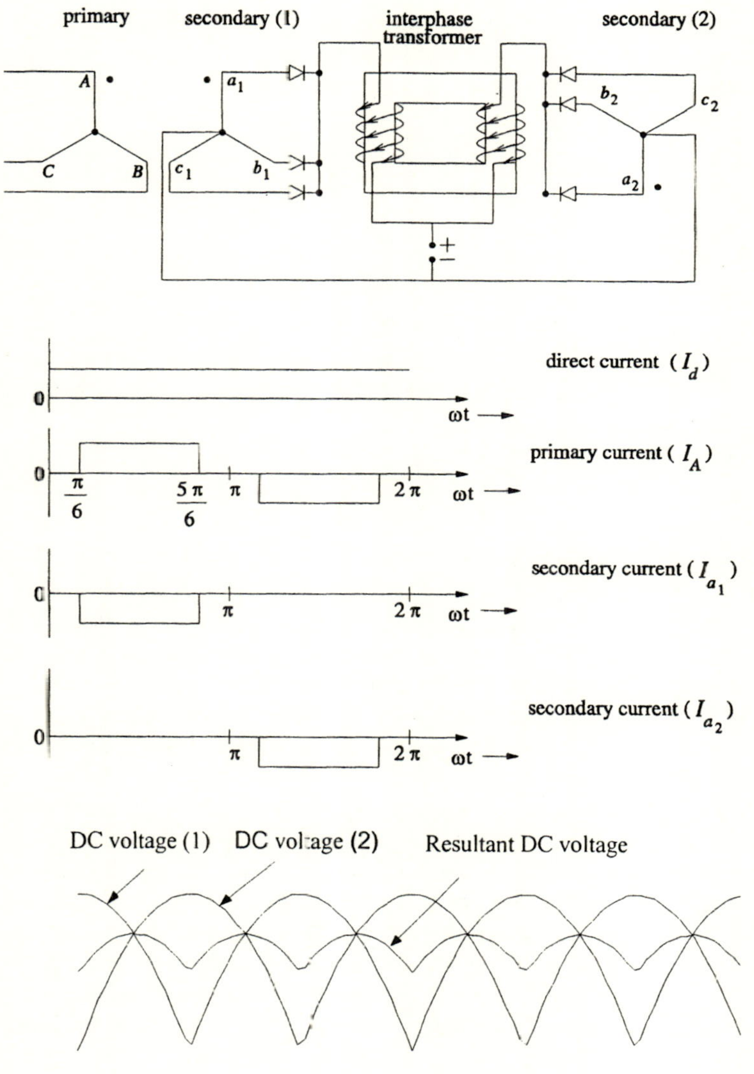

Figure 11.2 Arrangement with interphase transformer

The difference between the (instantaneous values of) direct voltages of two systems is balanced by the voltage induced in the windings of the interphase transformer, for which they are in series connection. Since both the windings are linked with the same magnitude of magnetic flux, the voltage difference is equally divided between them. The output DC voltage at any instant is the average value of DC voltages of the two systems. Thus, the paralleling of two 3-pulse systems results in a system with 6-pulse performance.

The r.m.s. value of the secondary current is given by

$$I_s = \sqrt{\frac{1}{2\pi}\left[\left(\frac{I_d}{2}\right)^2 \frac{2\pi}{3}\right]} = \frac{I_d}{2}\sqrt{\frac{1}{3}} \qquad (11.4)$$

where I_d is the total direct current (sum of the direct currents of two rectifier systems). Each secondary conducts for one-third of cycle, and it can be proved that the rating of two secondary windings considered together is $1.48\,P_d$. Since the primary winding carries the current pulses in both half cycles, it is utilized efficiently (compared to secondary windings). The r.m.s. value of its current is

$$I_p = \sqrt{\frac{1}{2\pi}\left[\left(\frac{I_d}{2}\right)^2 \frac{2\pi}{3} + \left(\frac{I_d}{2}\right)^2 \frac{2\pi}{3}\right]} = \frac{I_d}{2}\sqrt{\frac{2}{3}} \qquad (11.5)$$

The corresponding primary rating is $1.047\ P_d$, the minimum value which can be obtained for a 6-pulse performance.

Since the flux in the magnetic circuit of the interphase transformer is alternating with 3 times the supply frequency when two 3-pulse systems are paralleled or with 6 times the supply frequency when two 6-pulse systems are paralleled, the core losses are higher. Hence, the operating flux density in the interphase transformer is designed to be around 50 to 67% of the value used for the conventional transformer [1].

If a 12-pulse operation is desired, two 6-pulse rectifier systems operating with a phase displacement of 30° are combined through an interphase transformer. In this case, the time integral of the voltage to be absorbed is smaller as compared to the 6-pulse operation (due to smaller voltage fluctuation in the ripple). Also, the frequency of the voltage is 6 times the supply frequency. Hence, the size and cost of the interphase transformer is reduced. When the 12-pulse operation is obtained through one primary winding (usually star connected) and two secondary windings (one in star and other in delta connection), it may be difficult to get the ratio of turns of two secondary windings equal to $1/\sqrt{3}$ (because of low number of turns). In such a case, the 30° phase displacement is obtained by having two primary windings, one connected in star and other in delta, and two secondary windings both connected either in star or delta. One such arrangement is shown in figure 11.3.

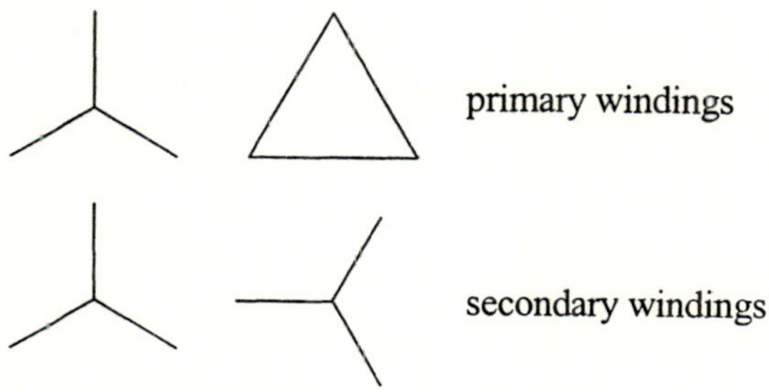

Figure 11.3 Twelve-pulse operation

Since the two primary windings are displaced by 30°, it is necessary to have an intermediate yoke [2] to absorb the difference between the two limb fluxes ϕ_1 and ϕ_2 (see figure 11.4). The intermediate yoke area should be corresponding to the difference of the two fluxes (which is about 52% of the main limb area).

Under the balanced condition of the two paralleled rectifier systems, the currents (average values) in both the windings of the interphase transformer are equal. This results in equal flux in the same direction in both the limbs forcing the flux to return through the high reluctance non-magnetic path outside the core (a substantial portion of DC ampere-turns is absorbed along the non-magnetic return path). Other way to explain it is that since net ampere-turns are zero in the window (currents are directed in opposite directions inside the window), flux lines in the closed magnetic path are absent. Hence, the flux density in the core is low under the balanced operation. A slight unbalance in currents of the two systems results in a non-zero value of ampere-turns acting on the closed magnetic path, which may drive the core into saturation [3]. Thus, the interphase transformer draws a high excitation current under the unbalanced conditions. This is one more reason (apart from higher core losses) for keeping the operating flux density lower in interphase transformers.

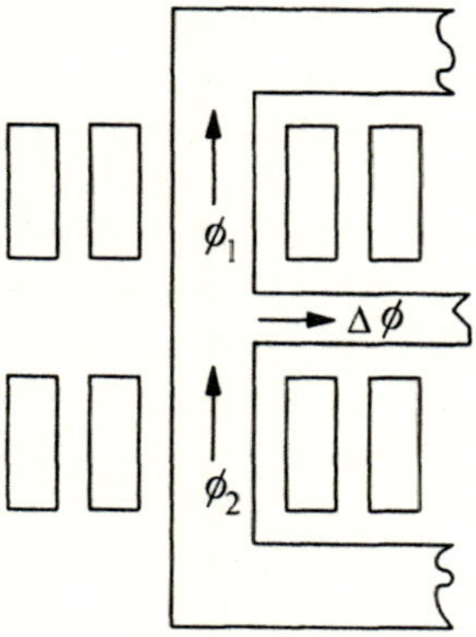

Figure 11.4 Intermediate yoke arrangement

Although the interphase transformer connection has some disadvantages, viz. higher rating of secondary winding and saturation of magnetic circuit due to unbalance between two paralleled systems, it competes well with the bridge connection in a certain voltage-current range. The application of the interphase transformers is not restricted to paralleling of two systems; for example with a three-limb core, three systems can be paralleled [1].

If the pulse number has to be further increased (e.g., 24-pulse operation), the required phase shift is obtained by using zigzag connections or phase shifting transformers [1, 2].

11.1.3 Features of rectifier transformers

Rectifier transformers are used in applications where the secondary voltage is required to be varied over a wide range at a constant current value. It is extremely difficult and uneconomical to have taps on the secondary winding because of its very low number of turns and high current value. The taps are either provided on the primary winding, or a separate regulating transformer (autotransformer) is used (feeding the primary of the main transformer) which can be accommodated in the same tank. Various circuit arrangements which can be used to regulate the secondary voltage are elaborated in [4].

For higher pulse operations, the extended delta connection is shown to be more advantageous than the zigzag connection, as it results into lower eddy losses and short circuit forces [5].

The output connections, which carry very high currents, increase the impedance of the transformer significantly. The increase in impedance due to these connections can be calculated for a single conductor as per equation 3.80. For go-and-return conductors of rectangular dimensions, the impedance can be calculated as per the formulae given in [6].

For large rating rectifier transformers, the field due to high currents causes excessive stray losses in structural parts made from magnetic steel. Hence, these parts are usually made of non-magnetic steel.

Rectifier transformers are subjected to harmonics due to non-sinusoidal current duty. Hence, sometimes the pulse number gets decided by harmonic considerations. Due to harmonics, more elaborate loss calculations are required for rectifier transformers as compared to the conventional transformers [7].

Sometimes the core of the rectifier transformer supplying power electronic loads is designed to have a small gap in the middle of each limb [5] to limit the residual flux and keep the magnetizing reactance reasonably constant. This feature also limits the inrush current thereby protecting the power electronic devices. Under normal operating conditions, the core flux fringing out in the gap between the two core parts hits the inner winding causing higher eddy losses. In order to mitigate this effect, the windings may also have to be designed with a gap at the location facing the core gap.

Because of possibilities of rectifier faults, special design and manufacturing precautions are taken for rectifier transformers. It is generally preferred to design the rectifier transformers with larger core area with the corresponding smaller number of turns to reduce short circuit forces [8]. Disk type windings are preferred since they have better short circuit strength compared to layer windings. Quality of drying / impregnation processes and integrity of clamping / support structures have to be very good. The paper insulation on winding conductors can also be strengthened.

11.2 Converter Transformers for HVDC

There has been a steady increase in High Voltage Direct Current (HVDC) transmission schemes in the world because of many advantages of HVDC transmission as compared to HVAC transmission [9, 10]. The converter transformer is one of the most important and costly components of HVDC transmission system. The converter transformer design has much in common with that of the conventional power transformer except a few special design aspects which are elaborated in this section.

11.2.1 Configurations

The standard 12-pulse converter configuration can be obtained using star-star and star-delta connections with one of the following arrangements, viz. 6 single-phase two-winding, 3 single-phase three-winding and 2 three-phase two-winding. The arrangements are shown in figure 11.5.

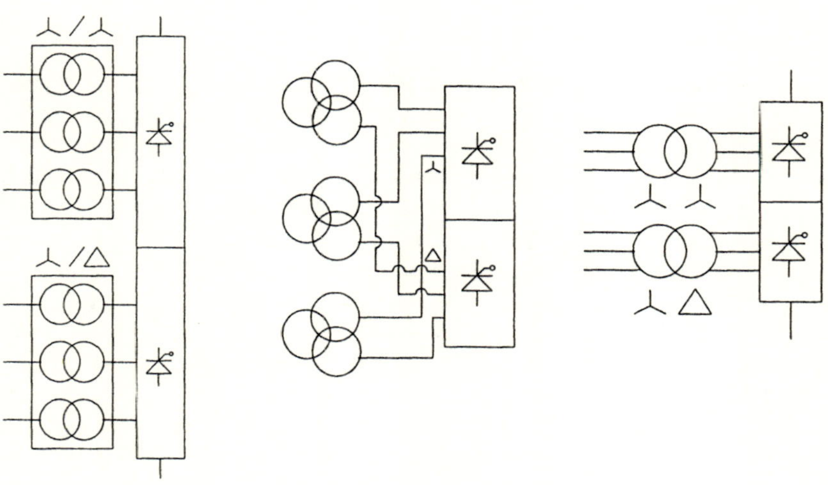

(a) 1-ph 2-winding (b) 1-ph 3-winding (c) 3-ph 2-winding

Figure 11.5 Configurations of converter transformers

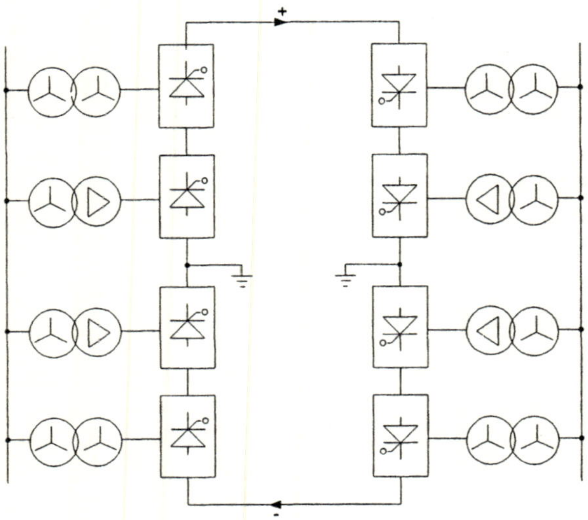

Figure 11.6 Schematic diagram of bipolar 12-pulse operation

The weight and size of individual transformer are highest and overall cost (with all transformers considered) is lowest in the three-phase two-winding configuration, whereas the weight and size of individual transformer are lowest and overall cost is highest in the single-phase two-winding configuration. Since the cost of spare transformer in the single-phase two-winding configuration is lowest (that of only one of the six transformers), it is more commonly used.

11.2.2 Insulation design

Simplified schematic diagram for bipolar (double) 12-pulse operation is shown in figure 11.6. The windings connected to converter and that connected to AC side are generally termed as valve and AC windings respectively. Since the potentials of the valve winding connections are determined by the combination of conducting valves at any particular instant, the entire valve winding has to be fully insulated. Also, unlike the AC winding, both the terminals of the valve winding experience the full DC voltage of the bridge to which it is connected. Hence, the end insulation is higher resulting into greater radial leakage field at the winding ends. The winding eddy loss due to radial leakage field can be much higher than the conventional transformer, if the conductor dimensions are not chosen properly.

Thus, in addition to the normal AC voltage, the valve windings are subjected to a direct voltage depending on their position with respect to ground. Under an AC voltage, potential distribution is in inverse proportion to capacitance or electric stress is inversely proportional to permittivity in a multi-dielectric system. Since the permittivity of oil is about half of solid insulation, the stress in oil is more under the AC voltage condition in the conventional transformers. Since the dielectric strength of the oil is quite less as compared to that of the solid insulation, the insulation design problem reduces mainly to designing of oil ducts as elaborated in Chapter 8. Contrary to AC conditions, under DC voltage conditions the voltage distribution is in direct proportion to resistance or electric stress is directly proportional to resistivity. At lower temperatures the resistivity of solid insulating materials used in transformers is quite high as compared to that of the oil. The ratio of resistivity of a high quality pressboard to that of the oil is about 100 at 20°C, which reduces to as low as 3.3 at 90°C [11]. This is because the fall in the resistivity of the pressboard with temperature is much higher than that of the oil [12]. Such a large variation in the ratio of the two resitivities increases the complexity of insulation design.

Thus, under DC conditions at lower temperatures, most of the voltage gets distributed across the solid insulation and stress in it greatly exceeds that in the oil. The oil ducts have practically only AC voltage across them, whereas solid insulations (barriers, washers, supporting and clamping components, etc.) generally have preponderance of DC voltage with a certain amount of superimposed AC voltage. Therefore, the pressboard barriers tend to be more in the converter transformers as compared to the conventional transformers. However, the proportion of solid cannot be higher than a certain value because the composite oil-solid system has to withstand AC voltage tests as well.

Let the symbols ε and ρ denote relative permittivity and resistivity. With a voltage V applied across two parallel plates shown in figure 11.7, under AC field conditions the stresses in oil and solid insulation are

$$E_1 = \frac{V}{d_1 + \frac{\varepsilon_1}{\varepsilon_2} d_2} \cdots \textit{for oil} \qquad E_2 = \frac{V}{d_2 + \frac{\varepsilon_2}{\varepsilon_1} d_1} \cdots \textit{for solid} \tag{11.6}$$

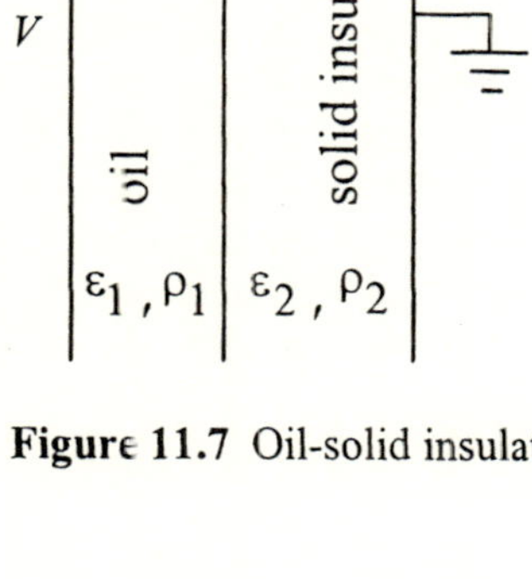

Figure 11.7 Oil-solid insulation system

and under DC field conditions the stresses are

$$E_1 = \frac{V}{d_1 + \frac{\rho_2}{\rho_1} d_2} \cdots \textit{for oil} \qquad E_2 = \frac{V}{d_2 + \frac{\rho_1}{\rho_2} d_1} \cdots \textit{for solid} \qquad (11.7)$$

For non-uniform field conditions involving complex electrode shapes, the techniques described in Chapter 8 should be used to calculate the stresses.

Under steady-state DC conditions, space charges get accumulated at the boundary of the oil and solid insulation. When there is a polarity reversal, in which the applied DC voltage changes from $+V$ to $-V$, an equivalent of the voltage difference $2V$ gets applied. As the time required for reversing the polarity of the applied voltage is much shorter than the space charge relaxation time [13], the voltage due to space charge is not affected during the time of polarity reversal. Therefore, using equations 11.6 and 11.7 the stresses in the oil and solid insulation under the polarity reversal condition can be given by

$$E_1' = \frac{V}{d_1 + \frac{\rho_2}{\rho_1} d_2} - \frac{2V}{d_1 + \frac{\varepsilon_1}{\varepsilon_2} d_2} \quad \cdots \textit{for oil} \qquad (11.8)$$

$$E_2' = \frac{V}{d_2 + \frac{\rho_1}{\rho_2} d_1} - \frac{2V}{d_2 + \frac{\varepsilon_2}{\varepsilon_1} d_1} \quad \cdots \textit{for solid} \qquad (11.9)$$

Thus, under the polarity reversal condition, the oil gap is stressed more (since $\rho_2 >> \rho_1$ and $\varepsilon_1/\varepsilon_2 \cong 0.5$). It can be easily seen from the above equations that the smaller the stress across the oil gap before the polarity reversal, the more the stress is across it after the polarity reversal. The voltage distribution under various conditions is shown in figure 11.8. The voltage across the oil gap is much higher during the polarity reversal condition (Case 3) as compared to Case 2 of steady-state DC voltage condition (prior to the polarity reversal).

For the oil-solid composite insulation system, a relatively low DC voltage superimposition on AC voltage has very little effect on the partial discharge inception voltage [12-15]; this is due the fact that most of the DC voltage gets dropped across the solid insulation, which has a high DC withstand voltage. If, however, the DC voltage magnitude is within the range of the breakdown DC voltage, the breakdown behaviour of the entire system is determined by the DC voltage. Although the converter transformers are stressed by combined AC and DC voltages during service conditions, it is not considered necessary to test them with superimposed voltages [16]. Conventional power frequency and impulse tests are generally sufficient besides pure DC voltage tests and the polarity reversal test.

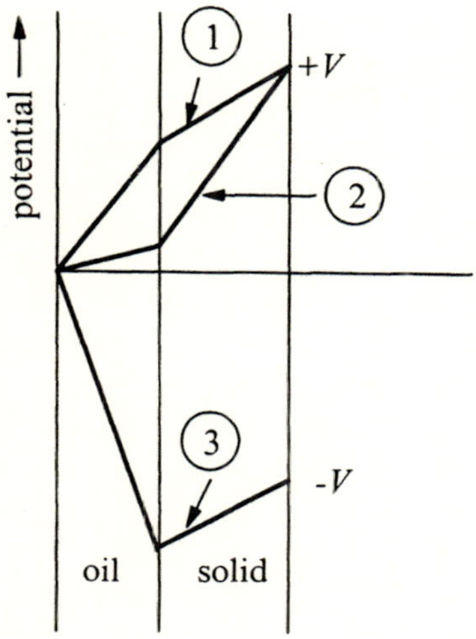

Figure 11.8 Voltage distribution under various conditions

Like in the case of AC insulation design (Chapter 8), the stresses under DC can be calculated accurately by numerical methods. The field distribution is generally calculated for the worst case situation, say at 20°C, since at this temperature the ratio of resistivity of the solid insulation to that of the oil is high resulting in high stress in the solid insulation. The density of equipotential lines in barriers / cylinders is much higher as compared to the oil, necessitating an increase in their thickness as compared to the conventional transformers. The shape and placement of barriers and the width of the resulting oil ducts would have already been decided by the requirements of AC design and the thermal considerations [11]. Hence, it is obvious that the high strength of the solid insulation cannot be fully utilized from the DC design consideration if the stress in the oil gap (having a lower strength) has to be kept low under the AC and polarity reversal conditions [17]. Discontinuities in solid insulations result into higher tangential (creep) stresses along the solid-oil interfaces, and hence these should be properly looked at while finalizing the insulation design.

The insulation design of the converter transformers is complicated by the fact that the ratio of resitivities of the solid insulation and oil, which varies considerably as explained earlier, is greatly influenced by a number of factors [16], viz. temperature, field strength, moisture, time and aging (contrary to the conventional transformers where the corresponding ratio of permittivities does not exceed about 2 and is practically independent of external factors).

The volt-time characteristics of the oil under DC voltage application are reported in [13, 18] for plane-plane electrode; the DC breakdown voltage shows a rapid rate of decrease for stressing time till 100 seconds, after which there is hardly any decrease. The long-time breakdown voltage ($t \rightarrow \infty$) is 70 to 80% of the one-minute breakdown voltage [18]. Under combined AC-DC voltage, the AC breakdown voltage of the oil decreases as the DC voltage increases. The DC 1-minute withstand voltage of the oil gap is about 20 to 30% lower than the AC 1-minute withstand voltage [14, 18, 19]; on the contrary the oil-impregnated paper / solid insulation withstands more DC voltage as compared to AC voltage [14]. The higher DC strength of the solid insulation can be partly explained by the absence of partial discharges which lower the strength in the case of AC voltage. Even if there are oil voids in the solid insulation, the stress in them is too low under DC voltage to initiate partial discharges.

It is reported in [20] that the dielectric strength of the converter transformer insulation under the polarity reversal condition is similar to that under switching impulse stresses. An equivalent AC power frequency voltage test has been suggested for the polarity reversal test.

From the typical schematic shown in figure 11.6, it is clear that when a number of converters are connected in series, the line (AC side) windings are connected in parallel across the same lines and the inductively transferred overvoltage to ground increases cumulatively from one converter bridge to the next higher bridge because the valve (DC side) windings are connected in series (although voltages across the valve windings remain almost the same for all the converters).

11.2.3 Other design aspects

***On-load tap changer (OLTC)*:** The OLTC of a converter transformer plays a crucial role in HVDC transmission system. The OLTC tap position is adjusted to get a voltage which minimizes the reactive power requirement of HVDC converters (firing angle of converters is kept as minimum as possible). Hence, the OLTC is an important constituent of the HVDC control scheme. The number of OLTC operations in the converter transformer is usually much higher than that in the conventional power transformer for the same reason. The OLTC is used for effective control of the power flowing through HVDC line and the DC voltage.

***Leakage impedance*:** The leakage impedance of the converter transformer is the principal component of commutating reactance, which limits the rate of rise of loop current during the small overlap period when the current is transferred from one valve (thyristor) to another. Thus, the leakage impedance helps in preventing instantaneous current transfer which otherwise would result in high di/dt value damaging the valves. The leakage impedance value has to be judiciously selected; a higher value reduces the rate of rise of the loop current during the current commutation process, but increases the overlap angle and reactive power demand of converters. The permissible tolerance on the impedance value of the converter transformers is usually lower than the conventional transformers so that the distortion in DC voltage waveform and the non-characteristic harmonics are reduced.

***DC Bushings*:** Creepage requirements of DC bushings are higher. The DC creepage withstand of an insulator can be about 30% lower than the AC withstand. While for conventional transformers, a creepage distance of about 31 mm/kV is specified for very heavily polluted areas, the creepage distance of as high as 40 mm/kV may be specified for DC bushings in converter transformers.

***Harmonics*:** One of the most severe duties of converter transformers is the presence of harmonics. Due to harmonics, eddy losses in windings and stray losses in structural components are higher in the converter transformers as compared to the conventional transformers. For a 6-pulse operation, the harmonics generated are $6k \pm 1$, where k is an integer. For a 12-pulse operation, the harmonics generated are $12k \pm 1$. Thus, the higher the pulse number the higher is the frequency of the lowest order harmonic produced. But as the pulse number increases, the number of transformers required is more and also the complexity of transformer connections increases. For a 24-pulse operation, use of zigzag windings or extended delta connection is required.

Under normal operating conditions, the current in the converter transformer windings is of stepped form. It is still AC current since it is symmetrical about the x axis. Non-uniformity or asymmetry of valve firing angles produces DC magnetization of the transformer core increasing the magnetizing current, and the windings carry a DC current corresponding to the level of DC magnetization. Although with modern firing controls, the DC magnetization is much lower, a careful design of the magnetic circuit is necessary to avoid excessive losses and noise in the core.

Due to harmonics and the possibility of some DC magnetization, the operating flux density in the converter transformer is less than the conventional transformer and its value is around 1.6 Tesla.

The higher harmonic content produces greater noise, and some special measures for noise control may be required as described in Chapter 10.

The leakage field in the converter transformer contains appreciable harmonic content which results into higher eddy loss in windings and structural parts. The eddy loss in windings can be controlled by using subdivided conductors or CTC (continuously transposed cable) conductor. If the CTC conductor is used, it is usually of epoxy bonded type for higher short circuit strength as explained in Chapter 6. The magnetic or eddy current shielding techniques, discussed in Chapter 5, have to be used to minimize losses and to eliminate hot spots in the tank and other structural components. Short circuit losses have been measured on a 213 MVA single-phase converter transformer at a number of frequencies between 60 Hz and 6 kHz and harmonic load loss factors are reported in [21] for accounting extra losses due to harmonics.

***Short circuit withstand*:** Due to possibility of valve mal-operation resulting into high currents, the short circuit withstand design of the converter transformer deserves more attention than the conventional transformer. Quality of processing (drying and impregnation) and integrity of clamping / support structures have to be ensured. The windings of the converter transformer need to be adequately braced. The current density of the windings can be lower to enhance the short circuit withstand strength.

11.3 Furnace Transformers

A transformer supplying arc furnace has to deliver an unusually high current over a wide range of voltage. The power ratings between 50 to 100 MVA are quite common now with the secondary currents of more than 50 kA. The furnace transformer has special features for handling very high currents as compared to the conventional transformer.

The arc furnace has three electrodes connected to the secondary terminals of the furnace transformer. The furnace transformer has to be specially designed to withstand frequent short circuits on the secondary side. Currents drawn in the arc furnace are characterized by wide fluctuations and unbalanced conditions, which lead to problems of voltage drops, harmonics, etc. These effects can be mitigated by supplying furnaces directly from a high voltage transmission line having high capacity (adequate fault level at the supply point) through a furnace transformer. In such a case, when the voltage ratio is quite large, suitable measures should be taken for protecting the secondary winding against the electrostatically transferred voltages from the high voltage primary winding. These measures are, viz. connection of a surge arrester or capacitor between the secondary terminals and ground, placement of electrostatic shield between the primary and secondary windings, etc.

The leakage reactance of the furnace transformer affects the furnace operation since it gets added to the reactance of the high current connections between the transformer secondary terminals and the electrode tips. The higher the reactance, the lower the useful service currents are, thereby reducing the efficiency of the operation. Hence, the leakage reactance needs to be kept as small as practically possible with due consideration to the mechanical design of windings and clamping / support structures. However, a certain minimum value of reactance is required in the furnace circuit to stabilize arc. In large furnace installations, the low voltage connections usually provide the necessary reactance. For smaller installations, a reactor may have to be added in series with the primary winding to give sufficient reactance value for the stability. These series reactors, which may be housed in the tank of furnace transformer, are usually provided with taps so that the reactance value can be varied for an optimum performance. Hence, depending on the rating of furnace installation and its inherent reactance, the leakage reactance of the furnace transformer has to be judiciously selected to meet the stability and efficiency requirements.

Although the core-type construction is common, the shell-type construction is also used [22] because one can get a desired low impedance value by suitably interleaving the primary and secondary windings. The furnace transformers are provided with a separate regulating (tap) winding. The variation of percentage reactance over the entire tapping range depends on the disposition of windings. The effects of various dispositions on the percentage reactance and the performance of furnace are analyzed in [23].

The melting process of a furnace requires initially more power to break down and melt the furnace charge. The power required afterwards for refining the molten metal is lower. The variable power input requirement is achieved by varying the supply voltage to electric arc furnace over a wide range continuously by use of OLTC. Its use is a must where temporary interruption of supply for changing taps (in off-circuit tap changer) is not desirable. Since the regulation required is quite fine, an OLTC with a large number of steps is required. Due to frequent operations, its oil quality should be regularly checked. It is preferable to place OLTC in a separate compartment so that its maintenance can be carried out without having to lower the oil to the extent that windings get exposed [24].

The commonly used arrangements for the voltage regulation are shown in figure 11.9. The arrangement (a), which consists of taps at the neutral end of the primary winding, is used for low rating furnace transformers (5 to 10 MVA). The cost of OLTC is minimum due to lower voltage and current values (the primary voltage may be of the order of 33 or 66 kV). The disadvantage of this arrangement is that the step voltage is not constant throughout the range of voltage regulation (for a fixed primary voltage, when the tap position is changed for varying the secondary voltage, the voltage per turn changes which results in a non-uniform change in the secondary voltage from one tap to another).

The arrangement (b), used for larger furnace applications, eliminates the disadvantage of the previous arrangement. A separate autotransformer is used for the voltage regulation. The step voltage is uniform throughout since the voltage per turn is independent of tap position for a fixed input voltage applied to the primary of autotransformer. The autotransformer, which may be supplied directly from a system at 66 or 132 kV, reduces the voltage down to the level of the primary winding of the furnace transformer. The OLTC voltage class is higher than that of arrangement (a) and three single-phase tap changers may have to be used. Also, the autotransformer and furnace transformer are usually housed in separate tanks thereby increasing the cost and size of the total system.

The most popular arrangement used for medium and large power furnace applications is the furnace transformer with a booster arrangement as shown in figure 11.9 (c). The booster transformer on the output side boosts or bucks the fixed secondary voltage of the main transformer. The primary winding of the booster transformer is supplied from the tap winding of the main transformer, and the supply voltage is selected such that it results in the least onerous operating conditions for the OLTC. Hence, the OLTC cost is quite low in this arrangement. Also, the variation of secondary voltage is same from one tap position to another throughout the range of regulation. Usually, the main and booster transformers are placed in the same tank minimizing the length of connections between the secondary windings of both the transformers. The amount of structural steel required is also reduced.

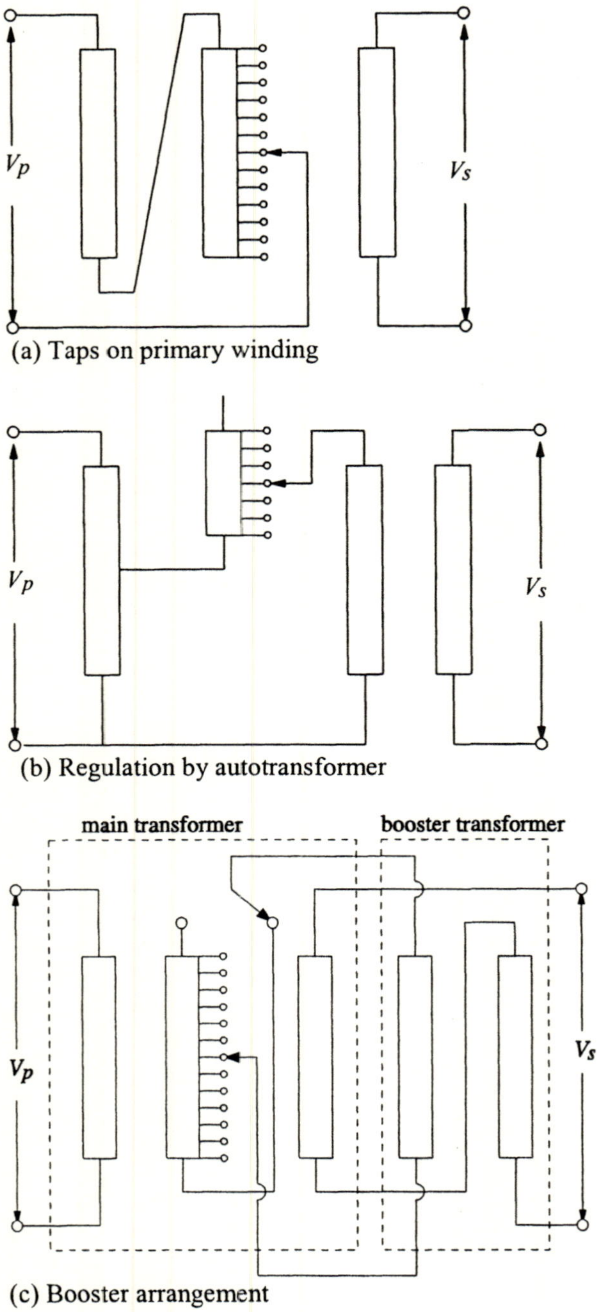

(a) Taps on primary winding

(b) Regulation by autotransformer

(c) Booster arrangement

Figure 11.9 Types of furnace transformer

The booster transformer rating is much smaller than that of the main transformer, being sufficient for the regulation purpose only. Although the core diameters (and core area) are different, the magnetic circuits of the two transformers have generally the same center-to-center distance and equal window heights to facilitate the connections between their secondary windings. If one wants to reduce the core material content, the center-to-center distance of the booster transformer can be lower, but the connections become a bit more difficult. Since the currents of the secondary windings of the main and booster transformers are equal, the same conductor type and size is generally used for both the windings. Also, the two windings are often connected by a *figure of eight* (figure 11.10) avoiding the extra connections between them. A special lifting arrangement is required to lift the two core-winding assemblies simultaneously.

Since the current carried by the secondary windings is quite high, a continuously transposed cable (CTC) conductor is used which minimizes the eddy losses, gets rid of the transposition problems and improves the winding space factor. The material of structural parts supporting high current terminations and tank parts in the vicinity of high current field should be non-magnetic steel. Analysis of a high current termination in a large furnace transformer is reported in [25] wherein excessive losses and hot spots observed during the tests have been analyzed by 3-D FEM analysis. Suitable modifications of magnetic clearances and material type eliminated the hot spots.

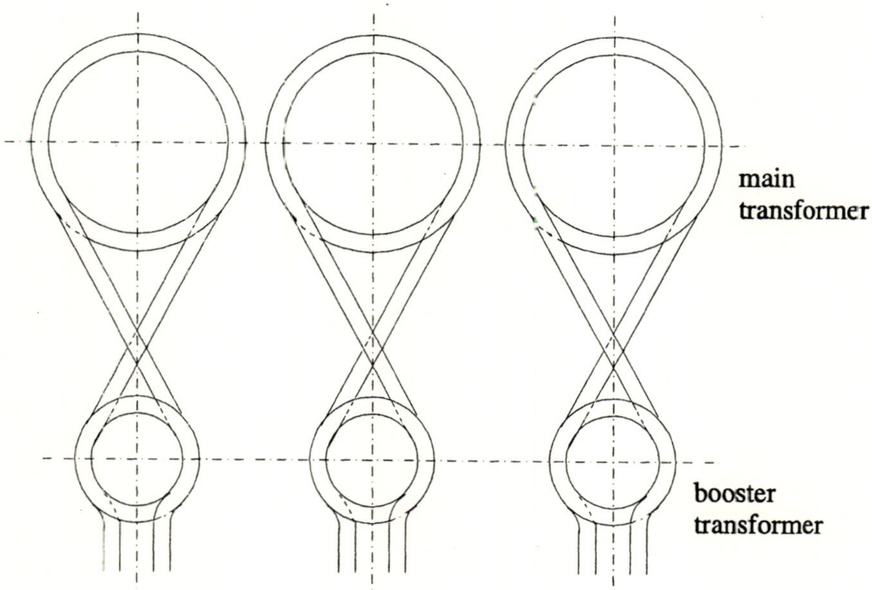

Figure 11.10 *Figure of eight* connection

The secondary winding of a furnace transformer is made up of a number of parallel coils arranged vertically and connected by vertical copper bars. The go-and-return arrangement is used for input and output connections (placed close to each other) reducing the magnetic field and associated stray losses in the nearby structural parts. A delta connected secondary winding is preferable since the current to be carried by it is reduced. Many times, both the ends of each phase of the secondary winding are brought out through the terminals and the delta connections are made at the furnace (get automatically formed by the metallic charge in the furnace). This minimizes the inductive voltage drops in the leads and can achieve a better phase balance between the electrode currents. Due to heavy connections, some unbalance may exist which has to be minimized by some specific arrangements [23]. The secondary winding terminals are usually located on the vertical side of the tank (instead of top cover) resulting in the reduced length of connections, stray losses and cost of the transformer. The LV (secondary) winding is invariably the outermost winding and the HV (primary) winding can be placed next to the core. In such a case, the regulating (tap) winding is in between the HV and LV windings. Such a winding disposition reduces the variation of percentage impedance (short circuit voltage) as the tap position is changed from the minimum to maximum value in variable flux designs.

Small furnace transformers are naturally cooled with radiators. For large ratings and where there are space restrictions, forced oil cooling with an oil-to-water heat exchanger can be used. The oil pressure is always maintained higher than the water pressure (so that water does not leak into oil if a leakage problem develops). The LV terminations may be of U shaped copper tubes of certain inside and outside diameters so that they can be water cooled from the inside. These copper tubes can be cooled by oil also [24].

11.4 Phase Shifting Transformers

A phase shifting transformer (PST) is used to control the active power flow in a complex power transmission network in a very efficient way. The PST has long been used to improve the transient stability of the power systems [26]. It has been successfully used to control and increase the power flow between two large systems [27]; in this case the option of using PST was finalized after its comparison with the other options, viz. HVDC link and series capacitors. The PST provides a well defined phase shift (advance or retard) between the primary (source) and secondary (load) terminals as shown in figure 11.11. In the phase-advance mode, the voltage vector at the output of the PST is made to lead the input voltage vector by adding a leading quadrature voltage to the source voltage. In the phase-retard mode, a lagging quadrature voltage is added to the source voltage so that the voltage vector at the output of the PST lags the input voltage.

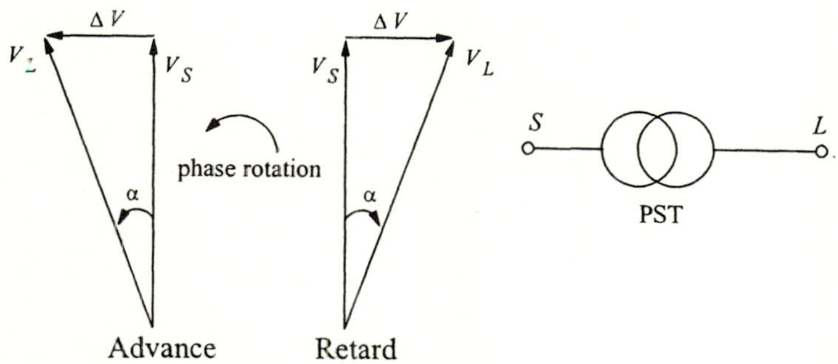

Figure 11.11 Advance mode and retard mode

Normally, this phase shift can be varied during operation in definite steps by use of an on-load tap changer (OLTC). The sign of the phase shift can be inverted (advance to retard mode and vice versa) with OLTC having the reversing switch. These transformers can be constructed with many different winding configurations depending on the rated voltage, power output and amount of phase shift.

The rated design (equivalent) power which decides the size of PST is given by [28]

$$S_{eq} = 3 \times \left\{ V_{ph} \times 2\sin(\alpha/2) \right\} \times I_{SL} = 3\, V_{ph}\, I_{SL} \times 2\sin(\alpha/2) \tag{11.10}$$

where V_{ph} is the line-to-ground voltage, I_{SL} is the line current flowing from source to load, and the term in brackets is the voltage across the winding (between S and L terminals) expressed in terms of the phase shift angle α. Hence, the required maximum value of phase shift angle decides the rating and size of PST. Depending on the voltage / power rating, phase shift angle requirements, connected system's short circuit capability and OLTC performance, two distinct designs of PST are used, viz. single-core design and double-core design.

The less complex single-core design is generally used at lower voltages for small phase shifts and small ratings of PST. The figure 11.12 shows the arrangement of PST with a delta connected exciting winding. In this configuration, the regulating windings are wound on the same core limb as the exciting winding. The phase shift between the source (S) and load (L) terminals is achieved by connecting the regulating winding as shown in the figure. Its voltage is in phase with that of the exciting winding between the other two phase terminals. The voltage magnitudes of S and L terminals are equal under the no-load condition. The phasor diagram shows the phase shift advancement obtained for the load terminal voltage with respect to the source terminal voltage. If the OLTC with reversing switch is used, one can get phase advance as well as phase retard.

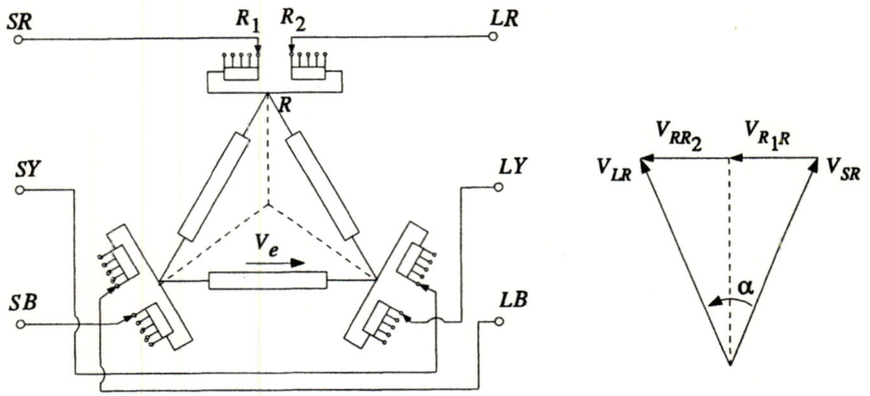

Figure 11.12 Delta configuration of PST and phasor diagram

A special case of the configuration shown in figure 11.12 is the design with only one tap winding and one OLTC with reversing switch, which can be used for smaller phase shifts.

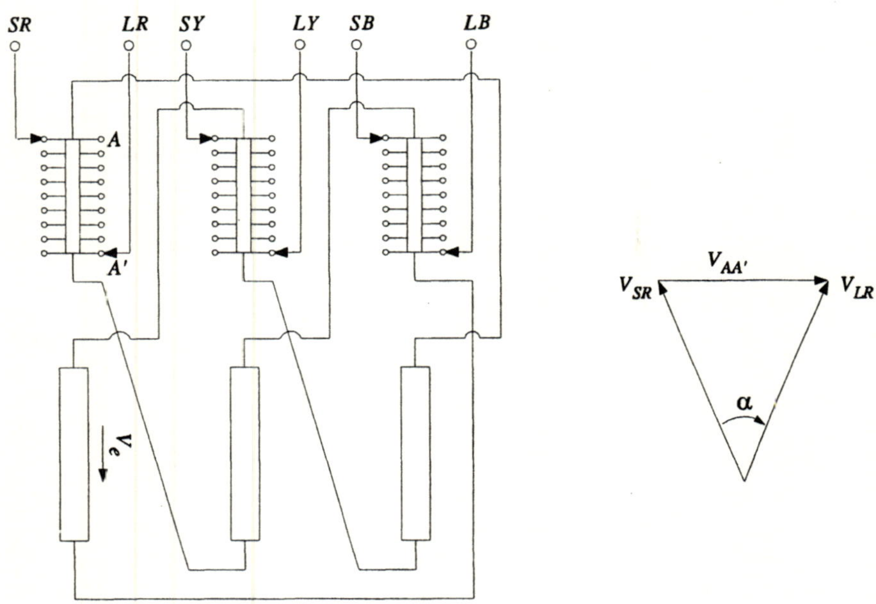

Figure 11.13 Delta-hexagonal PST (with retard phase shift)

Another configuration which uses the single-core design is shown in figure 11.13. The arrangement is known as delta-hexagonal PST which generally has OLTC with linear regulation, i.e., without a reversing switch. The regulating winding is wound on the same core limb as the main exciting winding. The regulating winding of R phase is located between phases Y and B, and produces a phase shift (retard) as shown in the figure.

The two configurations of the single-core design discussed are the cases of line end regulation. The tap changer and regulating (tap) winding are directly exposed to system disturbances (overvoltages and short circuit currents). Hence, the cost of OLTC increases. Additional impedances may have to be connected to the load side terminals to protect the tap changer from short circuit currents because no transformer impedance is present at phase angle of zero. If OLTC with reversing switch is used in the arrangement shown in figure 11.12, during the switch operation the tap winding gets momentarily disconnected from the main winding. Its potential is decided by the potentials of adjacent windings and the capacitances (between windings and between windings and ground). This problem of high recovery voltage can be tackled [28, 29] by use of shields between windings or by use of tie-in resistor (which connects the tap winding to a fixed potential during the reversing operation).

The two-core design shown in figure 11.14 is popularly used for large PST ratings and larger range of phase angle shift. This type of PST basically consists of the series unit and the exciting unit, which are enclosed in separate tanks. When the design is used for smaller ratings and lower voltages, both the units can be enclosed in the same tank.

The winding of the series unit, between the source and load terminals, is split into two halves, and the main winding of the exciting unit is connected to the connection point of these two split windings. The advantage of this arrangement is that the tap winding in the exciting unit and the winding $a\,a'$ in the series unit can be designed independently (windings AA' and BB' form a part of HV network). The voltage level of the tap winding and $a\,a'$ winding can be suitably chosen to reduce the tap changer cost.

It is clear from the phasor diagram that with phase angle α the voltage $V_{BB'}$ also changes because the exciting winding (BB') is located electrically in the middle of the series winding. At the phase angle $\alpha = 0°$, the voltage $V_{AA'}$ becomes zero and the three voltages V_{SR}, V_{LR} and $V_{BB'}$ become equal. Thus, the condition $\alpha = 0°$ gives the highest exciting voltage $V_{BB'}$, and the corresponding voltage at the tap winding also reaches a maximum value. The OLTC has to be designed for the step voltage corresponding to this highest voltage.

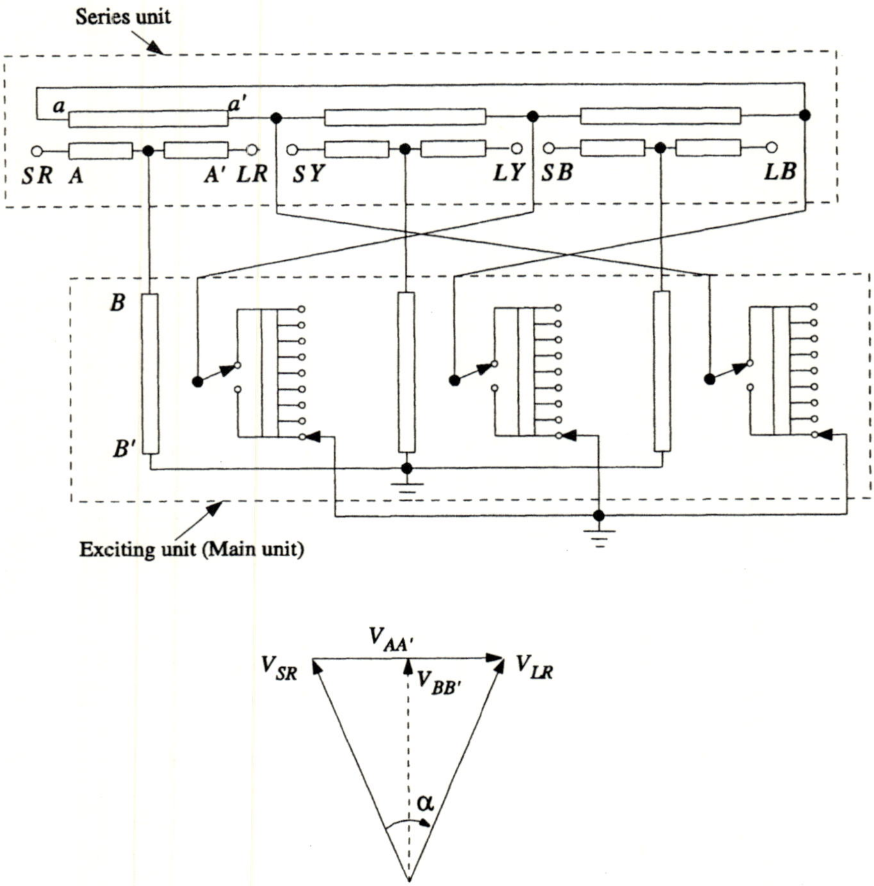

Figure 11.14 Commonly used configuration for two-core design PST

In order to get rapid and smooth control of power flow, static phase shifters can be used, which employ tap changers consisting of static devices like thyristors. One such scheme, reported in [30], is shown to minimize the subsynchronous resonance in large turbine-generators resulted due to capacitive series compensation of lines. Applications of static phase shifters in power systems are reported in [31]. The most of the dynamic characteristics of a static phase shifter can be achieved by augmenting an existing conventional phase shifter with a small size static phase shifter (thus giving a hybrid phase shifter).

The calculation of currents in the phase shifting transformers under system fault conditions requires more elaborate treatment as compared to that in the conventional transformers. The equivalent circuit model and the positive-sequence, negative-sequence and zero-sequence networks required for the fault analysis are given in [32]. The method is used to calculate currents under various fault conditions.

References

1. Schaefer, J. *Rectifier circuits: theory and design*, John Wiley and Sons, Inc., New York, 1965.
2. Pelikan, T. and Isler, J. Rectifier transformers for heavy currents, *The Brown Boveri Review*, Vol. 48, No.3/4, March/April 1961, pp. 215-228.
3. Wells, R. Rectifier interphase transformers and current balance, *Electrical Review*, Vol. 207, No. 14, October 1980, pp. 45-46.
4. Coppadoro, F. Voltage regulation of transformers for silicon rectifiers, *Brown Boveri Review*, Vol. 8-72, August 1972, pp. 416-421.
5. Andersen, O. W. Large transformers for power electronic loads, *IEEE Transactions on Power Delivery*, Vol. 12, No. 4, October 1997, pp. 1532-1537.
6. Copper Development Association, *Copper for busbars*, Publication 22, June 1996, p. 53.
7. Crepaz, S. For an improved evaluation of conventional losses of transformers for converters, *IEEE Transactions on Industry Applications*, Vol. IA-11, No. 2, March/April 1975, pp. 165-171.
8. Kasermann, P. Oerlikon rectifier transformers, *Bulletin Oerlikon*, No. 336/337, 1959, pp. 134-139.
9. Arrillaga, J. *High voltage direct current transmission*, Peter Peregrinus Ltd., London, 1983.
10. Kimbark, E. W. *Direct current transmission*, John Wiley and Sons, New York, 1971.
11. Tschudi, D. J. DC insulation design: paper-oil insulation systems, *WICOR Insulation Conference*, Rapperswil, Switzerland, September 1996.
12. Ganger, B., Falkenhagen, H., and Kury, H. Contributions to the problem of insulation stresses in HVDC converter transformers, *Direct Current*, Vol. 2, No. 1, February 1971, pp. 15-29.
13. Takahashi, E., Tsutsumi, Y., Okuyama, K., and Ogata, F. Partial discharge characteristics of oil-immersed insulation systems under DC, combined AC-DC and DC reversed polarity voltage, *IEEE Transactions on Power Apparatus and Systems*, Vol. PAS-95, No. 1, January/February 1976, pp. 411-420.
14. Kurita, K., Hyuga, S., Ikemoto, N., Oka, Y., Sudo, Y., Hirano, M., Kako, Y., and Kiwaki, H. Converter transformer, DC reactor and transductors for 125 kV 37.5 MW thyristor converter, *Hitachi Review*, Vol. 21, No. 4, pp. 146-156.
15. Hessen, P. and Lampe, W. Insulating problems in HVDC converter transformers, *Direct Current*, Vol. 2, No. 1, February 1971, pp. 30-44.

16. Wahlstrom, B. Voltage tests on transformers and smoothing reactors for HVDC transmission, *Electra*, Report No. 46, 1976, pp. 19-38.
17. Kull, U. The behavior of a laminated oil-paper dielectric with high direct voltages, *The Brown Boveri Review*, Vol. 55, No. 4/5, April/May 1968, pp. 215-221.
18. Ohshima, I., Motegi, S., Honda, M., Yanari, T., and Ebisawa, Y. HVDC breakdown of transformer oil and the effect of space charge on it, *IEEE Transactions on Power Apparatus and Systems*, Vol. PAS-102, No. 7, July 1983, pp. 2208-2215.
19. Ganger, B. E. The breakdown voltage of oil gaps with high DC voltage, *IEEE Transactions on Power Apparatus and Systems*, Vol. PAS-87, No. 10, October 1968, pp. 1840-1843.
20. Hasegawa, T., Yamaji, K., Hatano, M., Kouan, T., and Hosokawa, N. Dielectric strength of transformer insulation at DC polarity reversal, *IEEE Transactions on Power Delivery*, Vol. 12, No. 4, October 1997, pp. 1526-1531.
21. Forrest, J. Harmonic load losses in HVDC converter transformers, *IEEE Transactions on Power Delivery*, Vol. 6, No. 1, January 1991, pp. 153-157.
22. Grundmark, B. Large furnace transformers, *ASEA Journal*, December 1972, pp. 151-156.
23. Bonis, P. and Coppadoro, F. Transformers for arc furnaces, *Brown Boveri Review*, Vol. 60, No. 10/11-73, October/November 1973, pp. 456-467.
24. Hochart, B. *Power transformer handbook*, Butterworth and Co. Ltd., London, 1987.
25. Koppikar, D. A., Kulkarni, S. V., Khaparde, S. A., and Jha, S. K. Evaluation of eddy losses due to high current leads in transformers, *Proceedings IEE - Science Measurement and Technology*, Vol. 144, No. 1, January 1997, pp. 34-38.
26. O'Kelly, D. and Musgrave, G. Improvement of power-system transient stability by phase shift insertion, *Proceedings IEE*, Vol. 120, No. 2, February 1973, pp. 247-252.
27. Patel, B. K., Smith, H. S., Hewes, T. S., and Marsh, W. J. Application of phase shifting transformers for Daniel-Mcknight 500 KV interconnection, *IEEE Transactions on Power Delivery*, Vol. PWRD-1, No. 3, July 1986, pp. 167-173.
28. IEEE Standard C57.135-2001™, IEEE guide for the application, specification, and testing of phase-shifting transformers.
29. Kramer, A. and Ruff, J. Transformers for phase angle regulation considering the selection of on-load tap changers, *IEEE Transactions on Power Delivery*, Vol. 13, No. 2, April 1998, pp. 518-525.
30. Iravani, M. R. and Mathur, R. M. Damping subsynchronous oscillations in power systems using a static phase shifter, *IEEE Transactions on Power Systems*, Vol. PWRS-1, No. 2, May 1986, pp. 76-83.

31. Iravani, M. R., Dandeno, P. L., Nguyen, K. H., Zhu, D., and Maratukulam, D. Applications of static phase shifters in power systems, *IEEE Transactions on Power Delivery*, Vol. 9, No. 3, July 1994, pp. 1600-1608.
32. Del Vecchio, R. M., Poulin, B., Feghali, P. T., Shah, D. M., and Ahuja R. *Power transformer design principles: with applications to core-form power transformers*, Gordon and Breach Science Publishers, Canada, 2001, pp. 499-542.

12

Recent Trends in Transformer Technology

In the last one decade, rapid changes and developments are being witnessed in the transformer design, analysis, manufacturing and condition monitoring technologies. The technological leap is likely to continue for the forthcoming years with the simultaneous increase in the power rating and size of transformers. There is ongoing trend to go for higher system voltages for transmission which increase the voltage rating of the transformers. The phenomenal growth of power systems has put up tremendous responsibilities on the transformer industry to supply reliable and cost-effective transformers. Any failure of a transformer or its component will not only impair the system performance but it also has a serious social impact. The reliability of transformers is a major concern to users and manufacturers for ensuring a trouble-free performance during the service. The transformer as a system consists of several components such as core, windings, insulation, tank, accessories, etc. It is absolutely necessary that integrity of all these components individually and as a system is ensured for a long life of the transformer.

The chapter identifies the recent trends in research and development in active materials, insulation systems, computational techniques, accessories, diagnostic techniques and life estimation / refurbishment. The challenges in design and manufacture of the transformers are also identified.

12.1 Magnetic Circuit

There has been a steady development of core steel material in the last century from non oriented steels to scribed grain oriented steels. The trend of reduction in transformer core losses in the last few decades is related to a considerable increase in energy costs. One of the ways to reduce the core losses is to use better and thinner grades of core steels. Presently, the lowest thickness of commercially available steel is 0.23 mm. Although the loss is lower, the core-building time increases for the thinner grades. The price of the thinner grades is also higher. Despite these disadvantages, core materials with still lower thicknesses will be available and used in the future.

The commercial materials can be divided into three distinct groups: non oriented, grain oriented and rapidly quenched alloys [1]. The amorphous magnetic alloys, typically available in thickness of 0.025 to 0.05 mm, are part of the third group. The loss of amorphous materials is quite low; about 30% of cold rolled grain oriented (CRGO) steel materials, because of their high resistivity and low thickness. Due to their non-crystalline nature (low anisotropy), the flux distribution is more uniform in them as compared to the CRGO materials. However, they are costlier and have low saturation magnetization (~ 1.58 T as compared to 2.0 T for CRGO). The maximum operating flux density for amorphous cores is therefore limited to about 1.35 T. Hence, although the core (no-load) loss is substantially low, the size and cost of the core increases, and the load loss is also higher. Therefore, the use of amorphous material is attractive when users specify a high no-load loss capitalization ($ per kW). The space factor of the amorphous material is lower than the CRGO material. The amorphous materials are very sensitive to mechanical stresses; the core loss increases significantly with the stress. Also, they have a limited operating temperature range as compared to the CRGO materials. The properties of amorphous metals, viz. thinness, lower space factor, hardness and brittleness, pose design and manufacturing problems for the mass production of distribution transformers [2]. Distribution transformers up to 2.5 MVA have been made with amorphous core. Automation of core assembly process is desirable to make the amorphous core transformers cost-effective and to improve their performance.

12.2 Windings

There has been no significant change in the type of winding conductors used in distribution and power transformers. The rectangular strip or bunch conductors and continuously transposed cable (CTC) conductors are used for windings of power transformers. Foils of either copper or aluminum may find preference for the LV winding of distribution transformers. The CTC conductor is preferably of epoxy bonded type for greater short circuit strength. There have been some attempts [3] to improve the winding space factor significantly by using a cable in which a number of parallel rectangular insulated conductors are bonded edge-to-edge with epoxy. It is reported that there is significant reduction in transformer losses and weight when this type of cable conductor is used.

Recently, HV cable technology, used for power transmission and distribution, has been applied to transformers windings [4]. It results into a dry type transformer without oil with a current density lower than that of the conventional oil cooled transformer. The conductor consists of an innermost bundled conductor surrounded by a thin semi-conducting layer resulting into a more uniform field around the conductor. This semi-conductor layer is then surrounded by cross-linked polyethylene whose thickness depends on the voltage class. There is also an outermost semi-conducting layer which is earthed on each turn along the winding. Thus, the electric field is totally contained in the insulation. A special arrangement of forced air cooling is used. It is reported that the dielectric, mechanical and thermal design of windings can be done independently giving more flexibility for optimizing these functions. It is also claimed that the transformers manufactured by this technology will be more efficient, reliable and eco-friendly. The comparison of their cost with that of the conventional oil cooled transformers and their commercial viability are not yet reported.

Superconducting transformers: Advent of high-temperature superconducting (HTS) materials has renewed interest in research and development of superconducting transformers. Previously developed low-temperature superconductors (LTS) required cooling by liquid helium to about 4°K, which was quite expensive. The development of technology based on liquid nitrogen (LN_2) at temperatures up to 79°K has reduced the complexity and cost of the superconducting transformers [5]. Some of the most promising HTS materials are based on Bismuth compounds (BISCCO) and Yttrium compounds (YBCO). The principal advantages of HTS transformers are: much lower winding material content and losses (current density value of at least 10 times that of the conventional oil cooled transformers can be used), higher overload capacity up to about 2.0 per-unit current and possibility of coreless design [6].

Although HTS transformers have higher overload capacity, they have a very low through-fault sustaining capability due to small thermal mass. It is proposed in [7] that a conventional transformer can be operated in parallel with a HTS transformer. The HTS transformer is normally connected, and under through fault conditions it is disconnected and the conventional transformer is switched in immediately. The arrangement is shown to be more cost-effective (with lifetime costing) as compared to the parallel arrangement of two conventional transformers. In [8], it is suggested to use the HTS transformer as a current limiting device to limit the through-fault currents. During the fault conditions, the transition from the superconducting to normal conducting mode occurs increasing the resistance.

Due to greatly reduced conductor dimensions, the strength of the superconducting winding against radial and axial short circuit forces is inherently quite low. The series capacitance also reduces due to reduction in winding dimensions whereas the ground capacitance is not significantly affected. This results into a very non-uniform voltage distribution. Special countermeasures (e.g., interleaving) need to be taken which increase the complexity of construction. Although there is a possibility for optimization, certain minimum clearances between windings are required to get the specified leakage impedance. The main challenges of superconducting transformers are: short circuit withstand, through-fault recovery and withstand against high voltage tests (particularly the impulse test).

For efficient cooling, it is desirable to have a direct contact between LN_2 coolant and the conductor; hence in some designs the inter-turn insulation is arranged in such a way that the conductor edges are left as bare. Windings of each phase may be kept in a separate cryostat (made of fiberglass) and the tap winding is generally kept outside the cryostat to simplify the overall construction [5]. The tap winding and core may be cooled by forced gas cooling in which case it becomes oil-less, fire-hazard free and eco-friendly transformer.

There is a considerable amount of research and development work currently being done to make the superconducting transformers commercially viable. A development of three-phase 100 kVA superconducting transformer with amorphous core has been reported in [9]. A design feasibility study for a 240 MVA HTS autotransformer has been reported in [5]. With the rapid development in technology, the availability of commercial units is certainly on the horizon. The prototype HTS transformers of rating 30 / 60 MVA are being developed [10] for their use by utilities in the year 2005. The commercial units may be available thereafter.

12.3 Insulation

Low permittivity pressboard: If pressboard with low permittivity (around that of oil) is developed and if it is commercially made available, a more uniform electric stress distribution can be obtained opening avenues for insulation optimization as discussed in Chapter 8.

Gas insulated transformers: There is considerable progress in the technology of gas immersed transformers in the last one decade. Unlike the oil-immersed transformers, they have SF6 gas for the insulation and cooling purposes. Initially, SF6 transformers were manufactured in small ratings (10 to 20 MVA). Now, the ratings as high as 275 kV, 300 MVA are quite common in some parts of the world. SF6 gas has excellent dielectric strength and thermal / mechanical stability. It is non-flammable and hence the main advantage of SF6 transformers is that they are fire-hazard free. Hence, these are suited for operation in the areas with a high fire risk. Due to lower specific gravity of SF6 gas, the gas insulated transformer is generally lighter than the oil insulated transformer. The dielectric strength of SF6 gas is about two to three times that of air at atmospheric pressure and is comparable to that of the oil at about two to three atmospheric pressure. But as the operating gas pressure is increased, a tank with higher strength is required increasing its weight and cost.

Constructional features of SF6 transformer are not very much different than the oil-immersed transformer. The core of SF6 transformer is almost the same as that of the oil-immersed transformer. It usually has higher number of cooling ducts since the cooling is not as effective as that with the oil. Typical insulation over conductor is polyethylene terephthalate (PET). This material does not react with SF6 gas and permits higher temperature rise as compared to the oil-immersed transformer. The impulse strength ratio (strength for impulse test divided by strength for AC test) is lower for SF6 gas as compared to the oil-pressboard insulation system. Hence, the clearances in SF6 transformers get mostly decided by the impulse withstand considerations [11] and the methods have to be used which improve the series capacitance of windings. The ratio of the permittivity of SF6 to that of the solid insulation is lower than the corresponding ratio between the oil and solid insulation; this results into higher stress in SF6 gas than that in the oil. The duct spacers with lower permittivity may have to be used in the major insulation [12] to reduce the stress in the small SF6 gaps at the corners of winding conductors. The heat capacity of the gas is smaller than that of the oil and the thermal time constant is also smaller reducing the overload capacity of SF6 transformers as compared to the oil-immersed transformers. Due to the lower cooling ability of SF6 gas, a large volume of gas has to be circulated by gas blowers; this may increase the noise level of the transformers. For large capacity transformers, perflurocarbons may be used [13] for adequate cooling, and SF6 gas is used only as the insulating medium. But the construction becomes complicated; hence even for large capacity transformers, SF6 gas has been used as the insulation as well as the cooling medium [14]. Due to higher thermal stability of SF6 gas and quite a high value of temperature at which it decomposes, the dissolved gas analysis is not as easy as in the case of oil-immersed transformers to detect incipient faults [15]. The challenges which have to be overcome for the widespread use of SF6 transformers are viz. environmental concerns, sealing problems, lower cooling capability and present high cost of manufacture.

12.4 Challenges in Design and Manufacture of Transformers

Stray loss control: There is continuous increase in ratings of generator transformers and autotransformers. Hence, one of the challenges is to accurately evaluate stray losses for their optimization (to have competitive / compact designs) and for elimination of hot spots. Advanced 3-D numerical techniques are being used to optimize stray losses in the windings and structural parts of large transformers. These techniques along with the stray loss control methods are described in Chapter 5. Even in small distribution transformers, the shielding methods are being adopted to reduce the stray losses [16].

Short circuit withstand: A steady increase in unit ratings of transformers and simultaneous growth of short circuit capacities of networks have made *short circuit withstand* as one of the most important aspects of the power transformer design. The short circuit test failure rate is high for large transformers. In fact, the short circuit performance of transformers has been a preferential subject in a number of CIGRE conferences including the recent year 2000 conference. Although the static force and withstand calculations are well-established, efforts are being made to standardize and improve the dynamic short circuit calculations. The precautions that can be taken at the specification, design and manufacturing stages of transformers for improvement in short circuit withstand have been elaborated in Chapter 6.

Part winding resonance: There are a number of high voltage power transformer failures attributed to this phenomenon as described in Chapter 7. Factory and field tests with non-standard waveshapes and terminal conditions (simulating site conditions) reveal that the transient voltages could be developed across a section of a winding (e.g., tap section) significantly in excess of those during the standard tests. Switching operations and line faults at some distance from the transformer terminals are mainly responsible for such overvoltages within the transformer windings. Accurate simulation of transformers under such conditions by their designers and a greater cooperation between manufacturers and users are essential to avoid the part winding resonance.

Very fast transient overvoltages: Very fast transient overvoltages (VFTO) can be generated by switching operations and fault conditions in gas insulated substations (GIS). The behaviour of a transformer subjected to VFTO has been a topic of intensive research in the recent past. In the worst case, VFTO with a rise time of 10 ns and amplitude of 2.5 per-unit is possible. This steep fronted section of the wave is often followed by an oscillatory component having frequency in the range of 1 to 10 MHz [17]. It not only leads to severe inter-section / inter-turn voltages (due to highly nonlinear voltage distribution) but it may also produce a part-winding resonance. The knowledge of voltage distribution across inter-turn insulation is essential for transformers exposed to very fast transient overvoltages. For this, it is necessary to represent individual winding turns in the simulation models for the evaluation of very high frequency performance of the winding.

Geomagnetic disturbances: Although the effects of solar-geomagnetic activity on power system and equipment were known, the failure of large generator step-up transformer in 1989 during a solar-geomagnetic disturbance created a great concern and apprehension about the effects of geomagnetic currents on transformers. Magnitude and location of geomagnetic currents are very difficult to predict with any degree of accuracy [18]. Under normal excitation conditions, the exciting ampere-turns are less than 0.5% (of rated ampere-turns) for large transformers. Hence, even a small value of geomagnetically induced (excitation) current dramatically changes the field pattern and applies a DC bias to the core flux. During solar-geomagnetic disturbances, DC currents flow in low resistance paths via neutrals of transformers and transmission lines as a result of earth surface potentials. Because of the location of the north magnetic pole with respect to the north geographic pole, the regions of North America with low earth conductivity generally have high values of earth surface potential, and the transformers in these regions are vulnerable to the geomagnetic effects [19]. The DC currents may saturate the core completely which increases the excitation current drawn manifold (rich in even and odd harmonics). The stray losses in structural parts can increase to excessive values generating hot spots. Due to heavy field distortion, transposition scheme (decided based on usual leakage field which is predominantly axial) used in a winding with parallel conductors becomes ineffective resulting into unacceptable circulating current values. Due to different zero-sequence impedance characteristics, the three-phase three-limb transformers are less prone to geomagnetically induced saturation as compared to single-phase three-limb or three-phase five-limb transformers [18].

The duration of a geomagnetic activity can be quite high, lasting for repeat periods of several hours over a several day time span [20], which may result into long durations of overheating with a significant loss of transformer life. The transformer impedance changes drastically due to field distortion and harmonics. As a result, the reactive power associated with the transformer changes. This criterion is used in the monitoring program reported in [21] for protection of transformers from the geomagnetic effects. For mitigating the geomagnetic effects the methods which use active and passive devices are described in [19].

Static electrification: This phenomenon has been identified as the cause of failure of few power transformers with directed flow forced oil cooling. Considerable amount of work has been done in recent past to identify the factors influencing the charge separation phenomenon (see Chapter 9). The methods of avoiding / suppressing the static electrification are now known and practiced by the manufacturers.

Noise level prediction and control: Transformer noise is attracting attention as a result of the growing concern about the environment. While the trend of ever increasing ratings implies higher transformer noise, the noise-reducing measures can be adopted which make the transformer quieter. By using modern design methods and materials, noise emissions can be economically lowered to acceptable levels. The modal analysis, finite element method and sound intensity measurement provide the necessary know-why and know-how. The noise level prediction is a complex coupled field problem as highlighted in the next section. The noise reduction techniques have been discussed in Chapter 10.

12.5 Computer-Aided Design and Analysis

With the rapid development of computational tools, the routine design calculations can be efficiently programmed. Within a matter of few minutes, today's computer can work out thousands of designs to give the optimum design. With the ever increasing competition, there are continuous efforts to optimize the material cost of transformers. In most of the contracts, the transformers have to be delivered in a short period of time, and hence the speed of design and manufacture of the transformers is the key issue. Therefore, it should be ensured that the process of optimization does not lead to non-standard designs. The standardization not only reduces the engineering efforts but it also enhances the quality and reliability of the transformers. The standardization enables the use of drafting software packages which can generate manufacturing instructions and drawings thereby reducing the engineering design time drastically.

The main benefit to a transformer designer, due to modern computers, is in the area of analysis. The transformer is a multi-phase and three-dimensional structure having materials with nonlinear characteristics and anisotropic properties. Superimposition of various physical fields poses a real challenge to the designer. Many times, there are conflicting design requirements for these fields. For example, if the corner radius of a rectangular winding conductor is increased for reducing the dielectric stress, the short circuit strength may reduce. The design of conductor paper insulation (conflict between dielectric design and thermal design), design of supporting structures (conflict between stray loss control and structural design), etc. are some other examples. Due to geometric and material complexities, numerical methods are used for solution of such engineering problems (electrostatic, electromagnetic, structural, thermal, etc.). The Finite Element Method (FEM) is the most popular numerical method, and many commercial 2-D and 3-D FEM packages are available. Many manufacturers develop their own customized FEM programs for optimization and reliability enhancement of transformers. The 2-D FEM analysis, which is widely used for stray loss estimation / control, winding temperature rise calculations, short circuit force calculations, etc., can be integrated into the main electrical design optimization program. As the voltage / current rating of the transformer increases, it is very important to verify the new design using a tool such as FEM. Due to the three-dimensional (and asymmetrical) nature of the transformer structure, three-dimensional analysis is essential for more accurate calculations even though it may be computationally very time consuming and expensive.

The current research trends show that many of the complex design problems, involving more than one physical field, are increasingly being solved by using coupled field formulations [22]. The coupled field treatment is required for problems in which the involved fields (e.g., electromagnetic and thermal) interact either strongly or weakly. Hence, the coupled field problems can be broadly classified as strongly (directly) and weakly (indirectly or sequentially) coupled. This classification is mainly based on the degree of nonlinearity and the relative time constants of the involved fields. A weakly coupled problem is solved using cascaded algorithms; the fields (which are coupled) are solved in the successive steps. The coupling is performed by applying the results from the first analysis (involving only one field) as the loads for the second analysis (which involves the other field). Thus, the problem is divided into sub-problems which are solved sequentially in an iteration loop until the solution converges. When the method of indirect coupling is used, good properties like symmetricity, positive definiteness, etc. are preserved in the field coefficient matrix. The advantage of using this method is that the development of the formulation for each field can be done independently in a flexible and modular fashion. In transformers, the problem of estimation of temperature rise of conducting parts due to induced eddy currents is generally solved as a weakly coupled problem [23] since the thermal time constant is much higher than the electromagnetic time constant.

For strongly coupled problems, in which the interaction of coupled fields is highly nonlinear and the involved fields have comparable time constants, the governing equations are solved simultaneously with all the necessary variables. Thus, all the physical aspects of the coupled fields are simultaneously managed. The field-circuit coupling is also a type of strongly coupled problem. For example, a transient 3-D field analysis of a transformer under a short circuit condition is done in [24] by coupling magnetic field and electric circuit equations. Analysis of load-controlled noise is done in [25] by using a formulation in which the magnetic and structural fields are strongly coupled, whereas the structural and acoustic fields are weakly coupled.

12.6 Monitoring and Diagnostics

Due to severe competition and restructuring taking place in the power industry, there is need to reduce maintenance costs, operate transformers as much as possible and prevent forced outages. Hence, in the recent years the monitoring and diagnostics of transformers have attracted considerable attention. The monitoring can be either off-line or on-line. The trend is more towards on-line techniques due to continuous developments in computational / analysis tools and information technology.

The monitoring not only detects the incipient faults but it also allows a change from periodic to condition based maintenance [26]. It is important to identify the key parameters that should be monitored to reduce the cost of the overall monitoring system. The monitoring of transformers has several challenges, viz. cost of monitors, reliability of electronic equipment, performance under adverse field conditions and inadequate field expertise. There are a number of off-line and on-line monitoring / diagnostic techniques which are currently being used and developed further.

Dissolved gas analysis: It is the most established and proven method to detect incipient faults. Different faults produce different gases; for example arcing, overheated cellulose and partial discharges produce predominantly acetylene, carbon oxides and hydrogen gas respectively. There are a number of established dissolved gas interpretation methods / guidelines (IEEE, IEC, CIGRE, Rogers method, etc.). Sensors are being developed to detect gases. There is enough experience available already with on-line hydrogen sensors. Portable units are available which can detect presence of hydrogen in a sample of transformer oil, which indicates the occurrence of partial discharges.

Partial discharges: Measurement of partial discharge (PD) occurring inside a transformer is commonly done by either the acoustic technique or the electrical technique. The acoustical sensors based on the Piezoelectric effect are less expensive and the main advantage of the acoustic detection is that disturbing signals from electric network do not interfere with the measurement. But the PD detection is possible within a radius of about 200 to 300 mm from the source since the acoustic signals are attenuated by the medium / materials through which they travel. Hence, a number of acoustic sensors may have to be used which are distributed carefully around the transformer. Acoustic sensors can also be placed internally using wave guides (e.g., fiberglass rods) to enhance the strength of the received signal, but the system is expensive and difficult to install.

The PD detection range for the electrical method is larger. It covers a wider area, which includes for example tap changer and bushings. There is better correlation between the instrument reading and the actual PD magnitude as compared to that with the acoustic method. However, the measurements are generally hampered by electrical interference signals from surrounding equipment.

Direct hot spot measurement: The direct winding hot spot temperature measurement method, which uses fiber-optic sensors, is now increasingly used on critical transformers for on-line monitoring. Since the measuring instrument is costly, it can be used for transformers located in one substation / area by rotation (if the fiber-optic sensors are installed and brought out for these transformers). The technique is described in Chapter 9; the challenge for the transformer designer is to predict the hot spot temperature (and the location) very close to the measured values.

Degree of polymerization: This has a definite correlation with the mechanical strength of paper insulation. Its measured value is used to study the phenomena of aging and the corresponding influencing factors. The degree of polymerization (DP), which has a value of about 1000 to 1400 for a new paper, drops to just about 200 for a severely aged paper. Temperature, oxygen and moisture are the main degrading agents which reduce the value of DP. The main disadvantage of this technique is that a paper-insulation sample is required to be taken from inside of the transformer (generally from the lead insulation in the top portion of the transformer) necessitating shut-down of the transformer.

Furan analysis: When cellulose materials (paper and solid insulation) age due to thermal stresses, furanic compounds are generated. These compounds which are dissolved in oil can be detected. Since DP and concentration of furanic compounds are related the condition of insulation can be indirectly known.

Recovery voltage measurement: In this method the relaxation of insulation after excitation by a DC source is analyzed. It can be used to measure the moisture content of the insulation. It is based on the principle that water molecules present in the insulation get polarized in the direction of the applied field. After discharging (short circuiting) it, subsequently under the open circuit condition a finite voltage can be measured (voltage does not reduce to zero) which is a function of the energy stored in the water molecules.

Detection of winding displacement: Conventionally, reactance measurement is done to detect mechanical displacements and / or deformations of windings and structures. The loss of mechanical integrity might occur due to short circuit forces, winding shrinkage (causing release of clamping pressure) and transportation / relocation of transformers. Change in reactance of more than 2% for any phase indicates a winding deformation for transformers up to 100 MVA (particularly applicable to diagnostics after short circuit test). The corresponding value is 1% for transformers above 100 MVA as discussed in Chapter 6. There have been also some efforts to correlate winding looseness with the response of stray losses over a range of frequencies.

Low voltage impulse (LVI) tests are also popularly used to determine whether a transformer has passed a short circuit test (based on computed transfer function). The steepness of the applied voltage is adjusted to obtain a wide frequency band. It is difficult to obtain repeatable results because of test leads and ambient noise. The current trend is to use the frequency response analysis (FRA) technique for assessing the mechanical condition of windings.

Frequency response analysis: This technique is widely used nowadays [27] in which transfer function of transformer winding is determined by the swept frequency method (popularly called FRA). The transfer function essentially is a plot showing the poles (or natural frequencies) as a function of frequency. The transfer function is compared with the corresponding reference (fingerprint) taken earlier. The changes in overall shape and resonance frequencies (which occur due to change of winding inductances and capacitances) are the key indicators of any winding displacement after the short circuit test, shrinkage or transport. The FRA technique is more sensitive as compared to the method based on reactance measurement in detecting winding deformations. This is because, even a small change in winding geometry can have an appreciable effect on the characteristic frequencies. The repeatability of FRA results is also better. For the same reason, the FRA technique is preferred over the LVI method since the main problem with the LVI method is the difficulty in carrying out comparisons between original signatures and repeated measurements (since these are affected by ambient noise conditions).

The disadvantage of the FRA technique is the relatively long duration for each measurement as compared to the LVI method since the measurements are done at discrete frequencies to determine transfer function for full frequency range. It should be noted that both FRA and LVI techniques are used to determine the transfer function. In the former, the transfer function is measured by sweeping the frequency of an input sine wave, in which case, no Fourier transformation is required. In the latter time-domain method, the transfer function is estimated from low voltage impulse test data, for which the waveforms (i.e., input impulse voltage and neutral current response due to it) have to be converted to frequency domain using Fourier transformation (DFT or FFT).

Wavelet analysis: This technique, which is being applied to fault diagnosis and protection of transformers in recent years, is advancement over Fourier transform based methods. It allows study of each frequency component with any desirable time resolution. The inherent non-stationary nature of transformer neutral current waveforms (due to impulse excitations) during different fault conditions can be effectively classified, using its inherent frequency / time-selective feature. The conventional FRA based techniques cannot identify the time-localization of a particular frequency component in a time-dependent signal, whereas using wavelets this is possible, as they are not only localized in frequency but also in time. This simultaneous or joint time-frequency localization feature of the wavelet transform is used in [28] for pattern classification of impulse fault currents. Combined wavelet and neural network approach is used in [29] to discriminate between an internal fault and a magnetizing inrush current in a power transformer protection scheme.

On-load tap changers (OLTC): It is well-known that a maximum percentage of transformer failures are related to OLTC. Hence, the service reliability of OLTC is of vital importance. The OLTC problems are either of mechanical nature (faults related to drive mechanism, shaft, springs, etc.) or of electrical nature (contact wear and tear, burning of transition resistors, dielectric failures, etc.). Measurement of drive motor torque and current, recording of acoustic signals during tap changer operation, measurement of contact resistance, measurement of temperature difference between oil in the main tank and oil in the diverter switch, etc. are some of the methods that give good indications of these problems. The measurements are compared with the reference values. The important aspect in these monitoring systems is to gain a long-term experience for evaluating such data.

Bushings: On-line monitoring of power factor of bushings can be performed by vectorially adding currents of three phases obtained from the capacitance (test) tap. Since in reality the bushings are never identical and system voltages are never perfectly balanced, the sum current has a non-zero value which is unique for each set of bushings. Hence, generally the change in the sum current phasor is monitored. In [30], development of a fiber-optic instrument for on-line monitoring of dielectric dissipation factor of transformer bushings is reported. It consists of an electro-optic electric-field sensor (to sense the high voltage), capacitors (to sense the insulation current) and a signal processing unit.

12.7 Life Assessment and Refurbishment

Transformer life assessment is a process of reviewing the risks of failure for the given transformer and network conditions. Factors which determine the remaining life of a transformer can be categorized under three headings [31], viz. strategic (e.g., load increasing beyond rating), economic (high cost of losses or maintenance) and technical (aging, overstressing or contamination). A number of diagnostic tests are used to assess the overall technical condition of a transformer, which is then related to a criticality in terms of risk of failure or life expectancy. The remaining (residual) life of the transformer can be estimated by assessing the extent of deterioration of the paper insulation through its DP test or by knowing the content of furan compounds in the oil. Furan analysis gives an integrated overall health of cellulosic insulation whereas DP indicates the extent of degradation at a specific location [32]. Both these tests can also help in taking decision about the transformer refurbishment.

During the last one decade, refurbishment of transformers has been given a serious consideration by many users. It can slow down the insulation ageing and improve the short circuit strength. During refurbishment, critical components like on-load tap changer can be upgraded, repaired or replaced enhancing the transformer reliability. Re-clamping of windings can be done if there is looseness.

References

1. Moses, A. J. Electrical steels: past, present and future developments, *Proceedings IEE*, Vol. 137, Pt. A, No. 5, September 1990, pp. 233-245.
2. Harry, W. N., Hasegawa, R., Lee, A. C., and Lowdermilk, L. A. Amorphous alloy core distribution transformers, *Proceedings IEEE*, Vol. 79, No. 11, November 1991, pp. 1608-1623.
3. Girgis, R. S., Baldwin, C. J., Kennedy, W. N., Monaco, S., Lindgren, S. R., Holm, R. D., and Kothmann, R. E. Design and performance improvements in power transformers using ribbon cable, *IEEE Transactions on Power Delivery*, Vol. 10, No. 2, April 1995, pp. 869-877.
4. Jaksts, A., Leijon, M., Sasse, C., Frank, H., Andersson, T. H., Forsmark, S., Nilsson, L., Fogelberg, T., and Sjogren, A. A major breakthrough in transformer technology, *CIGRE 2000*, Paper No. 12-101.
5. Sykulski, J. K, Beduz, C., Stoll, R. L., Harris, M. R., Goddard, K. F., and Yang, Y. Prospects for large high temperature superconducting power transformers: conclusions from a design study, *IEE Proceedings – Electric Power Applications*, Vol. 146, No. 1, January 1999, pp. 41-52.
6. Yamaguchi, H., Sato, Y., and Katoka, T. Comparison between superconducting and conventional power transformers considering auxiliary facilities, *IEEE Transactions on Applied Superconductivity*, Vol. 5, No. 2, June 1995, pp. 937-940.

7. Sykulski, J. K, Stoll, R. L., Beduz, C., Power, A. J., Goddard, K. F., and Al-Mosawi M. K. The design, construction and operation of high temperature superconducting transformers - practical considerations, *CIGRE 2000*, Paper No. 12-203.
8. Serres, E., Levillain, C., Camescasse, F. X., Picard, J. F., Lakner, M., Rhyner, J., Chen, M., Paul, W., Zueger, H., Backes, J., Bonmann, D., Fleshler, S., Otto, A., Diehl, R., Christopherson, C., Mason, R., de Moranville, K., and Daniel, J. Current limitation in high temperature superconducting transformers and impact on the grid, *CIGRE 2000*, Paper No. 12-205.
9. Lee, J., Kim, W., Hahn, S., Choi, K., Cha, G., and Chang, S. Development of a three-phase 100 kVA superconducting power transformer with amorphous cores, *IEEE Transactions on Applied Superconductivity*, Vol. 9, No. 2, July 1999, pp. 1293-1296.
10. Reis, C. T., Mehta, S. P., McConnell, B. W., and Jones, R. H. Development of high temperature superconducting power transformers, *IEEE Power Engineering Society Winter Meeting*, Vol. 2, 2001, pp. 432-437.
11. Togawa, Y., Ikeda, M., Toda, K., and Esumi, K. Progress of gas-insulated transformers, *IEEE International Conference on Energy Management and Power Delivery*, Singapore, Vol. 2, November 1995, pp. 540-547.
12. Takizawa, A., Ono, Y., Inaka, S., Konno, K., Nozaki, Y., Suzuki, T., Yamazaki, S., Saito, S., and Shirone, T. Development of large capacity low-noise gas-insulated transformer, *IEEE Power Engineering Society Winter Meeting*, Vol. 2, 1999, pp. 1036-1041.
13. Muramoto, H., Yamazaki, T., Oshi, T., Uehara, K., Toda, K., Ikeda, M., and Yanabu, S. Development and application of the 275 kV 300 MVA separate cooling, sheet winding gas insulated transformer, *IEEE Transactions on Power Delivery*, Vol. 7, No. 2, April 1992, pp. 888-894.
14. Takahashi, E., Tanaka, K., Toda, K., Ikeda, M., Teranishi, T., Inaba, M., and Yanari, T. Development of large-capacity gas-insulated transformer, *IEEE Transactions on Power Delivery*, Vol. 11, No. 2, April 1996, pp. 895-902.
15. Chan, L. and Sum, C. L. Operating experience of SF6 gas-filled power transformers, *IEE International Conference on advances in power system control, operation and management*, November 1991, Hon Kong, pp. 410-414.
16. Olivares, J. C., Liu, Y., Canedo, J. M., Escarela-Perez, R., Driesen, J., and Moreno, P. Reducing losses in distribution transformers, *IEEE Transactions on Power Delivery*, Vol. 18, No. 3, July 2003, pp. 821-826.
17. Cornick, K., Filliat, B., Kieny, and Muller, W. Distribution of very fast transient overvoltages in transformer windings, *CIGRE 1992*, Paper No. 12-204.
18. Ringlee, R. J. and Stewart, J. R. Geomagnetic effects on power systems, *IEEE Power Engineering Review*, July 1989, pp. 6-9.

19. Albertson, V. D., Bozoki, B., Feero, W. E., Kappenman, J. G., Larsen E. V., Nordell, D. E., Ponder, J., Prabhakara, F. S., Thompson, K., and Walling, R. Geomagnetic disturbance effects on power systems, *IEEE Transactions on Power Delivery*, Vol. 8, No. 3, July 1993, pp. 1206-1216.
20. Kappenman, J. G. Geomagnetic storms and their impact on power systems, *IEEE Power Engineering Review*, May 1996, pp. 5-8.
21. Wrubel, J. N. Monitoring program protects transformers from geomagnetic effects, *IEEE Computer Applications in Power*, January 1992, pp. 11-14.
22. Hameyer, K., Driesen, J., Gersem, H. D., and Belmans, R. The classification of coupled field problems, *IEEE Transactions on Magnetics*, Vol. 35, No. 3, May 1999, pp. 1618-1621.
23. Driesen, J., Deliege, G., Belmans, R., and Hameyer, K. Coupled thermo-magnetic simulation of the foil-winding transformer connected to a non-linear load, *IEEE Transactions on Magnetics*, Vol. 36, No. 4, July 2000, pp. 1381-1385.
24. Renyuan, T., Shenghui, W., Yan, L., Xiulian, W., and Xiang, C. Transient simulation of power transformers using 3D finite element model coupled to electric circuit equations, *IEEE Transactions on Magnetics*, Vol. 36, No. 4, July 2000, pp. 1417-1420.
25. Rausch, M., Kaltenbacher, M., Landes, H., Lerch, R., Anger, J., Gerth, J., and Boss, P. Combination of finite and boundary element methods in investigation and prediction of load-controlled noise of power transformers, *Journal of Sound and Vibration*, 250 (2), 2002, pp. 323-338.
26. Bengtsson, C. Status and trends in transformer monitoring, *IEEE Transactions on Power Delivery*, Vol. 11, No. 3, July 1996, pp. 1379-1384.
27. Feser, K., Christian, J., Neumann, C., Sundermann, U., Leibfried, T., Kachler, A., and Loppacher, M. The transfer function method for detection of winding displacements on power transformers after transport, short circuit or 30 years of service, *CIGRE 2000*, Paper No. 12/33-04.
28. Purkait, P. and Chakravorti, S. Pattern classification of impulse faults in transformers by wavelet analysis, *IEEE Transactions on Dielectrics and Electrical Insulation*, Vol. 9, No. 4, August 2002, pp. 555-561.
29. Mao, P. L. and Aggarwal, R. K. A novel approach to the classification of the transient phenomenon in power transformers using combined wavelet transform and neural network, *IEEE Transactions on Power Delivery*, Vol. 16, No. 4, October 2001, pp. 654-660.
30. Jaeger, N. A. F., Polovick, G. S., Kato, H., and Cherukupalli, S. E. On-line dissipation factor monitoring of high-voltage current transformers and bushings, *CIGRE 1998*, Paper No. 12-102.
31. Breen, G. Essential requirements to maintain transformers in service, *CIGRE 1992*, Paper No. 12-103.
32. Shrinet, V., Patel, M. J., and Ramamoorthy, M. Role of furan and DP analysis for refurbishment of power transformer: few case studies, *CIGRE 2000*, Paper No. 12/33-09.

Appendix A: Fault Calculations

In this appendix, asymmetrical fault calculations for a transformer are given. A single-line-to-ground fault on LV side is considered. The neutrals of LV and HV windings are solidly grounded. Two distinct cases are considered, viz. with and without fault in-feed from the secondary (LV) side. For each of these two cases, two conditions are considered, viz. with and without stabilizing tertiary winding (TV). The effects of omitting tertiary winding are illustrated with these cases.

Transformer Data:

Rating (S) = 25 MVA, Core construction: three-phase three-limb
Voltage ratio: HV/LV/TV, 132/11/11 KV, star/star/delta
Leakage sequence impedances (based on 25 MVA base):

$$\text{HV/LV}: Z_1 = Z_2 = 12\%,\quad Z_0 = 10.2\%$$

$$\text{HV/TV}: Z_1 = Z_2 = 20\%,\quad Z_0 = 17.0\%$$

$$\text{LV/TV}: Z_1 = Z_2 = 6\%,\quad Z_0 = 5.4\%$$

132 KV system fault level (S_F) = 5000 MVA
The positive-sequence system impedance on the HV side is Z_{1HS} = 0.5% (calculated by equation 6.2). The HV and LV side systems are assumed as effectively grounded systems (X_0/X_1 not greater than 3). Here, in particular, the zero-sequence system impedances on the LV and HV sides are assumed to be equal to the corresponding positive-sequence system impedances for simplicity. As mentioned earlier, two alternatives are considered for 11 KV (LV) system:

(i) Without any fault in-feed from 11 KV system.

(ii) With fault in-feed from 11 KV system:

11 KV system fault level (S_F) = 833.33 MVA, which gives the positive-sequence system impedance on LV side: $Z_{1LS} = 3.0\%$ (by equation 6.2).

The equivalent impedances in the star equivalent circuit can be computed by using the theory given in Section 3.5 (Chapter 3) as

$$Z_H = (Z_{HL} + Z_{HT} - Z_{LT}) / 2$$

$$Z_L = (Z_{HL} + Z_{LT} - Z_{HT}) / 2$$

$$Z_T = (Z_{HT} + Z_{LT} - Z_{HL}) / 2$$

Therefore, the corresponding positive-, negative- and zero-sequence impedances of HV, LV and TV windings are computed as follows:

$$Z_{1H} = Z_{2H} = (12 + 20 - 6) / 2 = 13\%, \quad Z_{0H} = (10.2 + 17 - 5.4) / 2 = 10.9\%$$

$$Z_{1L} = Z_{2L} = (12 + 6 - 20) / 2 = -1\%, \quad Z_{0L} = (10.2 + 5.4 - 17) / 2 = -0.7\%$$

$$Z_{1T} = Z_{2T} = (20 + 6 - 12) / 2 = 7\%, \quad Z_{0T} = (17 + 5.4 - 10.2) / 2 = 6.1\%$$

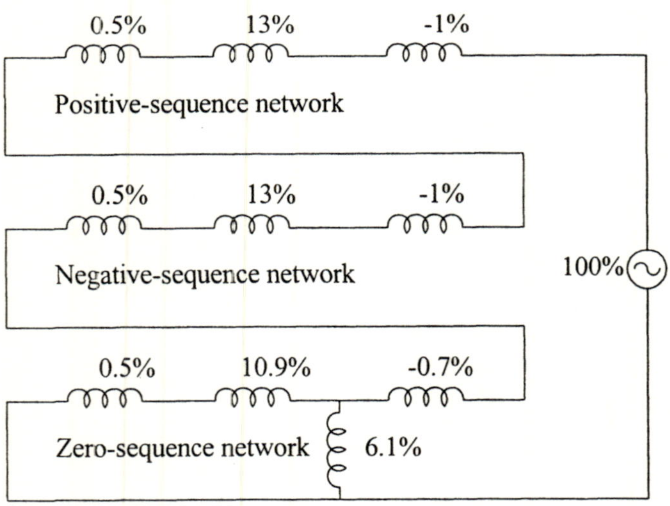

Figure A1 Sequence network for Case A1.1

A1 Asymmetrical Fault with No In-Feed from LV Side

A1.1 Delta connected tertiary winding is present

The equivalent network under the fault condition is shown in figure A1 for a single-line-to-ground fault on the LV side. All the impedances are in % values. The network can be derived from figure 6.3 (a) with impedances Z_{1LS}, Z_{2LS} and Z_{0LS} taken to be effectively infinite since there is no in-feed from the LV side.

Total impedance = 2 (0.5 + 13 – 1) – 0.7 + (11.4 // 6.1) = 28.27%

For a single phase-earth fault, using equations 6.6 and 6.7 we get the values of sequence components of currents as

$I_1 = I_2 = I_0 = 100/\,28.27 = 3.54$ p.u. (i.e., 3.54 times the rated current).

Total fault current= $I_1 + I_2 + I_0 = 3\ (3.54) = 10.62$ p.u.

The fault current in any of the windings is calculated by adding the corresponding sequence currents flowing in them in the three sequence networks. The neutral of LV and HV windings carries 3 times the corresponding zero-sequence current flowing in them. Therefore,

Current through HV winding = 3.54 + 3.54 + 3.54 (6.1/17.5) = 8.31 p.u.

Current through LV winding = 3.54 + 3.54 + 3.54 = 10.62 p.u.

Current through stabilizing winding = 3.54 (11.4/17.5) = 2.31 p.u.

Current through HV neutral = 3 × 3.54 (6.1/17.5) = 3.7 p.u.

Current through LV neutral = total fault current = 3 × 3.54 =10.62 p.u.

A1.2 Delta connected tertiary winding is absent

The equivalent network under the fault condition is shown in figure A2 for a single-line-to-ground fault on the LV side. Proceeding in the similar way as in Section A1.1 we get

Total Impedance = 2(0.5 + 13 – 1) – 0.7 + 0.5 + 10.9 = 35.7%

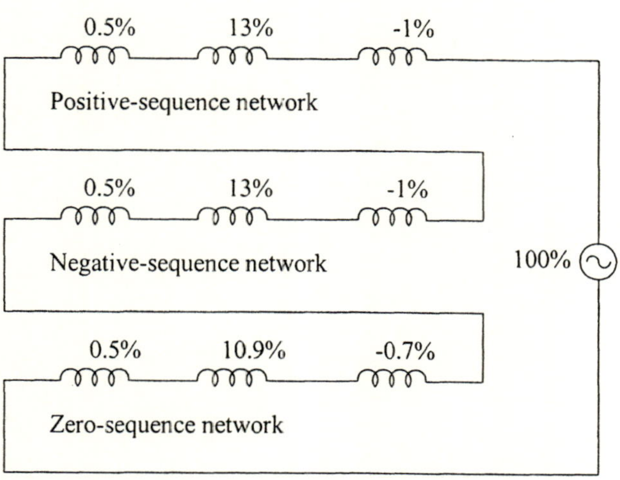

Figure A2 Sequence network for Case A1.2

$I_1 = I_2 = I_0 = 100/\,35.7 = 2.8$ p.u.

Total fault current $= I_1 + I_2 + I_0 = 3\ (2.8) = 8.4$ p.u.

Current through HV winding = 2.8 + 2.8 + 2.8 = 8.4 p.u.

Current through LV winding = 2.8 + 2.8 + 2.8 = 8.4 p.u.

Current through HV neutral = 8.4 p.u.

Current through LV neutral = 8.4 p.u.

It is to be noted that the single-line-to-ground fault current in this case (8.4 p.u.) is more than the three-phase fault current (i.e., $1/Z_1 = 1/0.12 = 8.33$ p.u.) due to two reasons. First, the zero-sequence leakage impedance of the transformer is less than the positive-sequence leakage impedance. Second, it is assumed that the positive- and zero-sequence system impedances of HV side are equal (0.5 p.u.). In many of the cases, the system zero-sequence impedance is more than the positive-sequence impedance, which will make the single-line-to-ground fault current lower than the three-phase fault current.

It can be observed from the above two case studies that the absence of the tertiary stabilizing winding reduces the total fault current and the current in the faulted winding (LV). However, the current in HV neutral increases significantly.

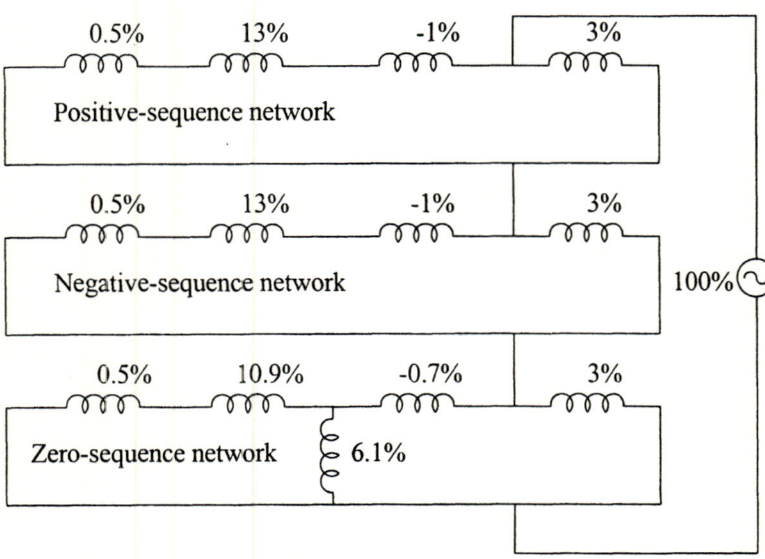

Figure A3 Sequence network for Case A2.1

A2 Asymmetrical Fault with In-Feed from LV Side

A2.1 Delta connected tertiary winding is present

The equivalent network under the fault condition is shown in figure A3 for a single-line-to-ground fault on the LV side. The network is same as that given in figure 6.3.

Total positive-sequence impedance = 12.5 // 3 = 2.42%

Total negative-sequence impedance = 12.5 // 3 = 2.42%

Total zero-sequence impedance = [(11.4 // 6.1) – 0.7] // 3 = 1.56%

Total impedance = 2.42 + 2.42 + 1.56 = 6.4%

$I_1 = I_2 = I_0 = 100/\ 6.4 = 15.62$ p.u.

Total fault current= $I_1 + I_2 + I_0 = 3\ (15.62) = 46.86$ p.u.

Current through HV winding

$$= 15.62\ (3/15.5) + 15.62\ (3/15.5) + 15.62\ (3/6.27)\ (6.1/17.5)$$
$$= 3.02 + 3.02 + 2.6 = 8.64 \text{ p.u.}$$

Current through LV winding = 3.02 + 3.02 + 15.62 (3/6.27) = 13.51 p.u.

Current through stabilizing winding = 15.62 (3/6.27) (11.4/17.5) = 4.87 p.u.

Current through HV neutral = 3 × 2.6 = 7.8 p.u.

Current through LV neutral = 3 × 7.47 = 22.41 p.u.

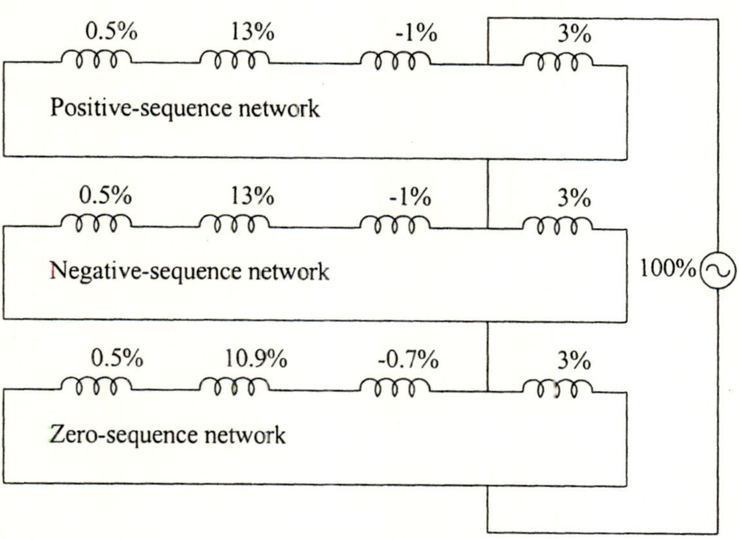

Figure A4 Sequence network for Case A2.2

A2.2 Delta connected tertiary winding is absent

The equivalent network under the fault condition is shown in figure A4 for a single-line-to-ground fault on the LV side. Proceeding in the similar way as in Section A2.1 we get

Total positive-sequence impedance = 12.5 // 3 = 2.42%

Total negative-sequence impedance = 12.5 // 3 = 2.42%

Total zero-sequence impedance = 10.7 // 3 = 2.34%

Total Impedance = 2.42 + 2.42 + 2.34 = 7.18%

$I_1 = I_2 = I_0$ = 100/ 7.18 = 13.93 p.u.

Total fault current = $I_1 + I_2 + I_0$ = 3 (13.93) = 41.79 p.u.

Current through HV winding

= 13.93 (3/15.5) + 13.93 (3/15.5) + 13.93 (3/13.7)
= 2.7 + 2.7 + 3.05 = 8.45 p.u.

Current through LV winding = 2.7 + 2.7 + 3.05 = 8.45 p.u.

Current through HV neutral = 3 × 3.05 = 9.15 p.u.

Current through LV neutral = 3 × 3.05 = 9.15 p.u.

It can be observed from the above two case studies (in-feed from LV side) that the absence of the tertiary stabilizing winding reduces the total fault current and the current in the faulted winding (LV). The current in HV neutral increases.

Appendix B: Stress and Capacitance Formulae

In this appendix, formulae are derived for electric stress and capacitance for commonly existing electrode configurations in transformers such as two round electrodes or round electrode and plane.

B1 Stress Calculations

The information about the electric field intensity and potential field between two parallel cylindrical electrodes can be found by considering the respective equivalent line charges. Consider two line charges $+\rho_L$ and $-\rho_L$ (charges per unit length) placed at $x = +m$ and $x = -m$ respectively as shown in figure B1. Now, due to single line charge ρ_L, the electric field intensity at a distance r is given by

$$E = \frac{\rho_L}{2\pi\varepsilon r} \tag{B1}$$

where ε is permittivity of medium. The potential reckoned from a distance R is

$$u = -\int_R^r E\,dr = \frac{\rho_L}{2\pi\varepsilon}\ln\left(\frac{R}{r}\right) \tag{B2}$$

The resultant potential at point A (figure B1) due to line charges $+\rho_L$ and $-\rho_L$ is

$$u = \frac{\rho_L}{2\pi\varepsilon}\ln\left(\frac{R}{r_2}\right) - \frac{\rho_L}{2\pi\varepsilon}\ln\left(\frac{R}{r_1}\right) = \frac{\rho_L}{2\pi\varepsilon}\ln\left(\frac{r_1}{r_2}\right) \tag{B3}$$

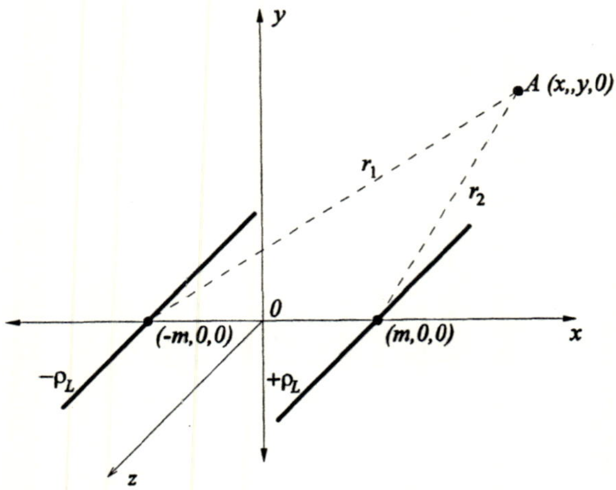

Figure B1 Two line charges placed at $x = -m$ and $x = +m$

Let us now find the nature of equipotential surface having potential of u. From equation B3 we get

$$\frac{r_1}{r_2} = e^{\frac{2\pi\varepsilon u}{\rho_L}}$$

But from figure B1 we have $\frac{r_1}{r_2} = \sqrt{\frac{(x+m)^2 + y^2}{(x-m)^2 + y^2}}$

$$\therefore \sqrt{\frac{(x+m)^2 + y^2}{(x-m)^2 + y^2}} = e^{\frac{2\pi\varepsilon u}{\rho_L}} \quad \text{or} \quad \frac{(x+m)^2 + y^2}{(x-m)^2 + y^2} = e^{\frac{4\pi\varepsilon u}{\rho_L}}$$

Solving by componendo and dividendo,

$$\frac{(x+m)^2 + y^2 + (x-m)^2 + y^2}{(x+m)^2 + y^2 - \left((x-m)^2 + y^2\right)} = \frac{e^{\frac{4\pi\varepsilon u}{\rho_L}} + 1}{e^{\frac{4\pi\varepsilon u}{\rho_L}} - 1}$$

$$\therefore \frac{x^2 + y^2 + m^2}{2mx} = \coth\left(\frac{2\pi\varepsilon u}{\rho_L}\right)$$

By algebraic manipulations we get

$$\left(x - m\coth\left(\frac{2\pi\varepsilon u}{\rho_L}\right)\right)^2 + y^2 = \frac{m^2}{\sinh^2\left(\frac{2\pi\varepsilon u}{\rho_L}\right)} \tag{B4}$$

This is the equation of a circle with radius $r = \dfrac{m}{\sinh\left(\frac{2\pi\varepsilon u}{\rho_L}\right)}$

and center $(s, 0) = \left(m\coth\left(\frac{2\pi\varepsilon u}{\rho_L}\right), 0\right)$

Thus, the equipotential surface is a cylinder which intersects the *x-y* plane in a circle with radius r and center at $(s, 0)$.

From the above expressions for radius and center we get

$$\frac{s}{r} = \cosh\left(\frac{2\pi\varepsilon u}{\rho_L}\right) \tag{B5}$$

Also, $s^2 - r^2 = m^2\coth^2\left(\frac{2\pi\varepsilon u}{\rho_L}\right) - \dfrac{m^2}{\sinh^2\left(\frac{2\pi\varepsilon u}{\rho_L}\right)} = m^2$ (B6)

By substituting the value of m in the equation for radius we have

$$\sinh\left(\frac{2\pi\varepsilon u}{\rho_L}\right) = \frac{\sqrt{s^2 - r^2}}{r} \tag{B7}$$

Now, from equations B5 and B7 we get

$$e^{\frac{2\pi\varepsilon u}{\rho_L}} = \cosh\left(\frac{2\pi\varepsilon u}{\rho_L}\right) + \sinh\left(\frac{2\pi\varepsilon u}{\rho_L}\right) = \frac{s + \sqrt{s^2 - r^2}}{r} = \frac{r}{s - \sqrt{s^2 - r^2}}$$

Thus, we get the expression for potential as

$$u = \frac{\rho_L}{2\pi\varepsilon}\ln\left[\frac{r}{s - \sqrt{s^2 - r^2}}\right] \tag{B8}$$

Now, we will consider two parallel cylindrical conductors of radii R_1 and R_2, placed such that the distance between their centers is $2s$. The electric field intensity and potential between the two conductors are calculated by considering the corresponding two equivalent line charges as shown in figure B2.

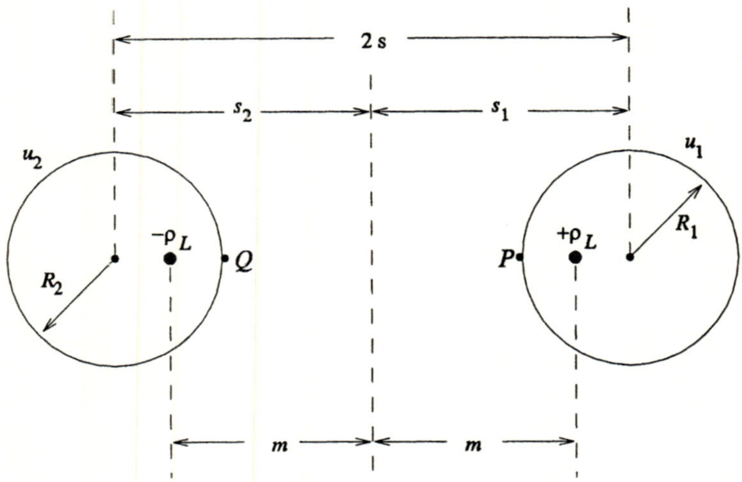

Figure B2 Configuration of two parallel cylindrical conductors

Now,

$$s_1 + s_2 = 2s \tag{B9}$$

Using equation B6 we can write

$$s_1^2 - s_2^2 = \left(m^2 + R_1{}^2\right) - \left(m^2 + R_2{}^2\right) = R_1^2 - R_2^2$$

$$\therefore\ s_1 - s_2 = \frac{R_1^2 - R_2^2}{2s} \tag{B10}$$

By solving equations B9 and B10 we get

$$s_1 = \frac{4s^2 + \left(R_1^2 - R_2^2\right)}{4s} \quad \text{and} \quad s_2 = \frac{4s^2 - \left(R_1^2 - R_2^2\right)}{4s}$$

The electric field intensity at point P on the surface of the conductor on the right side is given by

$$E_P = \frac{\rho_L}{2\pi\varepsilon}\left[\frac{1}{R_1 - (s_1 - m)} - \frac{-1}{2m - (R_1 - (s_1 - m))}\right] = \frac{\rho_L}{2\pi\varepsilon}\left[\frac{2m}{m^2 - (R_1 - s_1)^2}\right]$$

Now, by putting the value of $m = \sqrt{s_1^2 - R_1^2}$ in the above equation we get

$$E_P = \frac{\rho_L}{2\pi\varepsilon\ R_1}\sqrt{\frac{s_1 + R_1}{s_1 - R_1}}$$

By putting the value of s_1 obtained earlier in the above equation we get

$$E_P = \frac{\rho_L}{2\pi\varepsilon\ R_1}\sqrt{\frac{4s^2 + \left(R_1^2 - R_2^2\right) + 4s\,R_1}{4s^2 + \left(R_1^2 - R_2^2\right) - 4s\,R_1}} \tag{B11}$$

Now, by using equation B8 for potential, the potential difference between points P and Q is given as

$$u_1 - u_2 = \left(\frac{\rho_L}{2\pi\varepsilon}\ln\frac{R_1}{s_1 - \sqrt{s_1^2 - R_1^2}}\right) - \left(\frac{-\rho_L}{2\pi\varepsilon}\ln\frac{R_2}{s_2 - \sqrt{s_2^2 - R_2^2}}\right)$$

$$= \frac{\rho_L}{2\pi\varepsilon}\ln\left[\frac{R_1 R_2}{\left(s_1 - \sqrt{s_1^2 - R_1^2}\right)\left(s_2 - \sqrt{s_2^2 - R_2^2}\right)}\right] = \frac{\rho_L}{2\pi\varepsilon}\ln\left[\frac{R_1\left(s_2 + \sqrt{s_2^2 - R_2^2}\right)}{R_2\left(s_1 - \sqrt{s_1^2 - R_1^2}\right)}\right]$$

By putting the values of s_1 and s_2 in the above equation and simplifying,

$$\rho_L = \frac{\left(u_1 - u_2\right)2\pi\varepsilon}{\ln\left\{\frac{R_1}{R_2}\left[\frac{4s^2 - \left(R_1^2 - R_2^2\right) + \sqrt{\left[4s^2 - \left(R_1^2 - R_2^2\right)\right]^2 - 16s^2R_2^2}}{4s^2 + \left(R_1^2 - R_2^2\right) - \sqrt{\left[4s^2 + \left(R_1^2 - R_2^2\right)\right]^2 - 16s^2R_1^2}}\right]\right\}}$$

Putting this value in equation for E_P (equation B11) we have

$$E_p = \frac{\left(u_1 - u_2\right)}{\left(2s - R_1 - R_2\right)} f_P \tag{B12}$$

where

$$f_P = \frac{2s - R_1 - R_2}{R_1\ \ln\left\{\frac{R_1}{R_2}\left[\frac{4s^2 - \left(R_1^2 - R_2^2\right) + \sqrt{\left[4s^2 - \left(R_1^2 - R_2^2\right)\right]^2 - 16s^2R_2^2}}{4s^2 + \left(R_1^2 - R_2^2\right) - \sqrt{\left[4s^2 + \left(R_1^2 - R_2^2\right)\right]^2 - 16s^2R_1^2}}\right]\right\}} \times$$

$$\sqrt{\frac{4s^2 + \left(R_1^2 - R_2^2\right) + 4s\,R_1}{4s^2 + \left(R_1^2 - R_2^2\right) - 4s\,R_1}}$$

is called as non-uniformity factor.

Now, if both the electrodes have the same radius, i.e., $R_1 = R_2 = R$, then

$$f_P = \frac{2(s-R)}{R \ln\left[\frac{4s^2 + \sqrt{(4s^2)^2 - 16s^2 R^2}}{4s^2 - \sqrt{(4s^2)^2 - 16s^2 R^2}}\right]} \times \sqrt{\frac{4s^2 + 4sR}{4s^2 - 4sR}} = \frac{\sqrt{\gamma^2 - 1}}{\ln\left[\gamma + \sqrt{\gamma^2 - 1}\right]} \tag{B13}$$

where $\gamma = \dfrac{s}{R}$

Now, we will consider the other most commonly encountered geometry, i.e., cylindrical conductor - plane geometry as shown in figure B3. The ground plane at point G and the round conductor can be replaced by the configuration of the conductor and its image as shown in the figure. From equations B12 and B13 the electric field intensity, in this case, at point P is given as

$$E_p = \frac{(u-(-u))}{2(s-R)} f_P = \frac{u}{(s-R)} f_P \tag{B14}$$

and the non-uniformity factor is

$$f_P = \frac{\sqrt{\gamma^2 - 1}}{\ln\left(\gamma + \sqrt{\gamma^2 - 1}\right)}, \text{ where } \gamma = \frac{s}{R} \tag{B15}$$

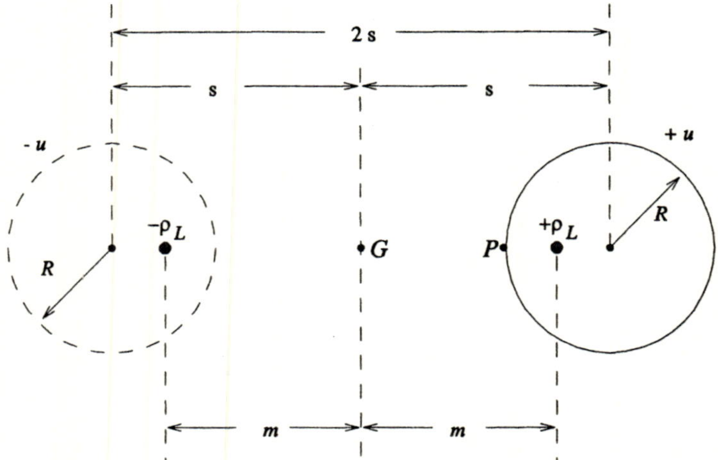

Figure B3 Cylindrical conductor – plane geometry

Now, the electric field intensity at point G is

$$E_G = \frac{\rho_L}{2\pi\varepsilon}\left[\frac{1}{m}+\frac{1}{m}\right] = \frac{\rho_L}{\pi\varepsilon\sqrt{s^2-R^2}} \tag{B16}$$

Using equation B8, the potential at the conductor surface is given as

$$u = \frac{\rho_L}{2\pi\varepsilon}\ln\left[\frac{R}{s-\sqrt{s^2-R^2}}\right] = \frac{\rho_L}{2\pi\varepsilon}\ln\left[\frac{s+\sqrt{s^2-R^2}}{R}\right] = \frac{\rho_L}{2\pi\varepsilon}\ln\left[\gamma+\sqrt{\gamma^2-1}\right]$$

$$\therefore \rho_L = \frac{2\pi\varepsilon\, u}{\ln\left(\gamma+\sqrt{\gamma^2-1}\right)} \tag{B17}$$

Putting this value in equation for E_G we get

$$E_G = \frac{2u}{\sqrt{s^2-R^2}\,\ln\left(\gamma+\sqrt{\gamma^2-1}\right)} = \frac{u}{s-R}f_G \tag{B18}$$

$$\text{where, } f_G = \frac{2(s-R)}{\sqrt{s^2-R^2}\,\ln\left(\gamma+\sqrt{\gamma^2-1}\right)} = \frac{2\sqrt{\dfrac{\gamma-1}{\gamma+1}}}{\ln\left(\gamma+\sqrt{\gamma^2-1}\right)} \tag{B19}$$

The non-uniformity factor f_x for any point x between the center of the conductor and ground in the x direction (figure B4) can now be found as below.

The electric field intensity at a point with distance of x from the conductor center is

$$E_x = \frac{\rho_L}{2\pi\varepsilon}\left[\frac{1}{x-(s-m)} - \frac{-1}{m+s-x}\right] = \frac{\rho_L}{\pi\varepsilon}\,\frac{m}{2s\,x-(x^2+R^2)} \tag{B20}$$

Now, from equation B8 we have

$$u = \frac{\rho_L}{2\pi\varepsilon}\ln\left(\frac{R}{s-m}\right), \text{ where } m=\sqrt{s^2-R^2}$$

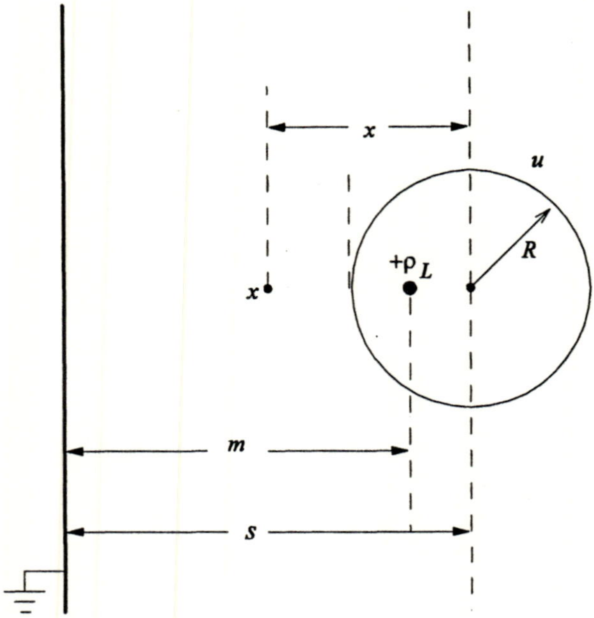

Figure B4 Stress and voltage at any point x from conductor center

$$\therefore\ \rho_L = \frac{2\pi\varepsilon u}{\ln\left(\dfrac{R}{s-m}\right)} \tag{B21}$$

Putting this value in the equation for electric field we get

$$E_x = \frac{2um}{\left[2sx - \left(x^2 + R^2\right)\right]\ln\left(\dfrac{R}{s-m}\right)} = \frac{u}{(s-R)}\ f_x \tag{B22}$$

where $$f_x = \frac{2m\,(s-R)}{\left[2sx - \left(x^2 + R^2\right)\right]\ln\left(\dfrac{R}{s-m}\right)} \tag{B23}$$

The voltage at point x can be calculated using equation B3 as

$$u_x = \frac{\rho_L}{2\pi\varepsilon}\ln\left[\frac{m+s-x}{x-s+m}\right] \tag{B24}$$

where, $\rho_L = \dfrac{2\pi\varepsilon u}{\ln\left(\dfrac{R}{s-m}\right)} = \dfrac{2\pi\varepsilon u}{\ln\left(\dfrac{s+\sqrt{s^2-R^2}}{R}\right)}$ and $m = \sqrt{s^2-R^2}$ (B25)

B2 Capacitance Calculations

B2.1 Capacitance between two parallel cylindrical conductors

From figure B2 for the conditions that $R_1 = R_2 = R$ and $s_1 = s_2 = s$, and by using equation B8, the capacitance between two parallel cylindrical conductors per unit length is given by

$$C = \frac{\rho_L}{u_1 - u_2} = \frac{\rho_L}{\dfrac{\rho_L}{2\pi\varepsilon}\ln\left(\dfrac{R}{s-\sqrt{s^2-R^2}}\right) - \dfrac{-\rho_L}{2\pi\varepsilon}\ln\left(\dfrac{R}{s-\sqrt{s^2-R^2}}\right)}$$

$$= \frac{2\pi\varepsilon}{\ln\left(\dfrac{s+\sqrt{s^2-R^2}}{R}\right) + \ln\left(\dfrac{s+\sqrt{s^2-R^2}}{R}\right)} = \frac{\pi\varepsilon}{\ln\left(\dfrac{s+\sqrt{s^2-R^2}}{R}\right)} \tag{B26}$$

Using equations B5 and B8 and simplifying we get the relation:

$$\cosh^{-1}\left(\frac{s}{R}\right) = \ln\left(\frac{R}{s-\sqrt{s^2-R^2}}\right) = \ln\left(\frac{s+\sqrt{s^2-R^2}}{R}\right) \tag{B27}$$

From the equations B26 and B27, we finally get the capacitance per unit length as

$$C = \frac{\pi\varepsilon}{\cosh^{-1}\left(\dfrac{s}{R}\right)} \tag{B28}$$

B2.2 Capacitance of cylindrical conductor and plane at ground potential

From figure B4 and by using the equation B8, the capacitance per unit length between a conductor and ground plane is given by

$$C = \frac{\rho_L}{u-0} = \frac{\rho_L}{\dfrac{\rho_L}{2\pi\varepsilon}\ln\left(\dfrac{R}{s-\sqrt{s^2-R^2}}\right)} = \frac{2\pi\varepsilon}{\ln\left(\dfrac{s+\sqrt{s^2-R^2}}{R}\right)} \tag{B29}$$

By using equation B27, we get the capacitance per unit length between the conductor and ground as

$$C = \frac{2\pi\varepsilon}{\cosh^{-1}\left(\frac{s}{R}\right)}. \tag{B30}$$

Index

Printed in the United Kingdom
by Lightning Source UK Ltd.
127125UK00001B/88-165/A